Bioinorganic Chemistry: Inorganic Elements in the Chemistry of Life

Inorganic Chemistry

A Wiley Series of Advanced Textbooks
ISSN: 1939-5175

Previously Published Books in this Series

Structural Methods in Molecular Inorganic Chemistry
David W. H. Rankin, Norbert W. Mitzel & Carole A. Morrison; ISBN: 978-0-470-97278-6

Introduction to Coordination Chemistry
Geoffrey Alan Lawrance; ISBN: 978-0-470-51931-8

Chirality in Transition Metal Chemistry
Hani Amouri & Michel Gruselle; ISBN: 978-0-470-06054-4

Bioinorganic Vanadium Chemistry
Dieter Rehder; ISBN: 978-0-470-06516-7

Inorganic Structural Chemistry 2nd Edition
Ulrich Müller; ISBN: 978-0-470-01865-1

Lanthanide and Actinide Chemistry
Simon Cotton; ISBN: 978-0-470-01006-8

Mass Spectrometry of Inorganic and Organometallic Compounds: Tools-Techniques-Tips
William Henderson & J. Scott McIndoe; ISBN: 978-0-470-85016-9

Main Group Chemistry, Second Edition
A.G. Massey; ISBN: 978-0-471-19039-5

Synthesis of Organometallic Compounds: A Practical Guide
Sanshiro Komiya; ISBN: 978-0-471-97195-5

Chemical Bonds: A Dialog
Jeremy Burdett; ISBN: 978-0-471-97130-6

The Molecular Chemistry of the Transition Elements: An Introductory Course
Francois Mathey & Alain Sevin; ISBN: 978-0-471-95687-7

Stereochemistry of Coordination Compounds
Alexander von Zelewsky; ISBN: 978-0-471-95599-3

For more information on this series see: www.wiley.com/go/inorganic

Bioinorganic Chemistry: Inorganic Elements in the Chemistry of Life

An Introduction and Guide

Second Edition

Written and Translated by

Wolfgang Kaim
University of Stuttgart, Institute of Inorganic Chemistry, Stuttgart, Germany

Brigitte Schwederski
University of Stuttgart, Institute of Inorganic Chemistry, Stuttgart, Germany

Axel Klein
University of Cologne, Institute of Inorganic Chemistry, Cologne, Germany

WILEY

Originally published in the German language by Vieweg+Teubner, 65189 Wiesbaden, Germany, as "Wolfgang Kaim and Brigitte Schwederski: Bioanorganische Chemie. 4. Auflage (4th Edition)".

Springer Fachmedien is part of Springer Science+Business Media

This edition first published 2013

First English language edition published 1994

Registered office
John Wiley & Sons Ltd, The Atrium, Southern Gate, Chichester, West Sussex, PO19 8SQ, United Kingdom

For details of our global editorial offices, for customer services and for information about how to apply for permission to reuse the copyright material in this book please see our website at www.wiley.com.

Library of Congress Cataloging-in-Publication Data

Kaim, Wolfgang, 1951–
[Bioanorganische Chemie. English]
Bioinorganic chemistry : inorganic elements in the chemistry of life : an introduction and guide / written and translated by Wolfgang Kaim, Brigitte Schwederski, Axel Klein. – Second edition.
pages cm
Translation of: Bioanorganische Chemie.
Includes bibliographical references and index.
ISBN 978-0-470-97524-4 (cloth) – ISBN 978-0-470-97523-7 (paper) 1. Bioinorganic chemistry. I. Schwederski, Brigitte, 1959– II. Klein, Axel, 1964– III. Title.
QP531.K3513 2013
572′.51–dc23 2013011894

A catalogue record for this book is available from the British Library.

HB ISBN: 9780470975244
PB ISBN: 9780470975237

Typeset in 10/12pt Times by Aptara Inc., New Delhi, India

1 2013

Contents

Preface to the Second Edition

The predictably enormous growth of bioinorganic chemistry has made a second edition of this text both necessary and difficult. While there are several extensive and often specialized reviews, major texts and handbooks on this subject, our experience in teaching it has suggested the provision of an updated overview of the classical, novel and applied sections of the field, which has not only become one of the major subdisciplines of inorganic chemistry but, due to its highly interdisciplinary nature, has also pervaded other areas of the life sciences.

The second edition contains updates of many kinds. New structure information on some intricate metalloproteins, such as water oxidase and the molybdopterin-based enzymes, has been included, replacing the earlier speculative models. Emerging developments are referred to at various points, covering such topics as bioorganometallic chemistry, nucleic acid ligation, gasotransmitters, nanoparticles and global cycles of the elements C, P and N. The vastly increased focus on the medical applications of inorganic compounds has required that more space be devoted to this particular aspect. Nonetheless, we have tried to keep the amount of material at a constant, manageable level suitable for an introductory overview, rather than the typical condensed fragments presented in general textbooks of inorganic chemistry or biochemistry. To achieve this, we have tried to concentrate on the facts and on descriptions of function, rather than on model compounds or mechanistic hypotheses (which may vary with time); excellent treatments of the reaction mechanisms of bioinorganic systems are available in T. D. H. Bugg's *Introduction to Enzyme and Coenzyme Chemistry*, third edition (John Wiley & Sons, 2012) and D. Gamenara, G. Seoane, P. Saenz Mendez and P. Dominguez de Maria's *Redox Biocatalysis: Fundamentals and Applications* (John Wiley & Sons, 2012). A basic knowledge of inorganic, organic, physical and biological chemistry remains necessary to make optimal use of this text.

Throughout this book, we have made reference to the RCSB Protein Data Bank for biological macromolecules. Each structure deposited therein is given a unique PDB code (e.g. 1SOD), and all information pertaining to that structure can be found using its code. For easy reference, we have included this code with all the structures in this book, so that the reader can refer to the original data online.

For comments and encouragement during the planning and completion of this edition, we thank many of our colleagues. We thank the publishers for their support and patience and Martina Bubrin for help in retrieving crystal structure files and drawing the structures. Most special thanks are due to Angela Winkelmann for her continued contributions to the preparation of the manuscript.

Wolfgang Kaim
Brigitte Schwederski
Axel Klein
Stuttgart and Cologne, January 2013

Instructors can access PowerPoint files of the illustrations presented within this text, for teaching, at: http://booksupport.wiley.com.

Preface to the First Edition

This book originated from a two-semester course offered at the Universities of Frankfurt and Stuttgart (W.K). Its successful use requires a basic knowledge of the modern sciences, especially of chemistry and biochemistry, at a level that might be expected after one year of study at a university or its equivalent. Despite these requirements we have decided to explain some special terms in a glossary and, furthermore, several less conventional physical methods are briefly described and evaluated with regard to their practical relevance at appropriate positions in the text.

A particular problem in the introduction to this highly interdisciplinary and not yet fully mature or definitively circumscribed field lies in the choice of material and the depth of treatment. Although priority has been given to the presentation of metalloproteins and the electrolyte elements, we have extended the scope to therapeutically, toxicologically and environmentally relevant issues because of the emphasis on functionality and because several of these topics have become a matter of public discussion.

With regard to details, we can frequently only offer hypotheses. In view of the explosive growth of this field there is implicit in many of the statements regarding structure and mechanisms the qualification that they are "likely" or "probable". We have tried to incorporate relevant literature citations up to the year 1993.

Another difficult aspect when writing an introductory and, at the same time, fairly inclusive text is that of the organization of the material. For didactic reasons we follow partly an organizational principle focused on the elements of the periodic table. However, living organisms are opportunistic and could not care less about such systematics; to successfully cope with a problem is all that matters. Accordingly, we have had to be "nonsystematic" in various sections, for example, treating the hemerythrin protein in connection with the similarly O_2-transporting hemoglobin (Chapter 5) and not under 'diiron centers' (Section 7.6). Several sections are similarly devoted to biological-functional problems such as biomineralization or antioxidant activity and may thus include several different elements or even organic compounds. The simplified version of the P-450 monooxygenase catalytic cycle which we chose for the cover picture illustrates the priority given to function and reactivity as opposed to static-structural aspects.

We regret that the increasingly available color-coded structural representations of complex proteins and protein aggregates cannot be reproduced here. General references to the relevant literature are given in the bibliography at the end of the book while specific references are listed at the end of each chapter in the sequence of appearance.

For helpful comments and encouragement during the writing and correction of manuscripts we thank many of our colleagues. Recent results have become available to us through participation in the special program "Bioanorganische Chemie" of the Deutsche

Forschungsgemeinschaft (DFG). We also thank Teubner-Verlag and John Wiley & Sons for their patience and support. Very special thanks are due to Mrs Angela Winkelmann for her continued involvement in the processing of the manuscript.

Wolfgang Kaim
Brigitte Schwederski
Stuttgart, December 1993

1 Historical Background, Current Relevance and Perspectives

The progress of an inorganic chemistry of biological systems has had a curious history.

R. J. P. WILLIAMS, *Coord. Chem. Rev.* **1990**, *100*, 573

The description of a rapidly developing field of chemistry as "bioinorganic" seems to involve a contradiction in terms, which, however, simply reflects a misconception going back to the beginning of modern science. In the early 19th century, chemistry was still divided into an "organic" chemistry which included only substances isolated from "organisms", and an "inorganic" chemistry of "dead matter".[1] This distinction became meaningless after Wöhler's synthesis of "organic" urea from "inorganic" ammonium cyanide in 1828. Nowadays, organic chemistry is defined as the chemistry of hydrocarbons and their derivatives, with the possible inclusion of certain nonmetallic heteroelements such as N, O and S, regardless of the origin of the material.

The increasing need for a collective, not necessarily substance-oriented designation of the chemistry of living organisms then led to the new term "biochemistry". For a long time, classical biochemistry was concerned mainly with organic compounds; however, the two areas are by no means identical.[2] Improved trace analytical methods have demonstrated the importance of quite a number of "inorganic" elements in biochemical processes and have thus revealed a multitude of partially inorganic natural products. A corresponding list would include:

- metalloenzymes (ca. 40% of the known enzymes, especially oxidoreductases (**Fe**, **Cu**, **Mn**, **Mo**, **Ni**, **V**) and hydrolases (e.g. peptidases, phosphatases: **Zn**, **Mg**; **Ca**, **Fe**);
- nonenzymatic metalloproteins (e.g. hemoglobin: **Fe**);
- low-molecular-weight natural products (e.g. chlorophyll: **Mg**);
- coenzymes, vitamins (e.g. vitamin B_{12}: **Co**);
- nucleic acids: (e.g. $DNA^{n-}(M^+)_n$, M = **Na**, **K**);
- hormones (e.g. thyroxine, triiodothyronine: **I**);
- antibiotics (e.g. ionophores: valinomycin/**K**);
- biominerals (e.g. bones, teeth, shells, coral, pearls: **Ca**, **Si**, . . .).

[1] There is increasing evidence that much of the "inorganic" material on the surface of the earth has undergone transformations during long-term contact with organisms and their metabolic products, such as O_2 [1].

[2] The term "bioorganic chemistry" is increasingly being used for studies of organic compounds that are directly relevant for biochemistry.

Bioinorganic Chemistry: Inorganic Elements in the Chemistry of Life – An Introduction and Guide, Second Edition.
Written and Translated by Wolfgang Kaim, Brigitte Schwederski and Axel Klein.

Some (by today's definition) "inorganic" elements were established quite early as essential components of living systems. Examples include the extractions of potassium carbonate (K_2CO_3, potash) from plants and of iron-containing complex salts $K_{3,4}[Fe(CN)_6]$ from animal blood in the 18th century, and the discoveries of elemental phosphorus (as P_4) by dry distillation of urine residues in 1669 and of elemental iodine from the ashes of marine algae in 1811.

In the middle of the 19th century, Liebig's studies on the metabolism of inorganic nutrients, especially of nitrogen, phosphorus and potassium salts, significantly improved agriculture, so that this particular field of science gained enormous practical importance. However, the theoretical background and the analytical methods of that time were not sufficient to obtain detailed information on the mechanism of action of essential elements, several of which occur only in trace amounts. Some very conspicuous compounds which include inorganic elements like iron-containing hemoglobin and magnesium-containing chlorophyll, the "pigments of life", were analyzed and characterized later within a special subfield of organic chemistry, the chemistry of natural products. It was only after 1960 that bioinorganic chemistry became an independent and highly interdisciplinary research area. The following factors have been crucial for this development:

1. Biochemical isolation and purification procedures, such as chromatography, and the new physical methods of trace element analysis, such as atomic absorption or emission spectroscopy, require ever smaller amounts of material. These methodical advances have made it possible not only to detect but also to chemically and functionally characterize trace elements or otherwise inconspicuous metal ions in biological materials. An adult human being, for example, contains about 2 g of zinc in ionic form (Zn^{2+}). Although zinc cannot be regarded as a true trace element, the unambiguous proof of its existence in enzymes was established only in the 1930s. Genuine bioessential trace elements such as nickel (Figures 1.1 and 1.2), (Chapter 9) and selenium (Chapter 16.8) have been known to be present as constitutive components in several important enzymes only since about 1970.

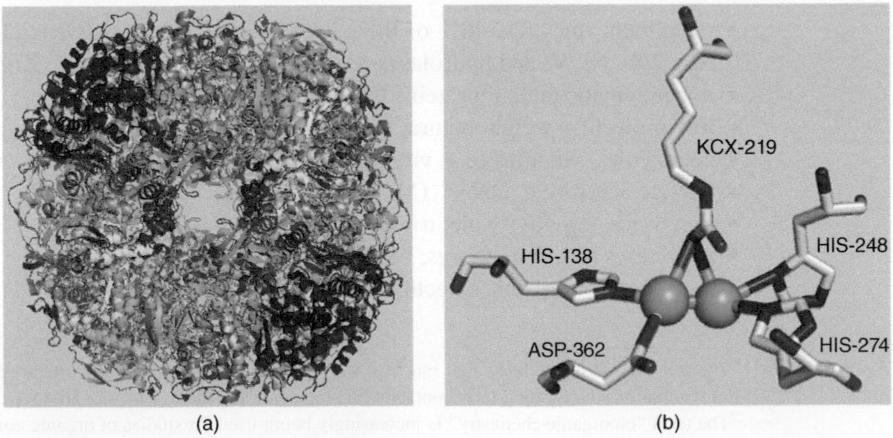

Figure 1.1
Nickel-containing urease, the first enzyme to be crystallized [2]. (a) Crystal structure of the full assembly of *Helicobacter pylori* urease, redrawn from [3] (PDB code 1E9Z). (b) Active site with two nickel centers (green spheres); histidine, aspartate, and a carbamylated lysine as ligands (Section 9.2).

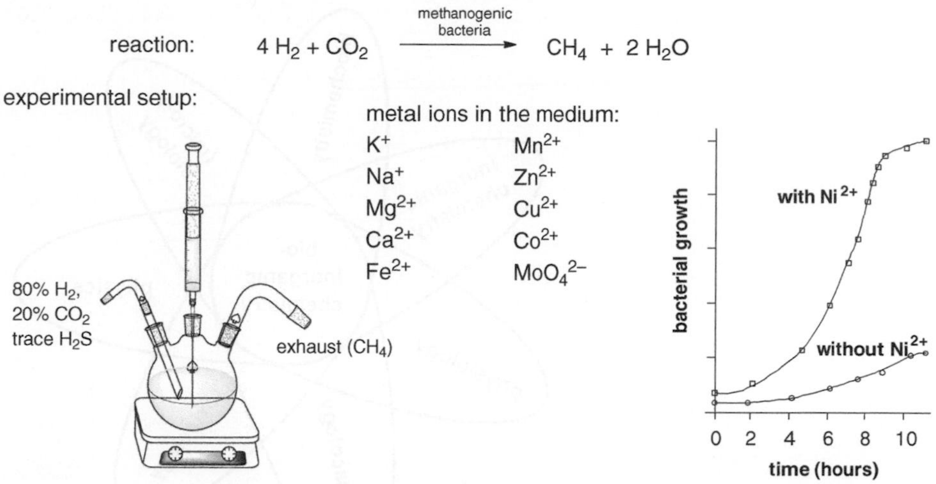

Figure 1.2
Discovery of nickel as an essential trace element in the production of methane by archaea.

In a desire "to accomplish something of real importance", the biochemist James B. Sumner managed to isolate and crystallize in 1926 a pure enzyme for the first time [2], much to the skepticism and disbelief of most experienced scientists. The chosen enzyme, urease (from jack beans), catalyzes the hydrolysis of urea, $O{=}C(NH_2)_2$, to CO_2 and 2 NH_3. It contains two closely associated nickel ions per subunit (Section 9.2). It was believed by many then that pure enzymes contained no protein, and only after other enzymes were crystallized was Sumner's discovery accepted. He was honored in 1946 with the Nobel Prize in Chemistry. However, Sumner's belief that urea contained *only* protein was corrected in 1975 when Dixon *et al.* proved that urease is a nickel metalloenzyme (Section 9.2).

In a very different research area, the biological reduction of carbon dioxide by hydrogen to produce methane has been investigated by studying the relevant archaebacteria, which are found, for example, in sewage plants. Even though the experiments were carried out under strictly anaerobic conditions and all "conventional" trace elements were supplied (Figure 1.2), the results were only partly reproducible. Eventually it was discovered that during sampling with a syringe containing a supposedly inert stainless steel (Fe/Ni) tip, minute quantities of nickel had dissolved. This inadvertent generation of Ni^{2+} ions led to a distinctive increase in methane production [4], and, in fact, several nickel containing proteins and coenzymes have since been isolated (see Chapter 9). Incidentally, a similar unexpected dissolution effect of an apparently "inert" metal led to the serendipitous discovery of the inorganic anti-tumor agent *cis*-$PtCl_2(NH_3)_2$ ("cisplatin", Section 19.2.1).

2. Efforts to elucidate the mechanisms of organic, inorganic and biochemical reactions have led to an early understanding of the specific biological functions of some inorganic elements. Nowadays, many attempts are being made to mimic biochemical reactivity

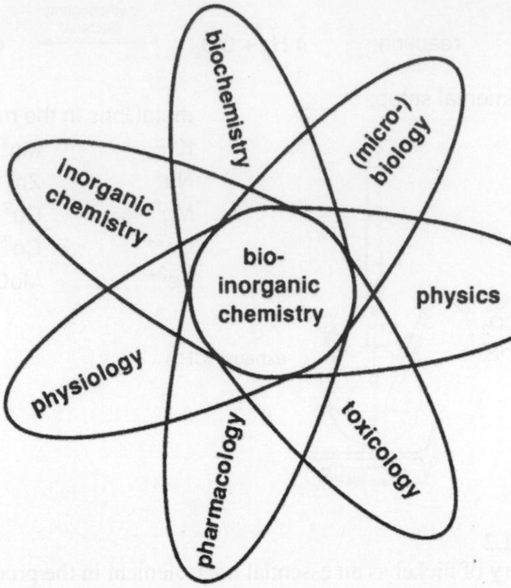

Figure 1.3
Bioinorganic chemistry as a highly interdisciplinary research field.

through studies of the reactivity of model systems, low-molecular-weight complexes or tailored metalloproteins (Section 2.4).

3. The rapid progress in bioinorganic chemistry, an interdisciplinary field of research (Figure 1.3), has been made possible through contributions from:
   - ○ physics (→ techniques for detection and characterization);
   - ○ various areas of biology (→ supply of material and specific modifications based on site-directed mutagenesis);
   - ○ agricultural and nutritional sciences (→ effects of inorganic elements and their mutual interdependence);
   - ○ pharmacology (→ interaction between drugs and endogeneous or exogeneous inorganic substances);
   - ○ medicine (→ imaging and other diagnostic aids, chemotherapy);
   - ○ toxicology and the environmental sciences (→ potential toxicity of inorganic compounds, depending on the concentration).

A list of examples illustrating the *application potential of bioinorganic chemistry* could include the following:

- Industrial sector:
  - ○ anaerobic bacterial degradation in sewage plants or in sediments: **Fe, Ni, Co**;
  - ○ biomining (bacterial leaching; ≈ 15% of the global copper production): **Cu, Au, Fe, U**.
- Environmental sector:
  - ○ agricultural trace element problems: nitrogen fixation (**Fe, Mo, V**); **Mo/Cu** antagonism; **Se** content of soil;
  - ○ pollution through metal species: **Pb, Cd, Hg, As, Al, Cr**;
  - ○ detoxification, for example via peroxidases: **Fe, Mn, V**.

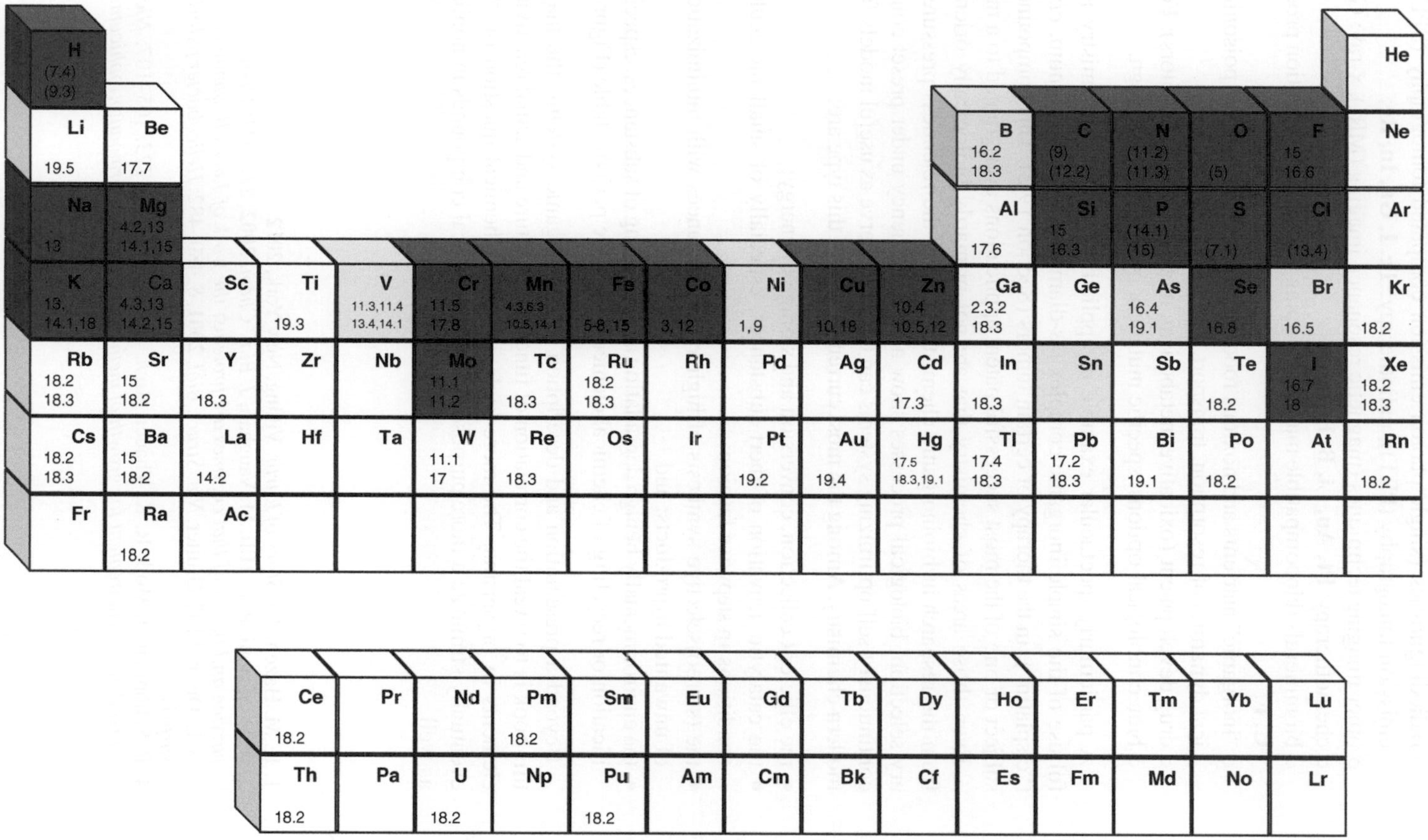

H (7.4) (9.3)
Li 19.5
Be 17.7
Na 13
Mg 4.2,13 14.1,15
K 13, 14.1,18
Ca 4.3,13 14.2,15
Sc
Ti 19.3
V 11.3,11.4 13.4,14.1
Cr 11.5 17.8
Mn 4.3,6.3 10.5,14.1
Fe 5-8, 15
Co 3, 12
Ni 1, 9
Cu 10, 18
Zn 10.4 10.5,12
B 16.2 18.3
C (9) (12.2)
N (11.2) (11.3)
O (5)
F 15 16.6
He
Ne
Al 17.6
Si 15 16.3
P (14.1) (15)
S (7.1)
Cl (13.4)
Ar
Ga 2.3.2 18.3
Ge
As 16.4 19.1
Se 16.8
Br 16.5
Kr 18.2
Rb 18.2 18.3
Sr 15 18.2
Y 18.3
Zr
Nb
Mo 11.1 11.2
Tc 18.3
Ru 18.2 18.3
Rh
Pd
Ag
Cd 17.3
In 18.3
Sn
Sb
Te 18.2
I 16.7 18
Xe 18.2 18.3
Cs 18.2 18.3
Ba 15 18.2
La 14.2
Hf
Ta
W 11.1 17
Re 18.3
Os
Ir
Pt 19.2
Au 19.4
Hg 17.5 18.3,19.1
Tl 17.4 18.3
Pb 17.2 18.3
Bi 19.1
Po 18.2
At
Rn 18.2
Fr
Ra 18.2
Ac
Ce 18.2
Pr
Nd
Pm 18.2
Sm
Eu
Gd
Tb
Dy
Ho
Er
Tm
Yb
Lu
Th 18.2
Pa
U 18.2
Np
Pu 18.2
Am
Cm
Bk
Cf
Es
Fm
Md
No
Lr

Figure 1.4

Periodic table of the elements. Indicated are the chapters and sections in which each element is discussed in this book. essential element; presumably essential element for human beings.

- Biomedical sector:
  - radiodiagnostics (single-photon emission computed tomography (SPECT), positron emission tomography (PET)), radiotherapy: **Tc, I, Ga, In, Re**;
  - other imaging techniques (magnetic resonance imaging (MRI), x-ray: **Gd, Ba, I**);
  - chemotherapy: **Pt, Au, Li, B, Bi, As**;
  - biominerals (biocompatible materials, coping with demineralization processes): **Ca, P, C, F**;
  - "inorganic" nutrients and noxious food components: deficiency, poisoning; physiological dynamics of resorption, transport, storage, excretion;
  - drug development (oxidative metabolism, metalloenzyme inhibitors): **Fe, Zn**;
  - biotechnological options: specific mutation, metalloprotein design.

A particularly spectacular example of applied bioinorganic chemistry is the successful use of the simple inorganic complex *cis*-diamminedichloroplatinum, *cis*-$Pt(NH_3)_2Cl_2$ ("cisplatin"), in the therapy of certain tumors (Section 19.2). This compound has been the subject of one of the most successful patent applications ever granted to a university.

Even those areas of chemistry that are not primarily biologically oriented can profit from the research in bioinorganic chemistry. Due to the relentless pressure of evolutionary selection, biological processes show a high efficiency under preset conditions. These continuously self-optimizing systems can therefore serve as useful models for problems in modern chemistry. Among the most current topics of this type are:

- the efficient collection, conversion and storage of energy;
- the catalytic activation of inert substances, especially of small molecules under mild conditions in stepwise fashion;
- the (stereo)selective synthesis of high-value substances with minimization of the yield of unwanted byproducts; and
- the environmentally benign degradation and recycling of substances, especially the detoxification or recycling of chemical elements from the periodic table (Figure 1.4).

Beyond a presentation and description of bioinorganic systems, the major purpose of this book is to reveal the correlation of function, structure and actual reactivity of inorganic elements in organisms. The more biological than chemical question of "Why?" should eventually stimulate a more purposeful use of chemical compounds in nonbiological areas as well.

References

1. R. M. Hazen, *The Story of Earth*, Viking, New York, **2012**.
2. R. D. Simoni, R. L. Hill, M. Vaughan, *J. Biol. Chem.* **2002**, *277*, e23: *Urease, the first crystalline enzyme and the proof that enzymes are proteins: the work of James B. Sumner*.
3. B. E. Dunn, M. G. Grutter, *Nat. Struct. Biol.* **2001**, *8*, 480–482: *Helicobacter pylori springs another surprise*.
4. P. Schönheit, J. Moll, R. K. Thauer, *Arch. Microbiol.* **1979**, *123*, 105–107: *Nickel, cobalt, and molybdenum requirement for growth of methanobacterium thermoautotrophicum*.

2 Some General Principles

2.1 Occurrence and Availability of Inorganic Elements in Organisms

"Life" is a process which, for an adult organism, can be characterized as a controlled stationary flow equilibrium maintained by energy-consuming chemical reactions ("dissipative system"). *Input* and *output* are essential requirements for such open systems. They differ very much from the more familiar and mathematically far more easily described "dead" thermodynamic equilibria (Figure 2.1).

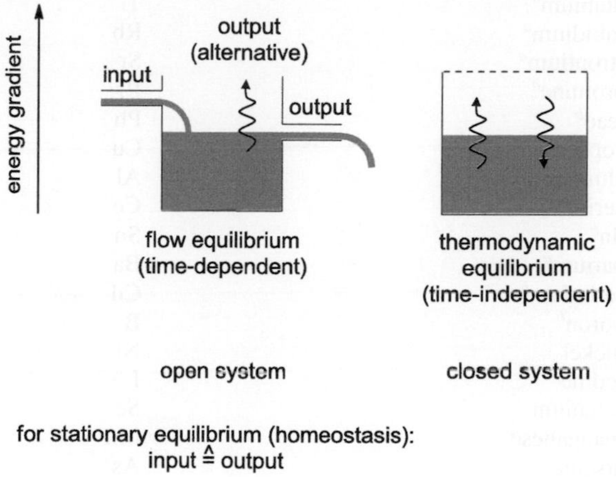

Figure 2.1
Two kinds of "equilibrium".

In addition to the energy flux, life requires a continuous material exchange which, in principle, includes *all* chemical elements (see Figure 1.4). The occurrence of these elements in organisms depends on external and endogeneous conditions; elements can be "bioavailable" to variable extents but can also be enriched ("bioaccumulated") by organisms using active, energy-consuming processes involving a local reduction of entropy. Some trends are obvious from the most familiar example, the elemental composition of the human body (Table 2.1).

The values for O and H in Table 2.1 reflect the high content of (inorganic) water; the "organic" element carbon only comes in third. Calcium as the first metallic element ranks fifth, its main quantitative use being the stabilization of the endoskeleton. Table 2.1 further shows relatively large quantities of potassium, chlorine, sodium and magnesium,

Bioinorganic Chemistry: Inorganic Elements in the Chemistry of Life – An Introduction and Guide, Second Edition.
Written and Translated by Wolfgang Kaim, Brigitte Schwederski and Axel Klein.

Table 2.1 Average elemental composition of a human body (adult, 70 kg) [1].

Element	Symbol	Mass (g)
oxygen	O	43 000
carbon	C	16 000
hydrogen	H	7000
nitrogen	N	1800
calcium	Ca	1200
phosphorus	P	780
sulfur	S	140
potassium	K	125
sodium	Na	100
chlorine	Cl	95
magnesium	Mg	25
fluorine	F	5.0 (var.)
iron	Fe	4.0
zinc	Zn	2.3
silicon	Si	1.0 (var.)
titanium[a]	Ti	0.70
rubidium[a]	Rb	0.68
strontium[a]	Sr	0.32
bromine[a]	Br	0.26
lead[b]	Pb	0.12
copper	Cu	0.07
aluminum[a]	Al	0.06
cerium[a]	Ce	0.04
tin[b]	Sn	0.03
barium[a]	Ba	0.02
cadmium[b]	Cd	0.02 (var.)
boron[b]	B	0.018
nickel	Ni	0.015
iodine	I	0.015
selenium	Se	0.014
manganese	Mn	0.012
arsenic[b]	As	0.007 (var.)
lithium[a]	Li	0.007
molybdenum	Mo	0.005
chromium	Cr	0.002 (var.)
cobalt	Co	0.002

[a]Not rated essential.
[b]Essential character not unambiguous.

the "mass" or "quantity elements" or "macronutrients". Iron, zinc and fluorine are distinctly less abundant inorganic elements. According to one definition, "trace elements" with regard to the human body involve a daily requirement of less than 25 mg (see Table 2.3), and some of them, such as boron, arsenic and tin, have not yet been unambiguously defined with regard to amount, essential character and function [2]. Since humans coexist with a host of supporting "lower" organisms – the "microbiome" – their requirements for trace elements will also have to be added. Elements are *essential* if their total absence in the organism causes severe, irreversible damage. Sometimes essentiality is invoked if the optimal functioning of organisms is impaired; in such instances the corresponding elements would be better referred to as "beneficial". Table 2.1 illustrates the occurrence of non-negligible quantities of obviously nonessential elements such as Ti, Rb, Sr, Br, Al and Li in the human body.

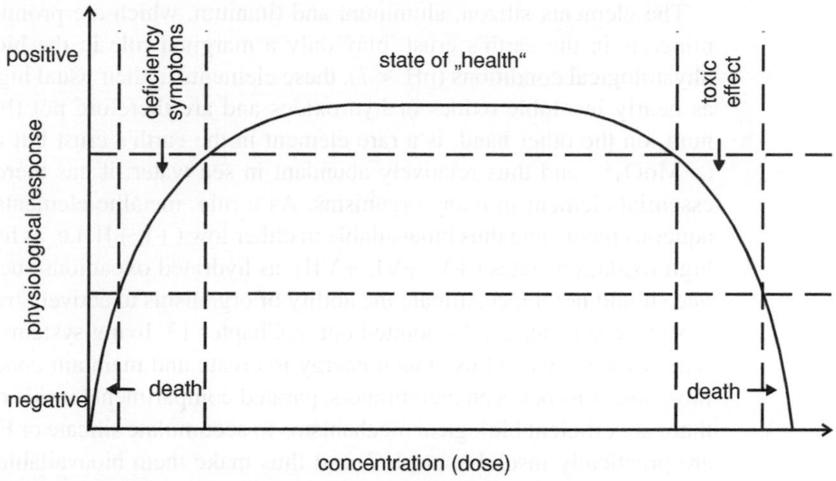

Figure 2.2
Schematic dose–response relationship (Bertrand diagram) for an essential element (compare Figure 17.1 for exclusively toxic elements).

These elements are probably incorporated due to a chemical similarity (indicated by ↔) with important essential elements (Li^+, Rb^+, Cs^+ ↔ Na^+, K^+; Sr^+, Ba^{2+} ↔ Ca^{2+}; Br^- ↔ Cl^-; Al^{3+}, Ti^{4+} ↔ Fe^{3+}). Elements that are known to be mainly toxic, such as As, Pb and Cd, deserve special attention; a positive effect of traces has been discussed for some of these elements, pointing to the ambivalence of many trace elements and to the problem of threshold values (see Figures 2.2 and 2.3). Possibly a physiological – if not always essential – function developed for all naturally occurring elements during the evolution of life [3].

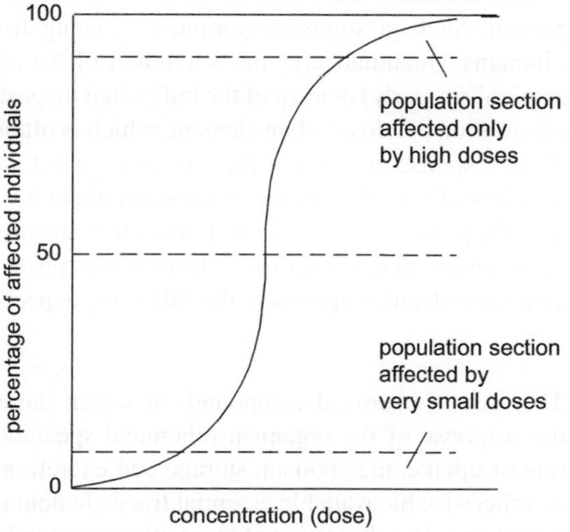

Figure 2.3
Typical variance of the (toxic) effect of a substance within a population.

The elements silicon, aluminum and titanium, which are prominent as components of minerals in the earth's crust, play only a marginal role in the biosphere. Under normal physiological conditions (pH ≈ 7), these elements in their usual high oxidation states exist as nearly insoluble oxides or hydroxides and are therefore not (bio)available. Molybdenum, on the other hand, is a rare element in the earth's crust but is quite soluble at pH 7 as MoO_4^{2-} and thus relatively abundant in sea water; it has therefore been found as an essential element in many organisms. As a rule, metallic elements are soluble in neutral aqueous media and thus bioavailable in either low (+I, +II, i.e. as hydrated cations) or very high oxidation states (+V, +VI, +VII), as hydrated oxoanions such as MO_4^{n-}. However, one should not underestimate the ability of organisms to actively transport and accumulate inorganic substances. As pointed out in Chapter 13, living systems have developed elaborate mechanisms and use much energy to create and maintain concentration gradients for inorganic ions between membrane-separated compartments inside of organisms. Similarly, there are efficient biological mechanisms to accumulate silicate or Fe^{3+} ions, both of which are practically insoluble at pH 7, and thus make them bioavailable for structural or other purposes (see Chapters 8 and 15). Not surprisingly, the elemental compositions are highly variable for different species and even different parts of higher organisms, depending *inter alia* on the kind of metabolism and on the biotope.

The flow equilibrium character of life processes (Figure 2.1) implies that the individual inorganic elements are continuously excreted and replenished even though their overall stationary concentration remains approximately constant ("homeostasis"). The rate of exchange is strongly dependent on the type of compound (chemical speciation) and on the site of action or storage in the individual organism. According to established principles of reaction kinetics, ions of low charge are exchanged relatively quickly (K^+, Mn^{2+}, MoO_4^{2-}), whereas more highly charged species such as Fe^{3+} have longer physiological half lives. It is not surprising that elements such as Ca^{2+}, which find their main quantitative use in the solid-state skeleton, are on the whole exchanged very slowly. The biological half life can then amount to several years; nevertheless, a continuous metabolism takes place even for "biominerals" (see Chapter 15).

Which elements are essential, which are beneficial and which are toxic for a certain organism? Such questions are continuously being discussed in popular science, particularly for humans. Quantitatively, this is a matter of the physiological state (i.e. of the ability to "function" properly) or even of the individual disposition of an organism, depending on the concentration or "dose" of an element, which is often related to its share in the food supply. A dose–response diagram of the type in Figure 2.2 can thus be discussed; this shows the ambivalent effects of many substances and illustrates the principle of Paracelsus: "The dose makes the poison". An important term here is the *therapeutic window*, which characterizes the concentration range causing advantageous physiological effects.

In a more detailed approach, the following aspects have to be considered:

- The type of chemical compound, of which the element is a part, is often crucial for the response of the organism (chemical speciation). The pathway, the extent and the rate of uptake, metabolism, storage and excretion can differ greatly; poor utilization of an otherwise bioavailable essential trace element may thus be responsible for deficiency symptoms. The absorption of inorganic compounds by the organism depends primarily on the solubility and therefore the charge of the system; humans resorb molybdate MoO_4^{2-} quite well, whereas slowly reacting Cr^{3+} is resorbed only to a small extent.

- Given the essentiality of biological compartmentalization within and between cells, the toxicity of an (inorganic) species may depend strongly on the site of action. Many bioessential elements are thus also potent carcinogens, depending on their location.
- It cannot be expected that higher organisms will react uniformly within a population or in the course of their individual development. Therefore, only average statements can be made with regard to a certain situation, such as for the adult state of a preferentially homogeneous population (Figure 2.3).
- The concentration variation of one particular element frequently affects the concentrations and physiological effects of other substances, including compounds of other inorganic elements. This multidimensional interdependence has been known qualitatively for a number of elemental nutrients since the experiments of Liebig. Two components can interact by mutually promoting corresponding effects (synergism) or by competing and suppressing each other's effects (antagonism).

An antagonistic relationship in a two-component system can be the result of displacement ($Zn^{2+} \leftrightarrow Cd^{2+}$, Pb^{2+}, Cu^{2+} or Ca^{2+}) or mutual deactivation ($Cu^{2+} + S^{2-} \rightarrow CuS$ (insoluble)). With three components, for example in the system Cu/Mo/S (Chapter 10 and Section 11.1.2), matters get more complicated, and in reality there is a multidimensional network of synergistic and antagonistic relationships, which is further complicated by the spatially unsymmetrical distribution of inorganic elements in organisms [4]. For instance, there are strongly contrasting distributions of soluble monocations (Na^+ vs K^+), dications (Ca^{2+} vs Mg^{2+}) and monoanions (Cl^- vs $H_2PO_4{}^-$) in the extra- and intracellular regions, respectively, as outlined in Chapter 13.

Despite this complexity, some deficiency symptoms of individual inorganic elements are quite familiar, particularly as they concern human beings [2,5] (see the incomplete list in Table 2.2). As far as causal connections are known for the single elements, these will be discussed in the relevant chapters within this book. A general syndrome of (trace) element deficiency is growth retardation: the number of truly essential elements seems to be smaller in fully developed organisms than during growth periods. This assumption was

Table 2.2 Some characteristic symptoms of chemical element deficiency in humans.

Deficient element	Typical deficiency symptoms
Ca	retarded skeletal growth
Mg	muscle cramps
Fe	anemia, disorders of the immune system
Zn	skin damage, stunted growth, retarded sexual maturation
Cu	artery weakness, liver disorders, secondary anemia
Mn	infertility, impaired skeletal growth
Mo	retardation of cellular growth, propensity for caries
Co	pernicious anemia
Ni	growth depression, dermatitis
Cr	diabetes symptoms
Si	disorders of skeletal growth
F	dental caries
I	thyroid disorders, retarded metabolism
Se	muscular weakness, esp. cardiomyopathy
As	impaired growth (in animals)

Table 2.3 Essential elements in food for adults and infants.

Inorganic constituent	RDA (adults),[a] in mg	AI (infants),[b] in mg
K	4700	400–700
Na	1200–1500	120–370
Ca	100–1300	200–260
Mg	310–420	30–75
Zn	8–11	2–3
Fe	8–18	0.27–11.0
Mn	1.6–2.3	0.003–0.6
Cu	0.9	0.2
Mo	0.045	0.002
Cr	0.02–0.035	0.0002–0.005
Co	~0.0024 (vitamin B_{12})	0.0004
Cl	1800–2300	180–570
PO_4^{3-}	700	100–275
F	3–4	0.01–0.5
I	0.15	0.11–0.13
Se	0.055	0.015–0.020

[a]Recommended Dietary Allowance (RDA) is derived from the Estimated Average Requirement (EAR [6]). 97.5% of the population meets or exceeds the EAR. Data taken for male and female age groups from 19 to 70 years.
[b]Adequate Intake (AI) is used when an EAR/RDA cannot be developed. The AI level is based on observed or experimental intakes. Data reflect mean values for infants 0–6 months and 6–12 months old (lower and higher values, respectively).
Sources: National Research Council. *Dietary Reference Intakes for Calcium, Phosphorus, Magnesium, Vitamin D, and Fluoride*, 1997; *Dietary Reference Intakes for Thiamin, Riboflavin, Niacin, Vitamin B_6, Folate, Vitamin B_{12}, Pantothenic Acid, Biotin, and Choline*, 1998; *Dietary Reference Intakes for Vitamin C, Vitamin E, Selenium, and Carotenoids*, 2000; *Dietary Reference Intakes for Vitamin A, Vitamin K, Arsenic, Boron, Chromium, Copper, Iodine, Iron, Manganese, Molybdenum, Nickel, Silicon, Vanadium, and Zinc*, 2001; *Dietary Reference Intakes for Water, Potassium, Sodium, Chloride, and Sulfate*, 2005; *Dietary Reference Intakes for Calcium and Vitamin D*, 2011. Washington, DC: The National Academies Press.

confirmed by pioneering experiments in the 1960s, which were designed to guarantee a nutritionally complete diet for astronauts during long space flights. The inorganic contents of such synthetic food are summarized in Table 2.3 in the form of the RDA (Recommended Dietary Allowance [6]) values of the US Food and Drug Administration (FDA). Whether such a composition is really sufficient or guaranteed in today's food supply and how far it can be exceeded via increased uptake or separate supplementation without detrimental consequences are still open questions in dietetics, particularly from the popular scientific point of view.

According to Figure 2.3, there are not only deficiency symptoms from the lack of essential elements but also toxic effects resulting from an excess of these, whether caused by insufficient excretion or by excessive uptake [7]. Such poisoning can be treated using "bioinorganic" measures, namely through the application of antagonists or a "chelate therapy" [8,9], which involves the complexation and excretion of acutely toxic metal ions using multidentate chelate ligands ((2.1), Table 2.4). Considering that there are a number of essential metal ions present in organisms, the problem of selectivity is obvious; selectively coordinating ligands have thus been developed for some specific heavy-metal ions. The most successful such ligands offer selectivity either (i) according to the preferred size of the coordinated ion or (ii) with respect to favored coordinating atoms (S for "soft" heavy metals, N especially for Cu^{2+}, O for "hard" metal centers). Furthermore, suitable

Table 2.4 Chelate ligands for detoxification after metal poisoning.

Ligand (2.1)	Trade or trivial name	Preferably coordinated metal ions	Detailed description in Chapter/Section
(a) 2,3-dimercapto-1-propanol	dimercaprol, BAL	Hg^{2+}, As^{3+}, Sb^{3+}, Ni^{3+}	17
(b) D-2-amino-3-mercapto-3-methylbutyric acid (D-β,β-dimethylcysteine)[a]	D-penicillamine	Cu^{2+}, Hg^{2+}	10, 17
(c) ethylenediaminetetraacetate	EDTA	Ca^{2+}, Pb^{2+}	
(d) deferrioxamine B	DFO, desferal	Fe^{2+}, Al^{3+}	8.2, 17.6
(e) 3,4,3-LICAMC		Pu^{4+}	18.1.3.3

[a]The L-enantiomer is toxic.

chelate ligands must (iii) form kinetically and thermodynamically stable complexes and (iv) facilitate rapid renal excretion, for example by containing hydrophilic hydroxyl groups.

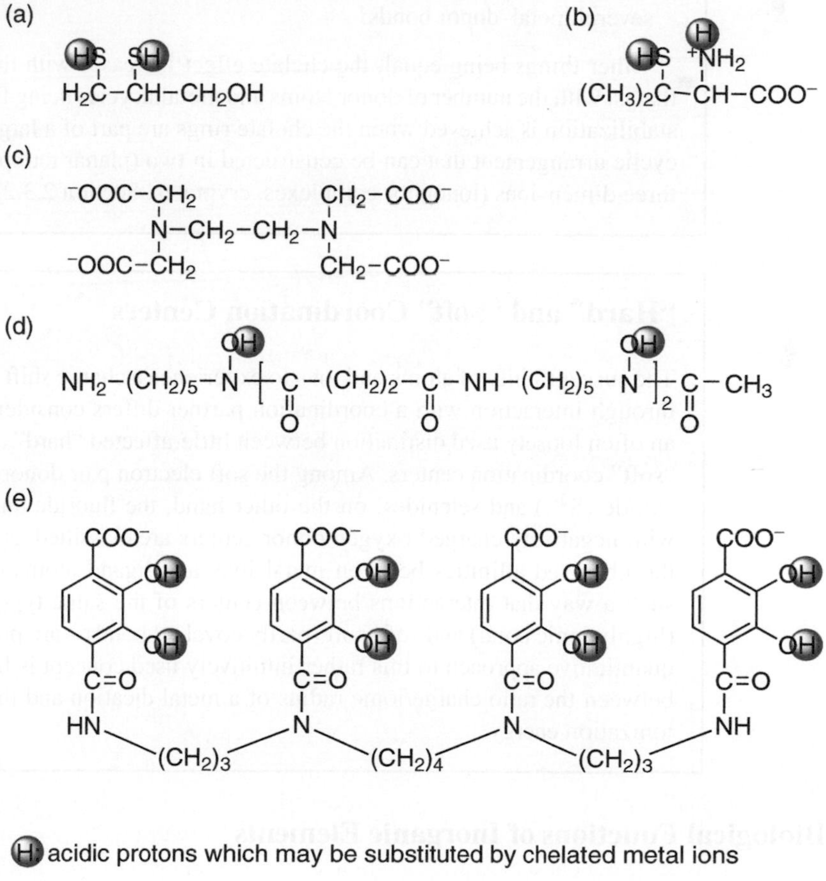

(2.1)

The Chelate Effect

Ligands that can use more than one nonadjacent donor atom for binding to a metal center are referred to as "chelate ligands". The corresponding chelate complexes contain at least one "metallacycle" ring structure, which restricts the torsional mobility of the system and contributes to enhanced selectivity. Due to the preference of sp^2- and sp^3-configured atoms for 120° and 109° bond angles, the optimum ring size of the metallacycles in chelate complexes is five, although four- and six-membered rings can also occur for very small and very large metal ions, respectively.

Chelate complexes can exhibit enhanced thermodynamic and kinetic stability (the "chelate effect").

- In addition to the enthalpy gain from optimum metal–donor interaction in suitable cyclic arrangements, an entropic factor favors the formation of chelate complexes because the number of free particles (including no-longer-coordinated individual solvent ligands) is increased.
- The kinetic stabilization of chelate complexes reflects the most unlikely complete dissociation of such species, which would require the simultaneous breaking of several metal–donor bonds.

Other things being equal, the chelate effect increases with the ligand "denticity"; that is, with the number of donor atoms and metallacycles being formed. An additional stabilization is achieved when the chelate rings are part of a larger preformed macrocyclic arrangement that can be constructed in two (planar tetrapyrrole complexes) or three dimensions (ionophor complexes, cryptates; Section 2.3.2).

"Hard" and "Soft" Coordination Centers

The susceptibility of atoms and ions to experience a charge shift in their electron shell through interaction with a coordination partner differs considerably. This has led to an often loosely used distinction between little-affected "hard" and easily polarizable "soft" coordination centers. Among the soft electron pair donors are thiolates (RS^-), sulfide (S^{2-}) and selenides; on the other hand, the fluoride anion (F^-) and ligands with negatively charged oxygen donor centers are classified as hard. In many cases, the observed affinities between metal ions and ligand atoms can be interpreted in such a way that interactions between centers of the same type – that is, hard–hard (highly ionic bond) and soft–soft (partly covalent bond) – are preferred. One possible quantitative approach to this rather intuitively used concept is based on a correlation between the ratio charge/ionic radius of a metal dication and the measurable second ionization energy.

2.2 Biological Functions of Inorganic Elements

The great efforts made by organisms to take up, accumulate, transport and store inorganic elements are justified only by their important and otherwise unguaranteed function. For living organisms, the arbitrary distinction between "organic" and "inorganic" compounds is irrelevant, since it is solely based on a historically grown definition. However, there are functions for which inorganic compounds or ions are particularly well suited:

1. The assembly of hard structures in the form of endo- or exoskeletons via biomineralization certainly falls into this category (Chapter 15). Another aspect of this *structural function* is that cell membranes require the presence of metal ions to crosslink the organic "filling material" and thus maintain the membrane integrity. Even the double helical structure of DNA is maintained only in the presence of sufficient cations to significantly reduce the otherwise dominating electrostatic repulsion forces between the negatively charged nucleotide phosphate groups. Solid state/structural functions are represented mainly by the elements Ca, Mg and Zn as dications and P, O, C, S, Si and F as parts of anions.
2. Simple atomic ions are superbly suited as *charge carriers* for very fast *information transfer*. Starting with a transmembrane concentration gradient that must be actively maintained by integral membrane ion pumps, information units in the form of sudden electrical potential changes can be created via diffusion, with maximum speed along ion channels. Electrical impulses in nerves, as well as more complex trigger mechanisms (e.g. in the control of muscle contractions), are thus initiated with the fastest possible effect by sudden fluxes (diffusion control) of atomic – that is, chemically and biologically nondegradable – inorganic ions of different sizes and charges (Na^+, K^+, Ca^{2+}; see Chapter 13 and Section 14.2).
3. *Formation, metabolism and degradation of organic compounds* in organisms often require acid or base catalysis. Since the physiological pH is generally limited to about 7, except for certain special compartments such as the stomach (which contains gastric acid), the rate enhancement of such reactions cannot be accomplished by simple proton or hydroxide catalysis, but requires *Lewis acid/Lewis base catalysis* involving metal ions. Many hydrolytically active enzymes thus contain the relatively small, dipositively charged metal ions Zn^{2+} and Mg^{2+} (Chapter 12 and Section 14.1).
4. The *transfer of electrons*, which is essential for short-term *energy conversion* in organisms, is mainly, but not exclusively, dependent on redox-active metal centers. A number of corresponding redox pairs have thus been found, some of which involve oxidation states that seem quite unusual under physiological conditions (marked **bold** in the following). Specific modifications induced by "bioligands" are largely responsible for the stabilization of such unusual oxidation states. Biologically relevant are the following oxidation states of redox-active metals: $\mathbf{Fe^{I}}$/Fe^{II}/Fe^{III}/$\mathbf{Fe^{IV}}$, $\mathbf{Cu^{I}}$/Cu^{II}, Mn^{II}/$\mathbf{Mn^{III}}$/Mn^{IV}, $\mathbf{Mo^{IV}}$/Mo^{V}/Mo^{VI}, $\mathbf{Co^{I}}$/Co^{II}/Co^{III}, $\mathbf{Ni^{I}}$/Ni^{II}/$\mathbf{Ni^{III}}$.
5. The redox *activation of small, highly symmetrical molecules* with large bond energies places stringent demands on the required catalysts. The ability of transition metal centers to provide unpaired electrons and to simultaneously accept and donate electronic charge (π back-bonding; Chapters 5 and 11) allows organisms to carry out energetically and mechanistically difficult multielectron-transfer reactions (2.2)–(2.5) under physiological conditions, such as:
   - the reversible conversion of the smallest molecule, H_2, to the oxidized form, H^+, through hydrogenases (Fe, Ni, Se; Section 9.3);

$$2\overset{+\mathrm{I}}{\mathrm{H}^+} \rightleftharpoons \overset{0}{\mathrm{H}_2} \qquad (2.2)$$

- the reversible uptake, transport, storage and conversion (Fe, Cu), or, conversely, the generation (Mn), of the paramagnetic dioxygen molecule, $^3O_2$ (Chapters 4–6 and 10);

$$\overset{0}{O_2} \rightleftharpoons \rightleftharpoons \overset{-I}{H_2O_2} \rightleftharpoons 2\,H_2\overset{-II}{O} \qquad (2.3)$$

- the nitrogen cycle, including the fixation of molecular nitrogen, N_2 (Sections 11.2 and 11.3), and its conversion to ammonia (Fe, Mo, V);

$$2\,\overset{+V}{NO_3^-} \rightleftharpoons \rightleftharpoons \rightleftharpoons \rightleftharpoons \rightleftharpoons \overset{0}{N_2} \rightleftharpoons \rightleftharpoons \rightleftharpoons 2\,\overset{-III}{NH_3} \qquad (2.4)$$

- the reduction of CO_2 with hydrogen to yield methane ("C_1 chemistry"; Ni, Fe; Section 9.5 and Figure 1.2).

$$\overset{+IV}{CO_2} \rightleftharpoons \underset{\text{or HCOOH}}{\overset{+II}{CO}} \rightleftharpoons \overset{0}{H_2CO} \rightleftharpoons \overset{-II}{CH_3OH} \rightleftharpoons \overset{-IV}{CH_4} \qquad (2.5)$$

6. Typical "*organometallic*" *reactivity*, such as reductive alkylation or the *facile generation of radicals* for rapid rearrangement of substrate molecules is found for cobalamin coenzymes that contain a σ bond between the transition metal cobalt and the primary alkyl groups (Chapter 3).

2.3 Biological Ligands for Metal Ions

The major part of bioinorganic chemistry is concerned with compounds of the metallic elements; nonmetallic inorganic "bio"elements will be discussed in Chapter 16. Within the biological context, metals mainly appear in oxidized form as formally ionized centers, which are therefore surrounded by electron-pair-donating ligands. Since the inorganic chemical elements cannot experience a biological evolution by themselves, it is their often highly complex coordination chemistry that is biologically relevant. What kinds of organic-biological material can serve as "natural" coordination partners for metal centers, in addition to simple or complex phosphates, **XPO_3^{2-}**, purely inorganic carbonate, **CO_3^{2-}** and sulfide, **S^{2-}**, or water, **H_2O**, together with its deprotonated forms, **OH^-** and **O^{2-}**? Relatively little is known about the relevance of metal coordination to **lipids** and **carbohydrates**, although the potentially negatively charged oxygen functions can bind cations electrostatically and even undergo chelate coordination via polyhydroxy groups [10,11]. Likewise, few molecular details are known about the *in vivo* interaction of low-molecular-weight **coenzymes** (vitamins; see (3.12)), **hormones** or metabolic products such as citrate with metal ions. The complexes formed are frequently labile, which makes their detection, isolation and structural characterization very difficult. Even so, it has long been known that the physiological function of ascorbate (vitamin C; see (3.12); [12]), for example, is connected with the Fe^{II}/Fe^{III} redox equilibrium. Examining the *in vitro* coordination chemistry of the redox-active isoalloxazine ring of flavins (2.6) more closely, one can find it coordinating via the atoms O(4) and N(5) to form chelate complexes with "soft" metal ions in its oxidized form and with "hard" metal centers in its half-reduced (i.e. semiquinone) form [13].

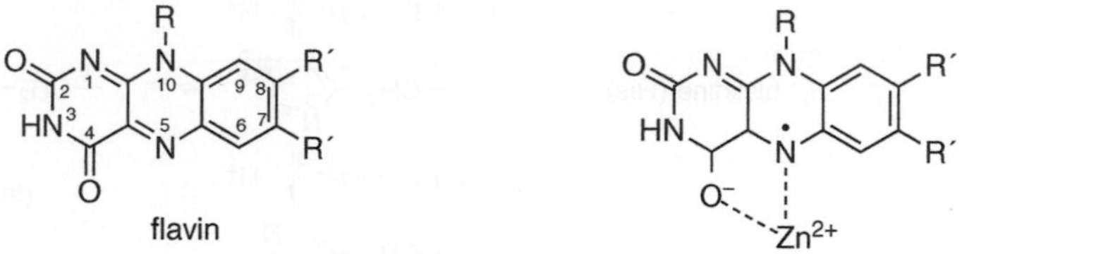

flavin

R′= CH_3, R = $CH_2(HCOH)_3CH_2OH$: riboflavin, vitamin B_2

flavosemiquinone complex of zinc(II)

(2.6)

In this section, the three most important classes of bioligand will be discussed in greater detail: **proteins** with amino acid side chains that can be used for coordination, specially biosynthesized **macrocyclic chelate ligands**, and **nucleic acids** with nucleobases as potentially coordinating components.

2.3.1 Coordination by Proteins-Comments on Enzymatic Catalysis

Proteins, including enzymes, consist of α amino acids, which are connected via "peptide bonds", —C(O)—N(H)—. This carboxamide function alone is only a rather poor metal coordination site. However, as in corresponding solvents such as N-methyl- and N,N-dimethylformamide, the high local concentration of amide functions leads to a relatively high dielectric constant (the "protein as medium") and therefore reduces ionic attraction and repulsion forces within proteins and protein complexes.

The functional groups in the side chains of the following amino acids are particularly well suited for metal coordination (see Table 2.5):

- **His**tidine coordinates mainly through the basic δ imine, sometimes through the ε imine nitrogen center of the imidazole ring (tautomeric forms). *Both* nitrogen atoms can become available for coordination after metal-induced deprotonation, producing a metal–metal bridging μ-imidazolate (see Cu,Zn-superoxide dismutase; Section 10.5).
- **Met**hionine, which is one of the limiting essential amino acids in animal feed, binds via the neutral δ sulfur atom of the weakly π-accepting thioether.
- **Cys**teine contains a negatively charged γ thiolate center after deprotonation ($pK_a \approx 8.5$) to σ- and π-donating "cysteinate". It can bridge two or more metal centers (see the P clusters of nitrogenase (7.11)).
- **Sele**no**cys**teine, the 21st proteinogenic amino acid, features a negatively charged σ- and π-donating selenolate center after deprotonation ($pK_a \approx 5$; see Section 16.8). Like cysteine (Sections 7.1–7.4), it can be oxygenated and still act as a ligand.
- **Tyr**osine coordinates mainly via the negatively charged phenolate oxygen atom after deprotonation to σ and π electron-rich tyrosinate ($pK_a = 10$; see Figure 8.4). However, metal binding can also occur via the neutral phenolic form or the simultaneously deprotonated and oxidized species, that is, the tyrosyl radical.

Table 2.5 The most important metal-coordinating amino acids.

α amino acid	Side chain, R $R\text{-}^{\alpha}CH(NH_3^+)CO_2^-$
histidine (His)	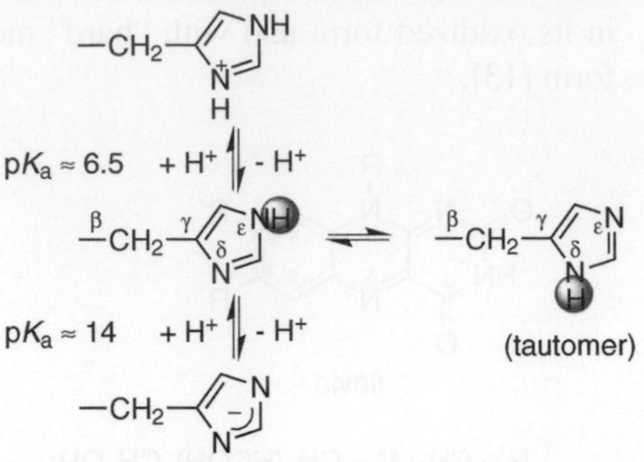
methionine (Met)	$—CH_2CH_2SCH_3$
cysteine (Cys)	$—CH_2SH$
selenocysteine (SeCys)	$—CH_2SeH$
tyrosine (Tyr)	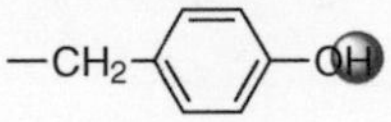
aspartic acid (Asp)	$—CH_2COOH$
glutamic acid (Glu)	$—CH_2CH_2COOH$

(H) acidic protons which may be substituted by metal cations

- **Glu**tamate and **Asp**artate bind via the negatively charged carboxylate functions (p$K_a \approx$ 4.5). Carboxylates can act as terminal (η^1), chelating (η^2) or bridging (μ-$\eta^1\eta^1$) ligands ((2.7); see also (7.14); [14]). A further distinction concerns the *syn* or *anti* positioning of the binding electron pairs. η^2-coordination involving four-membered chelate rings is found mainly in complexes with large metal ions, for example, in Ca^{2+}-containing proteins (Figure 14.4). The sometimes multiple μ-η^1:η^1 coordination of glutamate or aspartate has been observed in iron or manganese dimers, where the metal centers are often additionally bridged by μ-oxo or μ-hydroxo ligands ((7.14) or Figure 5.7). In all cases where the coordination is not η^1 (2.7), a strong deviation of the angle (C-O-M) from the ideal sp^2 angle of 120° is typical.

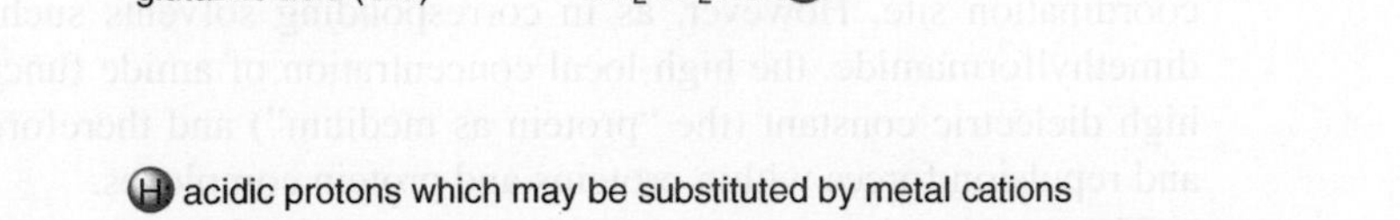

(2.7)

- Less common for metal-ion coordination are those amino acids with simple hydroxo or amino functions, such as **Ser**ine, **Thr**eonine, **Lys**ine and **Tryp**tophan.

The affinities of individual amino acid side chains to certain oxidation states of metals are often characteristic, resulting in a typical selectivity pattern. Metal complex formation constants of free amino acids, as well as observations made for proteins, point to the following preferred amino acid/metal ion combinations:

- **His:** Zn^{II}, Cu^{II}, Cu^{I} or Fe^{II};
- **Met:** Fe^{II}, Fe^{III}, Cu^{I} or Cu^{II};
- **Cys⁻:** Zn^{II}, Cu^{II}, Cu^{I}, Fe^{III}, Fe^{II}, $Mo^{IV\text{-}VI}$ or $Ni^{I\text{-}III}$;
- **Tyr⁻:** Fe^{III};
- **Glu⁻, Asp⁻:** Fe^{III}, Mn^{III}, Fe^{II}, Zn^{II}, Mg^{II} or Ca^{II}.

Conversely, individual oxidation states of the biometals favor the corresponding coordination environments. Two features are characteristic for these structural arrangements and unusual when compared to those of conventional coordination compounds:

1. The metal centers are often coordinatively unsaturated with regard to ligation by amino acid side chains; that is, one such residue is often missing in relation to a "regular" coordination number such as 4 (tetrahedron, square) or 6 (octahedron). For catalytic activity, however, such an open site is essential for coordinating the substrate; in most instances, that potentially open coordination site is temporarily occupied in the "resting state" by an easily replacable ligand such as H_2O. Proteins which exclusively transfer electrons do not have this requirement of coordinative unsaturation, since they do not have to coordinate the substrate directly at the metal center; however, other conditions apply for these proteins in terms of feature (2.) (see Sections 6.1 and 10.1).
2. While a certain amount of structure distortion must be expected for the metal coordination in proteins simply because of different amino acid ligands and the generally unsymmetrical environment, the coordination geometries of many protein-bound metal centers are quite irregular and deviate considerably from ideal structures. Figure 2.4 shows the typical structural representation of a relatively small metalloprotein resulting from x-ray crystallographic analysis.

The catalytic metal centers of enzymes are usually found deep within the more or less globular polypeptide, which thus acts as a huge chelate ligand, employing metal-binding amino acid residues. Access to substrates is often possible via specific channels in the tertiary structure, ensuring the desired substrate selectivity (lock-and-key model of enzymatic catalysis).

In many cases, however, the actual distortion of the metal coordination geometry is so pronounced in the enzyme that it cannot be regarded as coincidental. In 1968, Vallee and Williams rationalized this fact with their "theory of the entatic state" [16] of a catalytically active enzyme.

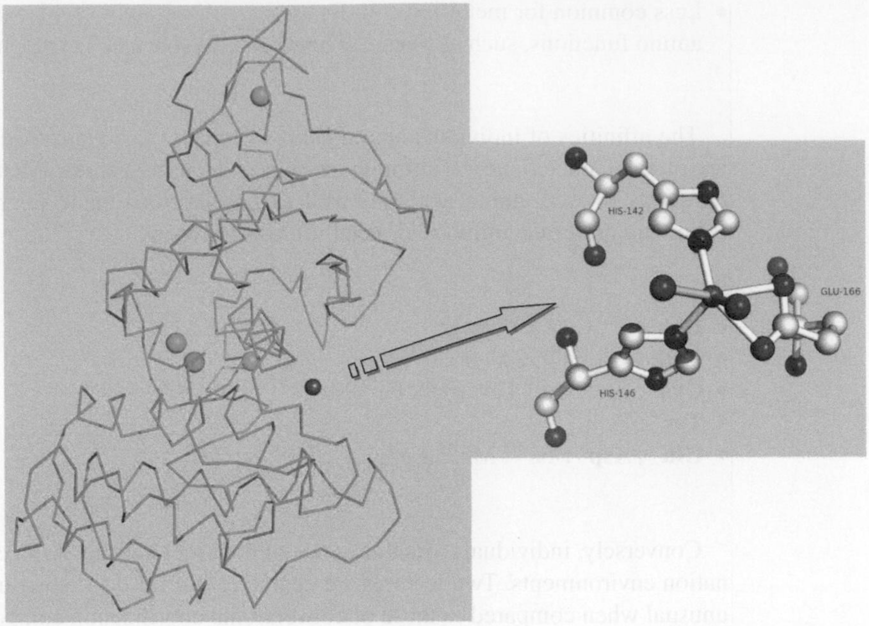

Figure 2.4
Structure of the proteolytic enzyme thermolysin (see Section 12.3) as determined by x-ray diffraction. The folding of the polypeptide chain of 316 amino acids (molecular mass 34 kDa) is shown in α carbon-backbone representation; that is, without depicting the side chains. Represented by spheres are the positions of four structure-stabilizing Ca^{2+} ions (green) and of the catalytic Zn^{2+} ion (dark grey), the detailed coordination of which (2 histidine, 1 glutamate, 2 H_2O) is shown in the insert (PDB code 1LNF) [15].

The "Entatic State" in Enzymatic Catalysis

Catalysts accelerate chemical reactions, inhibitors retard them. In an energy potential diagram (Figure 2.5), catalysis corresponds to a reduction in activation energy; in other words, the transition state to be overcome is changed in such a way that it is reached from the initial state with less energy. In metalloenzyme catalysis, a ternary complex is usually formed, consisting of substrate, enzymatically modified metal center and the second reactant, which might be a coenzyme or an acidic or basic group. The catalytic function of the metal center is at least twofold, namely to *electronically activate* one or both reacting species as intermediately bound ligands and to *position them in space, most often in a specific, unsymmetrical fashion* (three-point fixing). The latter role of the catalyst may be regarded simply according to statistical aspects: the probability for a successful encounter of the reactants increases greatly through coordination, which involves a restriction of the degrees of freedom for translation and rotation and thus produces a high "effective concentration". The energy difference between the initial and the final state is not affected at this stage.

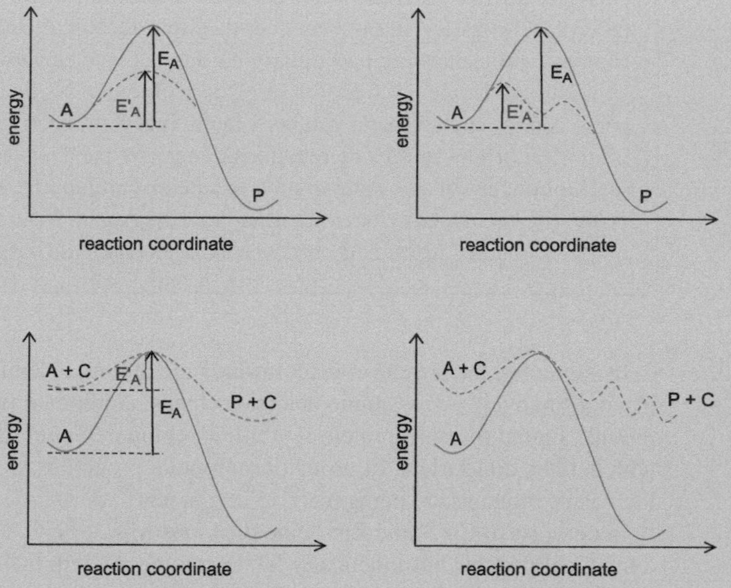

Figure 2.5
Energy profiles for catalyzed reactions at various degrees of sophistication. Upper left: conventional representation of the reduction of the activation energy $E_A \rightarrow E'_A$ upon transition from the initial state, A, to the products, P. Lower left: reduction of E_A by introduction of an "entatic" (strained, high-energy) enzymatic catalyst, C, which largely provides the preformed transition-state geometry of the substrate/catalyst complex. Upper right: realistic, multistep catalysis involving a new reaction pathway. Reprinted with permission from [19] © 1989, American Chemical Society. Lower right: realistic enzymatic catalysis.

According to the hypothesis of the entatic state, the often surprisingly efficient enzymatic catalysis can be largely explained by a "preformation of the transition state"; that is, the active center of the enzyme already features the (complementary) geometry necessary to reach the critical high-energy transition state of the substrate. In the "entatic" (strained) state of the enzyme, much of the energy necessary to reach that transition state is already stored and distributed over many chemical bonds. Small remaining geometrical changes between the initial and transition state of the *enzyme/substrate complex* then result in only a small activation energy (energy aspect), which means that productive encounters between reaction partners occur more often (statistical aspect) and that the reaction proceeds more rapidly (kinetic aspect).

For this reason, the active state of a metalloenzyme should not contain a regular (= low-energy, relaxed) coordination environment of the metal involved in the catalysis; on the contrary, the main goal should be a destabilization of the initial state [17]. Related concepts to that of the "entatic state" had been introduced previously by Haldane and Pauling and by Lumry and Eyring (1954) in the form of the "rack mechanism", implying that the protein serves as a structure-enforcing scaffold. The determination of enzyme structures can thus provide valuable information on otherwise experimentally inaccessible transition-state geometries of important types of chemical reactions. Since small changes in geometry are usually associated with low activation energies, the efficient enzymatic catalysis of reactions involving small molecules must take place *stepwise* in both biochemical and technical catalysis. Many small di- or triatomic molecules are particularly hard to activate because of their small number of geometrical degrees of freedom (e.g. the problem of N_2 fixation; see Section 11.2). In some cases, an enzymatic catalysis involving very reactive intermediates entails the inhibition of unwanted side reactions ("negative catalysis" [18]).

The concept of the entatic state has been particularly well demonstrated for hydrolyzing metalloenzymes (see Chapter 12). About 40% of all known enzymes are metal-dependent (especially oxidoreductases and hydrolases).

In summary, the protein can act towards metal ions as a multidentate chelate ligand via its coordinatively active amino acid side chains. Additional functions of the protein are to provide spatial fixation (protein as scaffold; compare Figure 2.4) and to serve as a medium with defined dielectrical, (a)protic or (non)polar properties [20]. Based on these principles, increasing attempts are being made to design new ("*de novo*") metal-coordinating proteins.

In contrast to Cu^{2+} and Zn^{2+}, the dications Mg^{2+}, Fe^{2+}, Ni^{2+} and Co^{2+} form thermodynamically stable but kinetically labile complexes with the amino acid side chains from Table 2.5. Despite favorable equilibria for complex formation, the activation energy for dissociation of these ions may be so small that exchange reactions are frequent. Such a situation would be detrimental to efficient catalytic functioning of a metalloenzyme; the kinetic stabilization of the aforementioned metal ions thus requires special measures: the use of macrocyclic chelate ligands.

2.3.2 Tetrapyrrole Ligands and Other Macrocycles

The tetrapyrroles [21] are fully or partially unsaturated tetradentate macrocyclic ligands (2.8) which, in their deprotonated forms, can tightly bind even substitutionally labile divalent metal ions.

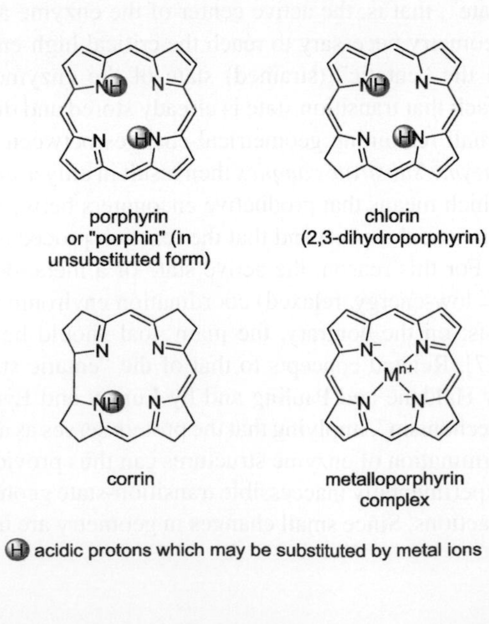

(2.8)

The resulting complexes [22–24] are among the most common and best known bioinorganic compounds (2.9):

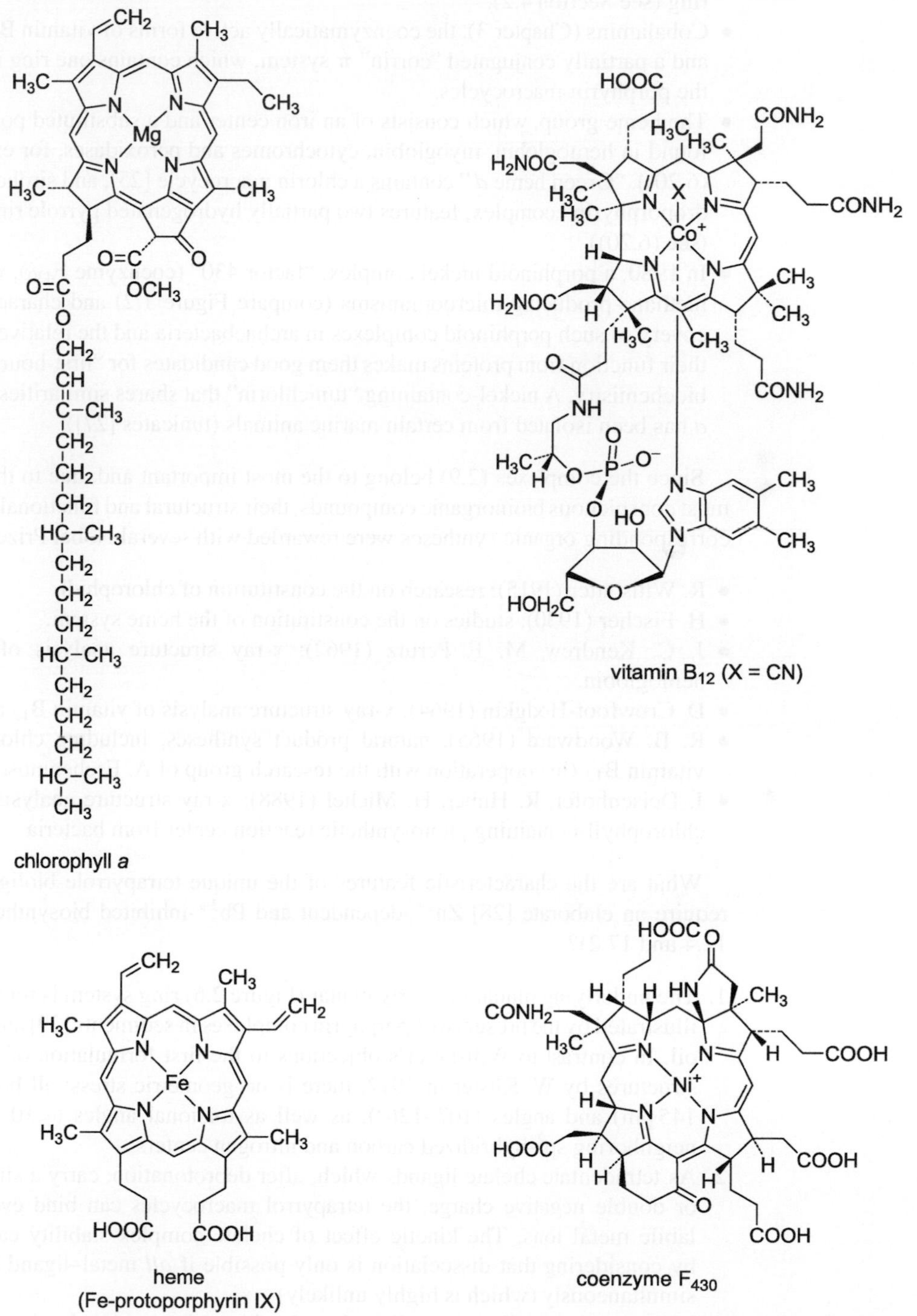

(2.9)

- Chlorophylls contain (otherwise labile) Mg^{2+} as the central ion and partially hydrogenated and substituted porphyrin ligands ("chlorins") with an annelated five-membered ring (see Section 4.2).
- Cobalamins (Chapter 3), the coenzymatically active forms of vitamin B_{12}, contain cobalt and a partially conjugated "corrin" π system, which contains one ring member less than the porphyrin macrocycles.
- The heme group, which consists of an iron center and a substituted porphyrin ligand, is found in hemoglobin, myoglobin, cytochromes and peroxidases, for example (compare (6.20)). "Green heme *d*" contains a chlorin macrocycle [25], and siroheme, another "hydroporphyrin" complex, features two partially hydrogenated pyrrole rings in *cis* position (see (6.20)).
- In 1980, a porphinoid nickel complex, "factor 430" (coenzyme F_{430}), was isolated from methane-producing microorganisms (compare Figure 1.2) and characterized. The discovery of such porphinoid complexes in archaebacteria and the relative independence of their function from proteins makes them good candidates for "first-hour catalysts" [26] in biochemistry. A nickel-containing "tunichlorin" that shares similarities with chlorophyll *a* has been isolated from certain marine animals (tunicates [27]).

Since the complexes (2.9) belong to the most important and, due to their intense color, most conspicuous bioinorganic compounds, their structural and functional investigation and corresponding organic syntheses were rewarded with several Nobel Prizes in Chemistry:

- R. Willstätter (1915): research on the constitution of chlorophyll.
- H. Fischer (1930): studies on the constitution of the heme system.
- J. C. Kendrew, M. F. Perutz (1962): x-ray structure analysis of myoglobin and hemoglobin.
- D. Crowfoot-Hodgkin (1964): x-ray structure analysis of vitamin B_{12} and derivatives.
- R. B. Woodward (1965): natural product syntheses, including chlorophyll and later vitamin B_{12} (in cooperation with the research group of A. Eschenmoser).
- J. Deisenhofer, R. Huber, H. Michel (1988): x-ray structure analysis of a heme- and chlorophyll-containing photosynthetic reaction center from bacteria.

What are the characteristic features of the unique tetrapyrrole bioligands (2.8) which require an elaborate [28] Zn^{2+}-dependent and Pb^{2+}-inhibited biosynthesis (see Sections 12.4 and 17.2)?

1. The underlying planar or nearly planar (Figure 2.6) ring system is remarkably stable, as illustrated by the presence of porphyrin complexes in sediments [29] and in crude mineral oil. In contrast to Willstätter's objections to the first formulation of such macrocyclic structures by W. Küster in 1912, there is no geometric stress; all bond lengths (134–145 pm) and angles (107–126°), as well as torsional angles (<10°), are normal for neighboring sp^2-hybridized carbon and nitrogen centers.
2. As tetradentate chelate ligands which, after deprotonation, carry a single (corrin, F_{430}) or double negative charge, the tetrapyrrol macrocycles can bind even coordinatively labile metal ions. The kinetic effect of chelate complex stability can be rationalized by considering that dissociation is only possible if *all* metal–ligand bonds are broken simultaneously (which is highly unlikely).
3. Macrocyclic ligands are usually quite selective with regard to the size of the coordinated ion [30]. This is especially true for the tetrapyrroles, because they are fairly rigid due to

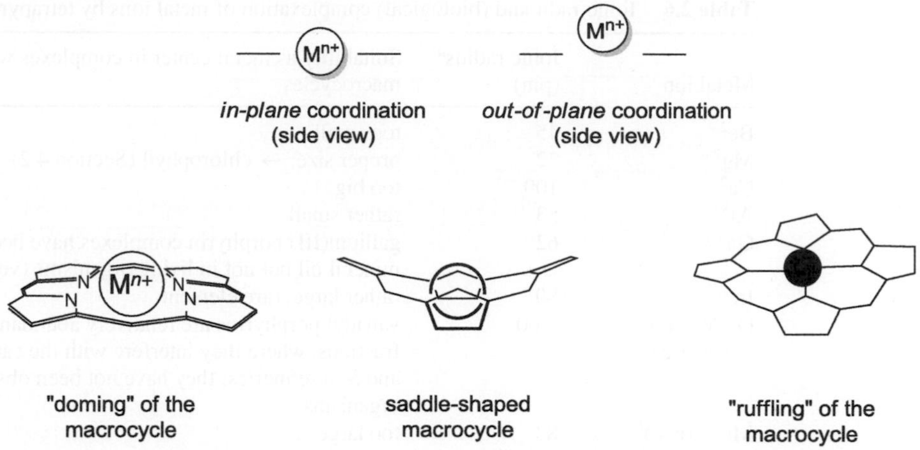

Figure 2.6
Typical geometrical deviations for complexes of tetrapyrrole macrocycles (cf. [31]).

the presence of many conjugated double bonds, although small, functionally significant distortions can occur (Figure 2.6). Structural data and model calculations show that spherical ions with a radius of 60–70 pm are best suited to fitting into the central cavity of tetrapyrrole macrocycles ("*in-plane*" coordination, Figure 2.6). A survey of different metal ions from the periodic table illustrates this fact (Table 2.6).

4. Most tetrapyrrole ligands contain an extensively conjugated π system. The Hückel rule for "aromatic" and thus particularly stabilized cyclic π systems is fulfilled for porphyrins as they feature $18 = 4n + 2$ π electrons in the inner 16-membered ring. This heteroaromaticity presumably contributes to the thermal stability of the ring system [22], which was mentioned earlier. As a consequence of the extensive π conjugation, the ligands and their metal complexes show intense absorption bands in the visible region of the electromagnetic spectrum, which led to their designation as tetrapyrrole "pigments" or "pigments of life". Furthermore, the uptake and release of electrons by these heterocycles (i.e. one-electron reduction or oxidation processes) are facilitated by the narrowing of the π frontier orbital gap. In fact, these ligands (Por^{n-}, n = 1,2,3) can behave "non-innocently" [32], and the resulting anion or cation radical complexes (2.10) can be quite stable. Both these properties, light absorption and redox behavior in terms of electron buffering and storage, render complexes of the tetrapyrrole macrocycles essential redox components in the most important biological energy transformations: photosynthesis and respiration (see Chapters 4–6).

$$[(Por^{\bullet -})M^{2+}]^{+} \underset{-\,e^-}{\overset{+\,e^-}{\rightleftharpoons}} [(Por^{2-})M^{2+}] \underset{-\,e^-}{\overset{+\,e^-}{\rightleftharpoons}} [(Por^{\bullet 3-})M^{2+}]^{-} \qquad (2.10)$$

Por^{n-}: porphinato ligand

5. Tetrapyrrole macrocycles are tetradentate chelate ligands that prefer a planar or nearly planar arrangement (Figure 2.6) around a metal center. Assuming a total coordination number of 6 in an approximately octahedral arrangement, this situation leaves two axial coordination sites, X and Y, available at the metal center (2.11). For a controlled

Table 2.6 Ionic radii and (biological) complexation of metal ions by tetrapyrrole ligands.

Metal ion	Ionic radius[a] (pm)	Suitability as metal center in complexes with tetrapyrrole macrocycles
Be^{2+}	45	too small
Mg^{2+}	72	proper size; → chlorophyll (Section 4.2)
Ca^{2+}	100	too big
Al^{3+}	53	rather small
Ga^{3+}	62	gallium(III) porphyrin complexes have been found in crude mineral oil but not in living organisms (very rare element)
In^{3+}	80	rather large, rare element
$O{=}V^{2+}$ (not spherical)	~60	vanadyl porphyrins are relatively abundant in certain crude oil fractions, where they interfere with the catalytic removal of N and S in refineries; they have not been observed in living organisms
Mn^{2+} (h.s.)[b]	83	too large
Mn^{3+}	~60	proper size; use in synthetic oxidation catalysts
Fe^{2+}(h.s.)	78	too large (*out-of-plane* structure; compare Figure 5.4)
Fe^{2+}(l.s.)[c]	61	proper size
Fe^{3+}(h.s.)	65	proper size
Fe^{3+}(l.s.)	55	rather small
average value for $Fe^{2+/3+}$	65	→ heme system with Fe^{n+} in various oxidation and spin states (Chapters 5 and 6)
Co^{2+}(l.s.)	65	proper size; → cobalamins (Chapter 3)
Ni^{2+}	69	proper size; → F_{430} (Section 9.5), tunichlorin
Cu^{2+}	73	relatively large; Cu porphyrins have not been found in organisms, strong bonds are formed mainly with histidine in proteins
Zn^{2+}	74	relatively large; Zn porphyrins have not been found in organisms, strong bonds are formed e.g. with histidine or cysteinate in proteins

[a]For coordination number 6.
[b]h.s., high spin.
[c]l.s., low spin.

stoichiometric or catalytic activation of substrates, there is indeed a need for two such open coordination sites: one for the actual binding of the substrate and another for the regulation of catalytic activity, using the "trans effect", for example. The following are some examples illustrating the useful functions of axial ligands in biological tetrapyrrole complexes:

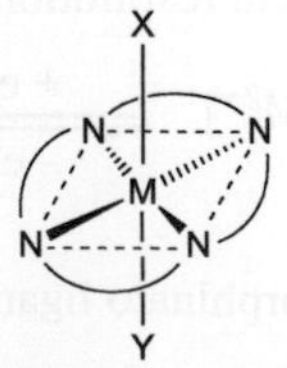

(2.11)

- The substrate of hemoglobin is molecular oxygen, O_2 = X, which has to be reversibly coordinated for transport to the tissue. A functionally useful "proximal" histidine is coordinated in the sixth position, Y, at the iron atom, determining the oxidation and spin state (Section 5.2).

- Coenzymatically active cobalamins contain primary alkyl groups, CH_2-R = X, coordinated directly to the cobalt center. The homolytic cleavage of these metal–alkyl bonds to yield radicals is effected in the enzyme via changes in the coordination of the sixth ligand, Y (a benzimidazole derivative in the isolated coenzyme (3.1)).
- Chlorophylls exist as highly aggregated and spatially oriented species in "antenna pigments" (Figure 4.2) or in the "special pair" of photosynthetic reaction centers (Figure 4.4). This well-controlled aggregation requires multiple coordinative interactions, in which the Mg^{2+} centers can act as bifunctional acceptors (Lewis acids) for electron-pair-containing ligands, X and Y, while the carbonyl groups of the chlorophyll molecules, for example, can function as electron pair donors (Lewis bases).

6. The tetragonal distortion of the octahedral symmetry through "strong" dianionic tetrapyrrole ligands causes a characteristic splitting of the d orbitals of coordinated transition metal centers (Figure 2.7). Some consequences of this effect for chemical reactivity are discussed in detail for vitamin B_{12} and cobalamins in Chapter 3. Although the equatorial ligand field strength of planar tetrapyrrole dianions should stabilize low-spin configurations versus their high-spin equivalents, there are species such as deoxy-hemoglobin and deoxy-myoglobin which feature a very critical high-spin iron(II) center. Due to the rather large size of high-spin Fe^{II} (see Table 2.6), this metal ion does not fit completely into the cavity of the macrocycle, which results in *out-of-plane* complexation (Figures 2.6 and 5.4) and hence a diminished ligand-field

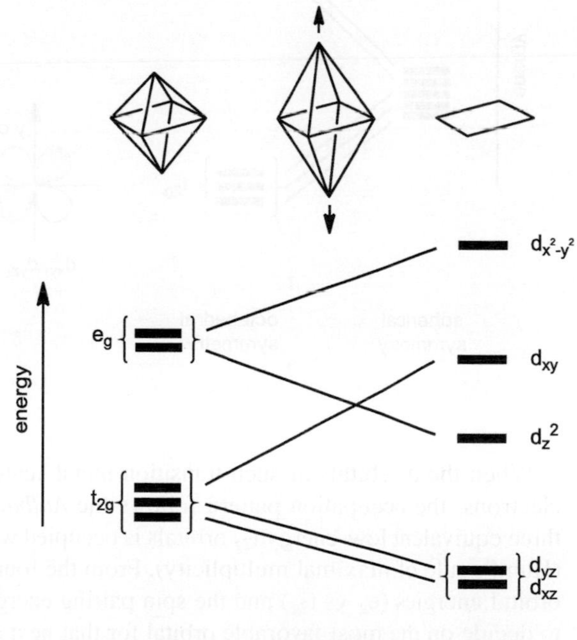

Figure 2.7
Correlation diagram for the splitting of transition metal d orbitals, depending on the extent of tetragonal distortion (here: axial elongation) of octahedral symmetry.

effect from the porphyrin. A similar relationship between spin state and the geometry of the tetrapyrrole ligand was found for the d^8 ion nickel(II) in coenzyme F_{430} (see Section 9.5).

Electron Spin States in Transition Metal Ions

The terms "high spin" and "low spin" are derived from **ligand field theory**. In a ligand field with octahedral symmetry, the five d orbitals (which are energetically equivalent in the free, spherically symmetrical transition metal ion) split into two energetically different groups of energy levels (2.12).

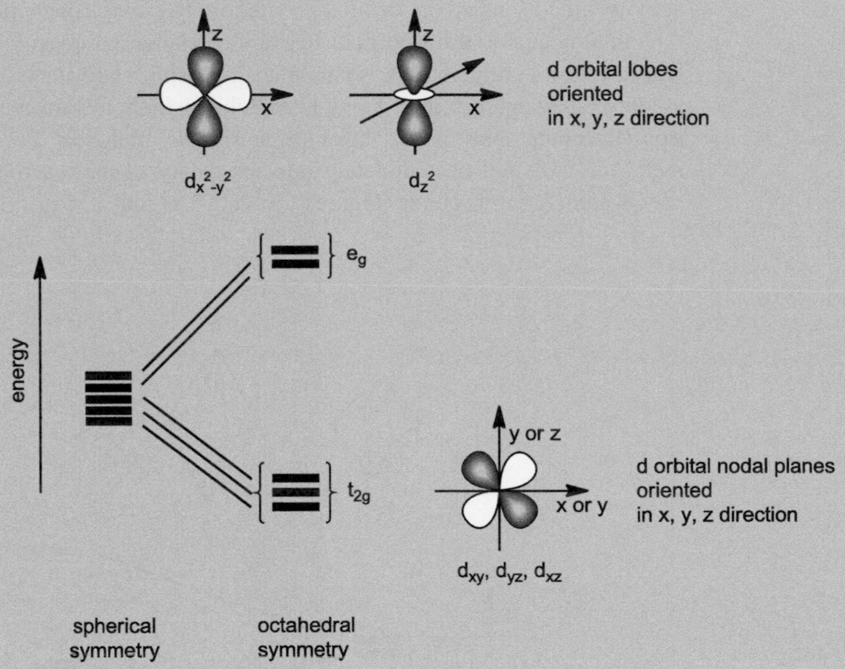

(2.12)

When the d orbitals of such transition metal centers are consecutively filled with electrons, the occupation pattern follows the *Aufbau* principle. At first, each of the three equivalent low-energy t_{2g} orbitals is occupied with one electron of the same spin (Hund's rule of maximal multiplicity). From the fourth electron on, the difference in orbital energies (e_g vs t_{2g}) and the spin pairing energy have to be compared in order to decide on the most favorable orbital for that next electron (2.13). If the difference in orbital energies is larger than the spin pairing energy, the electron will be placed in an already occupied orbital of the t_{2g} level such that those two spins are paired; two unpaired electrons of the same spin remain in a "low-spin" situation. If, on the other hand, the energy difference between e_g and t_{2g} orbitals is smaller than the spin pairing energy, the occupation of an e_g level becomes more favorable, which, according to Hund's rule, leads to four unpaired electrons in a "high-spin" configuration.

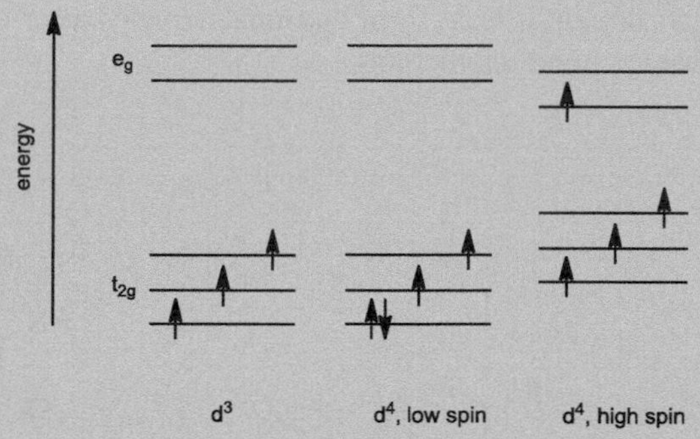

(2.13)

Octahedrally coordinated transition metal ions with an electronic configuration of d^4–d^7 can thus exist either as low-spin or as high-spin species. Which of these alternatives is realized in a complex is determined by the orbital splitting, as induced by the kind of metal and by the "strength" of the ligand field (see (2.14) for examples).

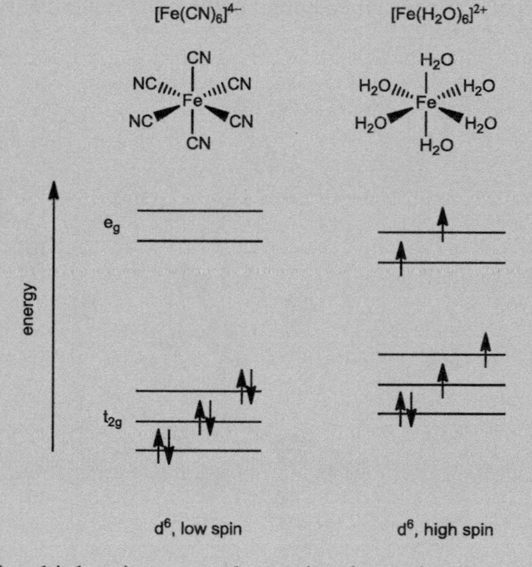

(2.14)

It is obvious that high-spin versus low-spin alternatives can also exist for less symmetrical coordination geometries and for other d^n configurations. For ions such as Fe^{2+} and Fe^{3+} in certain low-symmetry environments there is also the alternative of "intermediate-spin" states.

Despite the often slightly nonplanar structures (Figure 2.6) which permit some fine-tuning of the electronic structure and reactivity, the tetrapyrrole ligands essentially feature a planar two-dimensional ring system. For complexation of the extremely labile alkaline metal monocations, there are other, multidentate and much more three-dimensionally structured macrocycles available as biological ligands: the ionophores. These ligands, their complexes and their synthetic analoga are discussed in detail in Section 13.2; however, some important features shall be pointed out here.

The ionophores are multidentate (≥6) chelate ligands that either exist as macrocycles (2.15) or can at least form quasimacrocycles after coordination-induced ring closure via hydrogen bond interactions.

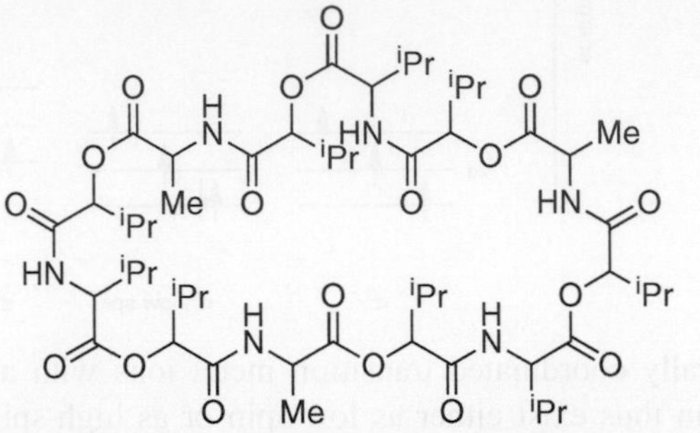

valinomycin (2.15)

The alkali metals, which generally form only highly labile complexes, and the rather labile Ca^{2+} ion can be bound in the polar inner cavity of such complex ligands (Figure 2.8)

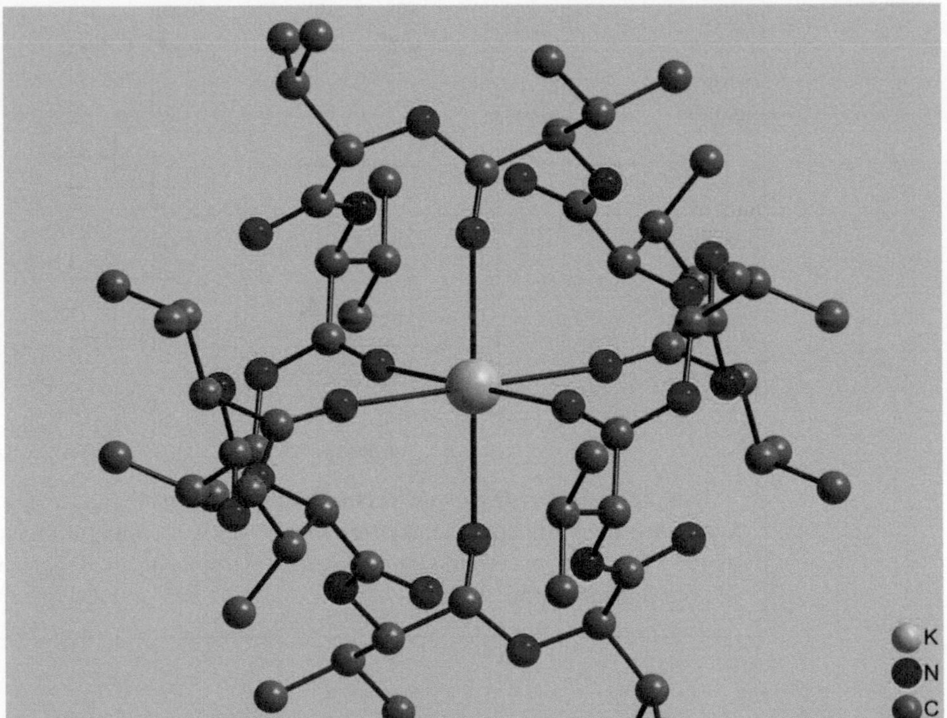

Figure 2.8
Molecular structure of the K^+/valinomycin complex. Reprinted with permission from [35] © 1995, Bioorganicheskaya Khim/Springer.

by several strategically positioned heteroatoms (N or O), according to their size (size selectivity [30]; chelate/crown ether/cryptate effect [33,34]). Dissociation is possible only when *all* coordinative bonds are broken simultaneously and when a substantial conformational change occurs. The often lipophilic outside of these complexes allows a transport of the corresponding ions through biological membranes. This ability renders such macrochelating natural products useful as antibiotics, because of the effect on the ion distribution between both sides of bacterial membranes (see Chapter 13). In 1987, the Nobel Prize in Chemistry was awarded to C. J. Pedersen, J.-M. Lehn and D. J. Cram for the development of synthetic analogues of such macrocyclic natural products, which greatly improved our understanding of the underlying principles of *molecular recognition* mechanisms.

2.3.3 Nucleobases, Nucleotides and Nucleic Acids (RNA, DNA) as Ligands

Suitable ligands for metal ions include not only proteins – the best known biochemical function carriers – but also information-carrying nucleic acids, oligo- and polynucleotides and ribozymes [36,37]. The negatively charged phosphate/carbohydrate backbone is the obvious first coordination site for monovalent and especially divalent cations. The formation, replication and cleavage of nucleic acid polymers (RNA, DNA) as well as their structural integrity (e.g. the double-helical arrangement of conventional DNA) require the presence of metal ions. The donor-rich nucleobases and the available regions for "supramolecular" interactions (→ secondary bonding) merit special comment.

As constituents of nucleic acids, the heterocyclic nucleobases (2.16) have long been recognized as potential metal coordination sites [38]. Nucleobases (2.16) are ambidentate ligands that offer several different coordination sites for metal ions. Depending on the characteristics of the donor (e.g. type of atom, hybridization, basicity, chelate assistance), on external conditions (e.g. pH) and on the size and nature of the metal center, monodentate and multidentate coordination to imine, amino, amido, oxo and hydroxo functions are possible (compare Figure 19.2). An important aspect is the ability of nucleobases to exist in different tautomeric forms, as shown in (2.17) by example of cytosine.

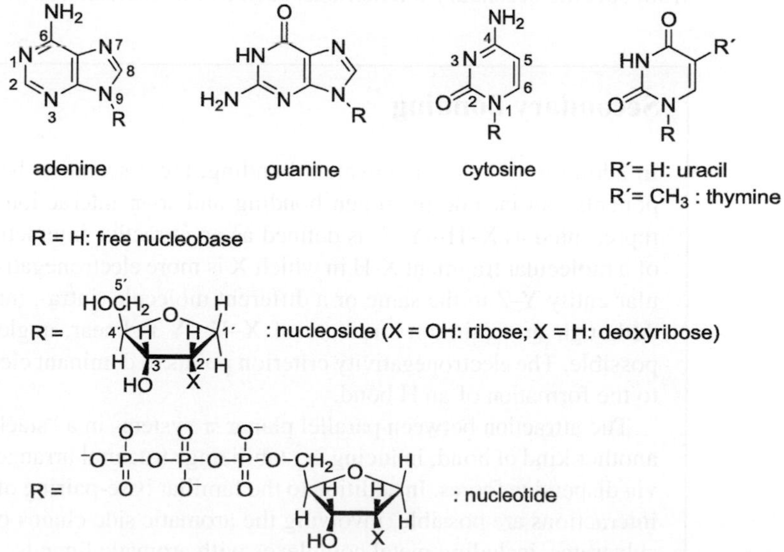

(2.16)

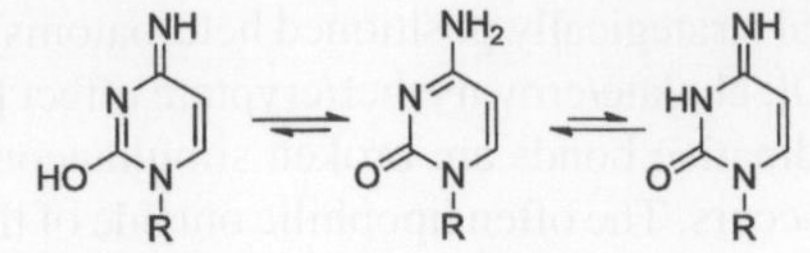

(2.17)

possible tautomers of N(1)-substituted cytosine

The presence of positively charged metal ions in the cell nucleus can thus affect the hydrogen-bond interactions essential for "natural" base pairing to such an extent that the intermediate formation of a "false" tautomer is favored. This can lead to a "mispairing" of nucleobases (2.18) and eventually, if not repaired, to altered genetic information transfer (cf. the potentially mutagenic or even carcinogenic effects of metal ions [39]).

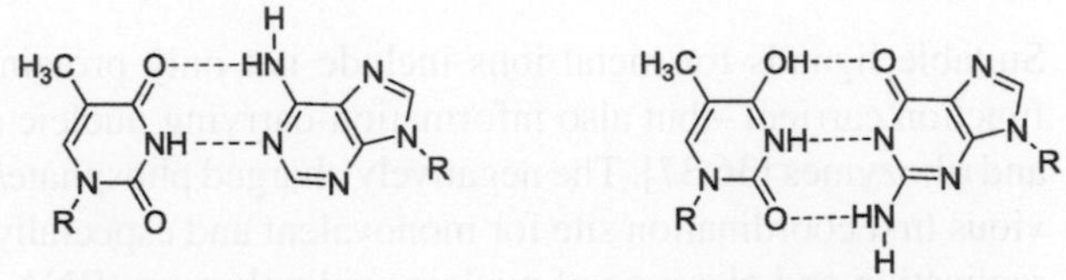

(2.18)

correct base pairing between thymine and adenine

mispairing between false thymine tautomer and guanine

Intensive studies aimed at understanding the antitumor activity of cisplatin (see Chapter 19.2) have shown that the coordination of metal complexes with nucleic acids can be very specific with regard to certain base-pair sequences in the double helix. These not necessarily covalent metal complex-nucleotide interactions ("secondary bonding") have received particular interest because relatively small, sequence-specific compounds can selectively target nucleic acids to act as chemical tools or potential medicinal agents, for example in tumor chemotherapy [40]. Sequence specificity, including the recognition of chiral structures, can result from direct metal coordination (see Figures 19.2 and 19.3), from specific coordination geometries of complexes and the resulting shape selection or from specific secondary interactions, such as the intercalation of ligands.

Secondary Bonding

In addition to ionic and covalent bonding, the association between molecular components can include hydrogen bonding and π–π interactions. The hydrogen bond, represented as X–H···Y–Z, is defined as an attractive interaction between an H atom of a molecular fragment X-H in which X is more electronegative than H, and a molecular entity Y–Z in the same or a different molecule (intra-, intermolecular H bonds). Although the optimum structure of X–H···Y is linear, angles lower than 180° are possible. The electronegativity criterion points to dominant electrostatic contributions to the formation of an H bond.

The attraction between parallel planar π systems in a "stacked" fashion represents another kind of bond, inducing or stabilizing structural arrangements in biomolecules via dispersion forces. In addition to the familiar base-pairing of DNA, other such π–π interactions are possible, involving the aromatic side chains of proteins and external substrates, including metal complexes with aromatic ligands.

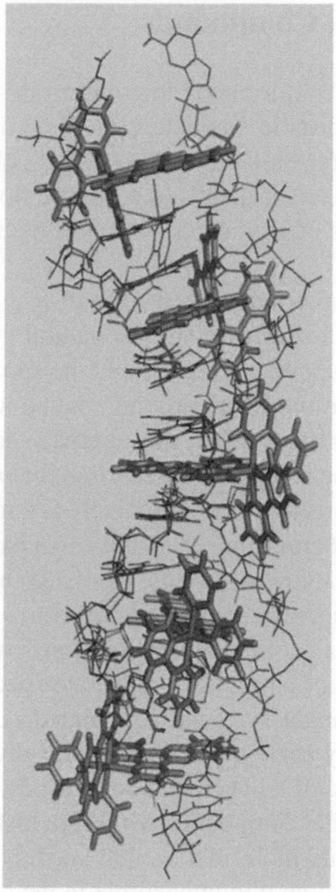

Figure 2.9
Example of a metal complex/oligonucleotide interaction (PDB code 4E1U) [41].

For instance, optically active (chiral) and light-responsive ruthenium(II) complexes such as $[Ru(phen)_3]^{2+}$ (2.19), phen = 1,10-phenanthroline and $[Ru(bpy)_2(dppz)]^{2+}$, with a potentially intercalating dppz ligand, can bind to oligonucleotide sequences (Figure 2.9) with base mismatch to effect intercalation, end-capping and insertion under rejection of mispaired base moieties [41].

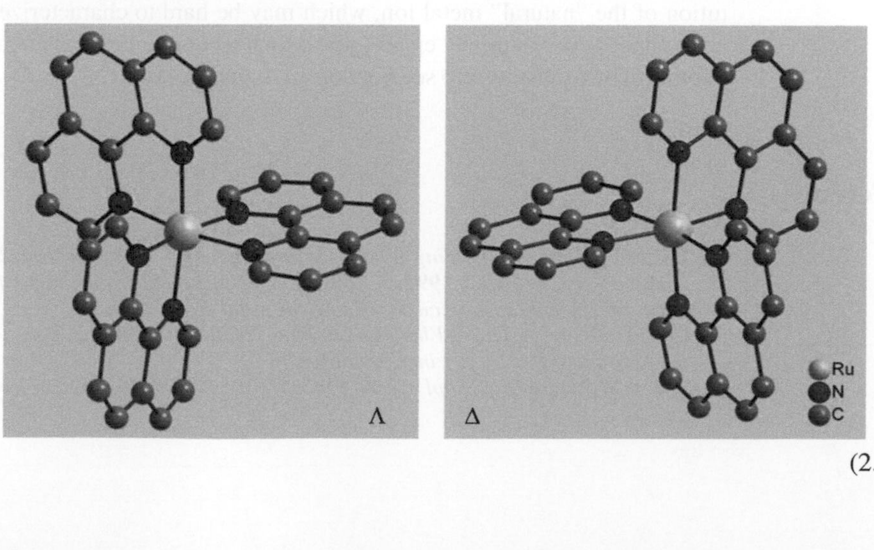

(2.19)

2.4 Relevance of Model Compounds

Research on "bioinspired" low-molecular-weight model compounds is widely used by inorganic chemists to simulate ("mimic") the main spectroscopic, structural and reactivity features of large bioinorganic systems. Such studies are particularly appropriate when structural or mechanistic details (e.g. of metalloproteins) are not yet available or are ambiguous. Several states of sophistication can be distinguished in modeling actual biological systems [42,43].

As a minimal requirement, *physical, in particular spectroscopic properties* and *basic structural characteristics* of the natural system should be adequately represented by the model. This approach implies that the spectroscopic behavior is determined mainly by the first coordination sphere around a metal ion. If very little is known about the structure of the system to be modeled, this approach can only help to exclude alternatives.

Even the next step, the *qualitative simulation of the reactivity* of the natural system, has been accomplished only in relatively few model studies. Such a simulation is very desirable because biochemical mechanisms often parallel technical processes, and efficient synthetic catalysts are attractive targets. Normally, however, synthetic enzyme models exhibit only a stoichiometric, noncatalytic reactivity towards the natural substrate.

The last stage of modeling, the approximately *quantitative simulation of reactivity* with respect to reaction rate and substrate specificity, is nearly impossible to attain with low-molecular-weight systems. As pointed out in Section 2.3.1, the desired high selectivity (lock-and-key analogy) *and* high reactivity (entatic state situation) require the highly complex structure of biochemical compounds. In this introductory text we can mention only a very few model compounds which are historically important.

Intermediate-molecular-weight models are those which provide a selected section of biopolymers – an oligonucleotide in the case of nucleic acids (Figure 2.9) or a peptide as part of a protein – in order to probe the interaction with metal ions.

It has now become possible to modify bioinorganic systems, particularly proteins, by specific recombinant DNA techniques; that is, via site-directed mutagenesis. The effects of altering a single amino acid in the peptide sequence can thus be elucidated, and new substrate specificities may result. Model studies of this type (see Section 6.1) have a promising potential for application in biotechnology, for example with respect to a transfer of N_2 fixation capability or to the microbial degradation of toxic waste.

A more inorganic and long-practiced modification of bioinorganic systems is the substitution of the "natural" metal ion, which may be hard to characterize spectroscopically by an isotope, or very similar "ersatz" ion, which is more suitable for certain physical methods (isomorphous substitution; see Section 12.1 and 13.1).

References

1. J. Emsley, *Nature's Building Blocks*, Oxford University Press, Oxford, **2001**.
2. F. H. Nielsen, *FASEB J.* **1991**, *5*, 2661–2667: *Nutritional requirements for boron, silicon, vanadium, nickel and arsenic: current knowledge and speculation.*
3. C. T. Horovitz, *J. Trace. Elem. Electrolytes Health Dis.* **1988**, *2*, 135–144: *Is the major part of the periodic system really inessential for life?*
4. R. J. P. Williams, *Pure Appl. Chem.* **1983**, *55*, 1089–1100: *Inorganic elements in biological space and time.*

5. E. J. Underwood and several authors in *Phil. Trans. Roy. Soc. London B* **1981**, *294*, 1–213: *Metabolic and physiological consequences of trace element deficiency in animals and man.*
6. National Academy of Sciences, National Research Council, Food and Nutrition Board, *Recommended Dietary Allowances*, 10th Edition, National Academy Press, Washington, **1989**.
7. H. G. Seiler, H. Sigel, A. Sigel (eds.), *Handbook on Toxicity of Inorganic Compounds*, Marcel Dekker, New York, **1988**.
8. R. A. Bulman, *Struct. Bonding (Berlin)* **1987**, *67*, 91–141: *The chemistry of chelating agents in medical sciences.*
9. M. M. Jones, *Comments Inorg. Chem.* **1992**, *13*, 91–110: *Newer chelating agents for in vivo toxic metal mobilization.*
10. W. I. Weis, K. Drickamer, W. A. Hendrickson, *Nature (London)* **1992**, *360*, 127–134: *Structure of a C-type mannose-binding protein complexed with an oligosaccharide.*
11. D. M. Whitfield, S. Stoijkovski, B. Sarkar, *Coord. Chem. Rev.* **1993**, *122*, 171–225: *Metal coordination to carbohydrates. Structures and function.*
12. M. B. Davies, *Polyhedron* **1992**, *11*, 285–321: *Reactions of l-ascorbic acid with transition metal complexes.*
13. M. J. Clarke, *Comments Inorg. Chem.* **1984**, *3*, 133–151: *Electrochemical effects of metal ion coordination to noninnocent, biologically important molecules.*
14. R. L. Rardin, W. B. Tolman, S. J. Lippard, *New J. Chem.* **1991**, *15*, 417–430: *Monodentate carboxylate complexes and the carboxylate shift: implications for polymetalloprotein structure and function.*
15. D. R. Holland, A. C. Hausrath, D. Juers, B. W. Matthews, *Protein Sci.* **1995**, *4*, 1955–1965: *Structural analysis of zinc substitutions in the active site of thermolysin.*
16. B. L. Vallee, R. J. P. Williams, *Proc. Natl. Acad. Sci. USA* **1968**, *59*, 498–505: *Metalloenzymes. Entatic nature of their active sites.*
17. R. J. P. Williams, *J. Mol. Catalysis – Review Issue* **1985**, *30*, 1–26: *Metallo-enzyme catalysis: the entatic state.*
18. J. Retey, *Angew. Chem. Int. Ed. Engl.* **1990**, *29*, 355–361: *Enzymic reaction selectivity by negative catalysis or how do enzymes deal with highly reactive intermediates?*
19. A. Haim, *J. Chem. Educ.* **1989**, *66*, 935–937: *Catalysis: new reaction pathways, not just a lowering of the activation energy.*
20. R. Huber, *Angew. Chem. Int. Ed. Engl.* **1989**, *28*, 848–869: *A structural basis for the transmission of light energy and electrons in biology* (Nobel address).
21. G. P. Moss, *Pure Appl. Chem.* **1987**, *59*, 779–832: *Nomenclature of tetrapyrroles.*
22. J. H. Fuhrhop, *Angew. Chem. Int. Ed Engl.* **1974**, *15*, 321–335: *Reactivity of the porphyrin ligands.*
23. D. Dolphin (ed.), *The Porphyrins*, Vol. I–VII, Academic Press, New York, from **1978**.
24. B. Kräutler, *Chimia* **1987**, *41*, 277–292: *The porphinoids – versatile biological catalyst molecules.*
25. R. Timkovich, M. S. Cork, R. B. Gennis, P. Y. Johnson, *J. Am. Chem. Soc.* **1985**, *107*, 6069–6075: *Proposed structure of heme d, a prosthetic group of bacterial terminal oxidases.*
26. A. Eschenmoser, *Angew. Chem. Int. Ed. Engl.* **1988**, *27*, 5–39: *Vitamin B_{12}: experiments concerning the origin of its molecular structure.*
27. K. C. Bible, M. Buytendorp, P. D. Zierath, K. L. Rinehart, *Proc. Natl. Acad. Sci. USA* **1988**, *85*, 4582–4586: *Tunichlorin: a nickel chlorin isolated from the Caribbean tunicate Trididemnum solidum.*
28. H. A. Dailey, C. S. Jones, S. W. Karr, *Biochim. Biophys. Acta* **1989**, *999*, 7–11: *Interaction of free porphyrins and metalloporphyrins with mouse ferrochelatase. A model for the active site of ferrochelatase.*
29. P. Schaeffer, R. Ocampo, H. J. Callot, P. Albrecht, *Nature (London)* **1993**, *364*, 133–136: *Extraction of bound porphyrins from sulfur-rich sediments and their use for reconstruction of palaeoenvironments.*
30. R. D. Hancock, *J. Chem. Educ.* **1992**, *69*, 615–621: *Chelate ring size and metal ion selection.*
31. J. A. Shelnutt, X.-Z. Song, J.-G. Ma, S.-L. Jia, W. Jantzen, C. J. Medforth, *Chem. Soc. Rev.* **1998**, *27*, 31–41: *Nonplanar porphyrins and their significance in proteins.*
32. W. Kaim, B. Schwederski, *Coord. Chem. Rev.* **2010**, *254*, 1580–1588: *Non-innocent ligands in bioinorganic chemistry – an overview.*
33. J. M. Lehn, *Angew. Chem. Int. Ed. Engl.* **1988**, *27*, 89–112: *Supramolecular chemistry – molecules, supermolecules, and molecular functional units* (Nobel address).

34. F. Vögtle: *Supramolecular Chemistry*, John Wiley & Sons, New York, **1993**.
35. V. Z. Pletnev, I. N. Tsygannik, Yu. D. Fonarev, I. Yu. Mikhailova, Yu. V. Kulikov, V. T. Ivanov, D. A. Langs, W. L. Duax, *Bioorganicheskaya Khim.* **1995**, *21*, 828–833: *Crystalline and molecular structure of the K^{+}-complex of meso-valinomycin, cyclo(-(D-Val-L-Hyi-L-Val-D-Hyi)$_3$)·$KAuCl_4$.*
36. E. Freisinger, R. K. O. Sigel, *Coord. Chem. Rev.* **2007**, *251*, 1834–1851: *From nucleotides to ribozymes – a comparison of their metal ion binding properties.*
37. A. Sigel, H. Sigel, R. K. O. Sigel (eds.), *Interplay between Metal Ions and Nucleic Acids; Metal Ions in Life Science Series, Vol. 10*, Springer, Berlin, **2012**.
38. L. G. Marzilli, *Adv. Inorg. Biochem.* **1981**, *3*, 47–85: *Metal complexes of nucleic acid derivatives and nucleotides: binding sites and structures.*
39. H. Schöllhorn, U. Thewalt, B. Lippert, *J. Am. Chem. Soc.* **1989**, *111*, 7213–7221: *Metal-stabilized rare tautomers of nucleobases.*
40. H.-K. Liu, P. J. Sadler, *Acc. Chem. Res.* **2011**, *44*, 349–359: *Metal complexes as DNA intercalators.*
41. H. Song, J. T. Kaiser, J. K. Barton, *Nature Chem.* **2012**, *4*, 615–620: *Crystal structure of Δ-[Ru(bpy)$_2$dppz]$^{2+}$ bound to mismatched DNA reveals side-by-side metalloinsertion and intercalation.*
42. J. A. Ibers, R. H. Holm, *Science* **1980**, *209*, 223–235: *Modeling coordination sites in metallobiomolecules.*
43. K. D. Karlin, *Science* **1993**, *261*, 701–708: *Metalloenzymes, structural motifs, and inorganic models.*

3 Cobalamins, Including Vitamin and Coenzyme B_{12}

3.1 History and Structural Characterization

For a number of reasons, coenzyme B_{12} and its derivatives (3.1), including vitamin B_{12}, are useful introductory examples in the field of bioinorganic chemistry. First, several milestones in the development of the field have been associated with vitamin B_{12}, and later with the coenzyme. These pertain to the immediate therapeutic benefit of cobalamins, the early use of chromatographic purification methods, structural elucidation by x-ray crystallography and the relationship between enzymatic and coenzymatic reactivity. Furthermore, modern natural product synthesis and both bioorganic and organometallic chemistry have strongly profited from studies of the B_{12} system.

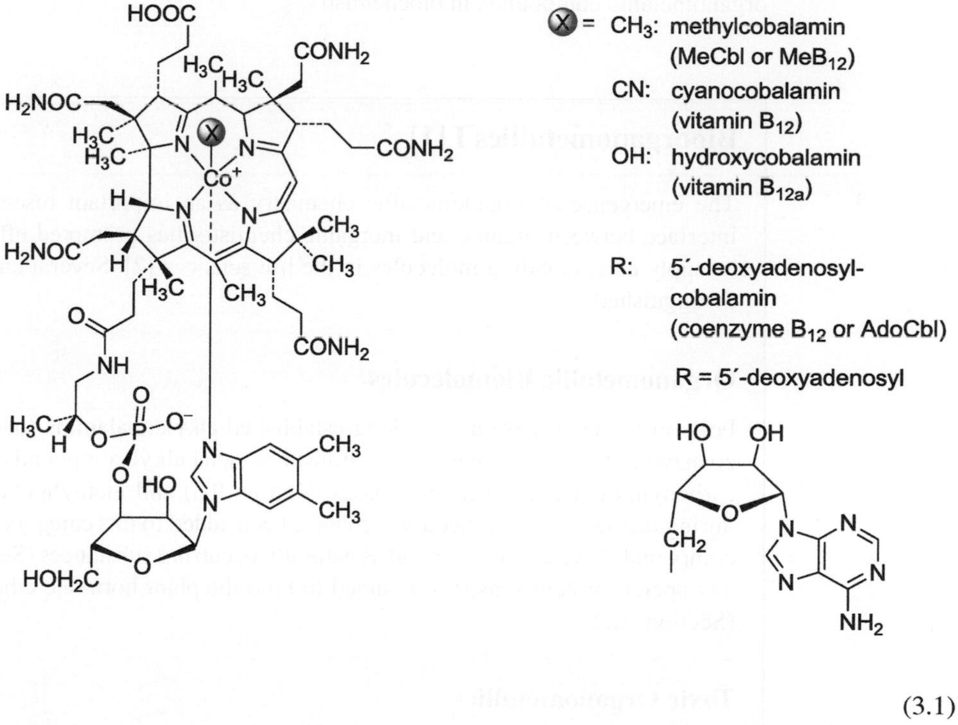

(3.1)

Bioinorganic Chemistry: Inorganic Elements in the Chemistry of Life – An Introduction and Guide, Second Edition.
Written and Translated by Wolfgang Kaim, Brigitte Schwederski and Axel Klein.

Coenzyme B_{12} (3.1) is a medium-sized molecule with a molecular mass of about 1350 Da that exhibits its characteristic specificity and high reactivity only in combination with corresponding apoenzymes (3.2).

coenzyme	+	**apoenzyme**	→	**holoenzyme**	
low molecular mass, determines the type of reaction		high molecular mass (protein), determines substrate specifity (selectivity) and the reaction rate		complete enzyme, fully functional	(3.2)

The incorporation of the element cobalt into the coenzyme is quite surprising, because cobalt is the least abundant first-row (3d) transition metal in the earth's crust and in sea water. Therefore, a very special functionality is to be expected. The corrin ligand (2.8) is also unique, particularly as regards its smaller ring size compared to the porphin systems. Cobalt-containing porphyrin complexes, although stable, are not suitable for mimicking the actions of coenzyme B_{12}.

The sixth, axial metal coordination site in coenzyme B_{12} and methylcobalamin features a primary alkyl group (3.1), which puts these complexes among the few examples of "natural" organometallic compounds in biochemistry.

Bioorganometallics I [1]

The emergence of organometallic chemistry as an important research area at the interface between organic and inorganic chemistry has prompted efforts to identify or apply corresponding molecules in the life sciences [2]. Several categories can be distinguished.

Organometallic Biomolecules

Foremost among these are the long-established alkylcobalamins (methylcobalamin, coenzyme B_{12}), as unusually stable transition metal alkyl compounds. More recently, carbonyliron centers of hydrogenases (Section 9.3) and methylnickel intermediates during methanogenesis (Section 9.5) have been added to this category. Organoarsenic compounds have also been found as naturally occurring substances (Section 16.4), and a copper(I) protein sensor is assumed to bind the plant hormone ethene, $H_2C{=}CH_2$ (Section 10.2).

Toxic Organometallics

Some of the most notorious toxic chemicals are found amongst the organoelement compounds, such as organomercury and organotin cations (Section 17.5). Membrane permeability and relatively inert M–C bonds contribute to their particularly harmful behavior.

Organometallics in Therapy

The toxicity of certain organometallics has been conducive to their use as antimicrobial agents. However, beyond this long-established application (organoarsenics: P. Ehrlich and S. Hata 1910) there has been much further progress in chemotherapy involving substitutionally reactive and nucleic acid-binding organometallics such as sandwich compounds for antitumor activity (Chapter 19). An essential aspect of these is their stability towards physiological conditions encountered. Carbon monoxide-releasing molecules have received attention because of the messenger function of CO (Section 19.4).

Organometallics for Diagnosis and Bioanalytics

Reasonably inert organometallic compounds have also been increasingly used in such areas as radioimaging, fluorescent imaging and vibrational monitoring. (See Chapter 9 for Bioorganometallics II.)

The Co–CH_2R configuration in alkylcobalamins is unusually stable towards hydrolysis in neutral aqueous solution. On the other hand, this cobalt–carbon bond shows a very special reactivity, namely the enzymatically controlled formation of reactive primary alkyl radicals. This unusual reactivity has prompted chemists from various fields beyond biochemistry to thoroughly study these remarkable complexes [3–8].

During the 1920s it was found that injections of extracts from animal liver were able to cure a very malignant ("pernicious") form of anemia that could otherwise be lethal. Improved methods of trace analysis soon showed that the essential component of these extracts contained cobalt. Since the substance was synthesized only by microorganisms and the trace element cobalt had to be supplied in any case, the factor was called "vitamin B_{12}". Enrichment and isolation turned out to be extremely laborious due to the low concentration of only 0.01 mg vitamin per liter of blood, so chromatographic separation methods had to be employed. The therapeutically useful but not directly active "vitaminic" form, cyanocobalamin (3.1), was obtained in pure form in 1948 [9].

As the complete determination of the molecular constitution of cobalamins was impossible using chemical means alone, its eventual structural elucidation required x-ray diffraction of single crystals, a method which could only be applied to relatively simple systems in the 1950s and early 1960s (see Insertion, Section 4.2). With approximately 100 nonhydrogen atoms, vitamin B_{12}, and later coenzyme B_{12}, posed a formidable crystallographic challenge [10]; for the solution of these problems, Dorothy Crowfoot Hodgkin was awarded the 1964 Nobel Prize in Chemistry.

The structure of the coenzyme (3.1) as obtained from crystallographic studies shows the cobalt–corrin macrocycle featuring an alkyl-bound 5′-deoxyadenosine group and an N(1)-coordinated 5,6-dimethylbenzimidazole ring (DMB) as axial ligands (Figure 3.1) [10]. The latter is connected to the corrin macrocycle via a long pendant chain, so that this corrin

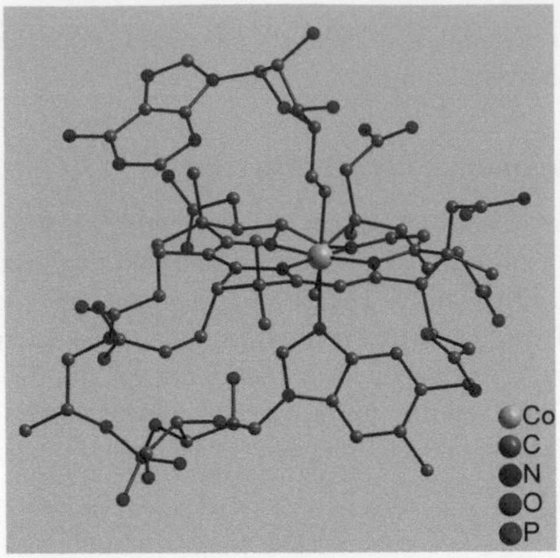

Figure 3.1
Molecular structure of coenzyme B_{12} AdoCbl. Reprinted with permission from [10c] © 2005, WILEY-VCH Verlag GmbH & Co. KGaA, Weinheim.

derivative effectively functions as a pentadentate chelate ligand. The resulting combination is unique, as the benzimidazole ligand can switch between the cobalt-coordinated "base-on" form and the decoordinated so-called "base-off" form, with possible replacement by a donor from the protein, as shown in B_{12} binding domains of methionine synthase (Figure 3.2; see Section 3.3.2) [11]. Coenzyme B_{12} can thus be considered a "molecular switch" [12,13]. In spite of extensive conjugation of π electrons, the unsaturated macrocycle is not completely flat but adopts a slightly bent "butterfly" or "saddle" conformation (Figures 3.1 and 2.6). Model studies have demonstrated the relevance of this structural feature for reactivity in terms of an entatic state structure. The nonplanarity results from the fact that the relatively large metal ion (Table 2.6) is encapsulated by a 15-membered instead of a 16-membered macrocycle.

In lieu of the 5′-deoxyadenosine moiety, a simple methyl group can be bound to the metal (methylcobalamin, MeCbl (3.1)); exchange with water, hydroxide or cyanide leads to the physiologically not directly active forms aqua-, hydroxo- and cyanocobalamin. The latter is called vitamin B_{12}, although this form represents only an artifact, that is, the result of the isolation procedure.

The existence of a relatively inert bond between a transition metal and a primary alkyl ligand is quite remarkable, especially since this true organometallic compound is stable under physiological conditions, that is, in aqueous solution at pH 7 and in the presence of oxygen. The corrin macrocycle creates a strong ligand field, resulting in a low-spin situation with considerable stabilization of the d^6 configuration (Co^{III}) in an approximately octahedral environment (2.14). However, the six-coordinate arrangement shows distinctive tetragonal distortion, and a corresponding splitting of the d orbitals according to Figure 2.7 is to be expected.

Considering the ideally suited low-spin d^6 situation for Co^{III} with accessible lower oxidation states Co^{II} and Co^{I}, it must be acknowledged that the equally d^6-configured neighbors of Co^{III} are less suitable: Fe^{II} is too labile with physiological ligands and tends to cross over to the less inert high-spin state, while Ni^{IV} will have a too-positive redox potential for a biological environment.

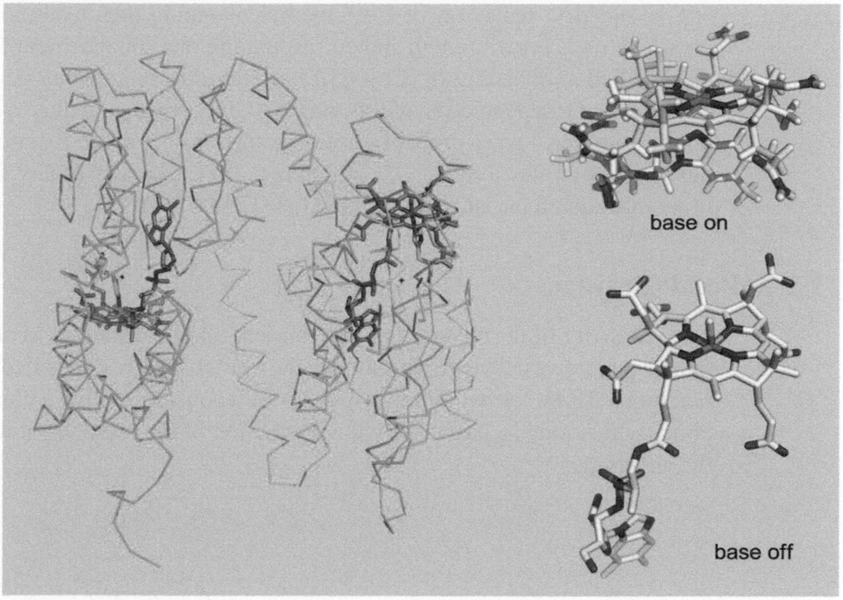

Figure 3.2
Representation of the structure of B_{12} binding domains of methionine synthase (PDB code 1BMT) [11]; base-on/base-off configurations shown for the coenzyme.

3.2 General Reactions of Alkylcobalamins

3.2.1 One-electron Reduction and Oxidation

In configuration (3.1), the trivalent cobalt ion exists as a six-coordinate metal center with the corrin monoanion, an anionic axial group (e.g. alkyl) and the neutral dimethylbenzimidazole base as ligands; a negatively charged phosphate in the side chain completes the charge balance. Starting from this configuration, two one-electron reduction steps are possible, the kind and number of axial ligands determining the actual redox potentials [14]. The reduction of the metal from a Co^{III} (d^6) (3.3, left) to a Co^{I} (d^8) configuration is accompanied by a tendency towards decreased axial coordination, with the ultimate case being a square planar arrangement (3.3, right).

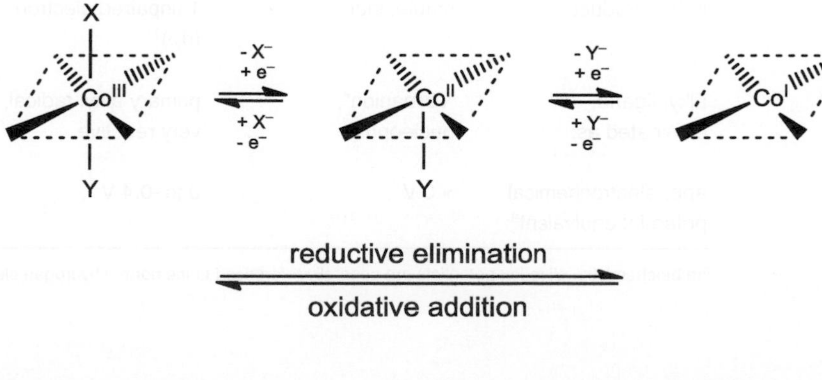

(3.3)

One-electron reduction of Co^{III}-methylcobalamin thus leads to a more than 50% decreased Co–C bond strength and corresponding rate enhancement for the homolysis due to a half-filled antibonding $\sigma^*(Co-CH_3)$ orbital (d_{z^2} component) [15]. Excitation with light can also lead to a population of this $\sigma^*(Co-CH_3)$ orbital and thus to a cleavage of the Co–C bond; however, this is probably not relevant for enzymatic reactions. Non-occupation of the strongly antibonding $d_{x^2-y^2}$ orbital in a d^8 system such as Co^I favors the sterically less favorable square planar configuration (see Figure 2.7).

3.2.2 Co–C Bond Cleavage

The reactivity of the physiologically relevant alkyl cobalamins is characterized by the fact that the reactive alkyl groups are made available in a controlled fashion for follow-up reactions. Three formal alternatives (3.4) are conceivable for a cleavage of the Co–CH_2R bond, which may be induced by the interaction of the coenzymes with the apoenzyme and the substrate.

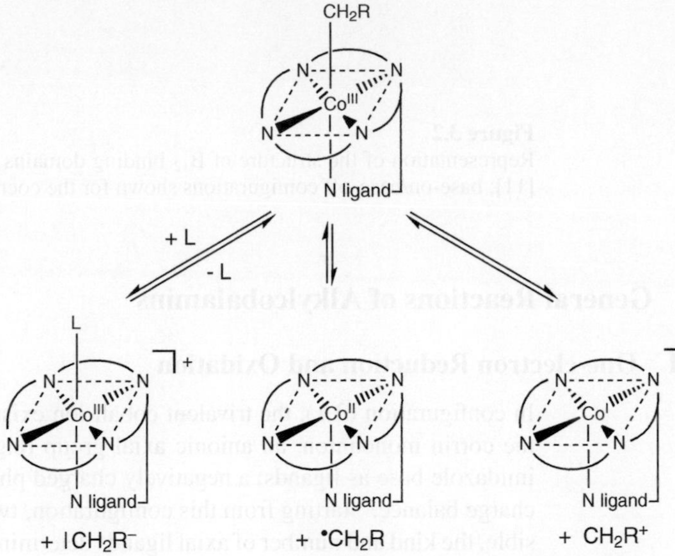

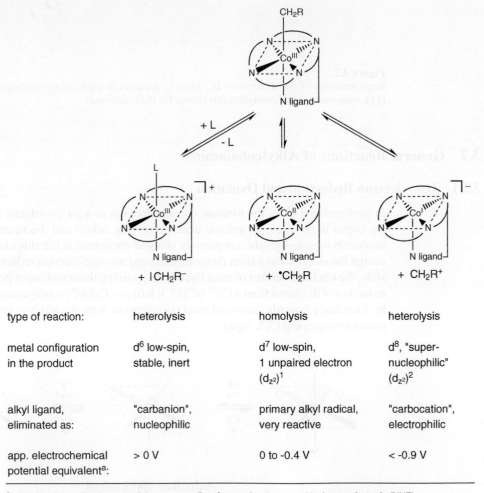

type of reaction:	heterolysis	homolysis	heterolysis
metal configuration in the product	d^6 low-spin, stable, inert	d^7 low-spin, 1 unpaired electron $(d_{z^2})^1$	d^8, "super-nucleophilic" $(d_{z^2})^2$
alkyl ligand, eliminated as:	"carbanion", nucleophilic	primary alkyl radical, very reactive	"carbocation", electrophilic
app. electrochemical potential equivalent[a]:	> 0 V	0 to -0.4 V	< -0.9 V

[a]In biochemistry, all redox potentials are generally referenced to the normal hydrogen electrode (NHE)

(3.4)

Heterolytic bond cleavage can lead either to low-spin Co^{III} and a carbanion equivalent $^{-}CH_2R$, involving substitution (e.g. by water), or to Co^{I} and a carbocationic alkyl moiety, $^{+}CH_2R$. In the latter case, a d^8-configurated metal center is formed, which behaves as a σ electron-rich "supernucleophile"; that is, its filled antibonding d_{z^2} orbital results in a high affinity towards σ electrophiles. A typical d^8 metal reactivity is thus the "oxidative addition" of organic compounds R–X. The carbanionic or carbocationic alkyl groups will not be produced as free ions but will be transferred in the presence of a reaction partner and a polar reaction medium in the transition state of the reaction.

The third alternative (3.4) is the homolytic bond cleavage, which leads to paramagnetic, electron paramagnetic resonance (EPR) spectroscopically detectable Co^{II} with a low-spin d^7 configuration (one unpaired electron) and a primary alkyl radical.

All three alternatives (3.4) are possible; axial coordination ("base-on") or noncoordination ("base-off") of donor ligands, the nature of the substrate, and the redox potential determine the actual reaction pathway (see Section 3.3). In the absence of a special base in the axial position, the carbanionic mechanism is realized at potentials above 0 V versus the normal hydrogen electrode (NHE). The Co^{I}/carbocation cleavage occurs only below approximately −0.9 V and thus beyond physiological conditions. Homolytic bond cleavage is a viable reaction in the physiologically interesting potential range between 0 and −0.4 V [3,14], resulting in EPR detectable radical intermediates [16–18].

Electron Paramagnetic Resonance I

EPR, or electron spin resonance (ESR), is an important spectroscopic method in bioinorganic chemistry [19–21]. The unpaired electron(s) in radicals or in complexes of transition metal centers with only partially filled d orbitals feature a spin whose orientation in a magnetic field can give rise (in the most simple case (3.5)) to two energetically different states (spin as a binary quantum mechanical feature of electrons). The transition from the energetically favorable and more populated state (**spin state I**) to the high-energy and thus less populated state (**spin state II**), the "resonance", can be induced by applying electromagnetic radiation, provided that certain spectroscopic selection rules are observed. With the magnetic fields of a few tenths of 1 Tesla (1 T = 10 000 G (Gauss)) normally used, this resonance requires microwave frequencies of approximately 10^{10} Hz.

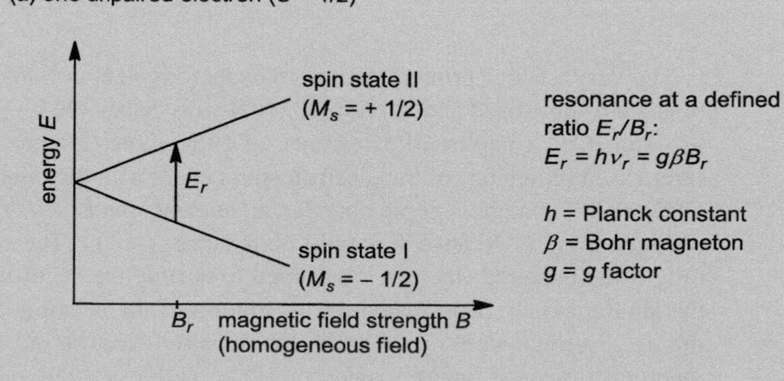

(3.5)

Both the magnitude of the resonance field strength at a given frequency and the hyperfine interaction(s) of the unpaired electron(s) with atomic nuclei that possess a nuclear spin $I \neq 0$ (e.g. $I = 1/2$ in (3.6)) are characteristic for the paramagnetic system. With this information, it is often possible to determine oxidation states, spin states and, under favorable circumstances, even details in the coordination sphere of metal centers. Since EPR spectroscopy is "blind" to the diamagnetic bulk of a protein, this method is often used for the initial characterization of a newly isolated metalloenzyme. The sensitivity of the EPR technique is such that small effective concentrations suffice to detect electron spin-bearing centers even in large proteins.

(b) one unpaired electron (electron spin $S = 1/2$) interacting with one proton (nuclear spin $I = 1/2$)

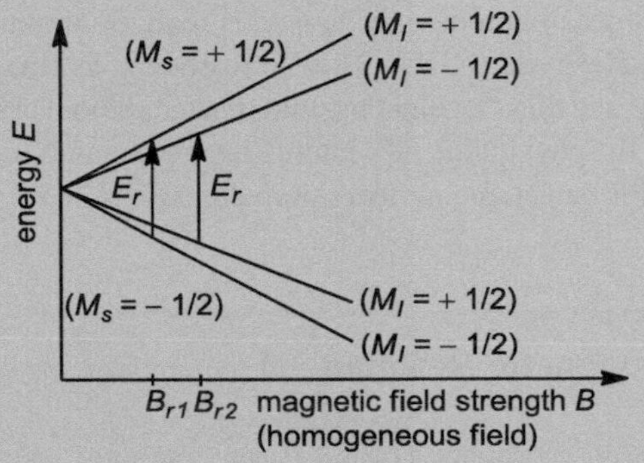

selection rules $\Delta M_I = 0$
$\Delta M_S = 0$

two resonances at E_r/B_{r1} and E_r/B_{r2}:
$B_{r2} - B_{r1} = a$

a = hyperfine coupling constant

(3.6)

Due to the complexity and asymmetry of many paramagnetic centers in biological systems, and because of an inherently high spectral linewidth, the simple EPR method often yields only unresolved signals. In such cases, the double-resonance technique ENDOR (Electron Nuclear DOuble Resonance) or pulsed EPR methods are used [21,22]. If several unpaired electrons are present, their exchange behavior and the resulting exited states have to be taken into account (see Section 4.3). The consequences of a not-spherically-symmetrical (i.e. anisotropic) distribution of the unpaired electron become evident in the solid state or when the mobility of the paramagnetic center is otherwise restricted. (See Chapter 10 for Electron Paramagnetic Resonance II.)

The very reactive primary carbon radicals formed after homolysis of $RH_2C{-}Co$(corrin) bonds are short-lived and, therefore, not always detectable by EPR [16,17]. The remaining low-spin Co^{II} complex (d^7) features an EPR signal for one unpaired electron [16,18]. Interaction (coupling) of the electron spin occurs with the nuclear spin of the metal center (^{59}Co: 100% natural isotopic abundance, nuclear spin $I = 7/2$) and the nuclear spin of one nitrogen atom (^{14}N: 99.6% natural abundance, $I = 1$). These findings can be explained only if the unpaired electron is assumed to occupy the d_{z^2} orbital, interacting mainly with the single, axially coordinated nitrogen atom of the benzimidazole ligand ("base-on"). If the $d_{x^2-y^2}$ orbital were occupied by the unpaired electron, all four nitrogen centers of the macrocyclic corrin ligand would contribute with essentially similar nuclear spin/electron

Table 3.1 Co(B_{12})dependent enzymes and their occurrence in organisms.

Co(B_{12})dependent proteins	Organisms
(a) adenosylcobalamin(AdoCbl)-dependent isomerases	
methylmalonyl-*CoA* mutase (MCM)	archaea, bacteria, eukaryotes
isobutyryl-*CoA* mutase (ICM)	archaea, bacteria, eukaryotes
ethylmalonyl-*CoA* mutase (ECM)	archaea, bacteria, eukaryotes
glutamate mutase (GM)	archaea, bacteria
methyleneglutarate mutase (MGM)	archaea, bacteria
D-lysine 5,6-aminomutase (5,6-LAM)	bacteria
diol dehydratase (DDH)	bacteria
glycerol dehydratase (GDH)	bacteria
ethanolamine ammonia lyase (EAL)	bacteria
(b) methylcobalamin(MeCbl)-dependent methyltransferases	
methionine synthase (MetH)	bacteria, eukaryotes
methyltransferases (Mta, Mtm, Mtb, Mtt, Mts, and Mtv)	archaea, bacteria
methyltetrahydromethanopterin *CoM* methyltransferase subunit A (MtrA)	archaea
(c) B_{12}-dependent reductive dehalogenase (CprA)	bacteria

spin coupling [20]. The thus-determined order of d orbitals corresponds to a relatively small distortion of octahedral symmetry (Figure 2.7) and is in accordance with the observed supernucleophilicity in axial direction after double occupation of the d_{z^2} orbital.

3.3 Enzyme Functions of Cobalamins

Three different enzyme functions (enzyme classes) are connected with B_{12} cofactors, reflecting different types of reactivity (3.4) [3,8,23]. The most important enzyme class contains the adenosylcobalamin-dependent isomerases (Table 3.1), which occur in all types of organisms. They rely on the homolytical splitting of the Co–C bond of AdoCbl to form highly reactive radicals. The second enzyme class involves methylcobalamin and exhibits alkylating activity. The methionine synthase is essential for the biosynthesis of methionine in many organisms, including humans. The third class, B_{12}-dependent reductive dehalogenases (CprA) of anaerobic microbes, plays an important role in the detoxification of aromatic and aliphatic organochloro compounds [23,24]. Most of these B_{12}-dependent reductive dehalogenases also contain Fe/S-clusters, confirming that the B_{12} cofactor plays the role of a redox catalyst [23]. Since many fundamental questions regarding the reaction mechanism of dehalogenases still remain unclear, the present discussion will be restricted to the first two classes of B_{12}-dependent enzymes.

3.3.1 Adenosylcobalamin (AdoCbl)-dependent Isomerases

This class of B_{12}-dependent enzymes can be divided into three subclasses, distinguished by the nature of the migrating group and the kind of substituent on the carbon atom to which the group migrates [8,25,26]. The Class Ia enzymes are mutases (3.7), which catalyze rearrangements of the carbon skeleton [26].

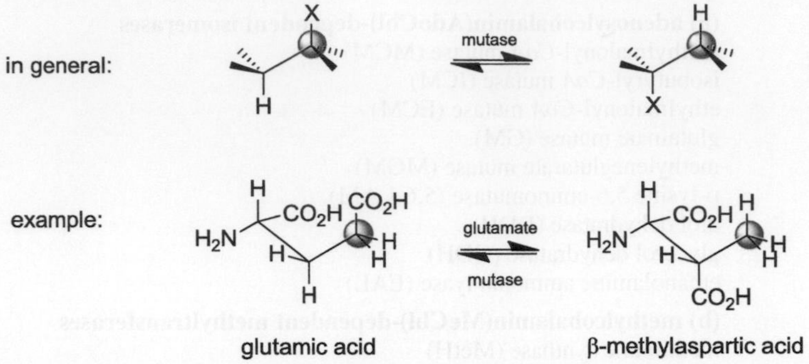

(3.7)

The eliminases (Class Ib) include dehydratases or lyases and are coenzyme B_{12}-dependent enzymes, since a 1,2-shift in 1,2-diols or 2-aminoalcohols leads to geminal (1,1)-isomers, which readily lose water or ammonia to form carbonyl compounds (3.8) [16]. In a larger context, the AdoCbl-dependent ribonucleotide reductases can be added to these Class Ib enzymes, catalyzing the reduction of ribonucleotides to their corresponding deoxy forms in certain bacteria [23,25]. The B_{12}-dependent ribonucleotide reductase requires coenzymatic dithiol, which is oxidized to disulfide (compare lipoic acid in (3.12) and Section 7.6.1). For this biochemically very important reaction (RNA→DNA), other organisms and bacteria such as *E. coli* use manganese- or iron-containing ribonucleotide reductases, which also require radicals for proper function (compare Section 7.6.1) [23,27].

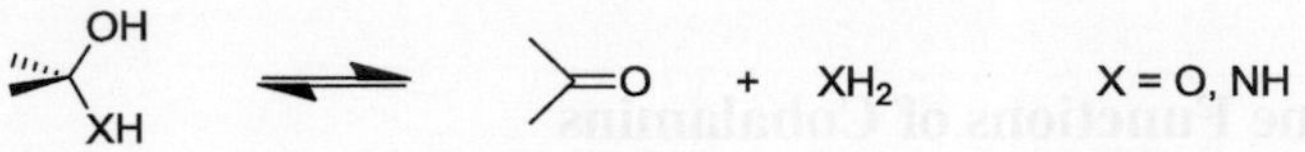

(3.8)

Class Ic enzymes are aminomutases, which catalyze amino group migrations and require pyridoxal phosphate in addition to AdoCbl. An overview and mechanistic interpretation of isomerase activity is given in (3.9) and (3.10).

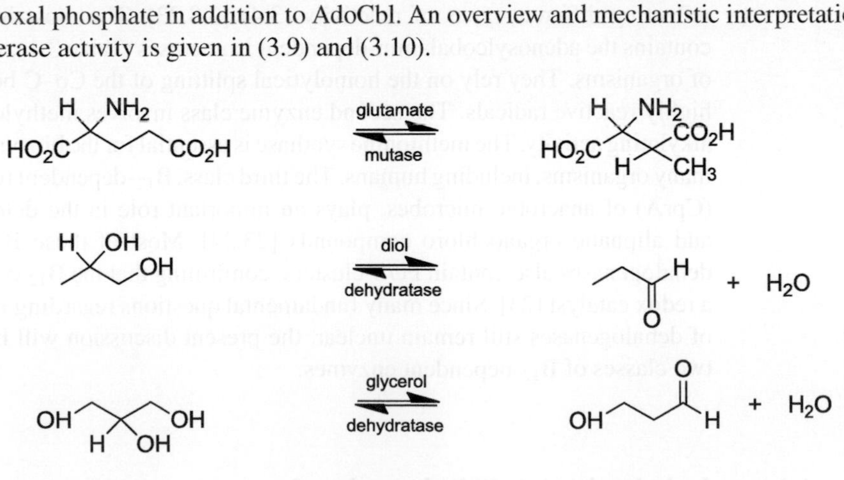

(3.9)

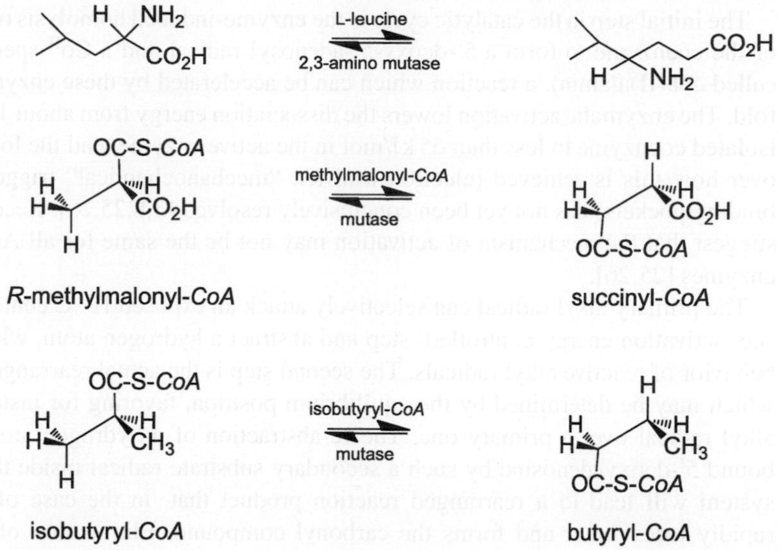

(3.9 *continued*)

Most reactions catalyzed by enzymes containing the AdoCbl cofactor are restricted to microorganisms that are also able to synthesize this coenzyme. For mammals, the methylmalonyl-coenzyme A (*CoA*) mutase is particularly important: it is required in the metabolism of amino acids in the liver and its absence due to genetic defects is lethal.

In synthetic organic chemistry, the deceptively simple 1,2-shifts (3.7) are not easily accomplished; there is thus much interest in the use of alkylcobalt corrin complexes and corresponding model compounds for organic synthesis, even if substrate- and stereospecificity are not guaranteed in the absence of the apoenzyme [28]. Numerous results from spectroscopic (EPR) and mechanistic (e.g. isotopic labeling) studies now point to the following, radical-based reaction cycle (3.10).

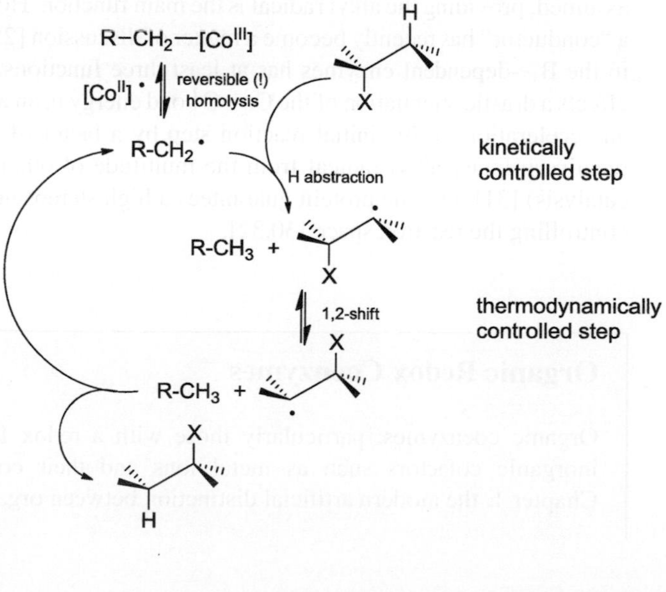

(3.10)

The initial step in the catalytic cycle is the enzyme-induced homolysis of the Co–C bond of the coenzyme to form a 5′-deoxy-5′-adenosyl radical and a Co^{II} species $[Co^{II}]^{\bullet}$ (also called cob(II)alamin), a reaction which can be accelerated by these enzymes 10^9- to 10^{14}-fold. The enzymatic activation lowers the dissociation energy from about 110 kJ/mole in the isolated coenzyme to less than 65 kJ/mol in the active enzyme, and the long-lasting debate over how this is achieved (electron transfer, "mechanochemical" triggering, adenosine-binding pockets) has not yet been conclusively resolved [3,8,25,26]. Recent investigations suggest that the mechanism of activation may not be the same for all AdoCbl-dependent enzymes [25,26].

The primary alkyl radical can selectively attack an exposed H–C center in a kinetically (i.e. activation energy-controlled) step and abstract a hydrogen atom, which is the typical behavior of reactive alkyl radicals. The second step is the actual rearrangement (1,2-shift), which may be determined by the equilibrium position, favoring for instance a secondary alkyl radical over a primary one. The re-abstraction of a hydrogen atom from enzyme-bound 5′-deoxyadenosine by such a secondary substrate radical inside the "radical cage" system will lead to a rearranged reaction product that, in the case of a geminal diol, rapidly loses water and forms the carbonyl compound. Elimination of water, however, can also be imagined for the radical itself; 1,2-dihydroxyalkyl radicals tend to lose OH^- or H_2O (after protonation) under the formation of carbonylalkyl radicals. Incidentally, the radical-induced transformation of glycols to aldehydes is not unknown in organic synthesis; such reactions can be triggered by hydroxyl radicals generated from Fenton's reagent (3.11).

$$Fe^{2+} + H_2O_2 \rightarrow Fe^{III}(OH^-) + OH^{\bullet} \tag{3.11}$$

If not immediately used, the 5′-deoxyadenosyl radical can recombine with the Co^{II} species, so alkylcobalamins can be described as reversibly-acting radical carriers [29]. A direct participation of the cob(II)alamin in the rearrangements of the substrates is thus not assumed, providing the alkyl radical is the main function. However, a role as a "spectator" or a "conductor" has recently become a matter of discussion [25,30]. The protein (apoenzyme) in the B_{12}-dependent enzymes has at least three functions. (1) After substrate binding, it effects a drastic attenuation of the Co–C bond energy in an as yet unknown manner, causing an acceleration of this initial reaction step by a factor of 10^9 to 10^{14}. (2) It protects the reactive primary alkyl radical from the multitude of other, undesired reactants (negative catalysis) [31]. (3) The protein guarantees a high stereoselectivity of the isomerization by controlling the reaction space [30,32].

Organic Redox Coenzymes

Organic coenzymes, particularly those with a redox function, often interact with inorganic cofactors such as metal ions and their complexes. As mentioned in Chapter 1, the modern artificial distinction between organic and inorganic molecules has no meaning for organisms; the important point is the relation between biosynthetic requirements and functional benefits.

Compounds (3.12) are among the most common organic redox coenzymes. They can be roughly divided into N-heterocycles, quinonoid systems and sulfur compounds. The 5,6,7,8-tetrahydro form of folic acid (which is related to the flavin systems) is an important carrier of C_1 fragments.

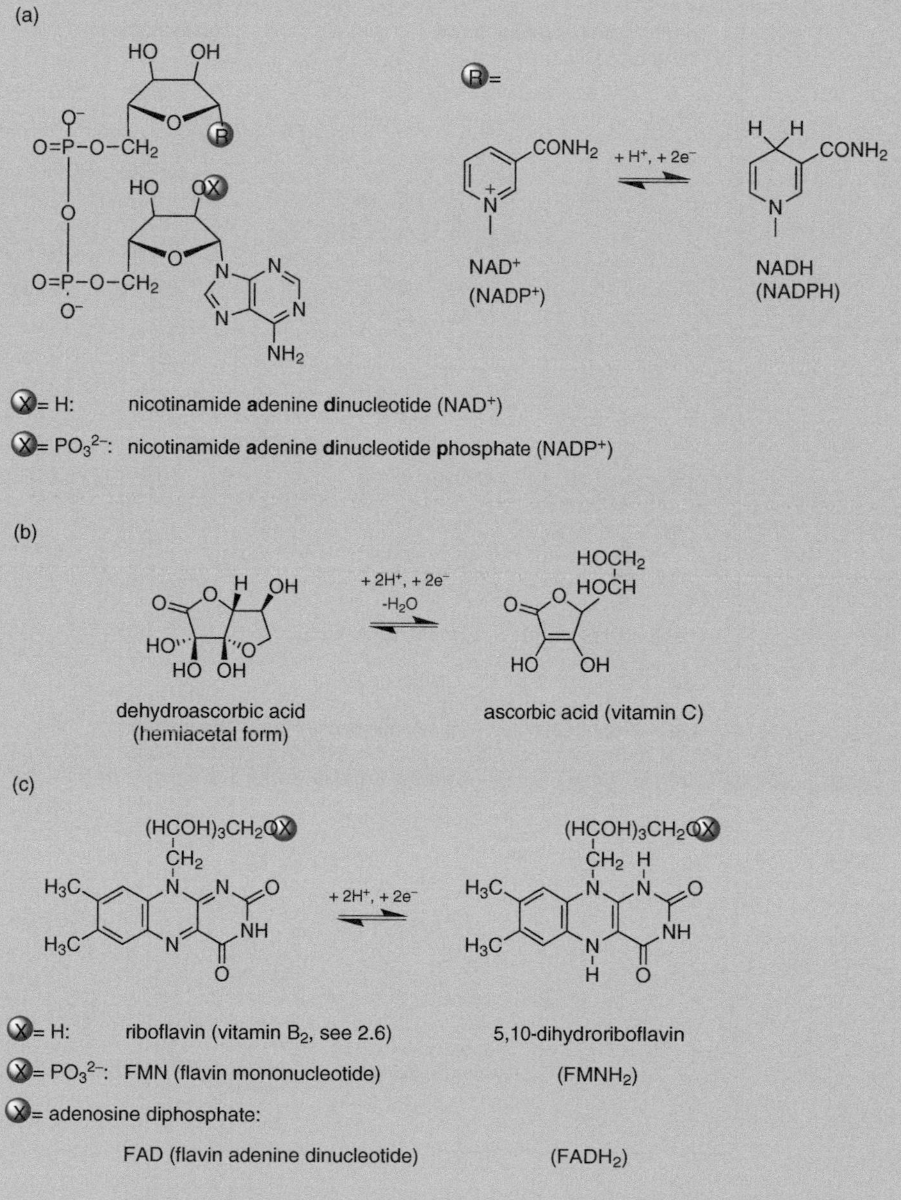

(3.12)

(d)

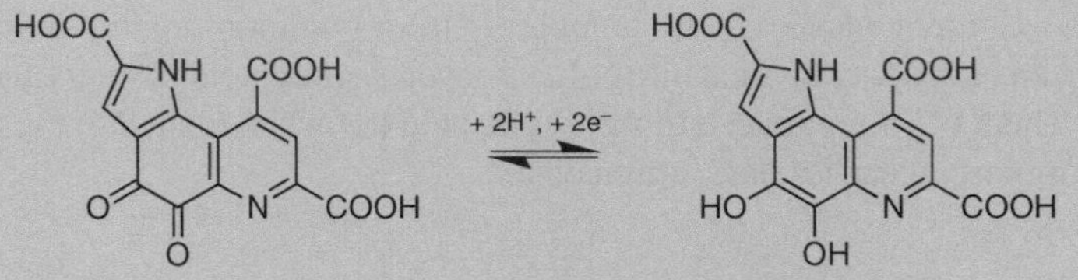

methoxatin (*o*-quinone form)
cofactor PQQ

(catechol form)

(e)

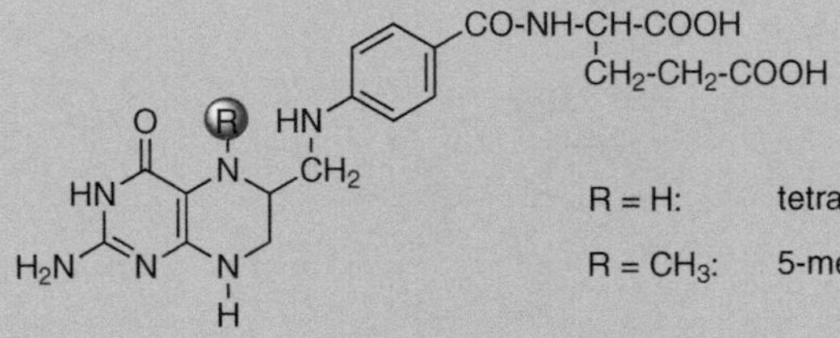

R = H: tetrahydrofolic acid (THFA)

R = CH_3: 5-methyl-THFA

(f)

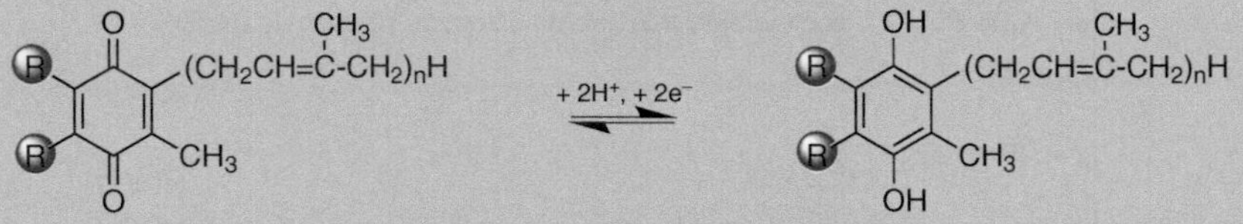

2 R = $(CH)_4$, n = 9: menaquinone (Q_a)

R = OCH_3, n = 2-10: ubiquinone (Q_b)

R = CH_3, n = 6-10 plastoquinones (PQ)

R = H, n = 4-7 vitamin K group

(g)

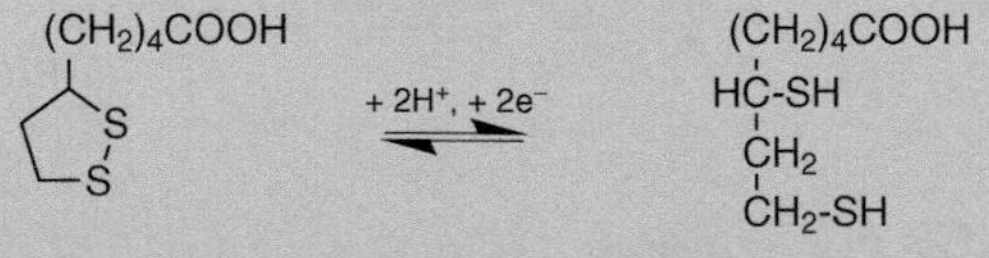

lipoic acid
(cyclic disulfide)

(dithiol form)

(3.12 *continued*)

3.3.2 Alkylation Reactions of Methylcobalamin (MeCbl)-dependent Alkyl Transferases

Methylcobalamin-induced "bio"methylations are of great importance for microbiology, biosynthesis and toxicology (compare Sections 17.3 and 17.5). While methyl groups with an electrophilic character, with a "positive partial charge", are biochemically available through the sulfonium species S-adenosyl methionine ("SAM"), (adenosyl)$(CH_3)S^+(CH_2CH_2CH(NH_3^+)CO^-)$, or through 5-methyltetrahydrofolic acid (3.12 and 3.13), the methylation of electrophilic substrates typically requires an organometallic compound that can react either in a carbanionic fashion (S_N2 reaction) or as a radical, involving a single-electron-transfer process (compare (3.4)). The methylation of compounds of less electropositive "soft" elements such as selenium or mercury (3.14) with oxidation potentials (E_0) $> 0\,V$ presumably occurs via a carbanionic mechanism, while less noble elements such as arsenic, tin and cadmium ($E_0 < 0\,V$) are alkylated in their compounds via radical pathways (see Section 17.3). In some instances, very toxic species like the methylmercury cation $(CH_3)Hg^+$ are formed by these reactions (3.14). If sufficiently stable under physiological conditions, such mixed hydrophilic/hydrophobic organometallic cations are able to penetrate the blood–brain barrier and deactivate sulfur-containing enzymes (Section 17.5).

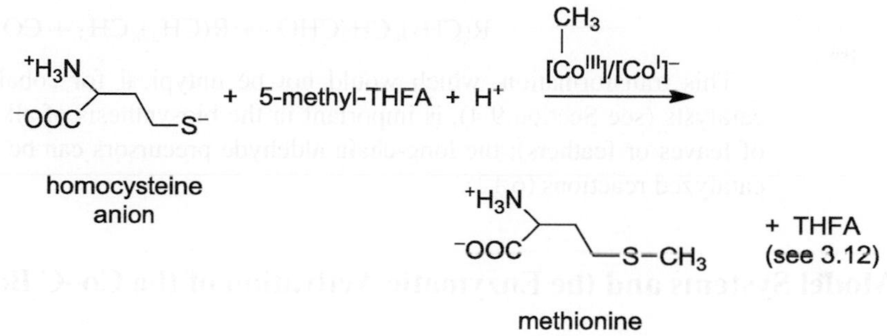

(3.13)

$$Hg^{2+} + H_3C{-}[Co^{III}] \longrightarrow (CH_3)Hg^+ + [Co^{III}]^+ \tag{3.14}$$

A biologically valuable methylation requiring cobalamin-dependent enzymes has been established for the substrate homocysteine ("homo": extended by one CH_2 chain link); the essential and often "limiting" amino acid methionine is biosynthesized by this reaction (methionine synthetase from *E. coli* (3.13; Figure 3.2)) [11,33].

In microorganisms, especially in "acetogenic" or "methanogenic" bacteria, which produce acetic acid and methane, respectively (Figure 1.2), methyl-transferring "corrinoid" (cobaltcorrin-containing) enzymes are of great importance. During bacterial CO_2 fixation, they participate in the catalytic formation of acetyl-*CoA* as "activated acetic acid" (3.15) [33,34], a process that involves a nickel enzyme-requiring carbonylation (acetyl-*CoA* synthase (ACS); compare Section 9.4) and a methyl-group transfer from a 5-methyltetrahydropterin (Pter-N_5)$-CH_3$ to *CoA* via a methylcobalt–corrin intermediate.

$$CO_2 + CO + 6\ \text{"H"} + \text{HS-}CoA \rightleftharpoons CH_3C(O)S\text{-}CoA + 2\ H_2O$$

HS-*CoA* = coenzyme A =

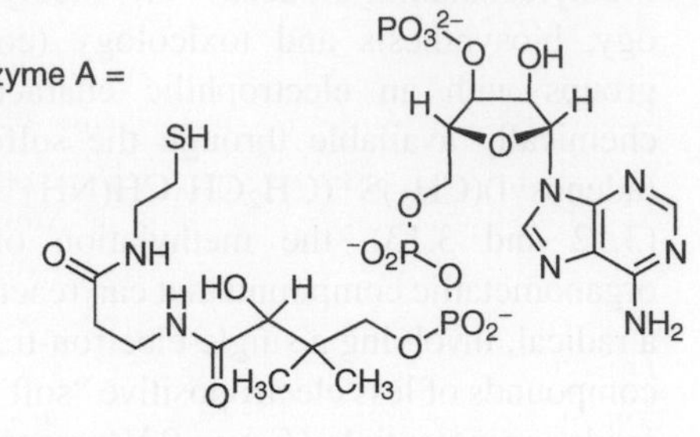

(3.15)

For archaea, the nickel-containing porphinoid "factor F_{430}" is directly involved in methane formation (see Section 9.5). Cobalt corrinoid-containing membrane proteins have been detected [34], which are presumably involved in the synthesis of coenzyme M ($H_3CSCH_2CH_2SO_3^-$; compare Section 9.5), the last methyl carrier before eventual methane production.

Very little is known about noncorrinoid cobalt enzymes. A possible cobalt–porphinoid factor has been described in the context of decarbonylation of long-chain aldehydes (3.16) [35].

$$R(CH_2)_nCH_2CHO \rightarrow R(CH_2)_nCH_3 + CO \quad (3.16)$$

This transformation, which would not be untypical for cobalt- or nickel-containing catalysts (see Section 9.4), is important in the biosynthesis of alkanes (→ waterproofing of leaves or feathers); the long-chain aldehyde precursors can be formed via peroxidase-catalyzed reactions (6.12).

3.4 Model Systems and the Enzymatic Activation of the Co–C Bond

Model systems of the cobalamins are of considerable interest because of the organic-synthetic attraction of coenzyme B_{12}-catalyzed reactions; the actual cobalamins are very sensitive and exhibit only limited solubility. It was realized quite early that simple bis(diacetyldioxime) complexes ("cobaloximes" (3.17)) represent surprisingly good models for B_{12} systems; these complexes contain a "quasimacrocyclic" chelate ring structure as a consequence of two strong hydrogen bonds [36]. Even better suited with respect to redox properties are the Costa complexes (3.17), with covalently linked α-diimine moieties [37]. Other model complexes contain the chelate ligand bis(salicylaldehyde)ethylenediimine ("salen" (3.17)), which also binds cobalt in an approximately square planar fashion.

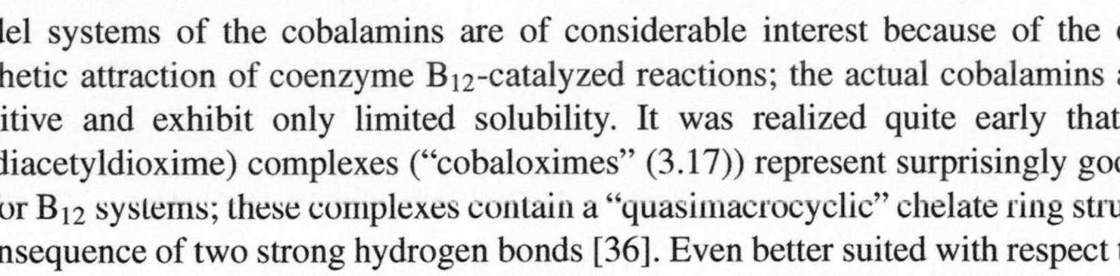

(3.17)

All model complexes, as well as the cobalt-containing porphyrins, have the disadvantage that their supernucleophilic Co^I state is much less stable under physiological conditions than that of the cobalt–corrin systems. The cob(III)yrinic acid heptamethyl ("cobester") complexes, which lack the nucleotide part of the B_{12} systems, fare better in this regard but are also harder to synthesize [3].

The value of model systems for mechanistic considerations has been evident from comparative studies of cobalt porphyrins and of cobaloximes as Co-corrin model compounds, with axially coordinated alkyl and organophosphine ligands (PR_3) of different sizes and basicities [38]. The results suggest that the enzymatically relevant activation, the Co–C bond cleavage, is influenced by electronic effects from the PR_3 ligands in porphyrin complexes, while in cobaloximes the steric bulk of the axial PR_3 ligands is more essential. This observation seems to confirm the significance of the nonplanarity of cobalt–corrin complexes and has lent support to the assumption that radical formation in the enzyme is sterically controlled. The term "mechanochemical triggering" was introduced to describe the assumption that base-on/base-off movements in conjunction with corrin deformations trigger the Co–C bond cleavage. Additional activation can occur through binding of the nucleotide part of the 5′-deoxyadenosyl group to the enzyme (adenosine-binding pockets) [3]. Structural data from Kräutler, Keller and Kratky have exhibited an essentially unchanged geometry for Co^{II} binding to the corrin ring in cob(II)alamin as compared to Co^{III} analogues [39], which enhances the electron-transfer rate due to a small reorganization energy and negligible conformational changes (compare Section 6.1). However, the bond to the axial base is shortened in the Co^{II} state, leading to a more pronounced out-of-plane situation for the metal center and a corresponding weakening of the Co–C bond. A similar, better-known example of the cooperativity of both axial ligands, X and Y (2.11 and 3.3), is the reversible binding of O_2 to heme–iron centers (see Chapter 5).

Despite extensive knowledge of the structure and reactivity of the B_{12} coenzymes and their model compounds, rather little is known about the conditions in the actual enzymes or about the reaction mechanisms; the catalytic cycle shown in (3.10) still has only model character [3,5,12,25]. Recent studies suggest a synergistic combination of electrostatic and large-scale protein structural changes causes the Co–C bond activation [40].

It seems clear, however, that the unique function of the element cobalt in B_{12} systems is its tolerance of a bond from the redox-active transition metal to primary alkyl groups, which can be set free through well-defined activation processes and then act as specific, reactive and reversibly transferable species.

References

1. *Organometallics* **2012**, *31*, 5671–5970: *Organometallics in Biology and Medicine Issue.*
2. C. G. Hartinger, P. J. Dyson, *Chem. Soc. Rev.* **2009**, *38*, 391–401: *Bioorganometallic chemistry – from teaching paradigms to medicinal applications.*
3. B. Kräutler, *Organometallic chemistry of B_{12} coenzymes*, in A. Sigel, H. Sigel, R. K. O. Sigel (eds.), *Metal Ions in Life Science, Vol. 6: Metal–Carbon Bonds in Enzymes and Cofactors*, RSC Publishing, Cambridge, **2009.**
4. R. G. Matthews, *Cobalamin- and corrinoid-dependent enzymes*, in A. Sigel, H. Sigel, R. K. O. Sigel (eds.), *Metal Ions in Life Science, Vol. 6: Metal–Carbon Bonds in Enzymes and Cofactors*, RSC Publishing, Cambridge, **2009.**
5. R. Banerjee, *Chemistry and Biochemistry of B_{12}*, John Wiley & Sons, Chichester, **1999.**
6. D. Dolphin (ed.), *B_{12}*, John Wiley & Sons, New York, **1982.**
7. Z. Schneider, A. Stroinski, *Comprehensive B_{12}*, de Gruyter, Berlin, **1987.**
8. K. L. Brown, *Chem. Rev.* **2005**, *105*, 2075–2149: *Chemistry and enzymology of vitamin B_{12}.*

9. K. Folkers, *J. Chem. Educ.* **1984**, *61*, 747–756: *Perspectives from research on vitamins and hormones.*
10. (a) P. G. Lenhert, D. Crowfoot Hodgkin, *Nature (London)* **1961**, *192*, 937–938: *Structure of the 5,6-dimethylbenzimidazolylcobamide coenzyme*; (b) P. G. Lenhert, *Proc. Roy. Soc. Series A* **1968**, *303*, 45–84: *The structure of vitamin B_{12}, VII. The x-ray analysis of the vitamin B_{12} coenzyme*; (c) S. Gschoesser, R. B. Hannak, R. Konrat, K. Gruber, C. Mikl, C. Kratky, B. Kräutler, *Chem. Eur. J.* **2005**, *11*, 81–93: *Homocoenzyme B_{12} and bishomocoenzyme B_{12}: covalent structural mimics for homolyzed, enzyme-bound coenzyme B_{12}.*
11. (a) C. L. Drennan, S. Huang, J. T. Drummond, R. G. Matthews, M. L. Ludwig, *Science* **1994**, *266*, 1669–1674: *How a protein binds B_{12}: a 3.0 Å X-ray structure of B_{12}-binding domains of methionine synthase*; (b) M. L. Ludwig, R. G. Matthews, *Annu. Rev. Biochem.* **1997**, *66*, 269–313: *Structure-based perspectives on B_{12}-dependent enzymes.*
12. B. Kräutler, C. Kratky, *Angew. Chem. Int. Ed. Engl.* **1996**, *35*, 167–170: *Vitamin B_{12}: the haze clears.*
13. S. Gschösser, K. Gruber, C. Kratky, C. Eichmüller, B. Kräutler, *Angew. Chem. Int. Ed. Engl.* **2005**, *44*, 2284–2288: *B_{12}-retro-riboswitches: constitutional switching of B_{12} coenzymes induced by nucleotides.*
14. D. Lexa, J.-M. Saveant, *Acc. Chem. Res.* **1983**, *16*, 235–243: *The electrochemistry of vitamin B_{12}.*
15. B. D. Martin, R. G. Finke, *J. Am. Chem. Soc.* **1992**, *114*, 585–592: *Methylcobalamin's full- vs. half-strength cobalt–carbon σ bonds and bond dissociation enthalpies: a > 10^{15} Co–CH_3 homolysis rate enhancement following one-antibonding-electron reduction of methylcobalamin.*
16. T. Toraya, *Chem. Rev.* **2003**, *103*, 2095–2127: *Radical catalysis in coenzyme B_{12}-dependent isomerization (eliminating) reactions.*
17. Y. Zhao, P. Such, J. Rétey, *Angew. Chem. Int. Ed. Engl.* **1992**, *31*, 215–216: *Detection of radical intermediates in the coenzyme-B_{12}-dependent methylmalonyl-CoA-mutase reaction by ESR spectroscopy.*
18. C. Michel, S. P. J. Albracht, W. Buckel, *Eur. J. Biochem.* **1992**, *205*, 767–773: *Adenosylcobalamin and cob(II)alamin as prosthetic groups of 2-methyleneglutarate mutase from Clostridium barkeri.*
19. (a) J. A. Weil, J. R. Bolton, J. E. Wertz, *Electron Paramagnetic Resonance: Elemental Theory and Practical Applications*, John Wiley & Sons, New York, **1993**; (b) M. Symons, *Chemical and Biochemical Aspects of Electron-spin Resonance Spectroscopy*, van Nostrand Reinhold, New York, **1978**.
20. W. R. Hagen, *Metallomics* **2009**, *1*, 384–391: *Metallomic EPR spectroscopy.*
21. M. Bennati, T. F. Prisner, *Rep. Prog. Phys.* **2005**, *68*, 411–448: *New developments in high field electron paramagnetic resonance with applications in structural biology.*
22. (a) B. M. Hoffman, *Acc. Chem. Res.* **1991**, *24*, 164–170, *Electron nuclear double resonance (ENDOR) of metalloenzymes*; (b) H. Kurreck, B. Kirste, W. Lubitz, *Electron Nuclear Double Resonance Spectroscopy of Radicals in Solution*, VCH, Weinheim, **1988**.
23. Y. Zhang, V. N. Gladyshev, *Chem. Rev.* **2009**, *109*, 4828–4861: *Comparative genomics of trace elements: emerging dynamic view of trace element utilization and function.*
24. J.-M. Savéant, *Chem. Rev.* **2008**, *108*, 2348–2378: *Molecular catalysis of electrochemical reactions. Mechanistic aspects.*
25. K. L. Brown, *Dalton* **2006**, 1123–1133: *The enzymatic activation of coenzyme B_{12}.*
26. R. Banerjee, *Chem. Rev.* **2003**, *103*, 2083–2094: *Radical carbon skeleton rearrangements: catalysis by coenzyme B_{12}-dependent mutases.*
27. (a) J. Stubbe, W. A. van der Donk, *Chem. Rev.* **1998**, *98*, 705–762: *Protein radicals in enzyme catalysis*; (b) M. D. Sintchak, G. Arjara, B. A. Kellogg, J. Stubbe, C. L. Drennan, *Nat. Struct. Biol.* **2002**, *9*, 293–300: *The crystal structure of class II ribonucleotide reductase reveals how an allosterically regulated monomer mimics a dimer.*
28. G. Pattenden, *Chem. Soc. Rev.* **1988**, *17*, 361–382: *Cobalt-mediated radical reactions in organic synthesis.*
29. J. Halpern, *Science* **1985**, *227*, 869–875: *Mechanisms of coenzyme B_{12}-dependent rearrangements.*
30. W. Buckel, C. Kratky, B. T. Golding, *Chem. Eur. J.* **2006**, *12*, 352–362: *Stabilisation of methylene radicals by Cob(II)alamin in coenzyme B_{12} dependent mutases.*
31. J. Rétey, *Angew. Chem. Int. Ed. Engl.* **1990**, *29*, 355–361: *Reaction selectivity of enzymes through negative catalysis or how enzymes work with highly reactive intermediates.*
32. W. Buckel, B. T. Golding, *Ann. Rev. Microbiol.* **2006**, *60*, 27–49: *Radical enzymes in anaerobes.*

33. R. G. Matthews, J. T. Drummond, *Chem. Rev.* **1990**, *90*, 1275–1290: *Providing one-carbon units for biological methylations*.
34. S. Ragsdale, *Chem. Rev.* **2006**, *106*, 3317–3337: *Metals and their scaffolds to promote difficult enzymatic reactions*.
35. (a) A. Bernard, J. Joubès, *Prog. Lipid Res.* **2013**, *52*, 110–129: *Arabidopsis cuticular waxes: advances in synthesis, export and regulation*; (b) F. Schneider-Belhaddad, P. Kolattukudy, *Arch. Biochem. Biophys.* **2000**, *377*, 341–349: *Solubilization, partial purification, and characterization of a fatty aldehyde decarbonylase from a higher plant, Pisum sativum*; (c) M. Dennis, P. E. Kolattukudy, *Proc. Natl. Acad. Sci. USA* **1992**, *89*, 5306–5310: *A cobalt-porphyrin enzyme converts a fatty aldehyde to a hydrocarbon and CO*.
36. G. N. Schrauzer, *Angew. Chem. Int. Ed. Engl.* **1977**, *16*, 233–244: *Recent developments in the vitamin B_{12} field: enzyme reactions depend on simple corrins and coenzyme B_{12}*.
37. G. Costa, G. Mestroni, L. Stefani, *J. Organomet. Chem.* **1967**, *7*, 493–501: *Organometallic derivatives of cobalt(III) chelates of bis(salicylaldehyde)ethylenediimine*.
38. (a) M. K. Geno, J. Halpern, *J. Am. Chem. Soc.* **1987**, *109*, 1238–1240: *Why does nature not use the porphyrin ligand in vitamin B_{12}?*; (b) S. M. Chemaly, L. A. Jack, L. J. Yellowlees, P. L. S. Harper, B. Heeg, J. M. Pratt, *Dalton Trans.* **2004**, 2125–2134: *Vitamin B_{12} as an allosteric cofactor; dual fluorescence, hysteresis, oscillations and the selection of corrin over porphyrin*.
39. B. Kräutler, W. Keller, C. Kratky, *J. Am. Chem. Soc.* **1989**, *111*, 8936–8938: *Coenzyme B_{12} chemistry: the crystal and molecular structure of Cob(II)alamin*.
40. J. Pang, X. Li, K. Morokuma, N. S. Scrutton, M. J. Sutcliffe, *J. Am. Chem. Soc.* **2012**, *134*, 2367–2377: *Large-scale domain conformational change is coupled to the activation of the Co-C bond in the B_{12}-dependent enzyme ornithine 4,5-aminomutase: a computational study*.

4 Metals at the Center of Photosynthesis: Magnesium and Manganese

In addition to their well-established functions in photosynthesis, the divalent forms of the main group element magnesium and of the transition metal manganese [1,2] are important as centers of hydrolytic and phosphate-transferring enzymes (Section 14.1). Moreover, higher-oxidized manganese plays a role as a redox center in several enzymes – including certain forms of ribonucleotide reductase (see Section 7.6.1), catalase and peroxidase (see Section 6.3) – and in the particular superoxide dismutase of mitochondria [3] (see Section 10.5). Iron and copper ions are also prominently involved in the overall photosynthetic process, contributing to directed electron transfer along or across membrane proteins. However, this chapter, which deals with the arguably most important chemical reaction for life on earth, restricts itself to two fundamental parts of photosynthesis: first, the absorption of light and the ensuing charge separation originating from magnesium-containing chlorophylls; and second, the manganese-catalyzed oxidation of water to oxygen ("dioxygen", O_2) in cyanobacteria, algae and higher plants.

4.1 Volume and Efficiency of Photosynthesis

The increasing public awareness of our dependence on fossil fuels and of the steadily rising accumulation of carbon dioxide in the atmosphere due to the combustion of those fuels has been responsible for enhanced efforts to understand the molecular workings of photosynthesis. This chemical process, often summarized as in (4.1), is fundamental for the existence of higher forms of life on earth; the production of reduced carbon compounds ("organic" material, including fossil fuels) on the one hand and the production of oxygen on the other are based on this radiation energy-consuming process.

$$H_2O + CO_2 \underset{\text{respiration (downhill catalysis)}}{\overset{\text{photosynthesis (uphill catalysis)}}{\rightleftharpoons}} 1/n\,(CH_2O)_n + {}^3O_2 \qquad (4.1)$$

$$\Delta H = +\,470\ \text{kJ/mole}$$

Progress in many diverse areas of science, such as protein crystallography [4], pico- and femtosecond laser spectroscopy [5] and high-resolution magnetic resonance [6], has

Bioinorganic Chemistry: Inorganic Elements in the Chemistry of Life – An Introduction and Guide, Second Edition.
Written and Translated by Wolfgang Kaim, Brigitte Schwederski and Axel Klein.

revealed several details of the many individual reaction steps involved in the highly complex photosynthetic process. Accordingly, Nobel Prizes in Chemistry have been awarded for the structural elucidation of a bacterial photosynthetic reaction center (J. Deisenhofer, R. Huber and H. Michel, 1988 [4]) and for the theoretical description of the underlying electron-transfer processes (R. A. Marcus, 1992 [7]). For the chemical sciences, an understanding of "photo-synthesis" would be particularly attractive within the context of model systems, as valuable (high-energy) substances are produced in these reactions from very simple, low-energy starting materials through the use of a readily available, innocuous and rather "diluted" form of energy: solar radiation. However, the hitherto constructed model systems have shown only moderate success in mimicking partial processes of photosynthesis. The reason lies in the very demanding requirements for an "uphill catalysis", which also explains the high degree of complexity of the photosynthetical "apparatus" in biology [8].

Certain bacteria and algae are photosynthetically active, as are green plants. Purple bacteria such as *Rhodopseudomonas viridis* possess a comparatively simple photosynthetical apparatus, without the ability to oxidize water. These bacteria use photosynthesis primarily to separate charges and thus create a transmembrane proton gradient (pH difference), which is used to synthesize high-energy adenosine triphosphate (ATP) from adenosine diphosphate (ADP phosphorylation, see (14.2)). Other, anaerobic bacteria use the redox equivalents to oxidize substrates such as hydrogen sulfide (H_2S) or dihydrogen H_2 instead of water.

In plants, the primary photosynthetical events take place in the highly folded, disk-shaped thylakoid membrane vesicles inside of chloroplasts (see Figure 4.6), and even in simple bacteria the process is membrane-spanning (compare Figure 4.4). Since immobilization and a defined orientation of pigments and of reaction centers are crucial for the success of photosynthesis, all chlorophyll molecules (which differ slightly with regard to some substituents (2.9 and 4.2)) feature a long aliphatic phytyl side chain by which they anchor these pigments in the hydrophobic phospholipid membrane, which has a thickness of about 5 nm.

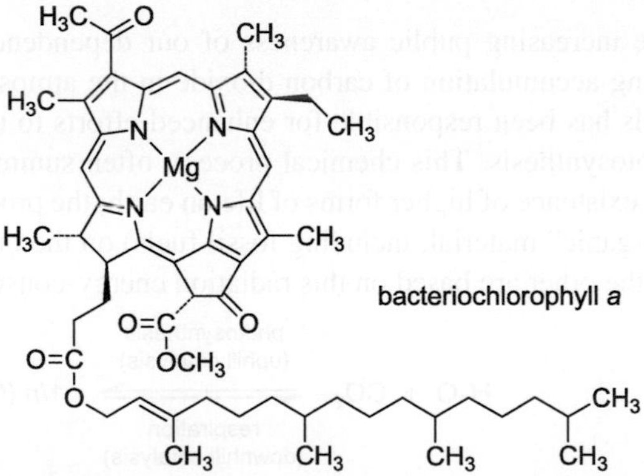

bacteriochlorophyll *a* (4.2)

The photosynthetic output of green plants in normal sunlight is usually assumed to be about 1 g of glucose per hour per 1 m^2 of leaf surface area. The photosynthetically active algae (phytoplankton) also play an important role on a global scale, since the water coverage of the earth is about 71%. Even though the total efficiency of photosynthesis is, on average, less than 1% if measured as production of fuel equivalents in comparison to the available radiation energy, the total global turnover is tremendous: about 200 billion tons of carbohydrate equivalents $(CH_2O)_n$ are produced from CO_2 each year. The efficiency of the primary, "physical" energy transformation in photosynthesis from incident light to transmembrane redox potential differences (about 20%) is comparable to that of very good photovoltaic elements.

4.2 Primary Processes in Photosynthesis

What are the main requirements for photosynthesis and what are the roles of the inorganic components?

4.2.1 Light Absorption (Energy Acquisition)

The sunlight available at the earth's surface includes the wavelength range visible to the human eye, from about 380 to 750 nm, but a considerable amount also features higher wavelengths in the near infrared region, up to more than 1000 nm. An efficient photosynthetic transformation of light of this energy (1.24–3.26 eV) requires the absorption of as many photons as possible. This condition is fulfilled through the presence of various organic pigments, including chlorophyll molecules, which are positioned in a highly folded membrane with an inherently large inner surface area and therefore a high cross-section for photon capture [8]. Chlorophylls themselves contain a conjugated tetrapyrrole π system (2.9 and 4.2), which shows a high absorptivity with molar extinction coefficients of about 10^5 M/cm at both the long- and short-wavelength ends of the visible spectrum. The complementary colors blue (after long-wavelength absorption) and yellow (after short-wavelength absorption) combine to form the typical green color seen in, for example, fresh leaves. Starting from a completely unsaturated porphyrin π system (2.8), the partial hydrogenation of pyrrole rings leads to a shift of photon absorption to longer wavelengths. Bacteriochlorophylls (BCs), which contain two partially hydrogenated pyrrole rings in contrast to the "normal" chlorophylls with only one dihydropyrrole ring (2.9), absorb light at particularly low energy in the near infrared region, an effect which can also be achieved by formyl substituents [9]. Carotenoids and open-chain tetrapyrrole molecules, such as phycobilins, complement the chlorophyll pigments [10] so that a broad spectral absorption range is covered (Figure 4.1); the separation of these "leaf pigments" marked the beginning of chromatography (by M. Tswett in 1906). At the end of each growth period, the non-green carotenoid leaf pigments became visible (autumn colors) following the disintegration of chlorophyll, which is quite unstable in its free, unprotected state.

4.2.2 Exciton Transport (Directed Energy Transfer)

The absorption of energy in the form of photons by pigments requires less than 10^{-15} seconds and yields short-lived electronically excited (singlet) states that, in principle, can produce a charge separation. In view of the rather low photon density in diffuse sunlight (rate of absorption less than one photon per pigment molecule per second) and the necessarily rapid charge separation process, it is more economical to use the major part (>98%) of the chlorophyll molecules to act as "antenna" devices and collect available photons ("light harvesting") [8,11]. This means, however, that there must be efficient and spatially oriented ("vectorial") transfer of the absorbed energy in the form of excited

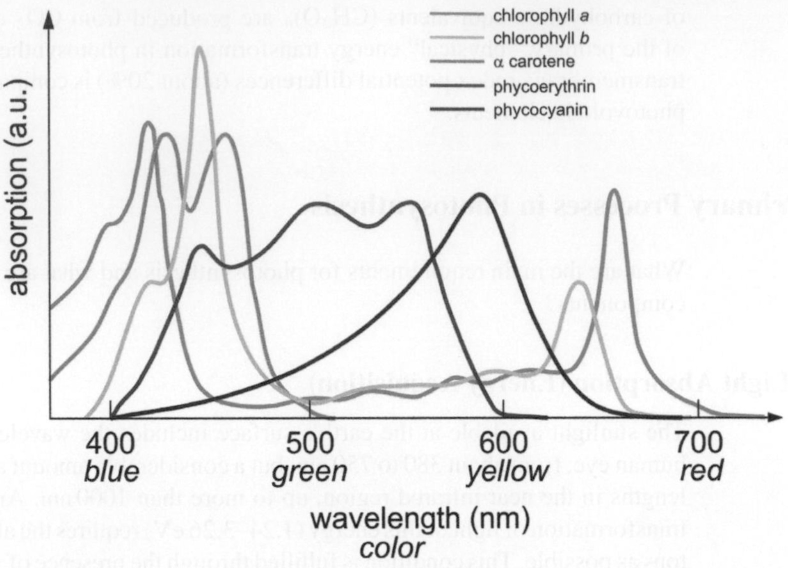

Figure 4.1
Absorption spectra of some pigments from algae and plants

states ("excitons") to the actual reaction centers, which contain less than 2% of the total chlorophyll content.

The energy transfer, which requires neither mass nor charge movement, is made possible by a special arrangement of many chlorophyll chromophors in a network of "antenna pigments" (Figure 4.2). These chromophors are arranged in spatial proximity and a certain, well-defined orientation; they are able to "tunnel" the light energy to the actual reaction centers with about 95% efficiency within 10–100 ps. In physical terms, this Förster mechanism of "resonance transfer" proceeds via spectral overlap of emission bands of the exciton source with the absorption bands of the exciton acceptor. This kind of mechanism also exists for the exciton transfer from other, higher energy-absorbing pigments to the reaction centers (energy transfer along an energy gradient), so that the light-harvesting complexes of the photosynthetic membrane feature a spatially as well as spectrally optimized cross-section for photon capture.

The role of magnesium in chlorophyll is to contribute to the particular arrangement of pigments. The virtually loss-free exciton transfer through a cluster network of antenna pigments requires a high degree of three-dimensional order (Figure 4.2). Therefore, a well-defined spatial orientation of the chlorophyll π chromophors with regard to each other is necessary. This orientation cannot be solely guaranteed by anchoring the chlorophyll molecules in the membrane via the phytyl side chains but must rely also on the coordination of polypeptide side-chain ligands to the two free axial coordination sites at the metal (three-point fixing for defined spatial orientation).

The more direct *in vitro* aggregation of the dihydrate of a chlorophyll derivative with an ethyl instead of a phytyl side chain is quite revealing (Figure 4.3), as a one-dimensional coordination polymer is thus formed. The coordinatively doubly unsaturated Lewis acidic

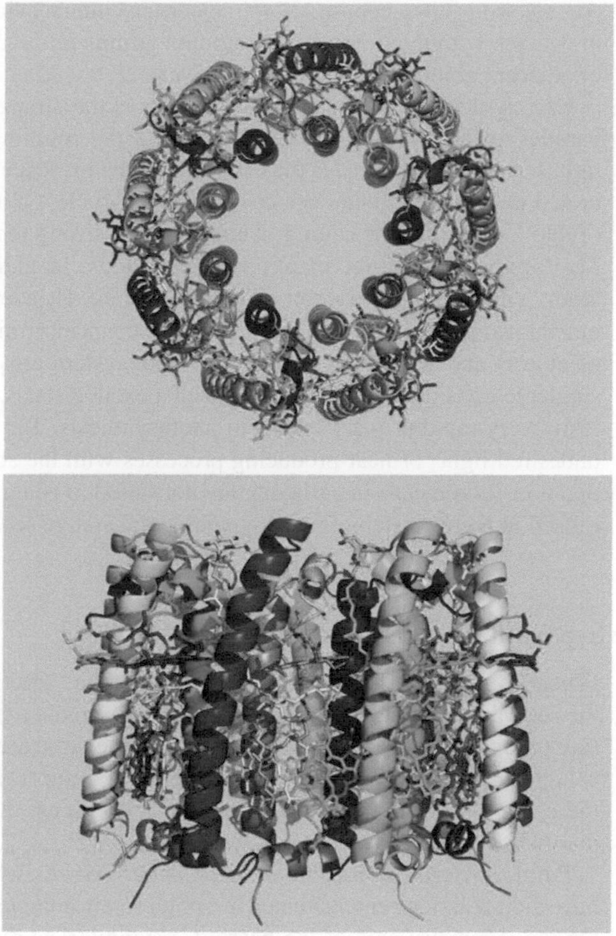

Figure 4.2
Light-harvesting architectures for *Rhodopseudomonas acidophila*, as seen from the top (top) and the side (bottom) (PDB code 1KZU).

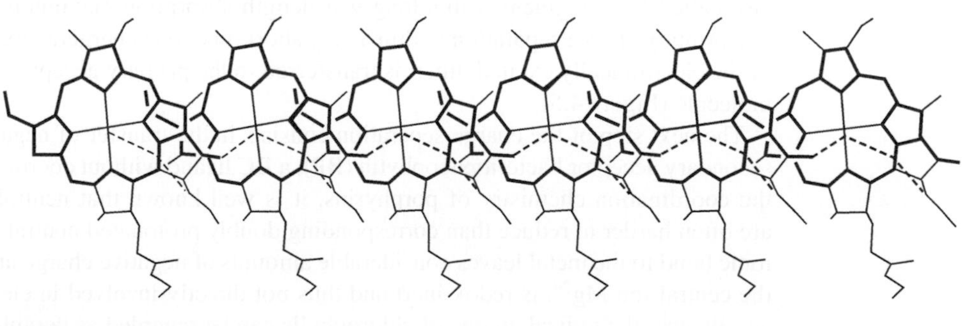

Figure 4.3
Structure of a one-dimensional aggregate occurring in crystals of ethyl chlorophyllide dihydrate. Reprinted with permission from [12] © 1975, American Chemical Society. The π electron conjugation is represented by thick lines, hydrogen bond links via water molecules by broken lines.

(i.e. electron pair-accepting) Mg^{2+} centers can interact via dipolar, hydrogen-bonding water molecules with the Lewis basic carbonyl groups in the characteristic cyclopentanone ring of adjacent chlorophyll molecules.

Exactly defined spatial orientation, such as the arrangement of tetrapyrrole ring planes parallel to the plane of the membrane and the resulting high degree of organization in light-harvesting systems, is thus ensured by the presence of a coordinatively doubly unsaturated electrophilic metal center in the macrocycle. Of all the metals with the proper size (Table 2.6), sufficient natural abundance and strong tendency for six-coordination, only Mg^{2+} remains of all the main-group metal ions. In addition, magnesium is a rather light atom, with a small spin–orbit coupling constant. Heavier elements, including main-group metal ions, have higher spin–orbit coupling constants (mixing spin and orbital momentums of atoms) and would thus enhance an intersystem crossing (ISC) from very short-lived singlet to considerably longer-lived triplet excited states, thereby slowing down the necessarily very rapid primary events in photosynthesis. The result would be a competition of undesired light- or heat-producing processes with the actual chemical reactions. The fact that transition metals in particular are not suited to being the central ions of chlorophylls is related to the next step of photosynthesis: the charge separation.

4.2.3 Charge Separation and Electron Transport

When excitonic energy reaches a photosynthetic "reaction center", the essential step for the separate production of an electron-rich (i.e. reduced) component and an electron-poor (i.e. oxidized) species can take place. Since the structural elucidation of bacterial reaction centers via x-ray diffraction of single crystals (Figure 4.4) has been accomplished [4,10,13], there is now a fairly solid basis from which to discuss the functions of the molecular units involved (see Table 4.1).

Purple bacteria, such as *Rhodopseudomonas viridis*, which contain only one photosystem have their reaction center situated in a polyprotein complex that *in vivo* spans the membrane. On the symmetry axis of the nearly C_2-symmetrical reaction centers is a BC dimer, the "special pair" BC/BC (Figure 4.5). Structure determination indicates that the aggregation can result from a coordinative interaction between acetyl substituents at the polypyrrole ring and the metal centers [10]. Because of the significant π–π orbital interaction, the special pair may function as an electron donor; electronic excitation of this pair, which is also called P_{960} (pigment with a long-wavelength absorption maximum at 960 nm), leads to a primary charge separation within a very short time: one energetically elevated electron of the electronically excited dimer is transferred to the primary acceptor, a monomeric BC molecule (Figure 4.5).

The next step of the charge separation consists in the transfer of negative charge to the secondary acceptor bacteriopheophytin (BP), a BC ligand without coordinated metal. From the coordination chemistry of porphyrins, it is well known that neutral M^{2+} complexes are often harder to reduce than corresponding doubly protonated neutral ligands; the more ionic bond to the metal leaves considerable amounts of negative charge at the ligand. Since the central ion Mg^{2+} is redox-inert and thus not directly involved in electron donation or acceptance, the radical anions of chlorophylls can be regarded as complexes of a divalent metal cation and the radical *trianion* of the macrocyclic ligand: $(Chl/Mg)^{\bullet-} = Chl^{\bullet 3-}/Mg^{2+}$. Correspondingly, radical cations of tetrapyrrole complexes can be formulated as compounds of metal dications with anion radical ligands (compare Sections 6.2–6.4).

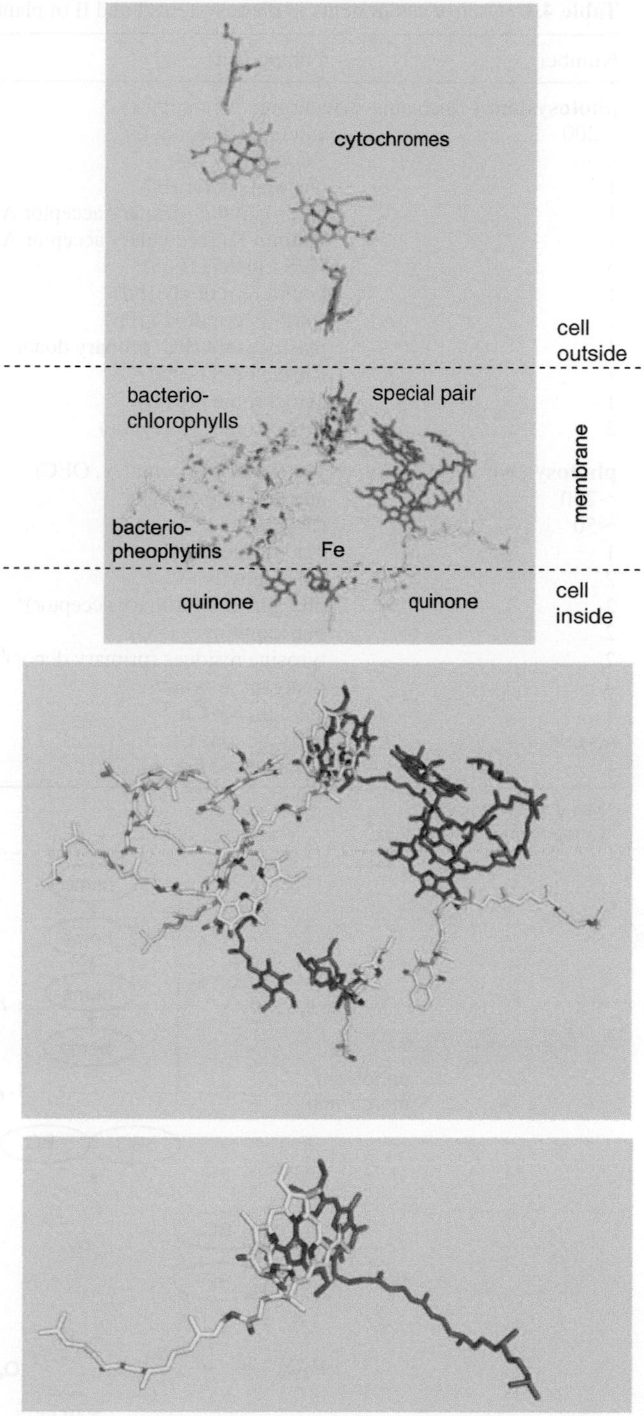

Figure 4.4
Photosynthetic reaction center of *Rhodopseudomonas viridis*. Top: arrangement of electron-transferring components in the membrane complex. Middle: extended view of the electron-transferring components around the special pair (mostly inside the membrane). Bottom: special-pair geometry (PDB code 1PRC).

Table 4.1 Active components in photosystems I and II of plants (compare Figures 4.6 and 4.7).

Number	Component	Reference
photosystem I (including cytochrome *b/f* complex)		
~200	antenna chlorophylls	(Figure 4.2)
~50	carotenoids	(compare Figure 4.1)
1	reaction center P_{700}	(compare Figure 4.4)
1	chlorophyll *a* (primary acceptor A_o)	(2.9)
1	vitamin K_1 (secondary acceptor A_1)	(3.12)
3	Fe/S-clusters (FeS)	(Sections 7.1–7.4)
1	bound ferredoxin (Fd)	(Sections 7.1–7.4)
1	soluble ferredoxin (Fp)	(Sections 7.1–7.4)
1	plastocyanin (PC, primary donor)	(Sections 10.1)
1	Rieske Fe/S center	(7.5)
1	cytochrome *f* (cyt *f*)	(Section 6.1)
2	cytochromes b_6 (cyt b_6)	(Section 6.1)
photosystem II (including oxygen-evolving complex, OEC)		
~200	antenna chlorophylls	(Figure 4.2)
~50	carotenoids	(compare Figure 4.1)
1	reaction center P_{680}	(compare Figure 4.4)
2	chlorophylls	(4.2)
2	pheophytins (primary acceptor)[a]	(Section 4.2.3)
2	plastoquinones (PQ)	(3.12)
2	tyrosine residues (primary donor)[b]	(compare Table 2.5)
4	manganese centers	(Figure 4.9)
1	calcium ion Ca^{2+}	(Section 14.2)
several	chloride ions Cl^-	
1	cytochrome b_{559}	(Section 6.1)

[a]Metal-free chlorophyll.
[b]Components of the protein.

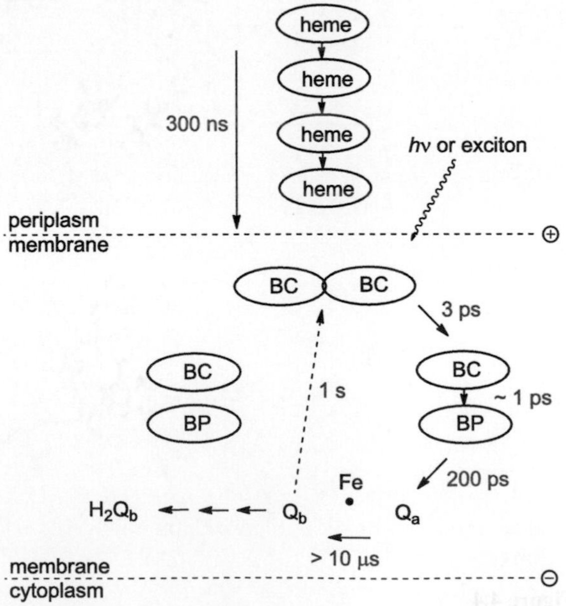

Figure 4.5
Schematic representation of the temporal and spatial sequence of the light-induced charge separation in the reaction center of bacterial photosynthesis (*Rps. viridis*, according to [4,10]). Heme: heme systems; BC: bacteriochlorophyll; BP: bacteriopheophytin; $Q_{a,b}$: quinones.

Structure Determination by X-ray Diffraction

In this context, and in the description of other physical methods, we can only provide very elementary presentations. The main purpose is to point out the principles and the usefulness of these methods for bioinorganic chemistry, as well as their limitations.

In structure determination by x-ray diffraction, uniformly crystallizing systems are examined with monochromatic x-rays. The mathematical analysis of the resulting diffraction pattern can then provide an idea of the three-dimensional distribution of electron density in the "unit cell" (i.e. the periodically repeated unit of a single crystal), revealing an illustration of molecular shapes (see Figure 2.4).

An initial requirement that is particularly difficult to achieve in the biochemical field is the growing of sufficiently large and qualitatively suited single crystals. Large proteins have many degrees of freedom and can often be arranged in a single-crystal setting only under very specific conditions (temperature, solvent, pH). Hydrophilicity and the propensity to form hydrogen bonds usually lead to the inclusion of water molecules during crystallization. A major problem, the structural determination of proteins which exist only within biological membranes, can be solved by using co-crystallizing, membrane-analogous detergent molecules [14]. It was this achievement that allowed for the first structural elucidation of a reaction center of bacterial photosynthesis by x-ray diffraction methods, which was honored by the Nobel Prize in Chemistry in 1988.

The computational effort necessary for a structure determination is much higher for proteins, with their very large number of atoms, than for low-molecular-weight compounds. Even with a good set of data, the resulting multiparameter problem can often be solved only to a certain extent, leaving a molecular resolution of 0.3 nm (3 Å) or higher, which is chemically not satisfactory and only provides for a rather diffuse structure, revealing the protein folding. In such cases, bond angles and bond lengths between small, little-diffracting atoms such as C, N and O are not available with sufficient precision. In general, hydrogen atoms cannot be localized by x-ray diffraction methods under the conditions of protein crystallography.

An important and not always easily dismissed objection to the relevance of single-crystal structural analyses concerns the mainly static aspect of crystalline systems. It is possible, and has been demonstrated in some cases, that important structural features (conformations) of biochemically active compounds are different in the crystal and in solution. Recent studies using high-resolution nuclear magnetic resonance (NMR) spectroscopy in solution (J. Fenn, K. Tanaka, K. Wüthrich, Nobel Prize in Chemistry 2002) and temperature-dependent diffraction of crystalline compounds have begun to contribute to an understanding of the molecular dynamics of biochemical systems.

Throughout this book we have included experimentally derived structures to illustrate structure–function relationships. Such data are available from specialized data banks, such as the Cambridge Structural Database for small molecule crystal structures and the RCSB Protein Data Bank for biological macromolecules.

The third detectable acceptor for the electron generated through light-induced charge separation in the reaction center is a para-quinone, Q_a, such as menaquinone (3.12), which is reduced to a semiquinone radical anion, $Q_a^{\bullet-}$, in this process. After "quenching" at low temperatures, the longer-lived paramagnetic states of the charge separation process can be examined by electron paramagnetic resonance (EPR) spectroscopy, as can the *in vitro*-generated radicals of the individual components [6]. Reduced Q_a may in turn reduce a different, more labile quinone, such as ubiquinone, Q_b (compare (3.12)), with a high-spin iron(II) center connecting the two quinones at the axis of the reaction center. The role of this six-coordinate Fe^{II}, which is equatorially bound by four histidine ligands and axially by carboxylate, is not fully clear; an active redox function in the sense of an *inner-sphere* electron transfer between Q_a and Q_b does not seem likely, since the exchange with redox-inert ZnII, for example, does not make a significant difference in the electron-transfer rate, at least in *Rps. sphaeroides*. Possibly, a divalent metal ion is required to guarantee the controlled electron transfer from reduced BP to the primary quinone through polarization of the hydrogen bond-forming histidine ligands or simply to maintain the necessary structure. Quinone Q_b, or its $2e^-/2H^+$ reaction product, the hydroquinone H_2Q_b (3.12), is not tightly bound to the protein but exchanges with quinones in the "quinone pool" of the membrane (compare Figure 4.6), so that electrons can now be transported further outside the protein. In simple bacteria, the electron gradient finally gives rise to a coupled H^+ gradient and photosynthetic phosphorylation takes place (ATP synthesis; compare Figure 4.6). In higher organisms there are further steps, the "dark reactions", which eventually lead to the production of the electron-rich coenzyme NADPH and to Mg^{2+}-requiring CO_2 reduction (Calvin cycle).

The radical cation of the "special pair" that remained after the initial charge separation is reduced after a relatively long time through regulated electron flow (Figure 4.5) via the heme centers of cytochrome proteins (Section 6.1). The electron deficiency or "hole" is thus further translocated and is finally filled up in purple bacteria through back electron transfer at a different site in the membrane. In more evolved organisms, the hole created at the special pair may be the starting point for substrate oxidation (Section 4.3), which requires an additional photosynthetic system consisting of another light-harvesting complex, a reaction center and an oxidase enzyme.

The ability of an electronically excited state to serve as effective reductant *and* oxidant is illustrated by the orbital-energy diagram (4.3). Electronic excitation creates an electron hole in a low-lying orbital, which invites electron transfer from an external donor (photooxidation). Simultaneously, the presence of the excited electron in a high-lying, previously unoccupied orbital allows for the photoreduction of an external acceptor. The photosynthetically undesired alternative of a simple radiative or radiationless recombination is obvious from (4.3).

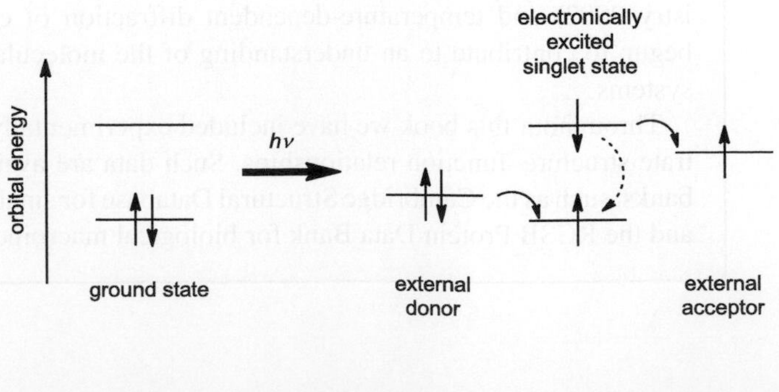

(4.3)

The formation of long-lived, storable oxidation and reduction products as a consequence of light-induced charge separation is very unusual, the normal course of events being a rapid recombination of charges to produce heat or light (emission, luminescence). According to results from structural, spectroscopic and magnetic resonance studies, the success of photosynthesis is based on the strong preference for the extremely rapid charge-separating steps rather than the slower, energy-releasing recombination processes, the ratio of corresponding rates amounting to about 10^8. The basis for this ratio, which is impossible to reach in "normal" chemical reaction systems, is the immobilization of the participating components in a special orientation to each other within *nonpolar* regions of proteins anchored in the membrane (Figure 4.3). Only then may a "vectorial" chemical uphill reaction prevail over the natural tendency for charge recombination, which would dominate if free diffusion were possible. In other words, the special arrangement of electron-transfer components results in a sizable reduction or even disappearance of the activation energies for the "forward" reaction process, while the reverse reaction (back electron transfer) falls in an "inverted region" of electron transfer. This means that the reaction rate *decreases* in spite of an increasing free reaction energy, ΔG^0, that is, in spite of a more favorable equilibrium [7]. Naturally, the electron-transfer rates should be highest for the initial steps (Figure 4.5). In summary, the redox potentials in the ground and excited states, the individual structural changes during the charge separation processes, and the orientations of the participating components with respect to each other have to be finely tuned in order to achieve a useful charge separation. Many functional details, such as the asymmetry of the electron-transfer pathway in the actually axially symmetrical reaction centers (Figures 4.4 and 4.5), require further investigation.

From the absolute necessity of preventing heat- or light-producing back electron-transfer reactions, it follows that the chlorophyll molecules must not contain a redox-active transition metal. Metal centers like $Fe^{II/III}$, which easily transfer electrons themselves, could accept or donate electrons in the ground or electronically excited states of an Fe-chlorophyll π system and thus preclude the photosynthesis, which requires a rapid spatial charge separation. An *intra*molecular instead of an *inter*molecular electron transfer would result, without chemical energy storage.

The role of the magnesium ion in chlorophylls is therefore to act as a *light, redox-inert, Lewis acidic coordination center* that contributes to a defined three-dimensional organization in light-harvesting systems and reaction centers. Monomers, π–π dimers and metal-free ligands each have separate, unique functions in the primary processes of photosynthesis. Redox-active transition metals or heavier metal centers with higher spin–orbit coupling constants would support additional, undesired reaction alternatives; of the remaining bioavailable metal ions, only Mg^{2+} fits exactly with regard to size and charge (Table 2.6).

4.3 Manganese-catalyzed Oxidation of Water to O_2

Although the photosynthetic fixation of CO_2 (carboxylation) requires polarizing Mg^{2+} ions in addition to the RuBisCO enzyme [15], the reductive side (4.4) of photosynthesis, which proceeds via the universal "hydride" carrier NADPH, is interesting mainly from an organic-biochemical point of view (Calvin cycle). The other side, the more "inorganic" water-oxidation part (4.5; e.g. of plant photosynthesis), has recently attracted much attention from coordination chemistry [16,17].

$$4\,e^- + 4\,H^+ + CO_2 \rightarrow 1/n\,(CH_2O)_n + H_2O \tag{4.4}$$

$$2\,H_2O \rightarrow O_2 + 4\,H^+ + 4\,e^- \tag{4.5}$$

The main reason for this interest lies in the crucial catalytic function of a polynuclear manganese complex in the mechanistically challenging oxidation of water to dioxygen, O_2 [18]. The efficient and long-term stable catalysis of dioxygen formation also poses a problem in the (photo)electrochemical water splitting used for the technical production of dihydrogen, H_2, as an energy carrier.

Since the production of NADPH *and* O_2 requires more potential difference than can be generated by *one* photosystem of the type depicted in Figure 4.4, especially when considering energy losses, the photosynthesis in plants and cyanobacteria differs from that in the more simple purple bacteria by featuring *two* separately excitable photosystems (Figure 4.6). These two systems can be connected in a redox potential diagram, the "Z scheme", which shows several, often metal-containing electron-transfer components (Table 4.1, Figure 4.7) [19]. In addition to photosystem I (PS I, absorption maximum 700 nm), a photosystem II (PS II, absorption maximum 680 nm) exists, which provides electrons for phosphorylation and for PS I. The electron holes remaining in PS II represent a very positive potential, which is used to oxidize two water molecules to dioxygen in an overall four-electron process (4.5).

Whereas PS II (without the "oxygen-evolving complex", OEC) shows structural similarity to the reaction centers of purple bacteria (Figures 4.4 and 4.8), PS I exhibits a different arrangement, according to diffraction studies [20]. Table 4.1 summarizes the nonprotein components for both photosystems.

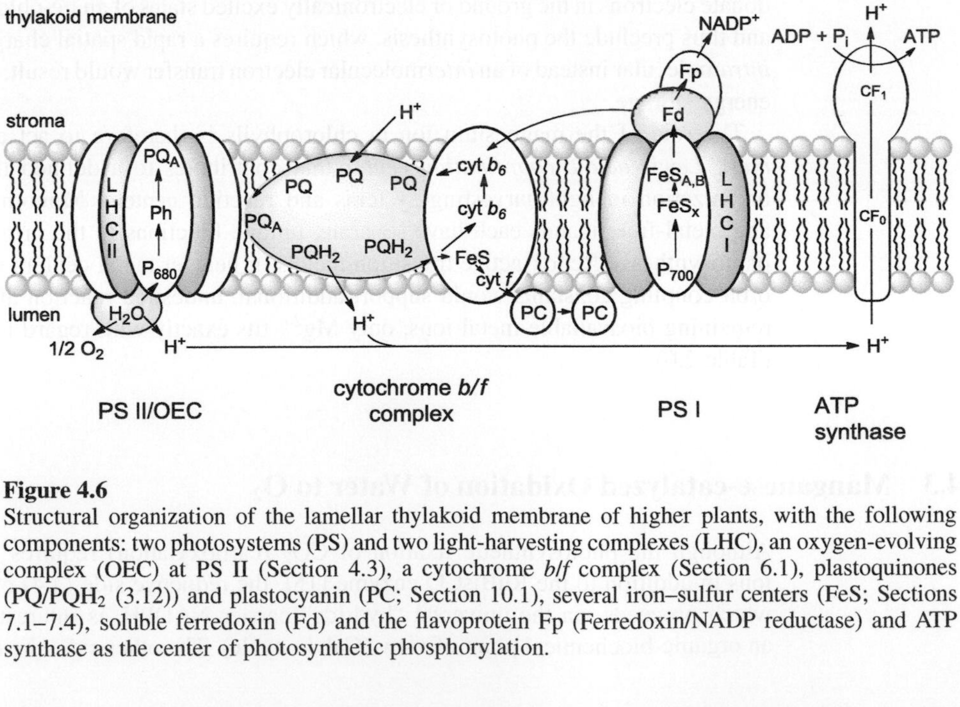

Figure 4.6
Structural organization of the lamellar thylakoid membrane of higher plants, with the following components: two photosystems (PS) and two light-harvesting complexes (LHC), an oxygen-evolving complex (OEC) at PS II (Section 4.3), a cytochrome *b*/*f* complex (Section 6.1), plastoquinones (PQ/PQH_2 (3.12)) and plastocyanin (PC; Section 10.1), several iron–sulfur centers (FeS; Sections 7.1–7.4), soluble ferredoxin (Fd) and the flavoprotein Fp (Ferredoxin/NADP reductase) and ATP synthase as the center of photosynthetic phosphorylation.

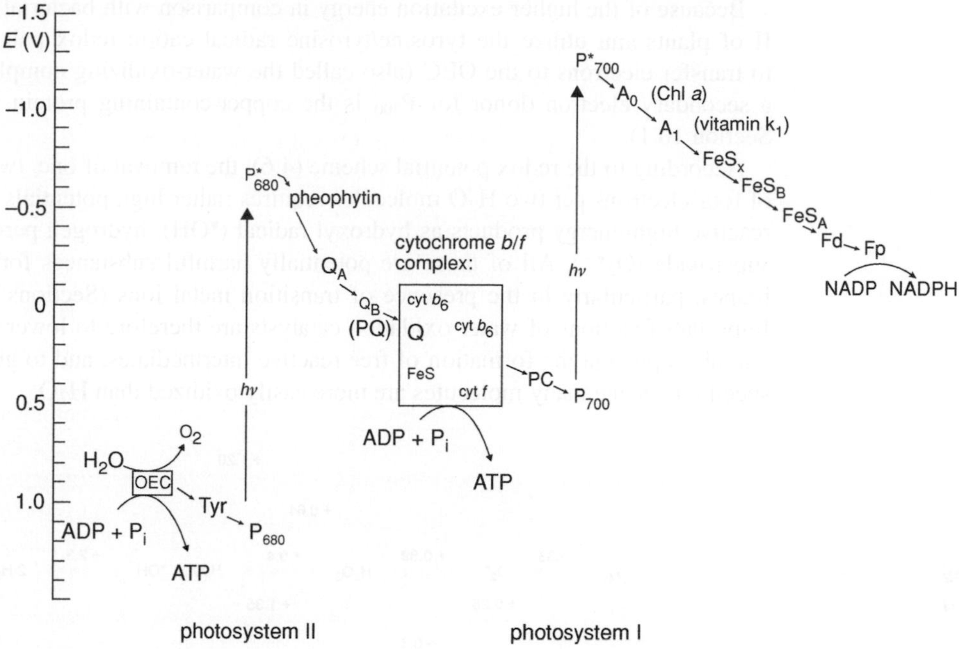

Figure 4.7
Z scheme of electron transfer in plant photosynthesis (see Figure 4.6). Tyr: tyrosine/tyrosine radical cation redox pair; for other abbreviations, see Table 4.1 (scheme modified according to [19]).

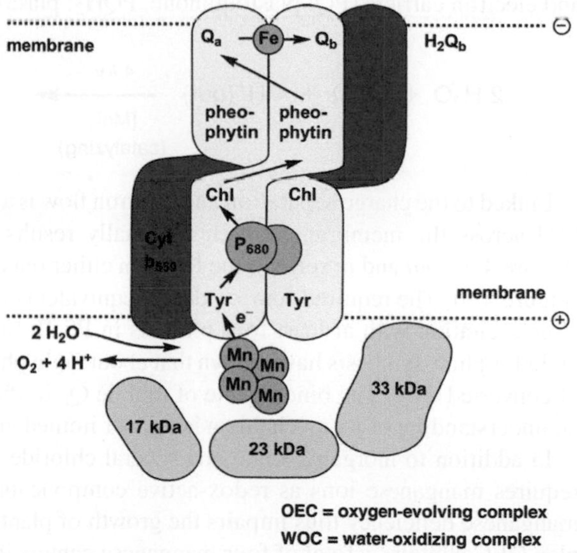

Figure 4.8
Schematic structure of PS II, including the OEC: protein subunits, electron-transferring components and electron-flow.

Because of the higher excitation energy in comparison with bacterial pigments, the PS II of plants can utilize the tyrosine/tyrosine radical cation redox pair ($E_0 = +0.95\,V$) to transfer electrons to the OEC (also called the water-oxidizing complex (PSII-WOC)); a secondary electron donor for P_{700} is the copper-containing protein plastocyanin (see Section 10.1).

According to the redox potential scheme (4.6), the removal of one, two or three instead of four electrons per two H_2O molecules requires rather high potentials and leads to such reactive high-energy products as hydroxyl radical ($^{\bullet}OH$), hydrogen peroxide (H_2O_2) and superoxide ($O_2^{\bullet-}$). All of these are potentially harmful substances for biological membranes, particularly in the presence of transition metal ions (Sections 10.5 and (16.8)). Important functions of water oxidation catalysts are therefore to lower the oxidation potential, to prevent the formation of free reactive intermediates, and to guarantee substrate specificity, since many molecules are more easily oxidized than H_2O.

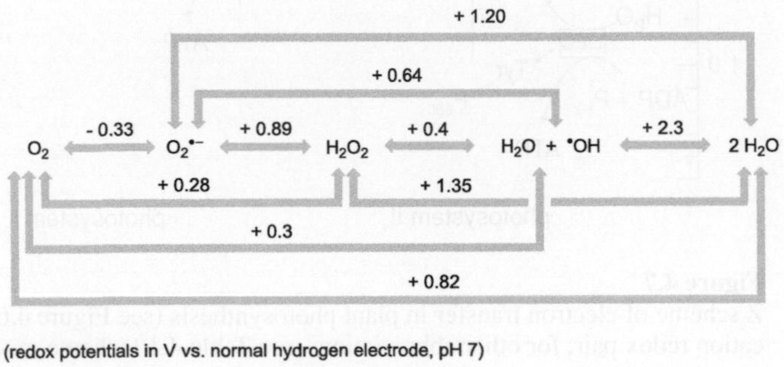

(4.6)

The overall reaction in PS II can be summed up as in (4.7), employing quinonoid proton and electron carriers (PQ: plastoquinone, PQH_2: plastohydroquinone; see (3.12)).

$$2\,H_2O + 2\,PQ + 4\,H^+(\text{out}) \xrightarrow[\substack{[Mn]_x\\(\text{catalyzing})}]{4\,h\nu} O_2 + 2\,PQH_2 + 4\,H^+(\text{in}) \qquad (4.7)$$

Linked to the charge separation and electron flow is a well-controlled transport of protons [21] across the membrane, which eventually results in energy-storing phosphorylation (Figure 4.7); *out* and *in* refer to the location either outside or inside the membrane vesicles (Figure 4.6). The required four oxidation equivalents for O_2 production are available only after excitation with at least four photons in PS II; measurements of the actual quantum yield for photosynthesis have shown that about eight photons are needed for every molecule of converted CO_2. The binding site of mobile Q_b in PS II is the target of many herbicides; an understanding of its mechanism is thus of immediate practical importance [22].

In addition to inorganic Ca^{2+} and several chloride ions, effective dioxygen generation requires manganese ions as redox-active components. EPR spectroscopically detectable manganese deficiency thus impairs the growth of plants. The catalyzing oligoprotein complex OEC contains a total of four manganese centers in one of its subunits (33 kDa). After having to rely on variants of x-ray absorption spectroscopy (XAS), such as extended x-ray absorption fine structure (EXAFS) or x-ray absorption near edge structure (XANES), for insufficiently crystallized preparations, it has been possible to establish the molecular arrangement to a satisfactory resolution (Figure 4.9) [23].

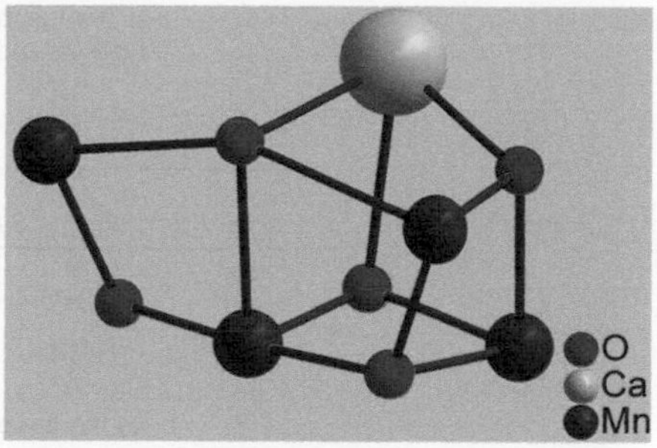

Figure 4.9
Metal coordination arrangement in the OEC from a 1.9 Å resolution x-ray crystal structure determination of PS II from the thermophilic cyanobacterium *T. vulcanus* (PDB code 3ARC) [23].

The structure of the metal cluster contains a distorted Mn_3CaO_4 heterocubane with one added ("dangling") Mn centre, connected to one of the cubane oxygen atoms and to one of the manganese sites via a bridging oxide. The first-shell coordination is complemented by carboxylates (glutamate, aspartate) and a histidine. The two non-metal-binding chloride ions can be located near the cluster and participate in structure stabilization and hydrogen bonding. The location of the substrate water (converted to O_2) and the details of the manganese oxidation states involved are being studied both experimentally and by computational methods.

In view of the intense x-ray irradiation needed to analyze the crystals of very large protein complexes, the assignment of structurally suggested oxidation states requires great caution, as radiation can cause redox reactions.

The charge-separation cascade [21] illustrated in (4.8) clearly shows how single oxidation equivalents become available for the polynuclear manganese cluster via a tyrosine radical cation ("$Y_z^{\bullet+}$") as a primary acceptor.

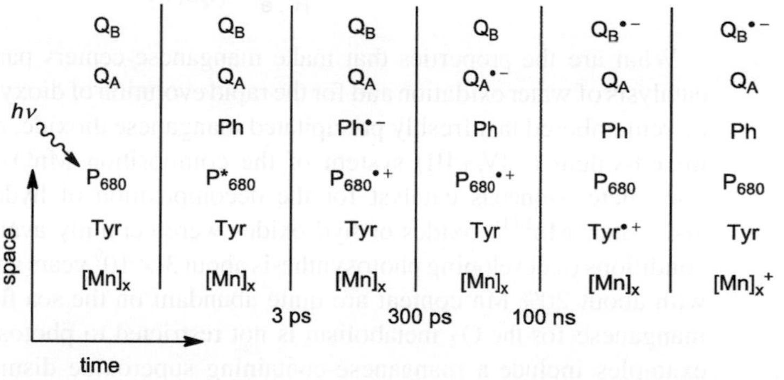

(4.8)

Until two molecules of water have been oxidized *stepwise* to dioxygen through successive light excitation/charge separation events, five exactly tuned (oxidation) states of the PS II, referred to as S_0 to S_4 (Kok cycle), are observed in the millisecond range by flash-photolysis techniques. Scheme (4.9) shows the charge-induced coupling of electron and proton flow and the different lifetimes of these five states under physiological conditions.

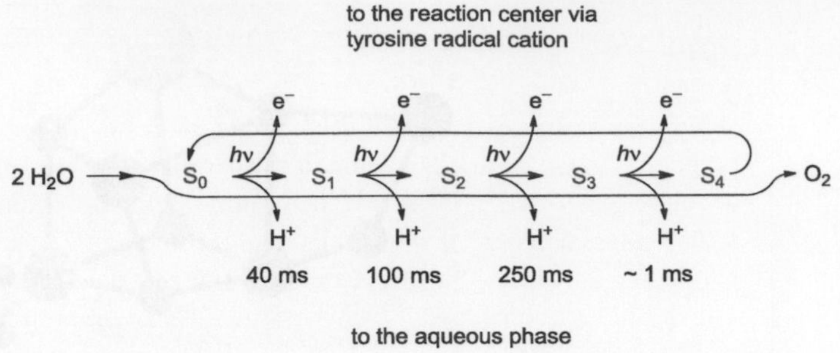

(4.9)

According to manganese XAS near edge studies, the metal oxidation state changes from S_0 via S_1 to S_2. The relatively long-lived intermediate state S_2 exhibits a most conspicuous, highly structured "multiline" EPR spectrum around $g = 2$, signifying an $S = 1/2$ ground state. Naturally occurring manganese contains exclusively the isotope ^{55}Mn, which has a nuclear spin of $I = 5/2$ and a large nuclear magnetic moment; a spin–spin hyperfine interaction of *one* unpaired electron with *two or more* different ^{55}Mn centers can thus be assumed for the S_2 state, attributed to an $Mn^{III}Mn^{IV}{}_3$ configuration.

Diagram (4.10) shows one of the alternatives for oxidation state assignment, including mixed-valent intermediates. Tetravalent manganese and an odd total number of electrons are postulated for the extremely short-lived O_2-evolving S_4 state. S_4 contains two electrons less than S_2.

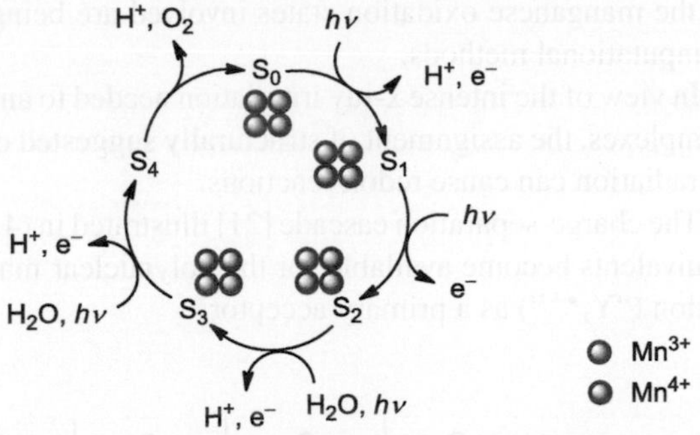

(4.10)

What are the properties that make manganese centers particularly well suited for the catalysis of water oxidation and for the rapid evolution of dioxygen? In this context, it should be remembered that freshly precipitated manganese dioxide, a typically nonstoichiometric mixed-valent (+IV,+III) system of the composition $MnO_{2-x} \times n\ H_2O$, may act as a good heterogeneous catalyst for the decomposition of hydrogen peroxide to dioxygen and water. $Mn^{III,IV}$ oxides or hydroxides were certainly available in sea water under the conditions of developing photosynthesis about 3×10^9 years ago; oxidic manganese nodules with about 20% Mn content are quite abundant on the sea floor. Also, the importance of manganese for the O_2 metabolism is not restricted to photosynthesis; further established examples include a manganese-containing superoxide dismutase (Section 10.5) [3], an azide-insensitive catalase and other peroxidases (see Section 6.3).

The remarkable features of manganese are:

- a large variety of stable or at least metastable oxidation states (+II, +III, +IV, +VI, +VII; compare Figure 18.3);
- the often very labile binding of ligands;
- a pronounced preference for high-spin states, due to inherently small d orbital splitting (2.13), resulting in an often complex magnetic behavior.

Spin–Spin Coupling

When two or more centers with unpaired electrons (↑) interact, the result may be a parallel "ferromagnetic" (↑↑) or an antiparallel "antiferromagnetic" (↑↓) coupling of the electron spins (4.11) [24]. If the orbital interaction is small, because of orthogonal arrangement of p or d orbitals for example, Hund's rule requiring maximal multiplicity in order to avoid the spin-pairing energy favors a parallel spin–spin coupled situation. The more frequent case, however, is antiparallel (antiferromagnetic) coupling, in which the energy gain from direct or indirect orbital interactions ("superexchange") compensates for the spin pairing.

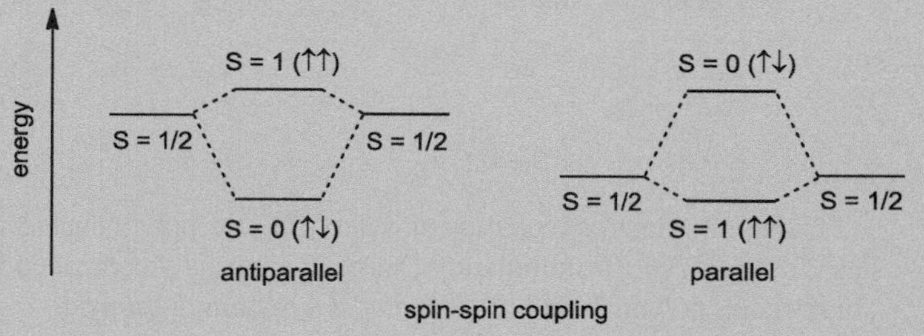

(4.11)

Magnetic states can be examined by measuring the paramagnetic component of the magnetic susceptibility, for example via the force experienced by a substance in an inhomogeneous magnetic field (Faraday balance, SQUID susceptometer). Theoretical models help to interpret the observed data; in particular, the simulation of the temperature dependence of the susceptibility provides information on the type and extent of couplings between electron spins at various centers. According to the Curie law, a higher susceptibility of paramagnetic systems should result at low temperatures, due to diminished averaging through thermal motion of the particles; however, this effect can be (over)compensated by antiferromagnetic behavior; that is, by the tendency for spin pairing.

Due to at least partial antiferromagnetic coupling between the individual high-spin metal centers, the total spin of the manganese cluster in the OEC is distinctly smaller than that of some synthetic polymanganese compounds with $S > 10$ ground states. Nevertheless, the water-oxidizing complex has *several odd-electron states* at its disposal. This fact, as well as the already mentioned availability of fairly high oxidation states and the high lability regarding coordinated ligands, makes manganese centers uniquely suited to catalyzing the generation and *release* of the molecule $^3O_2$ in its triplet ground state; that is, with an *even* number of unpaired electrons (compare Section 5.1). To appreciate this behavior, one must remember that the reaction of transition metals with dioxygen normally involves an irreversible binding.

It would require a "flipping" of electron spins to make the odd-electron catalyst and even-electron systems such as O_2 compatible; however, spin-flipping processes during chemical reactions often implicate a high activation energy because of their low probability (→ statistical aspect of the reaction rate). The best known example for a spin-inhibited reaction is the H_2/O_2 mixture, which reacts to give water only after activation by a bond-breaking catalyst or after ignition (= local bond breaking) in a radical chain reaction (4.12).

$$\begin{array}{ccccccccl}
\text{H–H} & + & \text{H–H} & + & \text{O=O} & \xrightarrow{\text{(spin-inhibited)}} & \text{H}^{\text{O}}\text{H} & + & \text{H}^{\text{O}}\text{H} \quad \text{(atom balance)} \\
\uparrow\downarrow & & \uparrow\downarrow & & \uparrow\downarrow\ \uparrow\uparrow & & \uparrow\downarrow\ \uparrow\downarrow & & \uparrow\downarrow\ \uparrow\downarrow \quad \text{(spin balance)} \\
 & & S = 1 & & & & & S = 0 &
\end{array} \qquad (4.12)$$

The hypothetical spin balance (4.13) shows the possible function of catalytic metal centers with variable spin quantum numbers $S = n/2$.

$$\begin{array}{ccccccccc}
2\,H_2O & + & (\text{Mn–Mn})^{n+} & \longrightarrow & {}^3O_2 & + & (\text{Mn–Mn})^{(n-4)+} & + & 4\,H^+ \\
\uparrow\downarrow & & \downarrow & & \uparrow\uparrow & & \downarrow\downarrow\downarrow & & \\
S = 0 & & S = 1/2 & & S = 1 & & S = 3/2 & &
\end{array} \qquad (4.13)$$

In the four-electron oxidation of water to O_2, the polymanganese system thus acts (1) as an electron reservoir, accumulating charge in an exactly controlled fashion at physiologically high redox potential, and (2) as a *non*-$^3O_2$-retaining catalyst.

With regard to synthetic polymanganese complexes, the emphasis has mainly been on *spectroscopic* and *structural* model compounds; functional model studies, for example of coupled H^+/e^--transfer reactions, have only begun to emerge [18]. For ruthenium, which is related to manganese via a diagonal relationship in the periodic system, there are functional models available (4.14) within attempts to approach artificial photosynthesis [25].

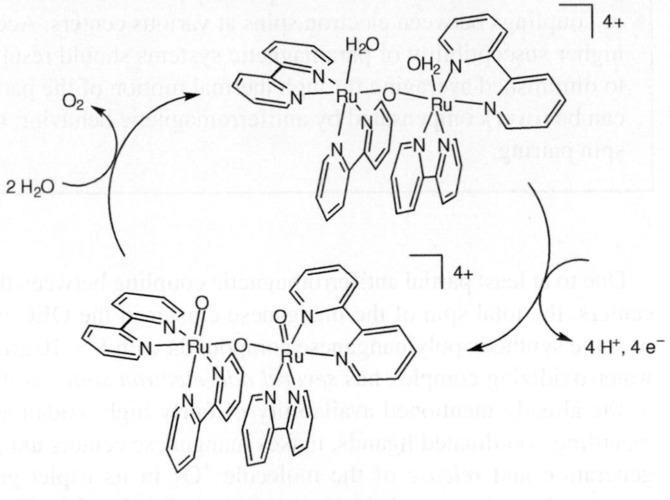

(4.14)

References

1. V. L. Pecoraro (ed.): *Manganese Redox Enzymes*, VCH, New York, **1992**.
2. K. Wieghardt, *Angew. Chem. Int. Ed. Engl.* **1989**, *28*, 1153–1172: *The active sites in manganese-containing metalloproteins and inorganic model complexes.*
3. G. E. O. Borgstahl, H. E. Parge, M. J. Hickey, W. F. Beyer, Jr., R. A. Hallewell, J. A. Tainer, *Cell* **1992**, *71*, 107–118: *The structure of human mitochondrial manganese superoxide dismutase reveals a novel tetrameric interface of two 4-helix bundles.*

4. J. Deisenhofer, H. Michel, *Angew. Chem. Int. Ed. Engl.* **1989**, *28*, 829–847: *The photosynthetic reaction center from the purple bacterium Rhodopseudomonas viridis* (Nobel lecture).
5. W. Holzapfel, U. Finkele, W. Kaiser, D. Oesterhelt, H. Scheer, H. U. Stilz, W. Zinth, *Chem. Phys. Lett.* **1989**, *160*, 1–7: *Observation of a bacteriochlorophyll anion radical during the primary charge separation in a reaction center.*
6. W. Lubitz, F. Lendzian, R. Bittl, *Acc. Chem. Res.* **2002**, *35*, 313–320: *Radicals, radical pairs and triplet states in photosynthesis.*
7. R. A. Marcus, *Angew. Chem. Int. Ed. Engl.* **1993**, *32*, 1111–1121: *Electron transfers reactions in chemistry – theory and experiment* (Nobel lecture).
8. A. V. Ruban, *The Photosynthetic Membrane: Molecular Mechanisms and Biophysics of Light Harvesting*, John Wiley & Sons, Chichester, **2012**.
9. M. Chen, M. Schliep, R. D. Willows, Z.-L. Cai, B. A. Neilan, H. Scheer, *Science* **2010**, *329*, 1318–1319: *A red-shifted chlorophyll.*
10. R. Huber, *Angew. Chem. Int. Ed. Engl.* **1989**, *28*, 848–869: *A structural basis for the transmission of light energy and electrons in biology* (Nobel address).
11. X. Hu, K. Schulten, *Phys. Today* **1997**, *50*, 28–34: *How nature harvests sunlight.*
12. H. C. Chow, R. Serlin, C. E. Strouse, *J. Am. Chem. Soc.* **1975**, *97*, 7230–7237: *The crystal and molecular structure and absolute configuration of ethyl chlorophyllide a dihydrate.*
13. J. P. Allen, G. Feher, T. O. Yeates, H. Komiya, D. C. Rees, *Proc. Natl. Acad. Sci. USA* **1987**, *84*, 5730–5734: *Structure of the reaction center from Rhodobacter sphaeroides R-26: The cofactors.*
14. H. Michel, *J. Mol. Biol.* **1982**, *158*, 567–572: *Three-dimensional crystals of a membrane protein complex. The photosynthetic reaction center from Rhodopseudomonas viridis.*
15. (a) I. Andersson, S. Knight, G. Schneider, Y. Lindqvist, T. Lundqvist, C.-I. Bränden, G. H. Lorimer, *Nature (London)* **1989**, *337*, 229–234: *Crystal structure of the active site of ribulose-bisphosphate carboxylase*; (b) I. Andersson, A. Backlund, *Plant Physio. Biochem.* **2008**, *46*, 275–291: *Structure and function of rubisco.*
16. *Inorg. Chem.* **2008**, *47*, 1697–1861: *Forum on making oxygen.*
17. H. Dau, C. Limberg, T. Reier, M. Risch, S. Roggan, P. Strasser, *Chem. Cat. Chem.* **2010**, *2*, 724–761: *The mechanism of water oxidation: From electrolysis via homogeneous to biological catalysis.*
18. R. J. Pace, R. Stranger, S. Petrie, *Dalton Trans.* **2012**, *41*, 7179–7189: *Why nature chose Mn for the water oxidase in Photosystem II.*
19. R. E. Blankenship, R. C. Prince, *Trends Biochem. Sci.* **1985**, *10*, 382–383: *Excited-state redox potentials and the Z scheme of photosynthesis.*
20. N. Krauss, W. Hinrichs, I. Witt, P. Fromme, W. Pritzkow, Z. Dauter, C. Betzel, K. S. Wilson, H. T. Witt, W. Saenger, *Nature (London)* **1993**, *361*, 326–331: *Three-dimensional structure of system I of photosynthesis at 6 Å resolution.*
21. A. Krauss, M. Haumann, H. Dau, *PNAS* **2012**, *109*, 16 035–16 040: *Alternating electron and proton transfer steps in photosynthetic water oxidation.*
22. W. Draber, J. F. Kluth, K. Tietjen, A. Trebst, *Angew. Chem. Int. Ed. Engl.* **1991**, *30*, 1621–1633: *Herbicides in photosynthesis research.*
23. Y. Umena, K. Kawakami, J.-R. Shen, N. Kamiya, *Nature (London)* **2011**, *473*, 55–60: *Crystal structure of oxygen – evolving photosystem II at a resolution of 1.9 Å.*
24. G. Blondin, J.-J. Girerd, *Chem. Rev.* **1990**, *90*, 1359–1376: *Interplay of electron exchange and electron transfer in metal polynuclear complexes in proteins or chemical models.*
25. J. W. Jurss, J. J. Concepcion, J. M. Butler, K. M. Omberg, L. M. Baraldo, D. G. Thompson, E. L. Lebeau, B. Hornstein, J. R. Schoonover, H. Jude, J. D. Thompson, D. M. Dattelbaum, R. C. Rocha, J. L. Templeton, T. J. Meyer, *Inorg. Chem.* **2012**, *51*, 1345–1358: *Electronic structure of the water oxidation catalyst cis,cis-*$[(bpy)_2(H_2O)Ru^{III}ORu^{III}(OH_2)(bpy)_2]^{4+}$*, the blue dimer.*

5 The Dioxygen Molecule, O_2: Uptake, Transport and Storage of an Inorganic Natural Product

5.1 Molecular and Chemical Properties of Dioxygen, O_2

The rather high concentration of potentially reactive dioxygen, O_2, in the earth's atmosphere (about 21 vol. %) is the result of continuous photosynthesis of organisms [1]. Thus, O_2 is a *natural product*, that is, a secondary metabolic product, just like alkaloids or terpenes. Initially, it even represented an exclusively toxic waste product. Studies of the atmospheres of planets and moons have generally shown O_2 contents of far less than 1 vol. %, and the primeval atmosphere of the earth until about 2.5 billion years ago is assumed to have been very similar. Because of the continuous growth of organisms and a corresponding need for reduced carbon compounds synthesized from fixated CO_2, the amount of photosynthesis increased to such an extent that the simultaneously produced oxidation equivalents could no longer be scavenged by auxiliary substrates such as sulfur(−II) and iron(II) compounds; eventually, a (Mn-)catalyzed oxidation (Section 4.3) of the surrounding water to dioxygen developed. For a while, the "environmentally" extremely harmful O_2 could be deactivated by reaction with reduced compounds, particularly soluble ions such as Fe^{2+} and Mn^{2+}, to form massive oxidic sediments, such as "banded iron formations" (5.1). However, about 2.5 billion years ago the concentration of O_2 in the atmosphere started to increase, and from about 700 million years ago another rise in O_2 concentration led to the current equilibrium between biogenic production and a general, biogenic and nonbiogenic consumption of O_2.

$$4\,Fe^{2+} + O_2 + 2\,H_2O + 8\,OH^- \rightarrow 4\,Fe(OH)_3 + 6\,H_2O \qquad (5.1)$$

The appearance of free **dioxygen, O_2, as a toxic waste product of an energy-producing process** was a true "eco-catastrophe". Most organisms living at that time probably perished as a consequence of this biogenic "self-poisoning", however, some may have survived in O_2-free niches as today's *anaerobic* organisms. In addition to the oxidizing character of O_2, partially reduced and highly reactive species (4.6) are easily formed through catalysis by transition metals such as Fe^{II}. Detoxification with regard to these species has made it necessary to develop a multitude of biological antioxidants (see Table 16.1). Therefore, only those *aerobic* organisms that were able to develop protective mechanisms against O_2 *and* the very toxic intermediates (often radicals) resulting from its partial reduction have survived in contact with the atmosphere (4.6), (5.2) [2].

Bioinorganic Chemistry: Inorganic Elements in the Chemistry of Life – An Introduction and Guide, Second Edition.
Written and Translated by Wolfgang Kaim, Brigitte Schwederski and Axel Klein.

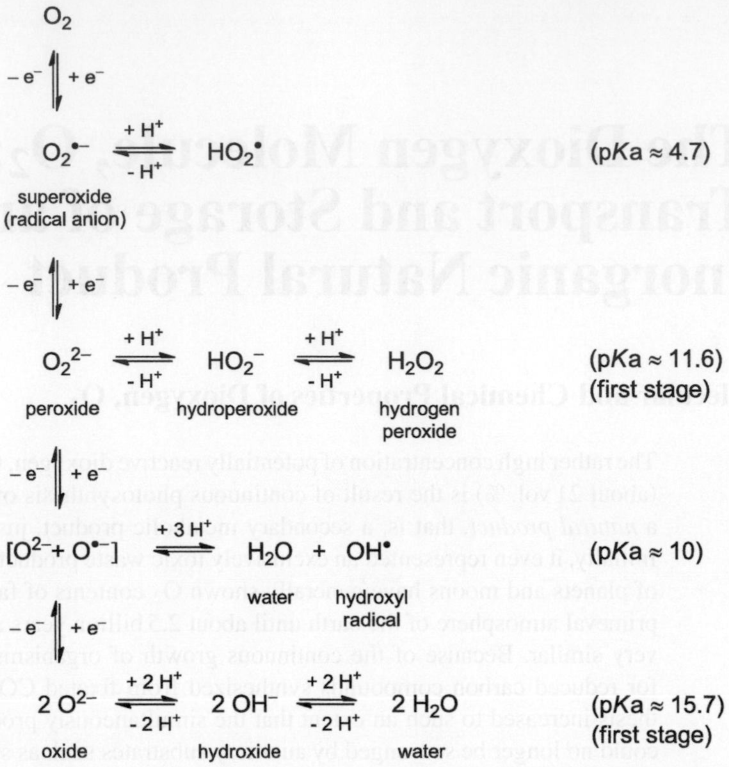

(5.2)

Through the process of evolution, organisms developed that were able to use the reverse process of photosynthesis (4.1) in an oxygen-containing atmosphere for a much more efficient and only indirectly light-dependent metabolic energy conversion and thus for a more dynamic form of life. This so-called "respiration" may be viewed as a "cold" (i.e. controlled and carefully catalyzed) combustion of reduced substrates ("food"). This remarkable development required not only a successful and eventually even biosynthetically useful degradation of partially reduced oxygen intermediates such as H_2O_2, $O_2^{\bullet-}$ and $^{\bullet}OH$ but also the ability to cope with a drastically changed bioavailability of some elements and their compounds [3]. The following changes resulted from the formation of an oxidized atmosphere at pH 7:

- Fe^{II} (soluble) $\rightarrow$ Fe^{III} (insoluble);
- Cu^{I} (insoluble) $\rightarrow$ Cu^{II} (soluble);
- S^{-II} (insoluble) $\rightarrow$ SO_4^{2-} (soluble);
- Se^{-II} (insoluble) $\rightarrow$ SeO_3^{2-} (soluble);
- MoS_x (insoluble) $\rightarrow$ MoO_4^{2-} (soluble);
- $CH_4 \rightarrow CO_2$;
- $H_2 \rightarrow H_2O$;
- $NH_3 \rightarrow NO_3^-$, NO_2^-, NO.

Incidentally, as a consequence of biogenic O_2 production, the ozone (O_3) layer in the stratosphere formed, which in turn contributed to a more controlled development of organisms by protecting them from the high-energy components of solar radiation.

The presence of free O_2 has been hazardous for many species but provided a great opportunity for other, adapting organisms, requiring a discussion of its molecular properties and coordination behavior [4]. According to its position in the periodic table and the resulting second highest electronegativity of all elements, O_2 is strongly oxidizing. Many substances react very exothermically with dioxygen, although activation is often required, due to inhibited reactivity. This frequently observed and quite characteristic activation energy barrier for many reactions involving O_2 can be rationalized by considering the triplet ground state of the O_2 molecule with two unpaired electrons. This situation, which is quite unusual for a small, stable molecule, is a consequence of the molecular orbital scheme (5.3) in connection with Hund's rule (compare (2.13)): when filling up degenerate orbitals such as $\pi^*(2p)$, the state with maximum multiplicity is favored; in the case of dioxygen, that is the triplet $^3O_2$ over the singlet form $^1O_2$.

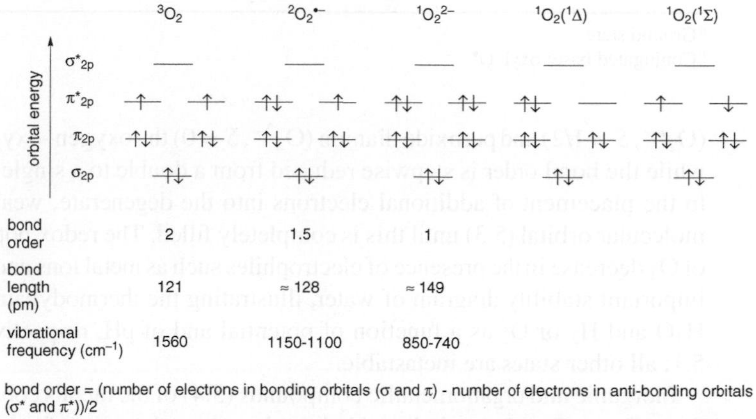

(5.3)

In fact, $^3O_2$ with two unpaired electrons in the ground state is favored over both singlet states, $^1O_2\ (^1\Delta)$ and $^1O_2\ (^1\Sigma)$, by more than 90 and 150 kJ/mole, respectively. The reactions of $^3O_2$ with most normal singlet molecules are thus inhibited because of the necessity for a statistically less likely "flipping" of spins (compare (4.12)). This phenomenon is responsible for the present *metastable* situation at the earth's surface, with the simultaneous presence of combustibles (wood, fossil fuels, carbohydrates etc.) and an oxygen-rich atmosphere without the instant formation (via fire) of the lowest-energy products, water and CO_2. Therefore, chemically or photogenerated singlet dioxygen $^1O_2$ represents another toxic form of oxygen in addition to partially reduced intermediates; the diamagnetic elemental modification ozone, O_3, also shows uninhibited and therefore uncontrolled oxidation behavior towards biomolecules.

Table 5.1 summarizes the biorelevant molecular oxygen species.

The "spin-verbot" (the impediment to a spin-forbidden process) is not valid for those reaction partners of $^3O_2$ that can easily undergo single electron-transfer reactions or which already contain unpaired electrons. The latter include:

- radicals ($S = 1/2$), either as stable free species or as intermediates produced by, for example, ignition;
- compounds with photochemically produced excited triplet states ($S = 1$); and
- paramagnetic transition metal centers ($S \geq 1/2$).

Almost all reactions between O_2 and metal complexes proceed irreversibly, as illustrated by equations (5.1) and (5.12). In most of these reactions, oxygen is eventually reduced to the (−II) state so that oxide (O^{2-}), hydroxide (OH^-) or water (H_2O) ligands result. During the first two one-electron reduction steps (compare (4.6) and (5.2)) to superoxide radical anion

Table 5.1 Biorelevant and other molecular oxygen species.

		S
$^3O_2$	dioxygen (triplet)[a]	1
$^1O_2$ (Δ)	dioxygen (singlet)	0
$^1O_2$ (Σ)	dioxygen (singlet)	0
$O_2^{\bullet-}$	superoxide	1/2
O_2^{2-}	peroxide	0
O_3	ozone	0
$O_3^{\bullet-}$	ozonide	1/2
$^{\bullet}OH$	hydroxyl[b]	1/2
O	oxygen atom	1

[a]Ground state.
[b]Conjugated base: oxyl, $O^{\bullet-}$.

($O_2^{\bullet-}$, $S = 1/2$) and peroxide dianion (O_2^{2-}, $S = 0$) the oxygen–oxygen bond remains intact, while the bond order is stepwise reduced from a double to a single bond. This corresponds to the placement of additional electrons into the degenerate, weakly antibonding $\pi^*(2p)$ molecular orbital (5.3) until this is completely filled. The redox potentials for the reduction of O_2 decrease in the presence of electrophiles such as metal ions and H^+. The biochemically important stability diagram of water, illustrating the thermodynamic equilibrium between H_2O and H_2 or O_2 as a function of potential and of pH, respectively, is shown in Figure 5.1; all other states are metastable.

Inorganic and organometallic compounds (5.4) of the metals cobalt, rhodium and iridium from Group 9 of the periodic table have long been known to form rather simple *reversibly* dioxygen-coordinating complexes [6,7]. Vaska's iridium complex shows reversible uptake and release of "side-on" (η^2) coordinated O_2, while pentacyano–, *salen*– (compare (3.17))

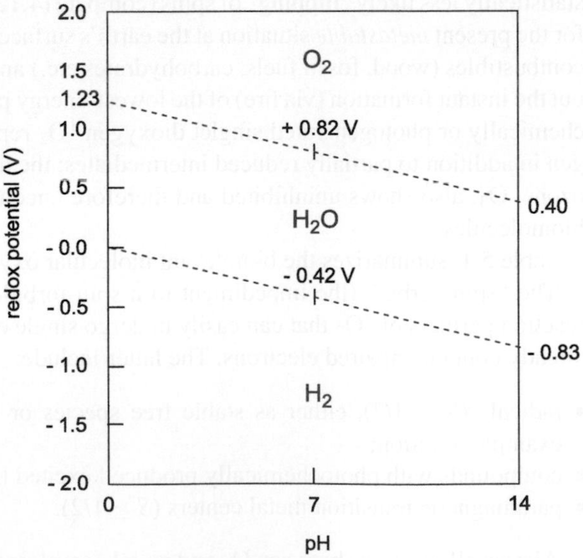

Figure 5.1
Stability diagram of water (- - -: equilibrium lines).

and pentaammine–cobalt complex fragments can coordinate "end-on" η^1-O_2 (5.4). Further possibilities of coordination, which are also discussed in biological O_2 utilization, are summarized in (5.5).

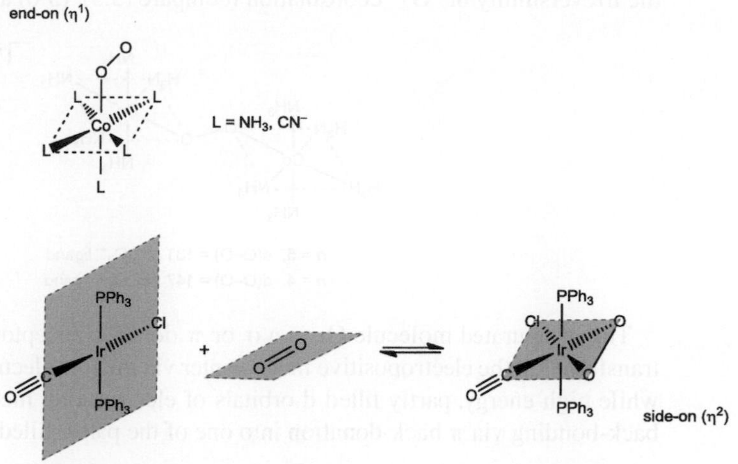

(5.4)

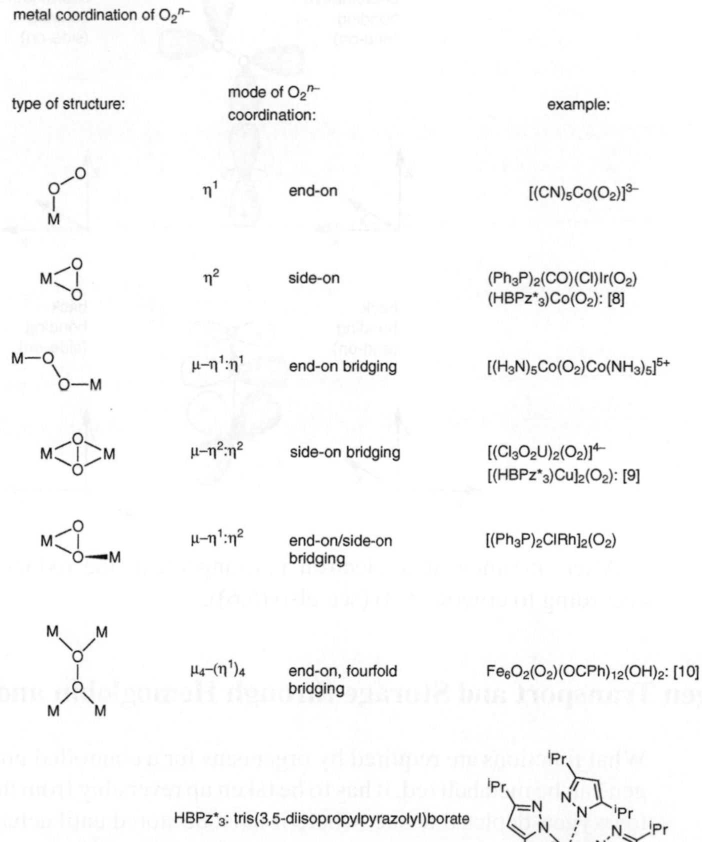

(5.5)

Structural and spectroscopic studies (electron paramagnetic resonance, EPR) suggest that the redox-active "non-innocent" dioxygen ligand [5] is mainly bound in singly or doubly reduced form (5.6). The ability of superoxide and peroxide, as well as of oxide and hydroxide, to frequently act as bridging (μ-) ligands between metal centers contributes to the irreversibility of "O_2" coordination (compare (5.5), (5.6) and (5.12)).

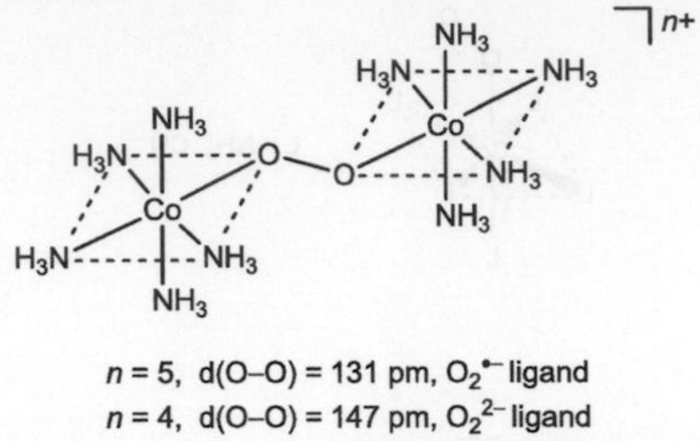

(5.6)

The unsaturated molecule O_2 is a σ or π donor/π acceptor ligand. Electron density is transferred to the electropositive metal center via the free electron pairs or the double bond, while high-energy, partly filled d orbitals of electron-rich metal centers can contribute to back-bonding via π back-donation into one of the partly filled π* orbitals of O_2 (5.7).

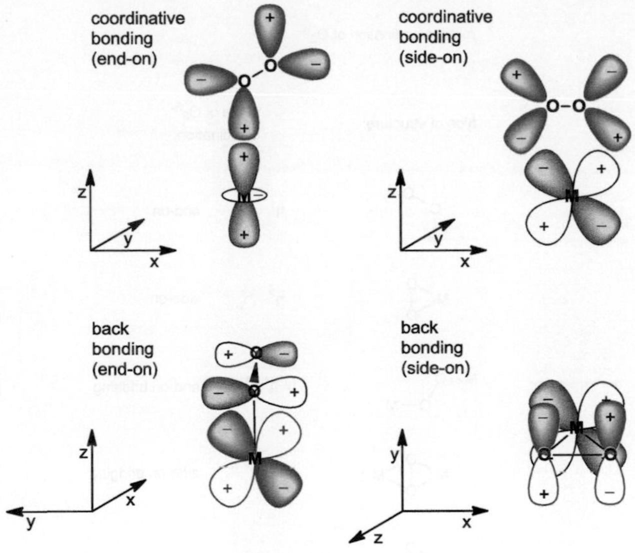

(5.7)

After intramolecular electron rearrangement, the oxidation states can be reassigned according to criteria (5.3) (see also (5.6)).

5.2 Oxygen Transport and Storage through Hemoglobin and Myoglobin

What functions are required by organisms for a controlled utilization of O_2? Before dioxygen can be metabolized, it has to be taken up reversibly from the atmosphere and transported to oxygen-depleted tissue, where it must be stored until actual use. Certain groups of mollusks, crustaceans, spiders and worms on one hand and the majority of other organisms, particularly vertebrates, on the other differ in their strategies in O_2 coordination. While the former contain dinuclear metal arrangements with amino acid coordination, namely the copper protein hemocyanin (*Hc*, Section 10.2) or the iron protein hemerythrin (*Hr*, Section 5.3), most other breathing organisms use the heme system, that is, monoiron complexes of a certain porphyrin macrocycle, protoporphyrin IX (see (2.9), (5.8), Chapter 6). The corresponding proteins are the tetramer hemoglobin (*Hb*, O_2 uptake in the lungs and transport in the blood stream) and the monomer myoglobin (*Mb*, O_2 storage in muscle tissue).

At this point, the versatile role of iron in general (Chapters 5–8) and of the heme group in particular for the biochemistry of humans (Table 5.2) [11] should be specified. Since dioxygen transport is not a catalytic but a "stoichiometric" function, about 65% of the iron present in the human body is confined to the transport protein hemoglobin (*Hb*) alone; the content of the oxygen-storage protein myoglobin is roughly 6%. After all, the share of O_2 in air is only about 21 vol. % and a sufficient level has to be maintained in the tissue even under unfavorable circumstances (e.g. at higher altitude); human blood has an approximately 30 times higher "solubility" for O_2 than water. Metal-storage proteins like ferritin (Section 8.4.2) constitute most of the rest of the iron in the human body. The catalytically active enzymes are present in only minute amounts.

Many, but not all (see Chapter 7), of the redox-catalytically active iron enzymes contain the heme group; peroxidases, cytochromes, cytochrome *c* oxidase and the P-450 system (see Chapter 6 and Section 10.4) belong to the heme proteins. Scheme (5.8) illustrates the determining role the protein environment can play for the quite variable biological functions of a tetrapyrrol complex.

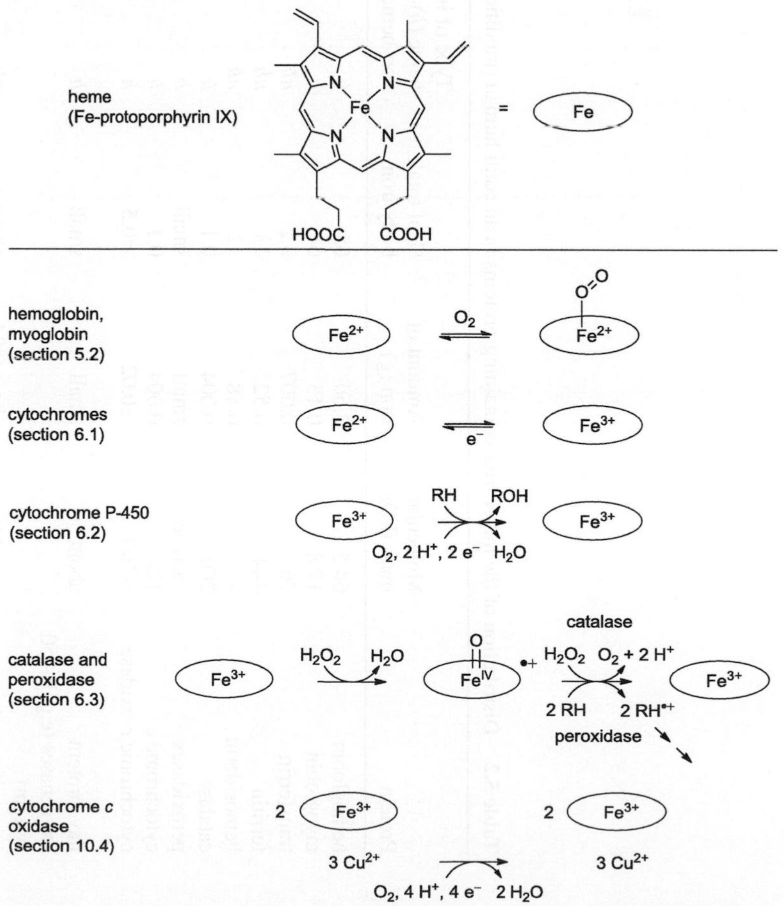

(5.8)

Table 5.2 Distribution of the major iron-containing proteins in an adult human (modified from [11]).

Protein	Molecular mass (kDa)	Amount of iron (g)	% of total body iron	Type of iron: heme (*h*) or non-heme (*nh*)	Number of iron atoms per molecule	Function	See Section
hemoglobin	64.5	2.60	65	*h*	4	O_2 transport in blood	5.2
myoglobin	17.8	0.13	6	*h*	1	O_2 storage in muscle	5.2
transferrin	76	0.007	0.2	*nh*	2	iron transport	8.4.1
ferritin	444	0.52	13	*nh*	up to 4500	iron storage in cells	8.4.2
hemosiderin	–	0.48	12	*nh*	–	iron storage	8.4.3
catalase	260	0.004	0.1	*h*	4	metabolism of H_2O_2	6.3
peroxidases	variable	small	small	*h*	1	metabolism of H_2O_2	6.3
cytochrome *c*	12.5	0.004	0.1	*h*	1	electron transfer	6.1
cytochromc *c* oxidase	>100	<0.02	<0.5	*h*	2	terminal oxidation ($O_2 \rightarrow H_2O$)	10.4
flavoprotein oxygenases (e.g. P-450 system)	about 50	small	small	*h*	1	incorporation of molecular oxygen	6.2
iron-sulfur proteins	variable	about 0.04	about 1	*nh*	2–8	electron transfer	7.1–7.4
ribonucleotide reductase	260 (*E. coli*)	small	small	*nh*	4	transformation of RNA to DNA	7.6.1

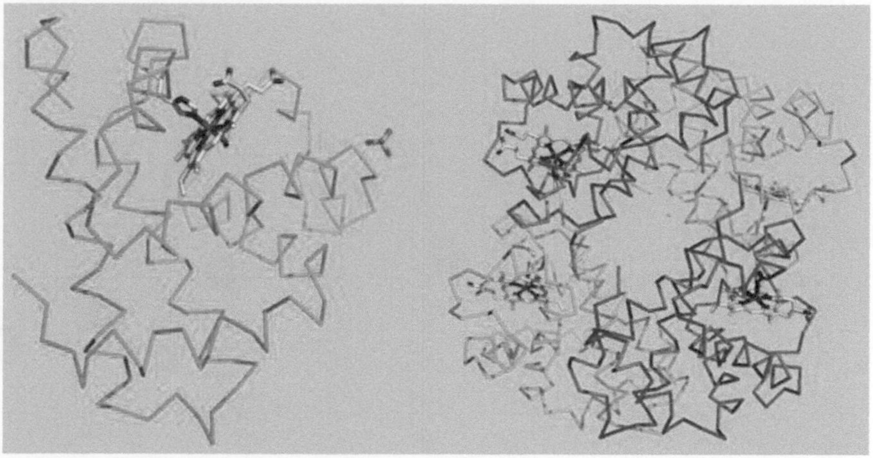

Figure 5.2
Structures of human myoglobin (left, oxy form, PDB code 1MBO) and of the tetrameric protein hemoglobin (right, deoxy form, PDB code 2HHB) [13].

The transport system for O_2 has to take up this molecule as effectively as possible in its ground-state form, $^3O_2$, from the gas phase, in order to transport it in specialized blood cells, the erythrocytes, via the circulatory system to an intermediate storage site and to release it there completely. This task is not easily solved, since both supply and demand of O_2 may change considerably; the storage system always has to have a higher affinity for O_2 than the nonetheless efficient transport system. In the case of hemoglobin (Figure 5.2), with its total of four heme sites, this efficiency is guaranteed by the cooperative effect [12]. In the course of loading with four molecules O_2, corresponding to 1 ml O_2 per 1 g *Hb*, the oxygen affinity increases, as graphically illustrated by a sigmoidal, nonhyperbolic saturation curve (Figure 5.3).

According to Figure 5.3, the cooperative effect guarantees an efficient transfer of O_2 to the storage site: the less O_2 is present in the transport system, the more completely it is released

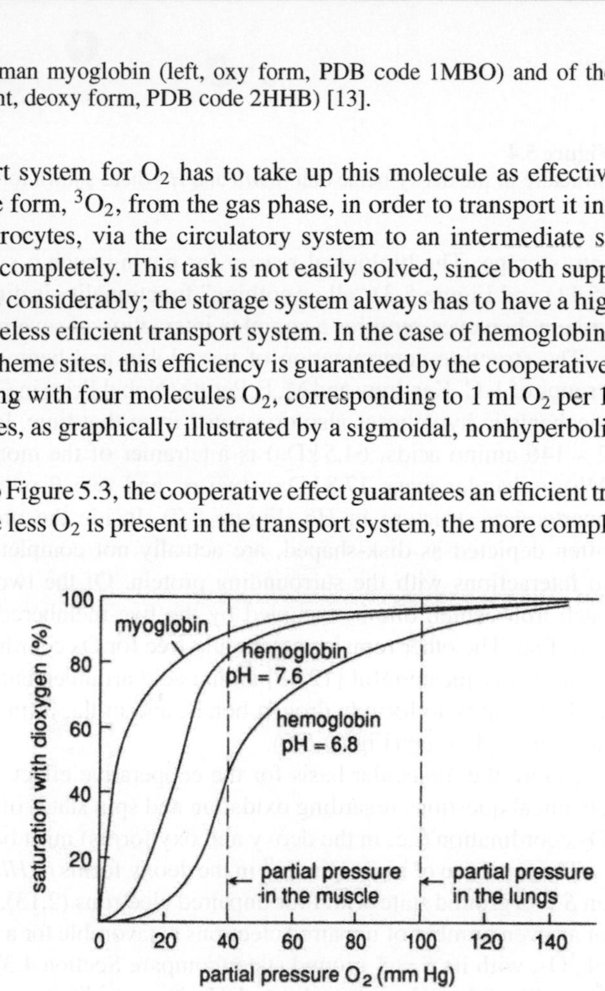

Figure 5.3
Oxygen saturation curves of myoglobin and hemoglobin at different pH values.

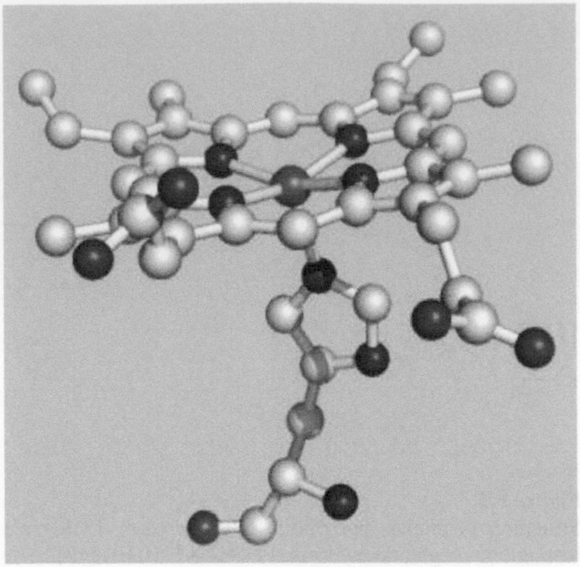

Figure 5.4
Structure of the deoxy heme unit in *Mb* and *Hb* (here shown for *Hb*, PDB code 2HHB) [13b].

into storage. The biological reason for having such a system of pH-regulated (compare (5.11) and Figure 5.3) "all or nothing" functionality is directly evident; its molecular realization, however, requires a complex interaction of several hemoprotein subunits [12,14].

The structure determination of myoglobin and hemoglobin was accomplished by the groups of J. C. Kendrew and M. F. Perutz (Nobel Prize in Chemistry, 1962), and some initial mechanistic hypotheses also originated from that time. In simple terms, *Hb* (2×141 and 2×146 amino acids, 64.5 kDa) is a tetramer of the monomeric hemoprotein myoglobin *Mb* (molecular mass 17.8 kDa); two α- and two β-peptide chains form a well-defined quarternary structure in *Hb* (Figure 5.2). Inside the protein, the deoxy heme systems, often depicted as disk-shaped, are actually not completely flat but slightly domed, due to interactions with the surrounding protein. Of the two free axial coordination sites at each iron center, one is occupied by the five-membered imidazole ring of the *proximal* histidine. The other remains essentially free for O_2 coordination (Figure 5.4), but there is a biologically meaningful [12,15] amino acid arrangement in the form of a *distal* histidine, with its ability to form hydrogen bonds, and in the form of a valine side chain containing an isopropyl group (Figure 5.5).

Before the molecular basis for the cooperative effect in *Hb* is discussed, the inorganic chemical questions regarding oxidation and spin states of the metal center before and after O_2 coordination (i.e. in the deoxy and oxy forms) must be answered.

The presence of high-spin Fe^{II} in the deoxy forms of *Hb* and *Mb* is firmly established and an $S = 2$ ground state with four unpaired electrons (2.13), (2.14) is observed. The presence of an even number of unpaired electrons is favorable for a rapid, non-spin-inhibited binding of $^3O_2$, with its $S = 1$ ground state (compare Section 4.3). For the diamagnetic oxy forms ($S = 0$) with end-on coordinated O_2 in a nonlinear arrangement and with an Fe-O-O angle of about 120° (compare (5.10) or Figure 5.5), the oxidation state assignment is less unambiguous because of the non-innocent nature of coordinated dioxygen; two alternatives (5.9) have been proposed [16].

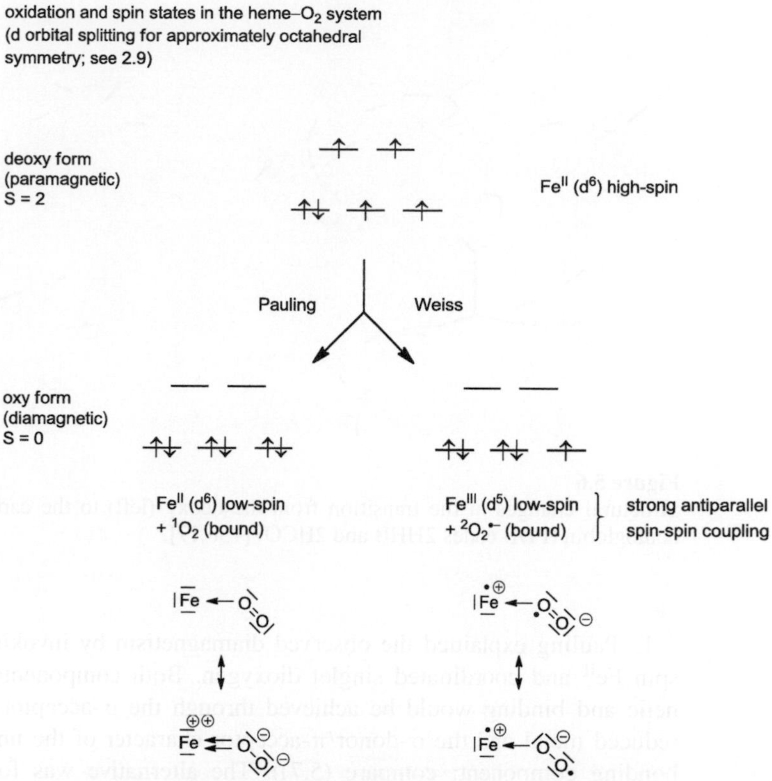

(5.9)

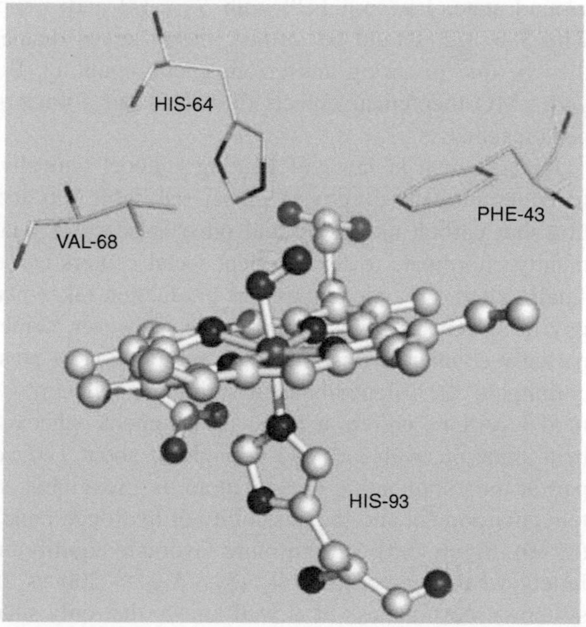

Figure 5.5
Arrangement of proximal (His$_{93}$) and distal (His$_{64}$, Val$_{68}$, Phe$_{43}$) amino acid residues with regard to oxy-myoglobin (PDB code 1MBO) [13a].

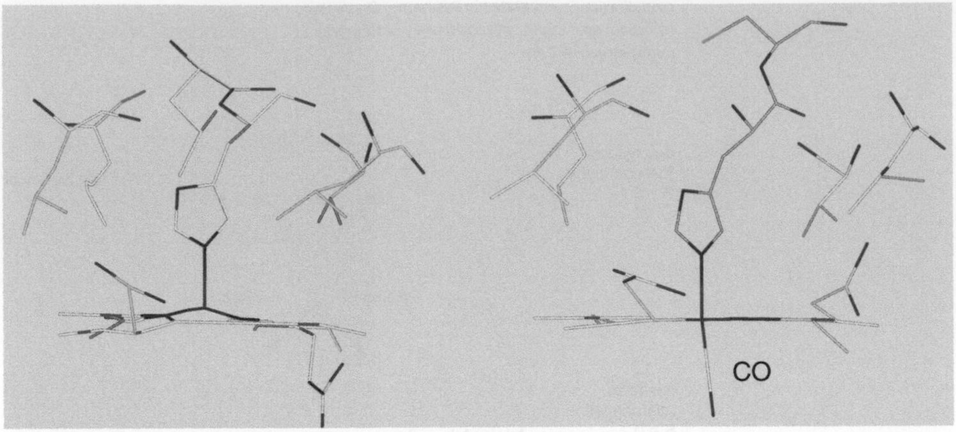

Figure 5.6
Structural changes in the transition from the deoxy (left) to the carbon monoxy form (right) of hemoglobin (PDB codes 2HHB and 2HCO) [13b,17].

L. Pauling explained the observed diamagnetism by invoking a combination of low-spin Fe^{II} and coordinated singlet dioxygen. Both components would each be diamagnetic and binding would be achieved through the σ-acceptor/π-donor character of the reduced metal and the σ-donor/π-acceptor character of the unsaturated ligand (π back-bonding component; compare (5.7)). The alternative was formulated by J. J. Weiss: following a single electron transfer from the metal to the approaching ligand in the ground state, low-spin Fe^{III} with $S = 1/2$ and a superoxide radical anion $O_2^{\bullet -}$, also with $S = 1/2$, should be formed; the observed diamagnetism at room temperature will then be due to strong antiferromagnetic coupling. Both alternatives have been studied under MO theoretical aspects (5.9) [16] but a universally accepted assignment has not been reached.

An argument in favor of Pauling's model with divalent low-spin iron (which would fit better into the porphyrin plane; see Table 2.6 and also Figure 5.6) is the observation that carbon monoxide and other π-acceptor ligands, such as NO, which predominantly coordinate to lower-valent metal centers can effectively replace dioxygen. To a small extent, an endogenous CO production takes place during the degradation of porphyrin systems of aged erythrocytes. However, some mechanisms exist which at least partially counteract this unwanted competition of poisonous carbon monoxide [15]. According to the valence-bond description (5.10), η^1(C)-coordinating CO in protein-free model systems prefers a linear arrangement, whereas end-on coordinated O_2 prefers a bent situation, with an Fe-O-O angle of about 120°, due to the remaining free electron pair at the coordinating oxygen atom. In myoglobin, spatial restrictions through the protein environment and the possibility of hydrogen bond formation with the distal histidine ((5.10), Figure 5.5) cause a more favorable equilibrium situation for binding of the definitely weaker π-acceptor O_2 ($K_{CO}/K_{O2} \approx 200$ vs 25 000 in protein-free heme model systems); nevertheless, it is well known that only small concentrations of CO in air can be tolerated.

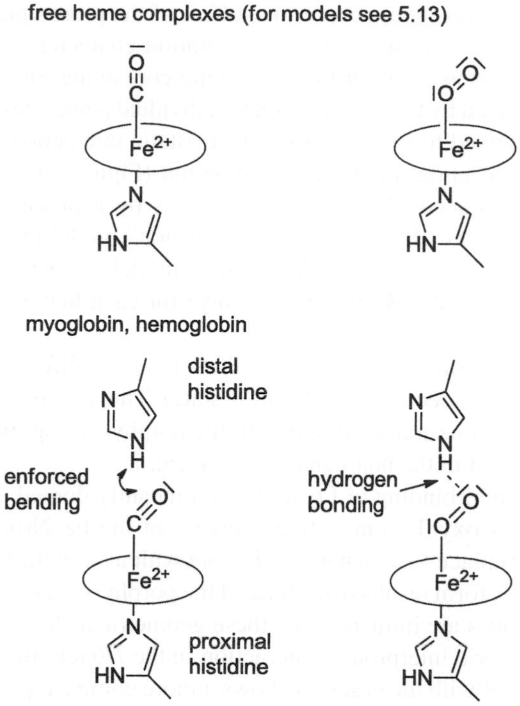

(5.10)

The imidazole ring of the distal histidine blocks an unhindered access to the sixth, "free" coordination site at the iron center, so that a controlled, rapid binding of small molecules may result only as a consequence of side-chain dynamics of the globin protein. Modifications of the distal histidine or of the valine site (Figure 5.5) led to a less favorable binding ratio O_2/CO [15]. Histidine is valuable as a distal amino acid residue because of its basic nature, which keeps protons away from the coordinated O_2 or from binding via N_ε and liberating them on the other side via N_δ (histidine as proton shuttle; compare Table 2.5). Protons act as electrophilic competitors in relation to the coordinating iron, weakening its bond to O_2 and thus favoring deleterious autoxidation processes.

It is remarkable that the coordination of an additional "weak" ligand, either O_2 or $O_2^{\bullet-}$ (the latter after inner-sphere electron transfer), induces a change from a high-spin to a low-spin state of the metal ("spin crossover"). The high-spin situation, which is nevertheless essential for the activation of $^3O_2$, is rather uncommon for tetrapyrrole complexes; its occurrence in the deoxy form is the result of a weakened metal–ligand interaction due to the incomplete fit into the cavity of the significantly domed macrocycle (out-of-plane situation; see Figure 5.4 and Table 2.6). For this apparent entatic state situation, the relatively light "pull" from coordinating O_2 or CO is sufficient to effect at least a partial charge transfer from the metal to the incoming ligand and a spin crossover of the metal center. The thus effected contraction of the metal (Table 2.6) and the relative movement of about 20 pm towards the now more strongly coordinating macrocycle (Figure 5.6) are probably essential factors for the cooperative effect [12,18]. Another significant structural change during O_2 coordination concerns the straightening of the Fe–N bond towards the proximal histidine with respect to the porphyrin plane.

5.3 Alternative Oxygen Transport in Some Lower Animals: Hemerythrin and Hemocyanin

Several groups of invertebrates, such as crustaceans, mollusks (e.g. snails), arthropods (e.g. spiders) and marine worms, possess nonporphinoid metalloproteins for reversible O_2 uptake. While iron-containing hemerythrin (*Hr*), which, contrary to its name, does *not* contain a heme system according to (5.8) (αιμα: blood) occurs mainly as an octameric protein with a molecular mass of 8 × 13.5 kDa, the even more complex copper-containing hemocyanins (*Hc*, Section 10.2) feature molecular masses of more than 1 MDa. For hemerythrin, many of the fundamental characteristics of the active center have been verified structurally (compare (5.17) and Figure 5.7); nevertheless, the contributions of various physical methods to an initially *indirect* structural elucidation of the O_2-coordinating centers will be summarized in this section [22].

5.3.1 Magnetism

Magnetic measurements of the virtually colorless deoxy form of hemerythrin indicate the presence of high-spin iron(II) with four unpaired electrons (2.14) at each center, just as in deoxy *Hb* and *Mb*. A weak antiparallel spin–spin coupling (4.11) between two apparently neighboring centers can be observed. For the red-violet oxy form with one dioxygen bound per Fe dimer, the susceptibility measurements indicate the presence of two strongly antiferromagnetically coupled ($S = 1/2$) centers, which leads to the conclusion that each monomeric oxyprotein contains *two* effectively interacting low-spin iron(III) centers (compare (5.9)). The strength of the spin–spin coupling can be inferred from the

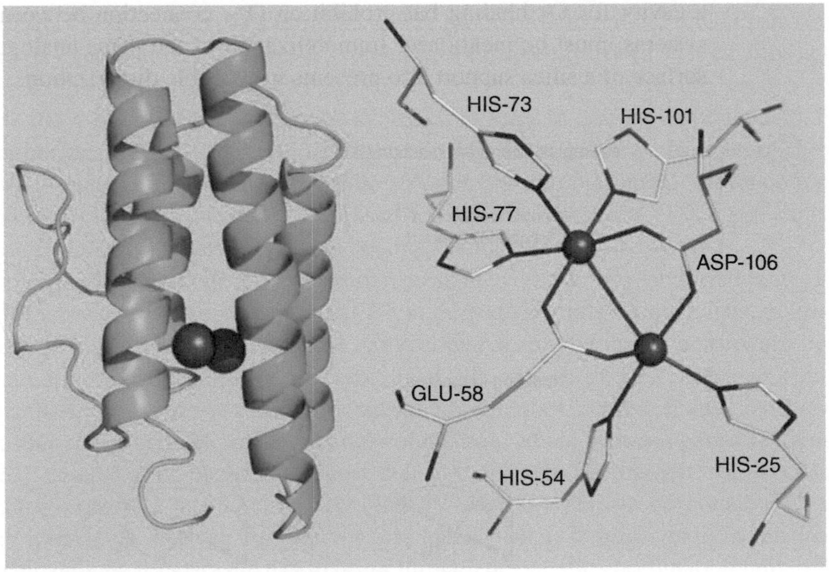

Figure 5.7
Protein folding in monomeric hemerythrin, with the positions of both iron atoms (red spheres, left) and the structure of the iron dimer center (right) in the deoxy form (PDB code 1HMD) [27].

temperature dependence of the magnetic susceptibility: the stronger this coupling, the more thermal energy is needed to observe normal paramagnetic behavior (i.e. uncoupled spins).

5.3.2 Light Absorption

The absence of strong light absorption in the deoxy form suggests protein-bound metal centers, because a pronounced color would be expected if a porphyrin π system were present. The color of the oxy form in the apparent absence of π-conjugated macrocyclic ligands can be attributed to a charge-transfer transition, which is "allowed" (compare (5.7)) because of compatible symmetry and effective overlap of ligand and metal orbitals (ligand-to-metal charge transfer, LMCT). This charge transfer in an electronically excited state occurs from an electron-rich peroxide ligand with a doubly occupied $\pi^*(2p)$ orbital (5.3) to electron-deficient, oxidized iron(III) with only partly filled d orbitals (5.14) [23].

$$\underset{\text{ground state}}{Fe^{III}(O_2^{2-})} \xrightarrow{h\nu} \underset{\text{LMCT excited state}}{{}^*\left[Fe^{II}(O_2)^{\bullet -}\right]} \qquad (5.14)$$

Many peroxo complexes of transition metals, particularly in higher oxidation states, are colored for that very reason; for instance, Ti^{IV}, V^{V} or Cr^{VI} may be analytically detected by color-forming reactions with H_2O_2. In contrast, the light absorption of the deoxy form is weaker by several orders of magnitude, since electronic transitions in the visible region can only occur between d orbitals at the *same* metal center. By definition, these d orbitals are of different symmetry (2.12) and the corresponding "ligand–field" transitions are therefore weak.

5.3.3 Vibrational Spectroscopy

If a "resonance" Raman experiment is carried out in the wavelength range of the LMCT absorption band of oxyhemerythrin, a resonance-enhanced vibrational band at 848 cm^{-1} is observed, which is characteristic for peroxides (5.3). When the isotopic combination ^{16}O–^{18}O is used for the bound dioxygen species, *two* signals are obtained for the O–O stretching vibration, suggesting a strongly asymmetric coordination (e.g. end-on) (Figure 5.5).

Resonance Raman Spectroscopy

In this method, molecular vibrations are excited in a scattering experiment (Raman effect) using an *absorption wavelength* [24]. Through coupling of electronic and vibrational transitions, a selective enhancement – the resonance – results for a limited number of vibrational bands, which are associated with the chromophor (e.g. a tetrapyrrole macrocycle). Those parts of the molecule experiencing a major geometry change following the electronic excitation respond most strongly. This highly selective kind of vibrational spectroscopy is thus suitable even for large proteins where normal infrared or Raman spectra do not provide very useful information because of the large number of atoms.

5.3.4 Mössbauer Spectroscopy

Mössbauer spectroscopy of oxyhemerythrin shows two distinctly different Fe^{III} resonance signals, whereas the Fe^{II} centers of the deoxy form cannot be distinguished.

Mössbauer Spectroscopy

Mössbauer spectroscopy is a nuclear absorption/nuclear emission spectroscopy (nuclear fluorescence) involving γ radiation. The strong line broadening expected at such high energies (several keV) is circumvented by embedding the absorbing nucleus in a solid-state matrix at low temperatures; the resulting distribution of recoil energy over many atoms makes resonance detection possible (5.15) [25,26].

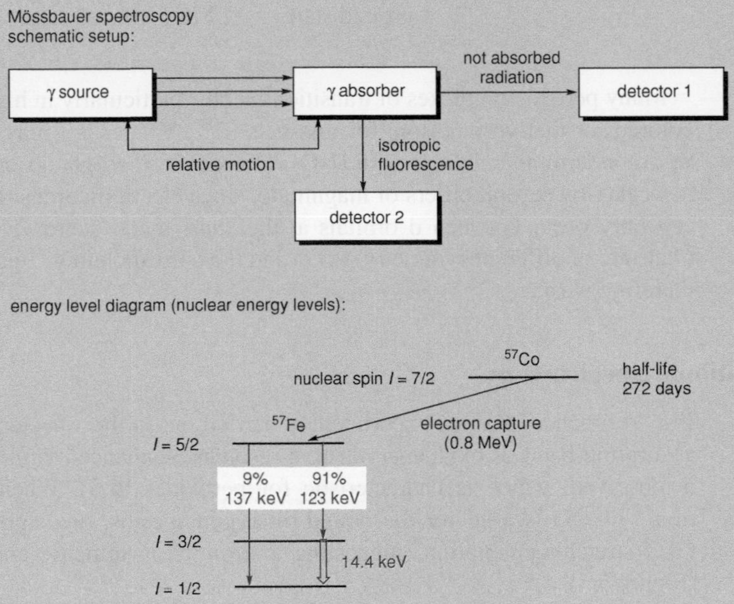

(5.15)

The very small relative line width of γ photons (ratio of about 10^{-13} between line width and energy) allows a detection of minute effects in the chemical environment (electron shell) of the absorbing nucleus. The actual measurement makes use of the Doppler effect, which occurs when the γ emitter and the absorber are moved relative to each other with a stepwise change of constant velocity. The resonance signal observed exhibits an "isomer shift" and a quadrupolar splitting, which reflect the chemical environment of the nucleus (symmetry, ligand field) and, in particular, the spin and ionization state. Unfortunately, only very few nuclei are well suited to this physical method; by far the most important isotope in the area of bioinorganic chemistry is ^{57}Fe, which is formed in a nuclear excited state during the radioactive decay of ^{57}Co, yielding a γ radiation of 14.4 keV energy.

5.3.5 Structure

Of the conceivable alternatives for the $O_2^{(2-)}$ coordination by the dinuclear metal arrangement in hemerythrin (5.16), only situation **2** and the strongly distorted arrangements **4** and **5** remain, considering the previously results. Scheme (5.17) shows the actual coordination environments for both forms [27], as suggested by crystal-structure analyses of various derivatives (Figure 5.7).

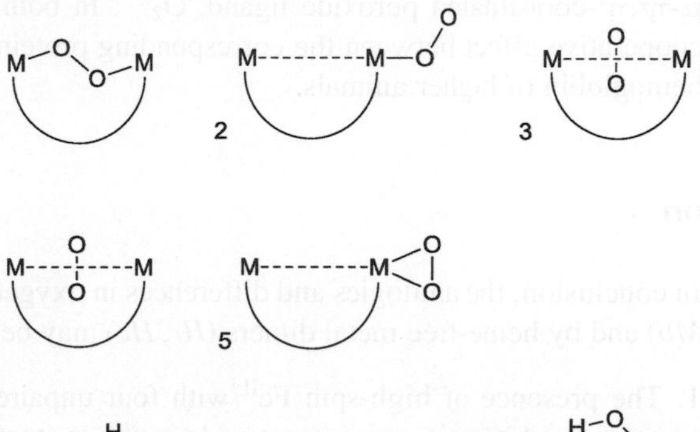

(5.16)

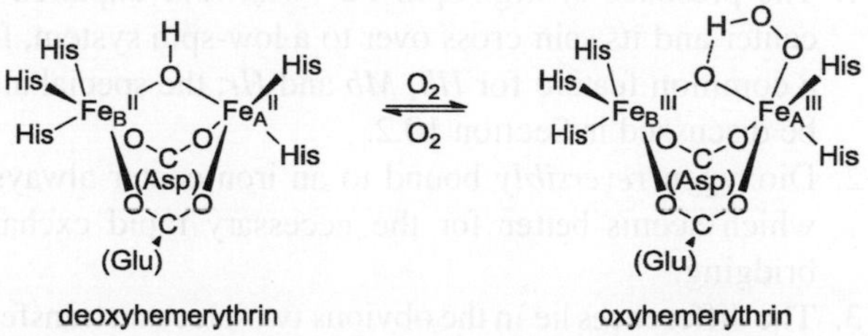

(5.17)

The doubly η^1:η^1 carboxylate- and singly hydroxide-bridged high-spin iron(II) centers (Fe_A, Fe_B) of the deoxy forms are coordinatively saturated with five histidine ligands, except for one position at Fe_A. When dioxygen is bound there, *both* centers are oxidized to Fe^{III} under simultaneous reduction of the substrate to the peroxide state and the metal centers are more tightly linked by formation of an oxo bridge ("superexchange" → antiferromagnetic coupling). The O_2 taken up presumably exists as hydroperoxo ligand, HOO^-, allowing a hydrogen bond to be formed with the bridging oxide ion (5.17). While this kind of reactivity between O_2 and Fe^{II} is not surprising (compare (5.1), (5.12)), it is astonishing that the O_2 binding in the protein is *reversible*. Structural model systems for hemerythrin have been synthesized with components (5.18) [28], and several pieces of physical data for the protein can be reproduced using these models.

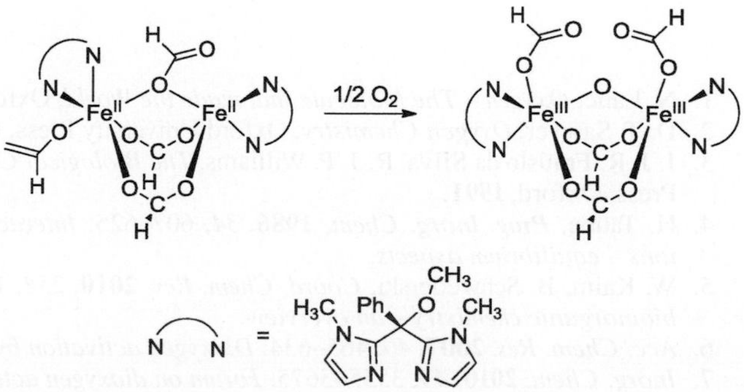

(5.18)

The copper-containing protein hemocyanin will be described in detail in Section 10.2, because of its structural relationship to copper-dependent oxygenases. However, some results from physical examinations parallel those for hemerythrin; for example, a dimeric arrangement of histidine-coordinated metal centers is assumed for each protein subunit (see (10.5)). The diamagnetic deoxy form features two copper(I) centers with formally closed (i.e. completely filled) 3d shells; the blue oxy form of *hemocyanin* contains antiferromagnetically interacting Cu^{II} dimers (d^9 configuration, $2 \times S = 1/2$) and an obviously μ-η^2:η^2-coordinated peroxide ligand, O_2^{2-}. In both non-heme systems, *Hr* and *Hc*, the cooperative effect between the corresponding protein subunits is smaller than that for the hemoglobin of higher animals.

5.4 Conclusion

In conclusion, the analogies and differences in oxygen coordination by heme systems (*Hb*, *Mb*) and by heme-free metal dimers (*Hr*, *Hc*) may be summarized as follows:

1. The presence of high-spin Fe^{II} with four unpaired electrons as the $^3O_2$-coordinating center and its spin cross over to a low-spin system, formulated either as Fe^{II} or Fe^{III}, is a common feature for *Hb*, *Mb* and *Hr*; the special alternative of the Cu system *Hc* will be discussed in Section 10.2.
2. Dioxygen *reversibly* bound to an iron center always shows *end-on* coordination (η^1), which seems better for the necessary rapid exchange than *side-on* coordination or bridging.
3. The differences lie in the obvious two-electron transfer through the metal dimers to create peroxo ligands versus a smaller extent of electron transfer in the oxy-heme species, which involve superoxo or dioxygen ligands, depending on the model used.
4. The electron "buffer capacity" necessary for the coordination of small, centrosymmetrical, unsaturated molecules is realized through metal–metal cooperation (cluster effect) in non-heme systems, whereas the heme-containing *Hb* and *Mb* exhibit interaction between only one redox-active iron center and the redox-active π system of the porphyrin ligand.

Both types of composite system are apparently better suited for reversible O_2 coordination than simple, isolated metal centers; the hemoproteins seem to represent the more elegant, more flexible and more efficient system, particularly with regard to cooperativity.

References

1. N. Lane, *Oxygen – The Molecule that made the World*, Oxford University Press, Oxford, **2003**.
2. D. T. Sawyer, *Oxygen Chemistry*, Oxford University Press, Oxford, **1991**.
3. J. J. R. Fraústo da Silva, R. J. P. Williams, *The Biological Chemistry of the Elements*, Clarendon Press, Oxford, **1991**.
4. H. Taube, *Prog. Inorg. Chem.* **1986**, *34*, 607–625: *Interaction of dioxygen species and metal ions – equilibrium aspects.*
5. W. Kaim, B. Schwederski, *Coord. Chem. Rev.* **2010**, *254*, 1580–1588: *Non-innocent ligands in bioinorganic chemistry – an overview.*
6. *Acc. Chem. Res.* **2007**, *40*, 465–634: *Dioxygen activation by metalloenzymes and models.*
7. *Inorg. Chem.* **2010**, *49*, 3555–3675: *Forum on dioxygen activation and reduction.*

8. J. W. Egan, B. S. Haggerty, A. L. Rheingold, S. C. Sendlinger, K. H. Theopold, *J. Am. Chem. Soc.* **1990**, *112*, 2445–2446: *Crystal structure of a side-on superoxo complex of cobalt and hydrogen abstraction by a reactive terminal oxo ligand.*
9. N. Kitajima, K. Fujisawa, Y. Moro-oka, *J. Am. Chem. Soc.* **1989**, *111*, 8975–8976: μ-η^2:η^2-*peroxo binuclear copper complex,* $[Cu(HB(3,5\text{-}iPr_2\,pz)_3)]_2\,(O_2)$.
10. W. Micklitz, S. G. Bott, J. G. Bentsen, S. J. Lippard, *J. Am. Chem. Soc.* **1989**, *111*, 372–374: *Characterization of a novel* μ_4*-peroxide tetrairon unit of possible relevance to intermediates in metal-catalyzed oxidations of water to dioxygen.*
11. F. A. Cotton, G. Wilkinson, *Advanced Inorganic Chemistry*, 5th Edition, John Wiley & Sons, New York, **1988**; p. 1337.
12. M. F. Perutz, G. Fermi, B. Luisi, B. Shaanan, R. C. Liddington, *Acc. Chem. Res.* **1987**, *20*, 309–321: *Stereochemistry of cooperative mechanisms in hemoglobin.*
13. (a) S. E. Phillips, *J. Mol. Biol.* **1980**, *142*, 531–554: *Structure and refinement of oxymyoglobin at 1.6 Å resolution*; (b) G. Fermi, M. F. Perutz, B. Shaanan, R. Fourme, *J. Mol. Biol.* **1984**, *175*, 159–174: *The crystal structure of human deoxyhaemoglobin at 1.74 Å resolution.*
14. R. E. Dickerson, I. Geis: *Hemoglobin: Structure, Function, Evolution and Pathology*, Benjamin Cummings, Menlo Park, **1983**.
15. M. F. Perutz, *Trends Biochem. Sci.* **1989**, *14*, 42–44: *Myoglobin and haemoglobin: role of distal residues in reactions with haem ligand.*
16. H. Chen, M. Ikeda-Saito, S. Shaik, *J. Am. Chem. Soc.* **2008**, *130*, 14778–14790; *Nature of the* $Fe{-}O_2$ *bonding in oxy-myoglobin: effect of the protein.*
17. J. M. Baldwin, *J. Mol. Biol.* **1980**, *136*, 103–128: *The structure of human carbonmonoxy haemoglobin at 2.7 Å resolution.*
18. J. Baldwin, C. Chothia, *J. Mol. Biol.* **1979**, *129*, 175–200: *Hemoglobin: the structural changes related to ligand binding and its allosteric mechanism.*
19. G. K. Ackers, J. H. Hazzard, *Trends Biochem. Sci.* **1993**, *18*, 385–390: *Transduction of binding energy into hemoglobin cooperativity.*
20. A. F. G. Slater, A. Cerami, *Nature (London)* **1992**, *355*, 167–169: *Inhibition by chloroquine of a novel haem polymerase enzyme activity in malaria trophozoites.*
21. J. P. Collman, *Acc. Chem. Res.* **1977**, *10*, 265–272: *Synthetic models for the oxygen-binding hemoproteins.*
22. D. M. Kurtz, *Chem. Rev.* **1990**, *90*, 585–606: *Oxo- and hydroxo-bridged diiron complexes: a chemical perspective on a biological unit.*
23. R. C. Reem, J. M. McCormick, D. E. Richardson, F. J. Devlin, P. J. Stephens, R. L. Musselman, E. I. Solomon, *J. Am. Chem. Soc.* **1989**, *111*, 4688–4704: *Spectroscopic studies of the coupled binuclear ferric active site in methemerythrins and oxyhemerythrins.*
24. R. J. H. Clark, T. J. Dines, *Angew. Chem. Int. Ed. Engl.* **1986**, *25*, 131–158: *Resonance Raman spectroscopy and its application to inorganic chemistry.*
25. E. Fluck, *Adv. Inorg. Chem. Radiochem.* **1964**, *6*, 433–489: *The Mössbauer effect and its application in chemistry.*
26. P. Gütlich, Y. Garcia, *Chemical applications of Mössbauer spectroscopy*, in Y. Yoshida, G. Langouche (eds.), *Mössbauer Spectroscopy*, Springer, Berlin, **2013**.
27. M. A. Holmes, I. Le Trong, S. Turley, L. C. Sieker, R. Stenkamp, *J. Mol. Biol.* **1991**, *218*, 583–593: *Structures of deoxy and oxy hemerythrin at 2.0 Å resolution.*
28. W. B. Tolman, A. Bino, S. J. Lippard, *J. Am. Chem. Soc.* **1989**, *111*, 8522–8523: *Self-assembly and dioxygen reactivity of an asymmetric, triply bridged diiron(II) complex with imidazole ligands and an open coordination site.*

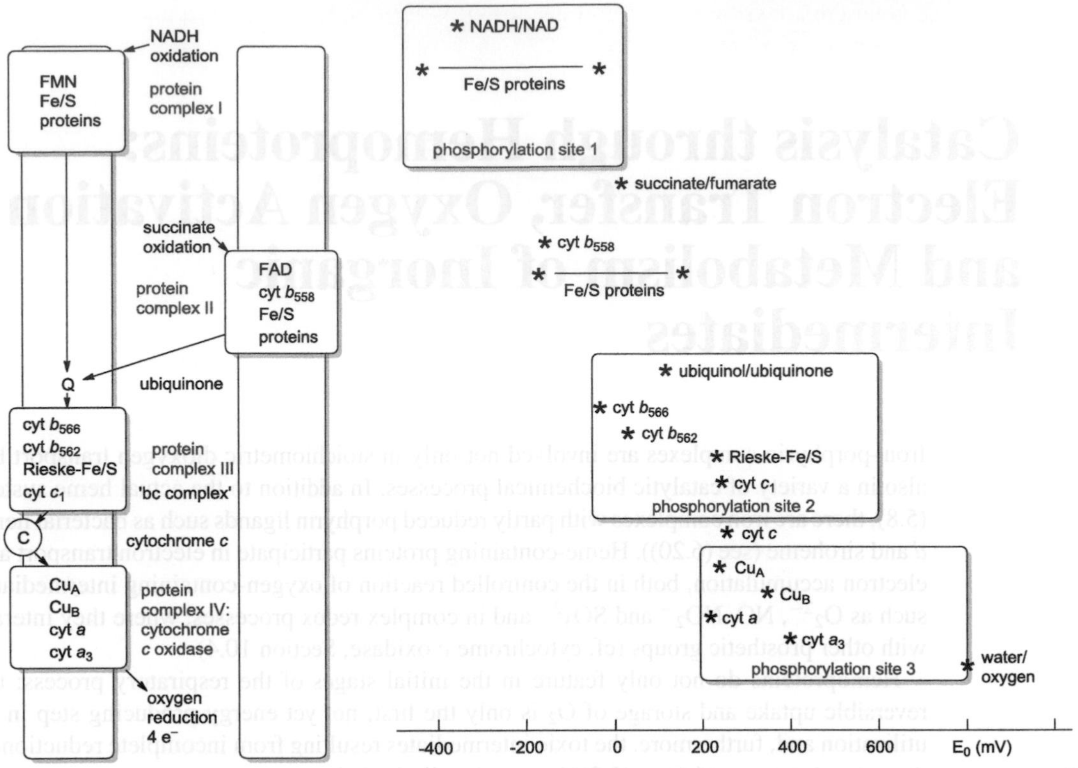

Figure 6.1
Schematic overview of the electron-transferring components in the respiratory chain. Representation of the functional assembly in the membranes (left; see also Figure 6.2) and of the redox potentials of individual components (∗). Modified with permission from [1] © 1980, Verlag Chemie, GmbH, Germany.

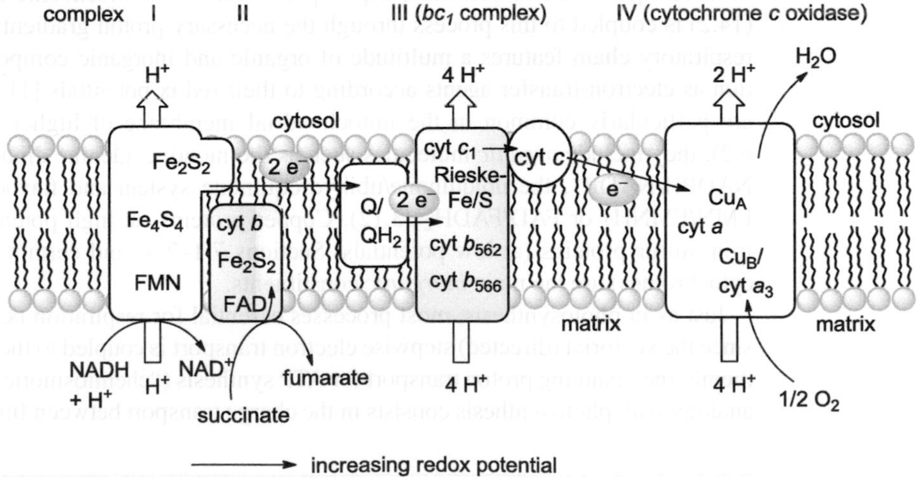

Figure 6.2
Schematic representation of interacting protein complexes in the mitochondrial membrane; the proton back-transport via ATP synthase is not shown here.

protein complexes via electron-transferring molecules, in particular quinones and heme-containing cytochromes.

6.1 Cytochromes

Cytochromes are hemoproteins which exert their electron-transfer function both in the respiratory chain and in photosynthesis (Figures 4.5–4.7) and other complex biological processes. Structure–reactivity relationships will be discussed in this section by example of the very thoroughly studied and relatively small cytochrome *c* [2].

The more than 50 known cytochromes ("cell pigments") can be divided into different groups according to their structural constitution and physical properties [3]. The cytochrome *a* type shows very high redox potentials; these species are important in the reduction of oxygen to water in the cytochrome *c* oxidase complex (see Section 10.4, Figure 10.7). Cytochromes of the *b* and *c* type (subscripts either serve for numbering or indicate the characteristic absorption maximum in nm) contain two tightly bound amino acid residues, histidine/histidine or histidine/methionine, as ligands at the heme iron (Figure 6.3). The heme group of cytochromes *c* is covalently bound via porphyrin/cysteine bonds. Histidine/lysine axial coordination is assumed for cytochrome *f* (Figure 4.6, Table 4.1), while methionine/methionine axial coordination occurs in bacterial ferritins (Section 8.4.2) [4]. The thus coordinatively *saturated* iron centers show quite variable redox potentials for the Fe^{II}/Fe^{III} transition, depending on the axial ligands and on the coordination environment (hydrogen bonding, electrostatic charge distribution, geometrical distortion); comparable effects of ligands and reaction media are well known from the "normal" coordination chemistry of iron (Table 6.1).

Cytochrome proteins may contain several heme groups, as for example in the mitochondrial complex III (Figure 6.2), in a nitrite reductase (Section 6.5), in cytochrome *c* oxidase

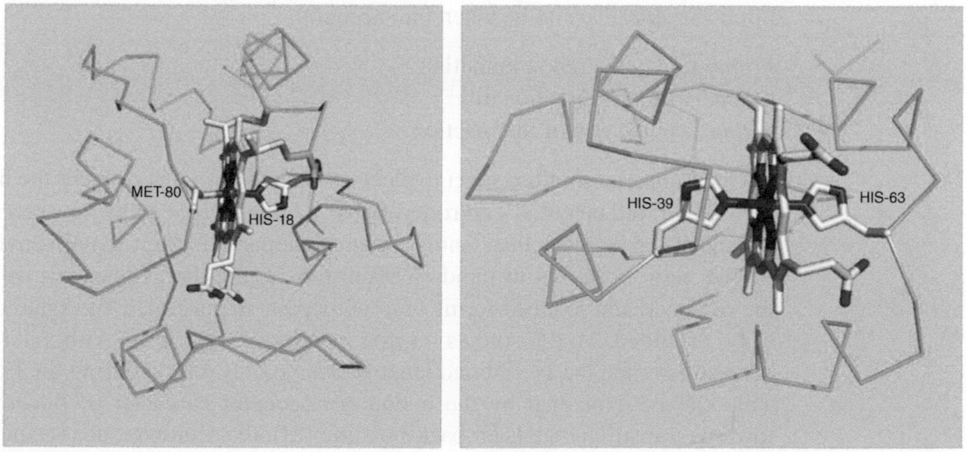

Figure 6.3
Schematic representation of protein folding, heme position and iron coordination in tuna cytochrome *c* (left; PDB code 3CYT) and bovine cytochrome b_5 (right; PDB code 1CYO) [5].

Table 6.1 Redox potentials for some chemical and biochemical Fe^{II}/Fe^{III} pairs.

Compound	E_0' (mV)	
hexaaquairon(II/III) $[(H_2O)_6Fe]^{2+/3+}$	771	
tris(2,2'-bipyridine)iron(II/III) $[(bpy)_3Fe]^{2+/3+}$	960	
hexacyanoferrate(II/III) $[(NC)_6Fe]^{4-/3-}$	358	
trisoxalatoiron(II/III) $[(C_2O_4)_3Fe]^{4-/3-}$	20	
hemoprotein iron(II/III)		Axial amino acid ligands
hemoglobin	170	His/–
myoglobin	46	His/–
horseradish peroxidase (HRP)	−170	His/–
cytochrome a_3	400	His/–
cytochrome *c*	260	His/Met
cytochrome b_5	20	His/His
cytochrome P-450	−400	Cys^-/–

(complex IV, Section 10.4) or in the subunit which participates in the bacterial photosynthesis of *Rps. viridis* (Figure 4.4). "Cytochrome P-450" (= **P**igment with an absorption maximum of the carbonyl complex at 450 nm) differs markedly from other cytochromes in its catalytic function; it shows monooxygenase activity and will be discussed separately in Section 6.2.

The best-examined cytochrome is cytochrome *c*, shown in Figure 6.3, which is typically obtained from the heart muscle tissue of tuna and horses. With only about 100 amino acids and a relatively low molecular mass of about 12 kDa, it is a rather small protein; structural determinations by protein crystallography can thus reveal several details. The amino acid sequences and tertiary structures of cytochrome *c* proteins isolated from very different organisms differ only slightly, suggesting that this is a very old protein in evolutionary terms and that its composition and structure have been well optimized. The protein is usually found at the outside of membranes (compare Figure 6.2) and for this reason it features a hydrophilic "surface".

Although the transfer of an electron is one of the simplest chemical reactions, at least three variables have to be taken into account:

- *energy*, i.e. the redox potential;
- *space*, i.e. the directionality;
- *time*, i.e. the rate of the reaction.

Open questions with respect to electron transfer proteins concern the molecular requirements for rapid potential-controlled electron transfer, especially between redox centers that are separated by more than 2 nm through an apparently inert protein environment.

The main obstacles to rapid electron transfer are the geometrical differences between the oxidized and reduced forms of a redox pair; in general, a "reorganization energy" has to be provided. Model studies on iron–porphyrin complexes with axial thioether ligands have shown that the Fe–S bond lengths change only a little during the Fe^{II}/Fe^{III} transition. This can be explained by the σ-donor/π-acceptor character of thioether sulfur centers towards transition metals (π back-bonding [2]), the geometric change being small through electron balancing according to $Fe \underset{\sigma}{\overset{\pi}{\rightleftharpoons}} S$. Unsaturated nitrogen ligands such as the extended porphyrin π system also show such an "electron buffer" capacity. According to the theory of the entatic state, the ground state is then situated *between* the typical structures for each of the individual redox states, that is, it is close to the transition state. In fact, the holoprotein of cytochrome *c* does not show a significant conformational change during electron transfer, and the same is true for other electron transfer proteins such as the Fe–S and blue copper systems (see Chapters 7 and 10).

In general, normal cytochromes are relatively simply structured proteins with clearly measurable activities (i.e. redox potential and electron-transfer rate); therefore, they have become attractive objects for deliberate enzyme modification via site-directed mutagenesis. In particular, they may feature different iron spin states depending on the coordination environment, including low-spin, intermediate-spin and high-spin alternatives, because porphyrin complexes of Fe^{II} and Fe^{III} often lie close to the spin crossover region [6]. For instance, "engineered" cytochrome b_5 in which basic His39 is exchanged for less basic Met exhibits high-spin ($S = 5/2$) iron(III) instead of a low-spin iron center in its oxidized state. That high-spin center is also characterized by a negatively shifted redox potential and additional substrate-catalytic (here: N-demethylase) activity [7].

6.2 Cytochrome P-450: Oxygen Transfer from O_2 to Nonactivated Substrates

Sulfur ligation and catalytic (enzymatic) activity come together in a special, ubiquitous "cytochrome" protein referred to as P-450 (because the CO complexes have a characteristic "pigment" absorption maximum at 450 nm). The corresponding hydroxylases (monooxygenases), an enzyme "superfamily" often referred to as "CYP", act in cooperation with reducing agents (NADPH, flavins, iron–sulfur proteins) and a P-450 reductase to produce oxygenated products from often rather inert substrates and dioxygen (6.1) [8,9].

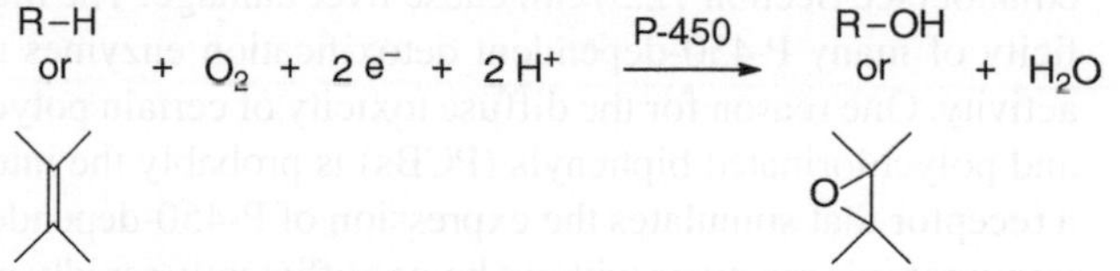

(6.1)

Several variants exist within the P-450 enzyme superfamily which play an essential role in the metabolism of endogeneous substances (e.g. steroids) and in the transformation of external, "xenobiotic" substances in animals, plants and microorganisms.

For instance, P-450-dependent monooxygenases are active as typical detoxification enzymes [10] in the microsomes of the human liver, where they often show relatively little substrate selectivity; on the other hand, fatty acids, amino acids and hormones (steroids such as the estrogenic and androgenic sex hormones or prostaglandines) are metabolized by more specific P-450 systems in a stereospecific fashion. Of great importance in pharmacology, medicine and toxicology are the monooxygenation reactions of xenobiotic substances with physiologically active metabolites, leading for example from calciferols (vitamin D group) to active 1,25-dihydroxycalciferols (6.2) or from β-naphtylamine to the carcinogenic α-hydroxy-β-aminonaphthalene (6.3). Drugs such as morphine and numerous other well-known pharmaceuticals (Table 6.2), as well as chlorinated and nonchlorinated hydrocarbons, are also transformed by P-450 monooxygenases [11]. In the absence of oxidizable aliphatic side chains, the cytochrome P-450 enzymes catalyze the epoxidation of benzene or benzo[*a*]pyrene to yield the mutagenic derivatives (6.4) and (6.5). Nitrosamines and polychlorinated methane derivatives are transformed by P-450 enzymes to reactive radicals

Table 6.2 Examples of the oxidative metabolism of pharmaceuticals by P-450 monooxygenation (according to [11]).

Reaction type	Equation	Examples
oxidation of aliphatic chains	$R{-}CH_2{-}CH_3 \longrightarrow R{-}CH(OH){-}CH_3$ and $R{-}CH_2{-}COOH$	barbiturates
oxidative N-dealkylation	$R^1{-}NH{-}CH_2R^2 \longrightarrow R^1{-}NH_2 + R^2{-}C(=O){-}H$	ephedrine
oxidative deamination	$R{-}CH_2{-}NH_2 \longrightarrow R{-}C(=O){-}H + NH_3$	histamine norepinephrine mescaline
oxidative O-de-alkylation		phenacetin codein mescaline
para-hydroxylation of aromatic compounds	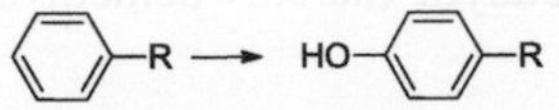	phenobarbital chlorpromazine
oxidation of aromatic amines	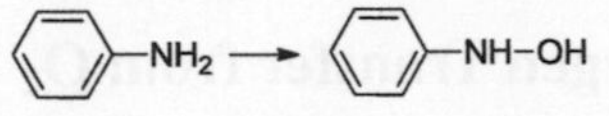	aniline derivatives
S-oxidation	$R^1R^2S \longrightarrow R^1R^2S{=}O \longrightarrow R^1R^2SO_2$	phenothiazine

and carbocations, and the large amounts of acetaldehyde formed in the oxidation of excess ethanol (see Section 12.5) can cause liver damage. The high reactivity and often low specificity of many P-450-dependent detoxification enzymes require a precise control of their activity. One reason for the diffuse toxicity of certain polychlorinated dibenzo-1,4-dioxines and polychlorinated biphenyls (PCBs) is probably the interaction of these substances with a receptor that stimulates the expression of P-450-dependent enzymes and thereby triggers immunotoxic reactions without being sufficiently rapidly metabolized. The oxidative degradation of such components is blocked if the most reactive sites, such as the *para*-positions of aromatic molecules or the 2,3,7,8-positions in dibenzo-1,4-dioxine heterocycles, are substituted by Cl instead of H; the C–Cl bond is not oxygenated by P-450 and Cl^+ would be a poor leaving group.

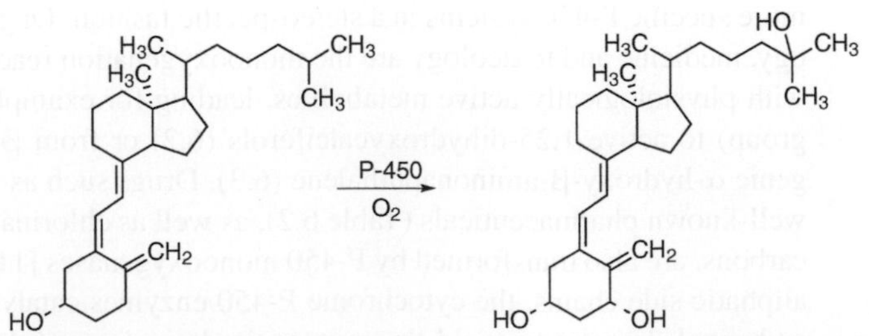

colecalciferol (vitamin D_3) 1,25-dihydroxy-colecalciferol (6.2)

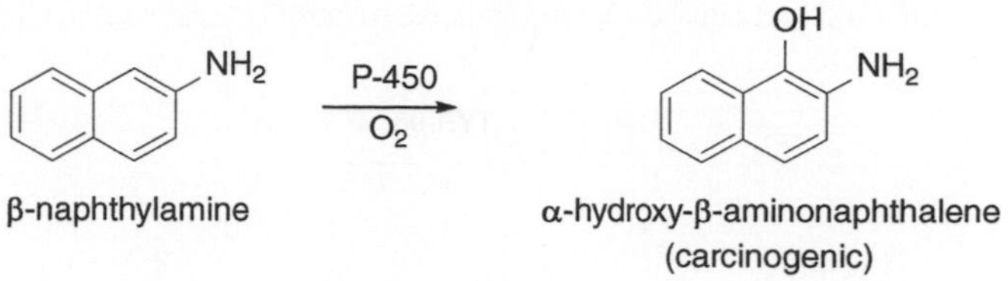

(6.3)

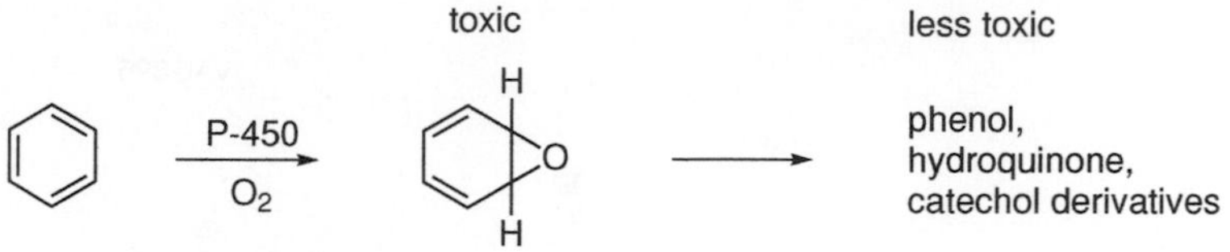

(6.4)

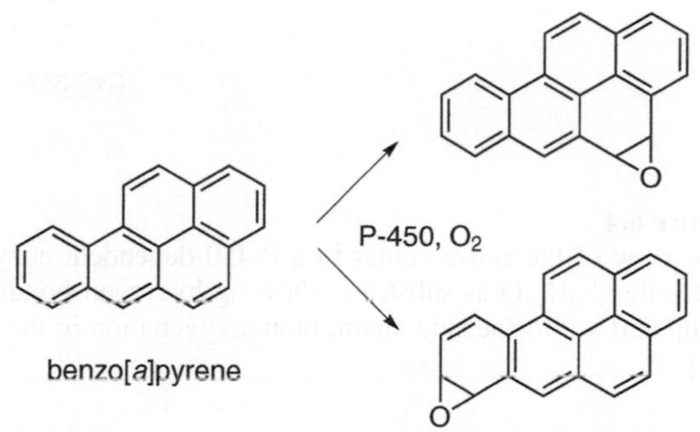

(6.5)

The interest in cytochrome P-450 arises not just from a desire to understand its physiological function but from the formidable challenge posed to synthetic chemistry by the controlled transfer of oxygen atoms from freely available O_2 to otherwise unactivated organic-chemical substrates, particularly hydrocarbons [12,13]. This is even more true since studies on the reaction mechanism of P-450 have shown a pattern that is also characteristic of many other metal-catalyzed processes; obviously, this catalysis involves the *controlled reaction of two ligands in the coordination sphere of the metal.* The enzyme thus functions both as a substrate-selective "template", bringing the reactants together in a spatially defined orientation (stereospecificity), and as an electronic activator, providing a new reaction pathway with lower activation energy. The activation of bound O_2 is presumably influenced by the axial ligand (2.11), which has been identified [14,15] as a cysteinate anion. In contrast to π-accepting thioethers such as methionine, the thiolates are strong σ and π electron donors and can therefore stabilize high oxidation states of metal centers.

Many P-450-dependent enzymes, which have a typical molecular mass of about 50 kDa, have been analyzed in terms of their amino acid sequences. Some forms have been characterized both with respect to the reaction mechanism and structurally, both with and without substrate [14] (compare Figure 6.4). According to such studies, the following mechanism is assumed for dioxygen activation and monooxygen transfer (6.6): Starting from the predominantly low-spin iron(III) state **1** with six-coordinate metal (porphyrin, cysteinate, water), the binding of the organic substrate occurs through hydrophobic and other (see Figure 6.4) interactions with the protein inside a cavity and close to the axial coordination site of the heme system, thereby causing a transition to the high-spin iron(III) form **2**. After the concomitant loss of the water ligand, **2** features

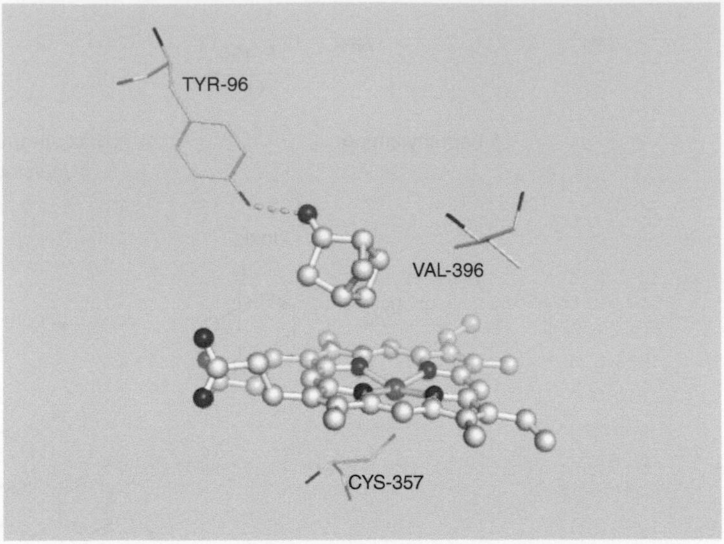

Figure 6.4
Side view of the active center in a P-450-dependent enzyme of *Ps. putida* with a bound camphor molecule $C_{10}H_{16}O$ as substrate, showing hydrogen bonding (yellow dotted line) between the keto group and a tyrosine side chain; monooxygenation in the 5-position of camphor (PDB code 7CPP) [14].

an open coordination site and an *out-of-plane* structure with a domed porphyrin ring (Figure 6.4).

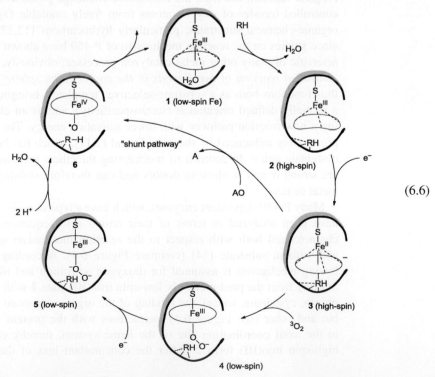

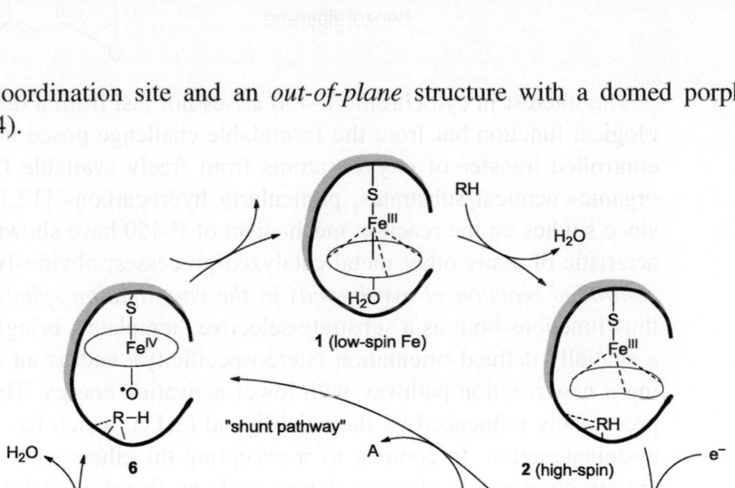

(6.6)

The next step is a one-electron reduction (via $FADH_2$; see (3.12)) to give a high-spin iron(II) complex **3**, which is predestined for $^3O_2$ binding because of its ($S = 2$) spin state and its pronounced *out-of-plane* situation (compare Section 5.2). In the presence of external oxygenation agents, AO, such as peracids, this iron(II) state **3** may directly yield compound **6**, the highly oxidized productive complex, via a "shunt" pathway. The physiological reaction, however, requires the uptake of dioxygen to form the coordinatively saturated low-spin oxy form **4** with a hydrogen bond to a threonine side chain and a possible $Fe^{III}/O_2^{\bullet-}$ oxidation state formulation (compare hemoglobin and myoglobin (5.9)). After a second one-electron reduction to form a very labile low-spin peroxo iron(III) complex **5**, this species adds two protons and releases one water molecule, thereby cleaving the O–O bond. This cleavage, however, requires two oxidation equivalents, which have to be made available intramolecularly; the result is a reactive complex **6**, which can be formulated with $Fe^{V}{=}O^{-II}$ or with $Fe^{IV}\text{-}O^{\bullet-I}$ oxidation state combinations and which collapses to yield the product and the initial state of the catalytic heme, possibly involving the transfer of a monooxygen ("oxyl") radical to the substrate ((6.7); for the overall reaction, see (6.1)).

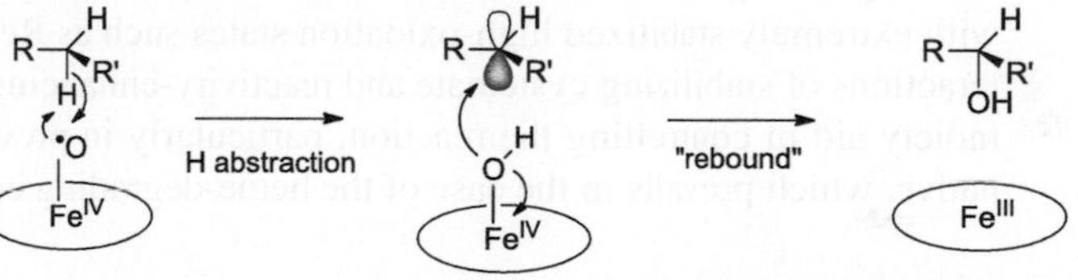

(6.7)

The high-valent oxo-iron center of the reactive heme group is stabilized by the electron-donating thiolate group, but it must also be regarded in context with the surrounding porphyrin π system. As has been shown for the chlorophylls in Section 4.2, the porphyrin system itself may add or release single electrons to form radical ions: In the presence of a dicationic oxo-metal ion $[Fe^{IV}{=}O]^{2+}$, which was detected here by resonance Raman spectroscopy, the formation of a radical cation complex via single-electron oxidation of the dianionic porphyrin ligand to a radical anion appears plausible. For the highest oxidized states of the oxo-heme systems in peroxidases and in the P-450 enzymes, such an electronic configuration is widely accepted [8,16]. A porphyrin radical anion, $Por^{\bullet-}$ ($S = 1/2$) will then coordinate to an oxoferryl(IV) fragment, $[Fe^{IV}{=}O]^{2+}$, with tetravalent iron (6.8).

$$[Fe^{II}(Por^{2-})]^0 \xrightarrow[-\,H_2O]{+\,O_2,\ 2\,H^+,\ e^-} [O^{-II}{=}Fe^{IV}(Por^{\bullet-})]^{\bullet+} \qquad (6.8)$$

The four d electrons of the rare [17] but biochemically very important oxidation state +IV of iron are not completely paired. From the ligand field splitting of an approximately D_{4h}-symmetrical complex (tetragonal compression; see correlation diagram in Figure 2.7) with oxide and cysteinate as the strongest ligands, one can deduce that an occupation with four electrons should lead to a triplet ground state ($S = 1$), in agreement with Hund's rule.

Interpretation (6.8) is based on magnetic measurements, which indicate a weak antiferromagnetic coupling between the assumed porphyrin radical anion ($S = 1/2$) and the iron center ($S = 1$).

Even considering the highly activated nature of compound **6**, it remains remarkable that this last state of the P-450 center in cycle (6.6) is able to transfer the iron-bound oxygen atom (6.7) to a substrate that is little or not at all activated; after release of the oxygenated substrate, the initial state of coordinatively unsaturated low-spin iron(III) is formed again. Formula (6.8) implies that the coordinated monooxygen exists as oxide ligand (O^{2-}) and that the oxidation number thus should be −II. This usually unquestioned convention, however, has been challenged by Sawyer [18], particularly with regard to the typical P-450 reactivity; a terminal monooxygen ligand can also be present as an oxyl radical anion ($O^{\bullet-}$), as the conjugated base of $^{\bullet}OH$ or, in analogy with carbenes and nitrenes, even as neutral "oxene" ligand (compare resonance structures (6.9)).

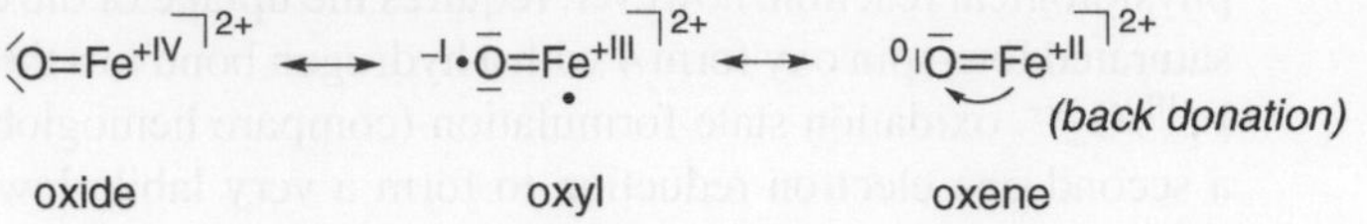

(6.9)

In fact, x-ray spectroscopic and mechanistic results suggest that in many "oxo" complexes of transition metals, and particularly in the reactive state **6** of the P-450 system [19] (compare (6.13)), the formulation with a weakly bound radical oxygen atom as reactive, hydrogen-abstracting ligand contributes significantly to the actual electronic structure. Adding the possible porphinate oxidation in alternative (6.9), there are then a multitude of conceivable resonance structures [20]. "Genuine" oxide (O^{2-}) ligands, which might tolerate even organic ligands in the metal coordination sphere, are found mainly in metal complexes with extremely stabilized high-oxidation states such as Re^{VII}. In the P-450 system, the interactions of stabilizing cysteinate and reactivity-enhancing distal ligands of the Fe–O(–O) moiety aid in controlling the reaction, particularly in preventing an autoxygenation alternative, which prevails in the case of the heme-degrading enzyme heme dioxygenase.

6.3 Peroxidases: Detoxification and Utilization of Doubly Reduced Dioxygen

Heme-containing peroxidases and catalases are enzymes that are related to cytochrome P-450; there are also manganese-, vanadium- and selenium-containing (Section 11.4 and 16.8) peroxidases [21], as well as manganese-dependent catalases. Unlike cytochrome P-450, peroxidases use the doubly reduced peroxidic form of O_2 to oxidize substrates of the type AH_2 to radical cations and their reaction products (6.10), (6.11), (6.12). The peroxide oxidation state of dioxygen can be produced as an undesired intermediate in the course of photosynthetic water oxidation or via incomplete oxygen reduction during respiration (4.6), (5.2); only about 80% of the dioxygen taken up by breathing is *completely* reduced. Therefore, peroxidases may be regarded at least partly as detoxification enzymes. This is especially true for the catalases, since their second substrate is also hydrogen peroxide (6.11); overall, the resulting reaction is the enzymatically catalyzed disproportionation of metastable H_2O_2, the equilibrium constant for (6.11) being about 10^{36}.

$$H_2O_2 + AH_2 \xrightarrow{\text{peroxidases}} 2\,H_2O + A \qquad (6.10)$$

$$H_2O_2 + H_2O_2 \xrightarrow{\text{catalases}} 2\,H_2O + O_2 \qquad (6.11)$$

$$R{-}CH_2{-}COOH + H_2O_2 \xrightarrow[\text{peroxidase}]{\text{fatty acid}} 3\,H_2O + R{-}CHO + CO_2 \qquad (6.12)$$

$$R{-}CHO \xrightarrow{\text{oxidation}} R{-}COOH$$

There are numerous not easily oxidized compounds, such as fatty acids, amines, phenols, chloride and xenobiotic substances (toxins), that can serve as substrates for peroxidases. For example, the controlled α oxidations of fatty acids during plant growth yield an α carbonyl carboxylic acid intermediate, which loses CO_2 (decarboxylation) to form an aldehyde with one less CH_2 group; its oxidation product is the correspondingly shorter fatty acid (6.12).

Other important reactions of heme peroxidases [21] concern the coupling of tyrosines and their iodination to the thyroid hormones by thyreoperoxidases [22] (see Section 16.7), the oxidation of cytochrome *c* by cytochrome *c* peroxidase (CCP), the oxidation of chloride to bactericidal hypochlorite, ^{-}OCl, by myeloperoxidase with cysteinate-coordinated iron, and the oxidative degradation of lignin from wood by lignin peroxidase [23]. A remarkable curiosity is the utilization of H_2O_2 and hydroquinone by certain beetles (*Brachymus*) to effect an explosive burst of O_2 and aggressive, oxidizing quinone in a peroxidase-catalyzed reaction.

The most thoroughly studied enzyme in the peroxidase group is horseradish peroxidase (HRP), which has been investigated since the days of R. Willstätter (1892–1942). Its low molecular mass (~40 kDa [24]) distinguishes it from the classical heme-containing catalases, which are associated proteins (tetramers) with total masses of 260 kDa and partially tyrosine-coordinated heme iron. Under physiological conditions, the resting state of most heme peroxidases contains high-spin iron(III) ($S = 5/2$, half-filled d shell, *out-of-plane* structure) and, in contrast to the P-450 system, an imidazole base from histidine as axial ligand (Figure 6.5). Oxidation, i.e. the protein-induced [25,26] internally catalyzed monooxygen transfer from H_2O_2 to the iron center under release of water (6.13), adds two oxidation equivalents and leads formally to an "iron(V)" species, again most probably involving a dicationic oxoferryl(IV) center with a coordinated porphyrin radical anion as ligand. This first, very electron-deficient cationic intermediate ("HRP I", $E_0 > 1$ V) can react with the substrate in a one-electron oxidation step; in fact, the follow-up products observed when phenols or certain strained hydrocarbons are used as substrates are otherwise known only from secondary reactions of chemically or electrochemically generated substrate radical cations. The resulting second enzymatic intermediate shows only *one* more oxidation

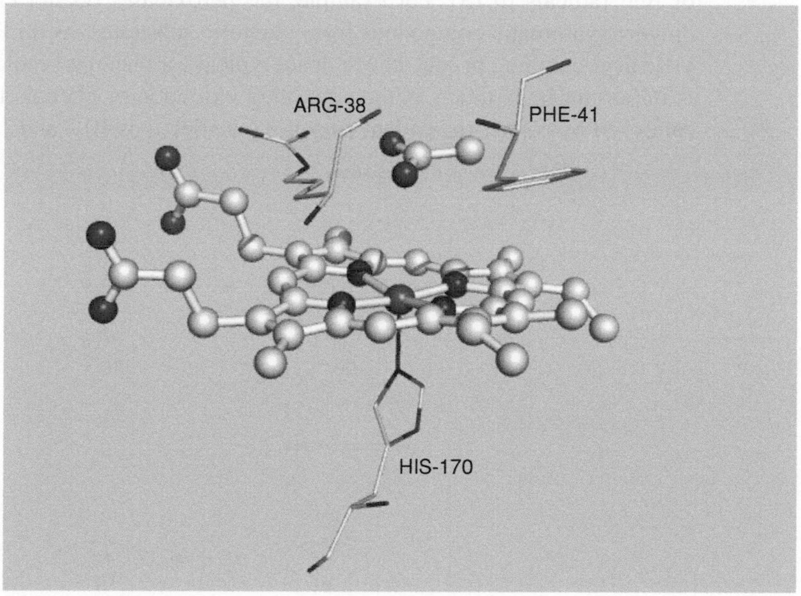

Figure 6.5
Active site structure of of horseradish peroxidase (HRP; crystallized with acetate; PDB code 1H5A) [21].

equivalent than the resting state but can still undergo a second one-electron oxidation reaction; according to physical measurements, this "HRP II" state contains an oxoferryl(IV) center ($S = 1$) coordinated to a normal (i.e. dianionic) porphyrin ligand (compare (6.9)) [16].

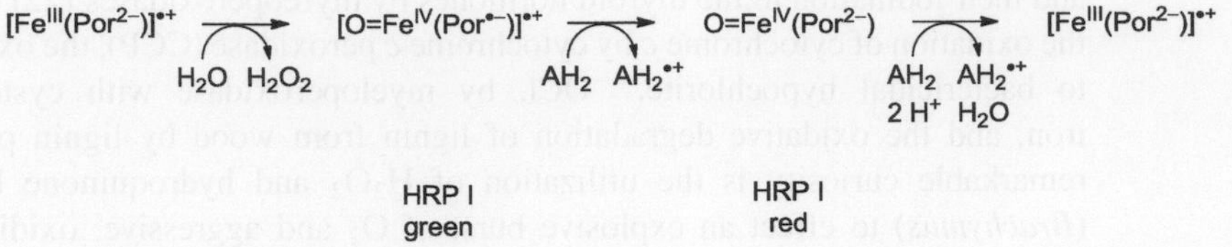

(6.13)

6.4 Controlling the Reaction Mechanism of the Oxyheme Group: Generation and Function of Organic Free Radicals

Cytochrome P-450 and the heme peroxidases go through reactive intermediate states with unusually high oxidation levels of the iron centers. Their rather different reactivities [19,24], namely monooxygenase activity (monooxygen transfer, one O from O_2) in one case and direct electron withdrawal (formation of a substrate radical cation) in the other, require explanation. Unlike P-450, most peroxidase iron centers feature a neutral histidine as axial ligand, which can get deprotonated. Compared to the anionic thiolate ligand of the P-450 systems, neutral histidine is a weaker electron donor, which may possibly effect a shift of the radical activity (spin density) from the iron-bound oxygen to the porphyrin π system (6.14). Electrophilic attack is then no longer connected to the oxygen transfer (6.6), (6.7) but consists of the extraction of one electron from the substrate and the conversion of the peroxidic oxygen to water (6.13). The different protein environments must be responsible for such different reactivities: in cytochrome P-450, the intermediate radicals that may be generated rapidly combine to yield the oxygenated products ("cage reaction"), while in peroxidases, a dissociation of the reactants can lead to the typical "escape" products of free radicals (6.15). For example, the peroxidases do not catalyze the formation of dihydroxyaromatic compounds from phenolic substrates, as do P-450 systems, but rather yield aryl-coupling products, which are typical for phenoxyl radicals [8]. This observation is important in as much as the controlled oxidation of phenols to catechols may also be catalyzed by copper-dependent enzymes (see Sections 10.2 and 10.3).

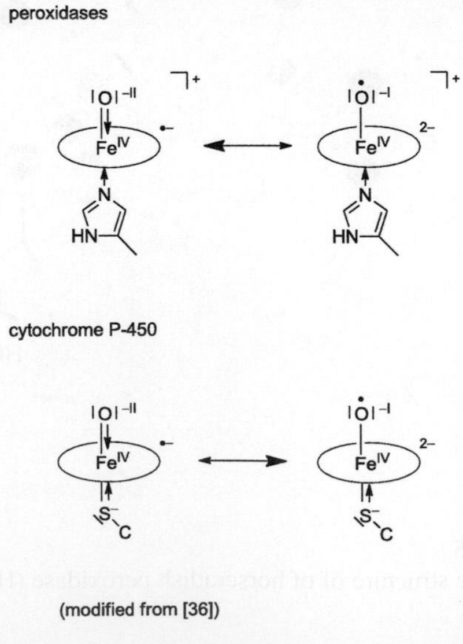

(6.14)

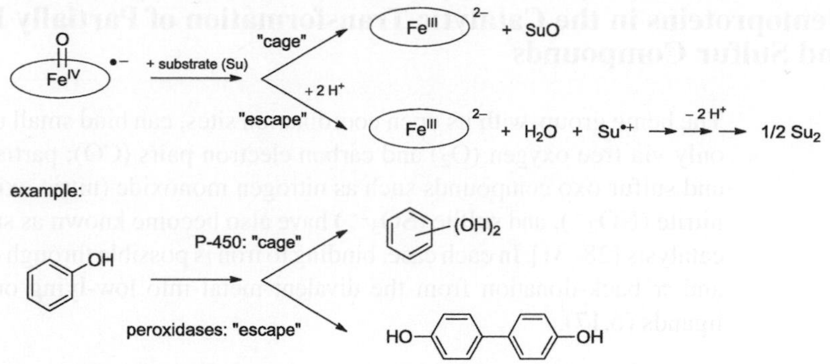

(6.15)

With their unusually high redox potentials of about +1.5 V, the reactive radical forms of oxidized peroxidases play an important role in the metabolism and degradation of the polymeric lignin of higher plants (about 25% of the world's biomass). Arylether bonds are formed and broken with the help of lignin peroxidases, which prefer rather low pH values [23]. Peroxidase-utilizing microorganisms are thus also being tested for the degradation of aromatic toxins, such as chlorinated polyaryls, phenols, dioxines and furanes, which can be relatively easily oxidized to unstable radical cations [27].

On the other hand, the potentially high reactivity of oxyheme iron centers with regard to substrates has to be avoided at all cost in myoglobin and hemoglobin; otherwise, an autoxidation of these exclusively O_2-transporting and -storing systems would result: a condition that only occurs in a pathological context. With this in mind, the requirement for the protein environment to inhibit such thermodynamically favorable oxidative processes in *Hb* and *Mb* should be particularly appreciated in retrospect (Section 5.2). The multiple functions of the heme group (5.8) are thus summarized in (6.16) and correlated with their electronic structure (according to [8]).

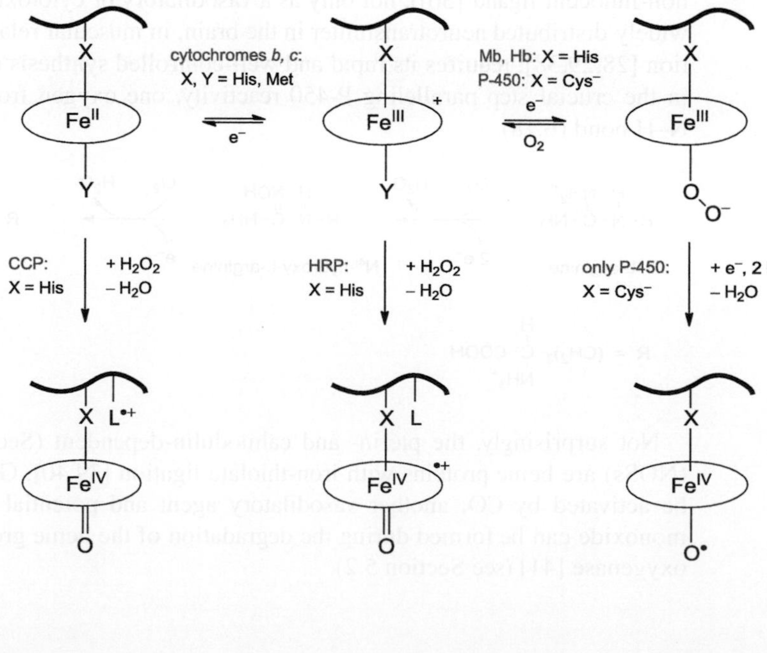

(6.16)

6.5 Hemoproteins in the Catalytic Transformation of Partially Reduced Nitrogen and Sulfur Compounds

The heme group, with its open coordination sites, can bind small unsaturated molecules not only via free oxygen (O_2) and carbon electron pairs (CO); partially reduced nitrogen oxo and sulfur oxo compounds such as nitrogen monoxide (nitric oxide, nitrosyl radical, NO), nitrite (NO_2^-), and sulfite (SO_3^{2-}) have also become known as substrates for hemoprotein catalysis [28–31]. In each case, binding to iron is possible through electron pair coordination and π back-donation from the divalent metal into low-lying orbitals of the unsaturated ligands (6.17).

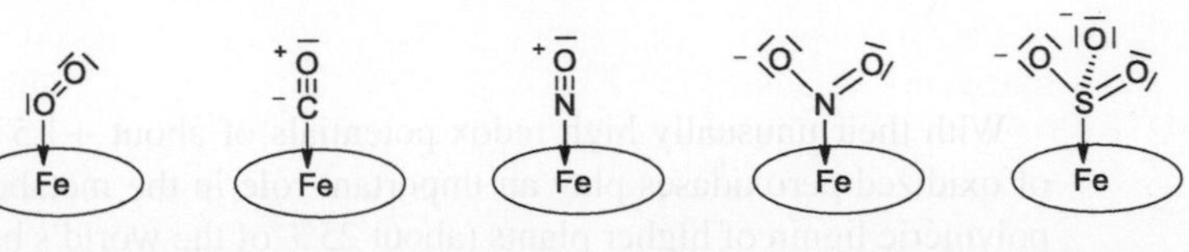

(6.17)

It has long been known that organic nitrites and nitrates, as well as the inorganic nitroprusside dianion $[Fe(CN)_5(NO)]^{2-}$, cause vascular relaxation in vertebrates as "nitrovasodilators" and may therefore be used against acute high blood pressure and angina pectoris (see also Section 19.4.5) [28,30]. Since vascular relaxation may be caused by an accumulation of the cyclic nucleotide guanosine monophosphate (cGMP; see Section 14.2), it is assumed that the enzyme guanylate cyclase is stimulated via binding of its heme component with endogeneous or exogeneous NO. While the binding to iron most probably involves the nitrosonium form, NO^+, of the non-innocent ligand (NO^+, being isoelectronic with the CO and CN^- ligands), the transport mode may be different ($NO^\bullet$, NO^+, thiol-bound form, metal complex?). As with O_2, the biological binding of NO (e.g. in the bacterial nitrogen cycle) can be effected not only through heme-iron but also through polynuclear copper centers [31] (see Section 10.3). The increasingly recognized essential role of NO ("molecule of the year 1992" [32]; Nobel Prize in Medicine 1998 for F. Murad, R. F. Furchgott and L. J. Ignarro [33–35]; non-innocent ligand [36]), not only as a vasodilatory or cytotoxic agent [37] but also as a widely distributed neurotransmitter in the brain, in muscular relaxation and in penile erection [28,29,38], requires its rapid and well-controlled synthesis (from arginine), whereby, in the crucial step paralleling P-450 reactivity, one oxygen from O_2 is inserted into an N–H bond (6.18).

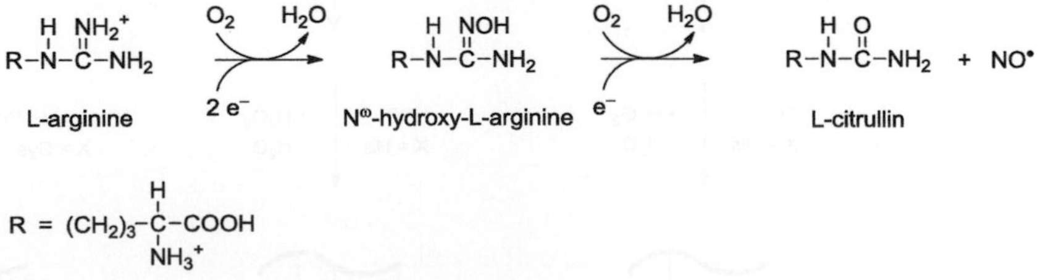

(6.18)

Not surprisingly, the pterin- and calmodulin-dependent (Section 14.2) NO synthases (NOSs) are heme proteins with iron-thiolate ligation [39,40]. Guanylate cyclase can also be activated by CO, another vasodilatory agent and potential neurotransmitter; carbon monoxide can be formed during the degradation of the heme group to biliverdin by heme oxygenase [41] (see Section 5.2).

Gasotransmitters

"Gasotransmitters" is the collective name given [42,43] to an emerging class of small gaseous molecules that includes NO, CO, H_2S and possibly others (C_2H_4, O_2). These molecules can act as information carriers (messengers, mediators); produced enzymatically, they can permeate membranes and induce specific responses by binding to receptors (signaling). Regulated generation in small concentrations and ready scavenging by receptors explain the biotolerance of these otherwise toxic substances.

The complete (i.e. six-electron) reduction (6.19) of nitrite, NO_2^-, another form of nitrogen(+III) that can be generated from nitrate, NO_3^-, via molybdoenzyme action (see Section 11.1), is part of the microbial denitrification and involves polyheme-containing nitrite reductases. These relatively large proteins contain several, partly unique (i.e. catalytic) heme centers.

$$NO_2^- + 6\,e^- + 8\,H^+ \xrightarrow[\text{nitrite reductase}]{\text{denitrification}} NH_4^+ + 2\,H_2O \qquad (6.19)$$

The four-electron oxidation of hydroxylamine, NH_2OH, to nitrite within the microbial nitrogen cycle (see Figure 11.1) is also catalyzed by a complex multiheme enzyme [44].

Another special heme group, heme d_1, is catalytically active in a dissimilatory bacterial nitrite reductase that reduces nitrite to NO and features an unusual porphyrin ligand with two carbonyl groups in two pyrrole rings [45] (6.20). Dissimilatorily nitrite-reducing bacteria may also contain copper enzymes; some assimilatorily NO_2^--utilizing microorganisms contain siroheme with a doubly hydrogenated porphyrin ring (6.20).

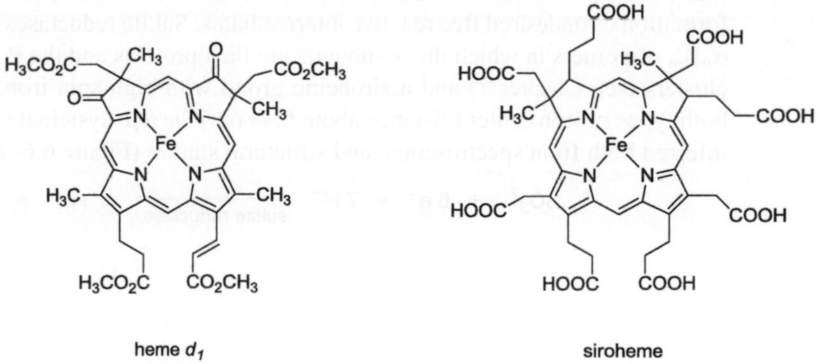

(6.20)

With the sulfite reductases, which most probably coordinate the partly reduced sulfur(+IV) atom of SO_3^{2-}, it is also possible to make a distinction between dissimilatory and assimilatory enzymes. The dissimilatory enzymes are found in simple bacteria in which SO_3^{2-} functions only as a terminal electron acceptor, whereas the assimilatory enzymes (e.g. of *E. coli*) serve to supply (hydrogen) sulfide for biosynthesis. Thermodynamically metastable sulfite can be formed from sulfate by molybdenum-containing oxotransferase enzymes (see Section 11.1.1).

The difficulty of reaction (6.21) is the catalysis of a six-electron process through several possible intermediates [31] using the normally available one-electron equivalents, a problem which is similar to that of reaction (6.19). Many-electron processes, such as the

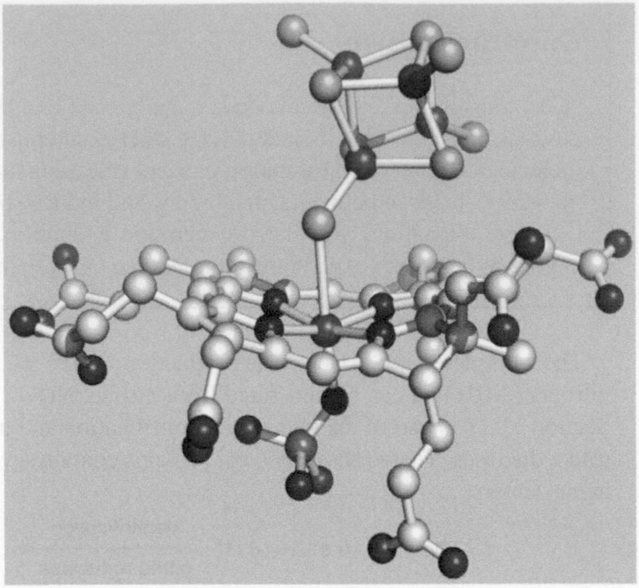

Figure 6.6
Active site structure of sulfite reductase, including connected Fe_4S_4 cluster (top) and heme macrocycle, with sulfate as inhibitor substrate (PDB code 1AOP).

transformation $2H_2O/O_2$ and the nitrogen fixation $8H^+ + N_2/2NH_3 + H_2$ (see (2.2), (2.3) and (2.4)), require the combined action of several redox-active, often inorganic, centers (see Sections 4.3, 10.4 and 11.2) in order to utilize favorable potentials and to avoid the formation of undesired free reactive intermediates. Sulfite reductases are therefore complex $\alpha_n\beta_m$ oligomers in which the α subunits are flavoproteins and the β subunits contain Fe/S clusters (see Chapter 7) and a siroheme group with high-spin iron. A direct bridging of both types of iron center (distance about 0.44 nm) via a μ-cysteinate sulfur center has been inferred both from spectroscopic and structural studies (Figure 6.6) [46].

$$SO_3^{2-} + 6\,e^- + 7\,H^+ \xrightarrow[\text{sulfite reductase}]{} HS^- + 3\,H_2O \qquad (6.21)$$

References

1. G. von Jagow, W. D. Engel, *Angew. Chem. Int. Ed. Engl.* **1980**, *19*, 659–675: *Structure and function of the energy converting systems of mitochondria.*
2. G. R. Moore, G. W. Pettigrew, *Cytochromes c*, Springer-Verlag, Berlin, **1990**.
3. G. Palmer, J. Reedijk, *Eur. J. Biochem.* **1991**, *200*, 599–611: *Nomenclature of electron-transfer proteins.*
4. G. N. George, T. Richards, R. E. Bare, Y. Gea, R. C. Prince, E. I. Stiefel, G. D. Watts, *J. Am. Chem. Soc.* **1993**, *115*, 7716–7718: *Direct observation of bis-sulfur ligation to the heme of bacterioferritin.*
5. (a) T. Takano, R. E. Dickerson, *Proc. Natl. Acad. Sci. USA* **1980**, *77*, 6371–6375: *Redox conformation changes in refined tuna cytochrome c*; (b) C. Lange, C. Hunte, *Proc. Natl. Acad. Sci. USA* **2002**, *99*, 2800–2805: *Crystal structure of the yeast cytochrome bc1 complex with its bound substrate cytochrome c.*

6. W. R. Scheidt, C. A. Reed, *Chem. Rev.* **1981**, *81*, 543–555: *Spin-state/stereochemical relationships in iron porphyrins: implications for the hemoproteins.*
7. S. G. Sligar, K. D. Egeberg, J. T. Sage, D. Morikis, P. M. Champion, *J. Am. Chem. Soc.* **1987**, *109*, 7896–7897: *Alteration of heme axial ligands by site-directed mutagenisis: a cytochrome becomes a catalytic demethylase.*
8. P. R. Ortiz de Montellano, *Cytochrome P450: Structure, Mechanism, and Biochemistry*, 3rd Edition, Kluwer Academic/Plenum Publishers, New York, **2005**.
9. T. D. Porter, M. H. Coon, *J. Biol. Chem.* **1991**, *266*, 13 469–13 472: *Cytochrome P-450.*
10. W. B. Jakoby, D. M. Ziegler, *J. Biol. Chem.* **1990**, *265*, 20 715–20 718: *The enzymes of detoxification.*
11. E. Mutschler, G. Geisslinger, H. K. Kroemer, S. Menzel, P. Ruth, *Arzneimittelwirkungen, Pharmakologie – Klinische Pharmakologie – Toxikologie*, 10th Edition, WVG, Stuttgart, **2013**.
12. D. Mansuy, P. Battioni, J. P. Battioni, *Eur. J. Biochem.* **1989**, *184*, 267–285: *Chemical model systems for drug-metabolizing cytochrome-P-450-dependent monooxygenases.*
13. H. Patzelt, W.D. Woggon, *Helv. Chim. Acta* **1992**, *75*, 523–530: *O-insertion into nonactivated C-H bonds: the first observation of O_2 cleavage by a P-450 enzyme model in the presence of a thiolate ligand.*
14. R. Raag, T. L. Poulos, *Biochemistry* **1989**, *28*, 917–922: *The structural basis for substrate-induced changes in redox potential and spin equilibrium in cytochrome P-450CAM.*
15. K. G. Ravichandran, S. S. Boddupalli, C. A. Hasemann, J. A. Peterson, J. Deisenhofer, *Science* **1993**, *261*, 731–736: *Crystal structure of hemoprotein domain of P450BM-3, a prototype for microsomal P450's.*
16. J. Rittle, M. T. Green, *Science* **2010**, *330*, 933–937: *Cytochrome P450 compounds I: capture, characterization, and C-H bond activaton kinetics.*
17. K. L. Kostka, B. G. Fox, M. P. Hendrich, T. J. Collins, C. E. F. Richard, L. J. Wright, E. Münck, *J. Am. Chem. Soc.* **1993**, *115*, 6746–6757: *High-valent transition metal chemistry. Mössbauer and EPR studies of high-spin (S=2) iron(IV) and intermediate-spin (S=3/2) iron(III) complexes with a macrocyclic tetraamido-N ligand.*
18. D. T. Sawyer, *Comments Inorg. Chem.* **1987**, *6*, 103–121: *The nature of the bonding and valency for oxygen in its metal compounds.*
19. P. M. Champion, *J. Am. Chem. Soc.* **1989**, *111*, 3433–3434: *Elementary electronic excitations and the mechanism of cytochrome P450.*
20. H. Sugimoto, H. C. Tung, D. T. Sawyer, *J. Am. Chem. Soc.* **1988**, *110*, 2465–2470: *Formation, characterization, and reactivity of the oxene adduct of [tetrakis(2,6-dichlorophenyl)porphinato]iron(III) perchlorate in acetonitrile. Model for the reactive intermediate of cytochrome P-450.*
21. N. C. Veitch, *Phytochem.* **2004**, *65*, 249–259: *Horseradish peroxidase: a modern view to a classic enzyme.*
22. S. Hashimoto, R. Nakajima, I. Yamazaki, T. Kotani, S. Ohtaki, T. Kitagawa, *FEBS Lett.* **1989**, *248*, 205–209: *Resonance Raman characterization of hog thyroid peroxidase.*
23. H. E. Schoemaker, *Recl. Trav. Chim. Pays-Bas* **1990**, *109*, 255–272: *On the chemistry of lignin biodegradation.*
24. J. H. Dawson, *Science* **1988**, *240*, 433–439: *Probing structure-function relations in heme-containing oxygenases and peroxidases.*
25. K. Yamaguchi, Y. Watanabe, I. Morishima, *J. Am. Chem. Soc.* **1993**, *115*, 4058–4065: *Direct observation of the push effect on the O-O bond cleavage of acylperoxoiron(III) porphyrin complexes.*
26. G. I. Berglund, G. H. Carlsson, A. T. Smith, H. Szöke, A. Hendriksen, J. Hajdu, *Nature (London)* **2002**, *417*, 463–468: *The catalytic pathway of horseradish peroxidase at high resolution.*
27. G. Winkelmann (ed.), *Microbial Degradation of Natural Products*, VCH Publishers, New York, **1992**.
28. A. R. Butler, D. L. H. Williams, *Chem. Soc. Rev.* **1993**, 233–241: *The physiological role of nitric oxide.*
29. J. S. Stamler, D. J. Singel, J. Loscalzo, *Science* **1992**, *258*, 1898–1902: *Biochemistry of nitric oxide and its redox-activated forms.*
30. M. J. Clarke, J. B. Gaul, *Struct. Bonding (Berlin)* **1993**, *81*, 147–181: *Chemistry relevant to the biological effects of nitric oxide and metallonitrosyls.*

31. P. M. H. Kroneck, J. Beuerle, W. Schumacher, *Metal-dependent conversion of inorganic nitrogen and sulfur compounds*, in H. Sigel (ed.) *Metal Ions in Biological Systems, Vol. 28*, Marcel Dekker, New York, **1992**.
32. E. Culotta, D. E. Koshland, *Science* **1992**, *258*, 1862–1865: *NO news is good news.*
33. F. Murad, *Angew. Chem. Int. Ed. Engl.* **1999**, *38*, 1856–1868: *Discovery of some of the biological effects of nitric oxide and its role in cell signaling* (Nobel lecture).
34. R. F. Furchgott, *Angew. Chem. Int. Ed. Engl.* **1999**, *38*, 1870–1880: *Endothelium-derived relaxing factor: discovery, early studies, and identifcation as nitric oxide* (Nobel lecture).
35. L. J. Ignarro, *Angew. Chem. Int. Ed. Engl.* **1999**, *38*, 1882–1892: *Nitric oxide: a unique endogenous signaling molecule in vascular biology* (Nobel lecture).
36. *Inorg. Chem.* **2010**, *49*, 6223–6365: *Inorganic chemistry forum: the coordination chemistry of nitric oxide and its significance for metabolism, signaling, and toxicity in biology.*
37. M. A. Tayeh, M. A. Marletta, *J. Biol. Chem.* **1989**, *264*, 19 654–19 658: *Macrophage oxidation of l-arginine to nitric oxide, nitrite and nitrate.*
38. A. L. Burnett, C. J. Lowenstein, D. S. Bredt, T. S. K. Chang, S. H. Snyder, *Science* **1992**, *257*, 401–403: *Nitric oxide: a physiologic mediator of penile erection.*
39. M. A. Marletta, *J. Biol. Chem.* **1993**, *268*, 12 231–12 234: *Nitric oxide synthase structure and mechanism.*
40. T. Doukov, H. Li, M. Soltis, T. L. Poulos, *Biochemistry* **2009**, *48*, 10 246–10 254: *Single crystal structural and absorption spectral characterizations of nitric oxide synthase complexed with* N^{ω}*-hydroxy-l-arginine and diatomic ligands.*
41. A. Verma, D. J. Hirsch, C. E. Glatt, G. V. Ronnett, S. H. Snyder, *Science* **1993**, *259*, 381–384: *Carbon monoxide. A putative neural messenger?*
42. R. Wang, *FASEB J.* **2002**, *16*, 1792–1798: *Two's company, three's a crowd: can* H_2S *be the third endogenous gaseous transmitter?*
43. R. Wang, *Sci. American*, **2010**, March, 50–55: *Toxic gas, lifesaver.*
44. M. P. Hendrich, M. Logan, K. K. Anderson, D. M. Arciero, J. D. Lipscomb, A. B. Hooper, *J. Am. Chem. Soc.* **1994**, *116*, 11 961–11 968: *The active site of hydroxylamine oxidoreductase from Nitrosomonas: evidence for a new metal cluster in enzymes.*
45. C. K. Chang, R. Timkovich, W. Wu, *Biochemistry* **1986**, *25*, 8447–8453: *Evidence that heme d1 is a 1,3-porphyrindione.*
46. D. E. McRee, D. C. Richardson, J. S. Richardson, L. M. Siegel, *J. Biol. Chem.* **1986**, *261*, 10 277–10 281: *The heme and* Fe_4S_4 *cluster in the crystallographic structure of Escherichia coli sulfite reductase.*

7 Iron–Sulfur and Other Non-heme Iron Proteins

Three main groups of iron-containing proteins can be distinguished according to the ligation of the metal. Iron centers that are exclusively coordinated by amino acid residues, components of water (H_2O, HO^-, O^{2-}) or oxoanions occur in photosynthetic reaction centers (Figures 4.5–4.7), in hemerythrin (Section 5.3), in non-heme iron enzymes (Section 7.6) and in iron transport and storage proteins (see Chapter 8). In addition to these often polynuclear systems and the heme species (5.8) in which the iron centers are chelated through a porphinoid macrocycle (see Section 5.2 and Chapter 6), the *iron–sulfur (Fe/S) centers* represent another major and important class of iron species in proteins [1].

7.1 Biological Relevance of the Element Combination Iron–Sulfur

The majority of the ubiquitous Fe/S centers in proteins are involved in electron transfer at typically negative redox potentials (Tables 7.1 and 7.2).

Fe/S centers have essential functions in photosynthesis (Figure 4.7), cell respiration (Figures 6.1 and 6.2), nitrogen fixation (see Section 11.2) and in the metabolism of H_2 (hydrogenases with and without nickel; Section 9.3), NO_2^- and SO_3^{2-} (sulfite oxidation and reduction; Sections 6.5 and 11.1.2). In addition to the electron-transfer function, Fe/S centers have been recognized as sites for redox and nonredox catalysis; examples include the exclusively Fe/S-dependent hydrogenases and nitrogenases, as well as (de)hydrolase/isomerase enzymes of the aconitase type (see 7.10) [2]. A further role of coordinatively open Fe/S centers would be to serve as iron sensors [3], whereas the [4Fe-4S] center found in the DNA repair enzyme endonuclease III [4] seems to have "only" a structural/polarizing function.

Fe/S centers in proteins sometimes occur as isolated clusters, for example in the small, electron-transferring "ferredoxins", [Fd]; however, they can also interact with other prosthetic groups such as other metal centers (Ni, Mo, V, heme-Fe) and flavins [5]. A characteristic feature of Fe/S proteins is the coordination of iron ions with protein-bound cysteinate sulfur (RS^-) and, in polynuclear Fe/S centers, with "inorganic" acid-labile sulfide, (S^{2-}). Sulfide and iron ions are often extractable and in many cases the remaining apoenzymes can be reconstituted with external S^{2-} and $Fe^{2+/3+}$ ions (Figure 7.1).

Approximately 1% of the iron content of mammals is present in the form of Fe/S proteins (compare Table 5.2). The facile formation and thermal robustness of such proteins, their

Bioinorganic Chemistry: Inorganic Elements in the Chemistry of Life – An Introduction and Guide, Second Edition.
Written and Translated by Wolfgang Kaim, Brigitte Schwederski and Axel Klein.

Table 7.1 Some reactions catalyzed by Fe/S center-containing redox enzymes.

Enzymes	Catalyzed reaction	Further explanations in chapter
hydrogenases	$2\,H^+ + 2\,e^- \rightleftharpoons H_2$	9.3
nitrogenases	$N_2 + 10\,H^+ + 8\,e^- \rightleftharpoons NH_4^+ + H_2$	11.2
sulfite reductase	$SO_3^{2-} + 7\,H^+ + 6\,e^- \rightleftharpoons HS^- + 3\,H_2O$	6.5
aldehyde oxidase	$R\text{-}CHO + 2\,OH^- \rightleftharpoons R\text{-}COOH + H_2O + 2e^-$	12.5
xanthine oxidase	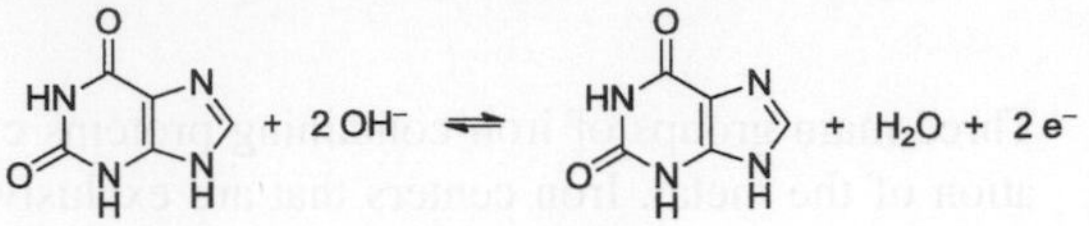	11.1.1
NADP oxidoreductase	$NADP^+ + H^+ + 2\,e^- \rightleftharpoons NADPH$	(Fig. 4.7)

Table 7.2 Redox potentials of representative Fe/S proteins (see also Table 6.1).

Protein	Typical origin	Type of Fe/S center	Molecular mass (kDa)	E (mV)
rubredoxin	*Clostridium pasteurianum*	$[Rd]^{2+;3+}$	6	−60
2Fe ferredoxin	spinach	$[2Fe\text{-}2S]^{1+;2+}$	10.5	−420
adrenodoxin	adrenal mitochondria	$[2Fe\text{-}2S]^{1+;2+}$	12	−270
Rieske center	adrenal mitochondria	$[2Fe\text{-}2S]^{1+;2+}$	250 (*bc₁* complex)	+280
4Fe ferredoxin	*Bacillus stearothermophilus*	$[4Fe\text{-}4S]^{1+;2+}$	9.1	−280
8Fe ferredoxin	*Cl. pasteurianum*	$2[4Fe\text{-}4S]^{1+;2+}$	6	−400
High-potential iron–sulfur protein (HiPIP)	*Chromatium vinosum*	$[4Fe\text{-}4S]^{2+;3+}$	9.5	+350
ferredoxin II	*Desulfovibrio gigas*	$[3Fe\text{-}4S]^{n+}$	24	−130
ferredoxin I	*Azotobacter vinelandii*	$[3Fe\text{-}4S]^{n+}$		
		$[4Fe\text{-}4S]^{n+}$	14	−460

distribution in nearly all organisms, particularly in evolutionarily "old" species, and the remarkable conservation of critical amino acid sequences (7.7) suggest that they may have played an important role very early in evolution, that is, in the absence of free O_2 [6]. The typically negative redox potentials (Table 7.2), the occurrence in highly temperature-resistant (>100 °C) "hyperthermophilic" microorganisms ("extremophiles" [7]) and the general oxygen sensitivity of the reduced states also support this hypothesis.

Extremophiles and Bioinorganic Chemistry

Organisms which live under uncommon ("extreme") physical or chemical conditions such as very high/very low temperatures or in apparently "aggressive" environments (acidic, basic, halide-rich, heavy-metal-rich etc.) are referred to as "extremophiles".

Their role at the beginning (prebiotic conditions) and (presumably) at the end of life on earth, their possible occurrence on other planets (→ astrobiology) and their possible commercial applications have directed scientific attention to this kind of microbial life, revealing frequently unconventional biochemical metabolism, including the unorthodox use of inorganic substances by these organisms. Examples include hyperthermophilic and hyperbaric life at hydrothermal sites based on the conversion of sulfur compounds with the help of metal-rich biomolecules, the tolerance of acidic conditions by mineral-dissolving thiobacilli and inhibition of the crystallization of potentially cell-rupturing ice by the cryophilic organisms.

There are experimentally supported [8] theories according to which inorganic iron sulfides (FeS) or the disulfide (S_2^{2-})-containing pyrite (FeS_2) might have been involved in the beginning of chemoautotrophic metabolism through reduction of CO_2, according to (7.1), and thus possibly in the evolution of life [8,9].

$$H_2S + 2\,FeS + CO_2 \longrightarrow 2\,FeS_2 + HCOOH$$
$$H_2O + 2\,FeS + CO_2 \xrightarrow{h\nu} 2\,FeO + 1/2\,(CHOH)_n + 2\,S \qquad (7.1)$$

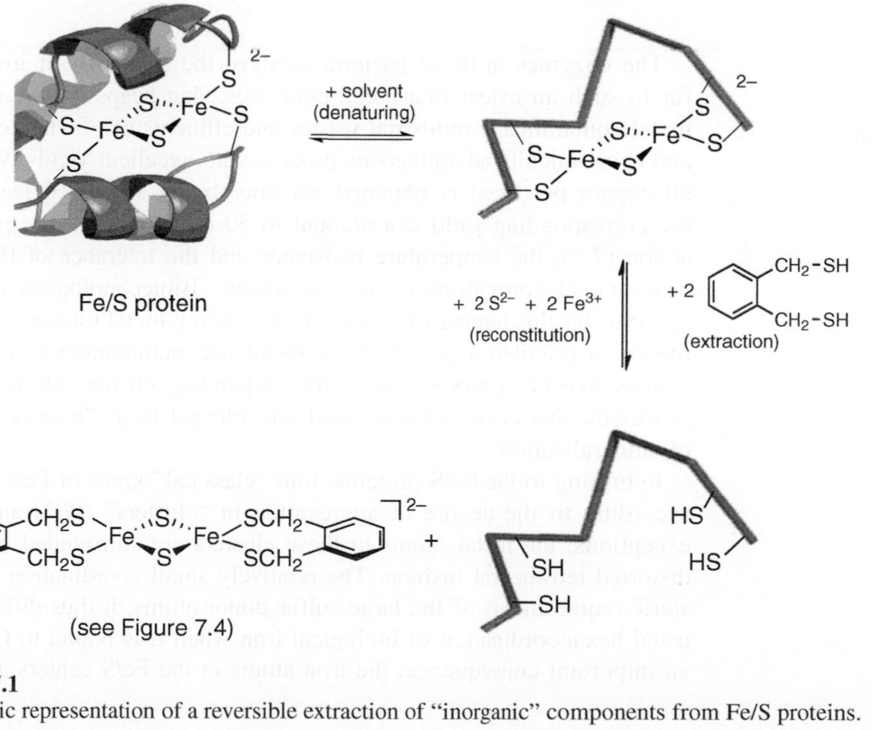

Figure 7.1
Schematic representation of a reversible extraction of "inorganic" components from Fe/S proteins.

Robust "chemolithotrophic" sulfur bacteria, which obtain their energy from the transformation of inorganic compounds, have become important in geobiotechnology as the organisms essential for the process of "bacterial leaching" or "biomining" [10]. This leaching process of ores or slag heaps with the help of the ubiquitous bacteria *Thiobacillus thiooxidans* and *T. ferrooxidans* is directed at hardly soluble sulfides such as CuS and $CuFeS_2$ and at oxides such as UO_2 which become transformed into soluble sulfates (7.2), at the same time releasing enclosed noble metals such as gold.

$$S + 3/2\,O_2 + H_2O \xrightarrow{\textit{T. thiooxidans}} H_2SO_4$$

$$2\,FeSO_4 + 1/2\,O_2 + H_2SO_4 \xrightarrow{\textit{T. ferrioxidans}} Fe_2(SO_4)_3 + H_2O$$

$$MS + Fe_2(SO_4)_3 \longrightarrow MSO_4 + 2\,FeSO_4 + S$$

overall reaction:

$$MS\ (\text{insoluble}) + 2\,O_2 \longrightarrow MSO_4\ (\text{soluble})$$

M: e.g. Cu, Zn, Ni, Co

(7.2)

The enzymes in these bacteria catalyze the oxidation of iron(II) and elemental sulfur to such an extent that metal-poor ores, slag heaps of primary ore mining and even metal-contaminated industrial wastes and effluents can be treated using this ecologically and economically advantageous process with excellent yields. Worldwide, about 15% of all copper produced is obtained via microbially supported leaching; in a single mine, the corresponding yield can amount to 50 tons per day. The pH optimum (2–3) of reaction (7.2), the temperature resistance and the tolerance of thiobacilli with respect to heavy-metal concentrations are remarkable. Biotechnological modifications to increase and transfer this temperature stability and heavy-metal tolerance are being attempted with regard to possible applications in metal decontamination and recycling. In nature, reactions like (7.2) proceed reversibly depending on the external conditions; the reverse processes also occur naturally and are referred to as "bioweathering" or "geochemical biomineralization".

Returning to the Fe/S proteins: four "classical" kinds of Fe/S center are distinguished, according to the degree of aggregation in "clusters" ((7.3) and Figure 7.2); with few exceptions, the metal atoms in these clusters are surrounded by four sulfur atoms in a distorted tetrahedral fashion. The relatively small coordination number is caused by the steric requirements of the large sulfur donor atoms; it thus differs significantly from the usual hexacoordination of biological iron when it is bound to O or N donor ligands. As an important consequence, the iron atoms in the Fe/S centers are exclusively high-spin,

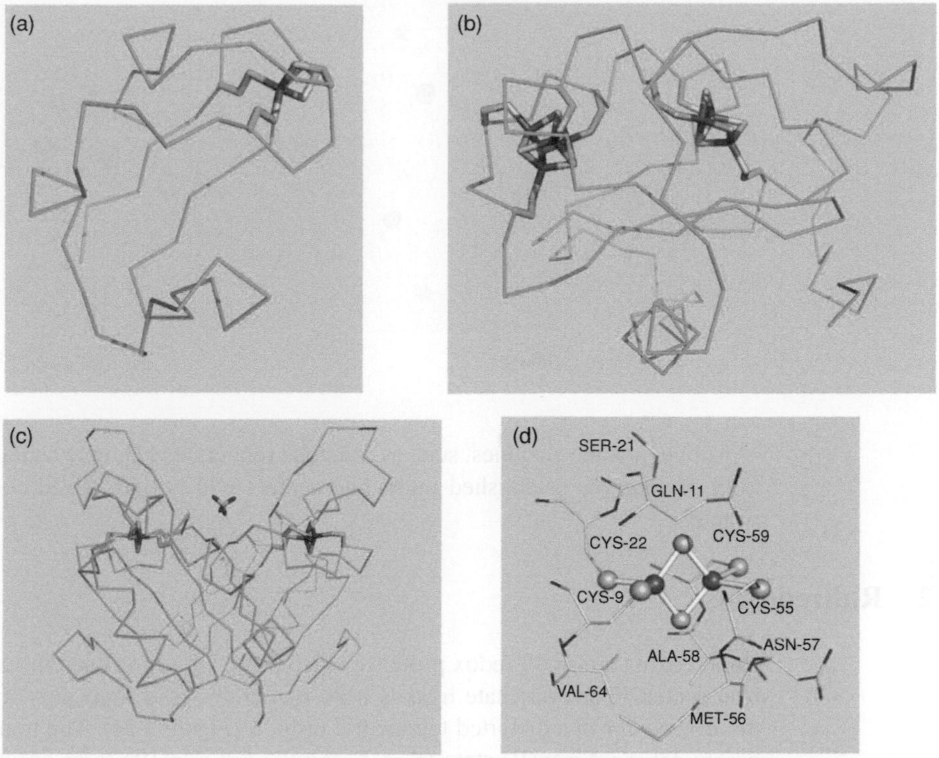

Figure 7.2
Structures of Fe/S centers in proteins. (a–c) Ribbon representations of the protein folding with (a) rubredoxin (PDB code 2DSX), (b) ferredoxin from *Azotobacter vinelandii* (PDB code 1FDD) and (c) [2Fe-2S] ferredoxin from *Aquifex aeolicus* (PDB code 1M2A). (d) Detailed close-up of the immediate protein environment of (c) the [2Fe-2S] ferredoxin from *Aquifex aeolicus* [12].

because the ligand field splitting in tetrahedral symmetry is less than half of what it would be in an octahedral arrangement (see (12.4)).

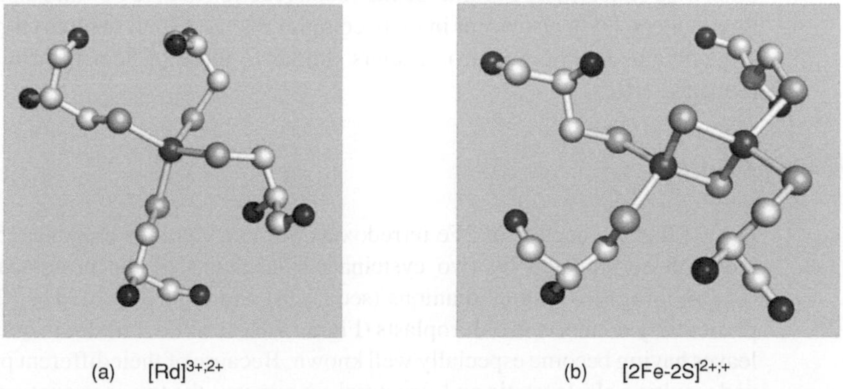

(7.3)

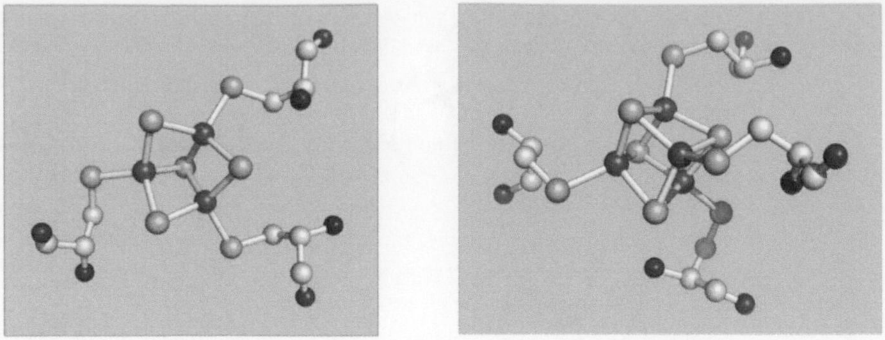

(c) $[3Fe\text{-}4S]^{+;0}$ (d) $[4Fe\text{-}4S]^{3+;2+;+}$

(7.3 *continued*)

Although some enzymes, such as fumarate reductase [11], may contain several different Fe/S centers, the established individual centers will be introduced consecutively in this chapter.

7.2 Rubredoxins

Rubredoxins are small redox proteins which occur in certain bacteria and contain just one iron center. Four cysteinate ligands from two amino acid sequences -Cys-X_2-Cys- ligate the iron center in a distorted tetrahedral fashion (Figure 7.2a). The transition between the nearly colorless iron(II) state ($S = 2$) and the red iron(III) form occurs without a major change in Fe–S distances; larger changes and, therefore, slower electron transfer (compare Section 6.1) are observed for simple model complexes (see Section 7.5). The intensely red color responsible for the name of these proteins results from a ligand-to-metal charge transfer transition (7.4) from the σ and π electron-rich thiolate ligands to the oxidized (i.e. electron-poor) iron(III) center [13]. Comparably intense colors are known from the analytically important thiocyanate(NCS^-) complexes of iron(III).

$$Fe^{III}(^{-}S{-}R) \xrightarrow{h\nu} [Fe^{II}(^{\bullet}S{-}R)]^* \quad \text{*represents the excited state} \tag{7.4}$$

An important function of the protein is to stabilize the trivalent form with respect to an otherwise conceivable intramolecular redox reaction to give Fe^{II} and disulfide. Rubredoxin-type centers, [*Rd*], also occur in more complex proteins such as ruberythrin, which additionally contains non-heme diiron centers similar to those of hemerythrin and ribonucleotide reductase (RR).

7.3 [2Fe-2S] Centers

In the [2Fe-2S] centers of 2Fe ferredoxins or more complex enzymes [5], two iron centers are each coordinated by two cysteinate side chains of the protein and by two shared (i.e. bridging) (μ-)sulfide dianions (see (7.3b) and Figure 7.2b). The [2Fe-2S] centers are particularly common in chloroplasts (Figure 4.8), the 2Fe ferredoxin obtained from spinach leaves having become especially well known. Because of their different protein environment and resulting electrostatic and structural asymmetry, the two iron centers are not equivalent; the question is whether or not the same is true for their redox behavior. The biologically relevant electron transfer consists in a transition from the Fe^{III}/Fe^{III} state, with compensating (i.e. strongly antiferromagnetically coupled) spins, to a one-electron reduced mixed-valent form. Does the unsymmetrical formulation Fe^{II}/Fe^{III}, implicating valence localization, or rather the symmetrical description $Fe^{2.5}/Fe^{2.5}$, corresponding to delocalization in the ground state, conform with experimental results? Mössbauer spectroscopy (insertion Section 5.3) and other physical methods suggest a localized description with fixed valences Fe^{II} and Fe^{III} for reduced [2Fe-2S] centers in proteins and in some model complexes (Section 7.5). Despite antiferromagnetic coupling of the high-spin centers via the sulfide ions (which are capable of effecting superexchange) one EPR-detectable unpaired electron remains for the reduced form.

Some [2Fe-2S] proteins contain centers with unusual spectroscopic properties and relatively high redox potentials (Table 7.2). These "Rieske centers" are found in the cytochrome-containing membrane protein complexes of mitochondria ("*bc*$_1$-complexes"; compare Figures 6.1 and 6.2) and in chloroplasts (*b*/*f*-complexes; compare Figure 4.7). The Rieske proteins contain two markedly different iron centers, due to an unsymmetrical coordination (7.5) involving the neutral *non*-sulfur ligand histidine [14].

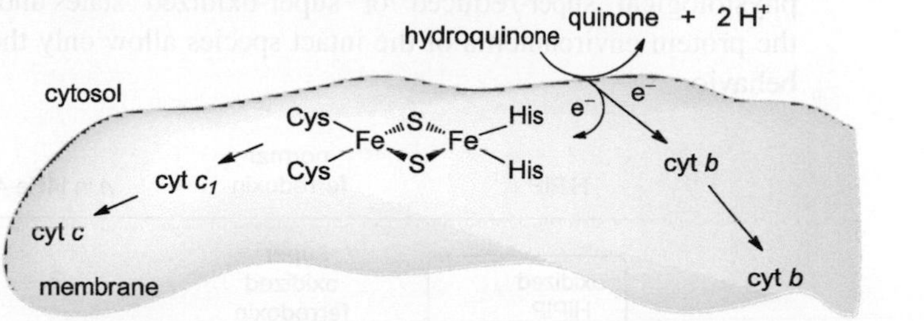

(7.5)

In combination with a cytochrome *b*, the function of the Rieske centers is to guarantee a splitting of the electron flow in the intramembrane electron transport chain [14]: starting from the two-electron-donating hydroquinones, there is one electron pathway *along* the membrane at high potential and another *across* the membrane at low potential (7.5).

7.4 Polynuclear Fe/S Clusters: Relevance of the Protein Environment and Catalytic Activity

The most common and most stable Fe/S centers are of the [4Fe-4S] type. These "clusters" are found in numerous complex enzymes and in electron-transfer proteins such as the 4Fe, 7Fe and 8Fe ferredoxins, the latter of which contain two rather isolated [4Fe-4S] centers within the protein (distance > 1 nm; compare Figure 7.2d). The 4Fe and 8Fe ferredoxins with exclusively S-coordinated iron are relatively small, ubiquitous electron-transfer proteins. As the term "[4Fe-4S]" suggests, four iron centers and four sulfide ions are arranged in what may be described as a distorted cuboidal arrangement with approximate D_2 symmetry, each four-coordinate iron center being anchored in the protein by one cysteinate residue ((7.3), (7.8), Figure 7.2c,d). According to a first approximation, the four iron centers form a tetrahedron with the triply bridging μ_3-sulfide ions above the triangular planes, so that these too are tetrahedrally arranged. [4Fe-4S] centers participate in nearly all complex biological redox reactions, such as photosynthesis, respiration and N_2 fixation, in which they act mainly as electron-transfer centers at negative potentials. However, they may also have nonredox catalytic or noncatalytic functions [4].

Mössbauer spectroscopy indicates that there are two pairs of iron dimers present in the diamagnetic (2−) charged state of the system ($[4Fe\text{-}4S]^{2+}$ nucleus and 4 Cys^{-}); both pairs show approximately the same isomer shift, corresponding to an oxidation state of +2.5, but different quadrupole splitting. The fact that two mixed-valent Fe^{II}/Fe^{III} pairs do not show any resulting paramagnetism is not self-evident despite the even number of electrons. Extensive electron delocalization, including the sulfur ligands, and effective spin-coupling seem to operate in this configuration. Starting from this (2–) state, electron paramagnetic resonance (EPR)-active forms are created in the "normal" 4Fe ferredoxins through one-electron reduction to $3Fe^{II}/1Fe^{III}$ species with spin ground states of $S = 1/2$ or higher. However, there are also protein types in which the diamagnetic [4Fe-4S] center can be reversibly oxidized at relatively high potentials to a paramagnetic ($S = 1/2$) form with a $1Fe^{II}/3Fe^{III}$ oxidation state distribution; these proteins are referred to as "high potential iron–sulfur-proteins" (HiPIPs) (Table 7.2).

A "three-state hypothesis" has been applied to describe the fact that HiPIP and normal 4Fe ferredoxins are distinguished by different stabilities of their one-electron oxidized and reduced forms, respectively (7.6) [15]. Although it is possible to obtain the respective unphysiological "super-reduced" or "super-oxidized" states under denaturation of the protein, the protein environments of the intact species allow only the biologically intended redox behavior.

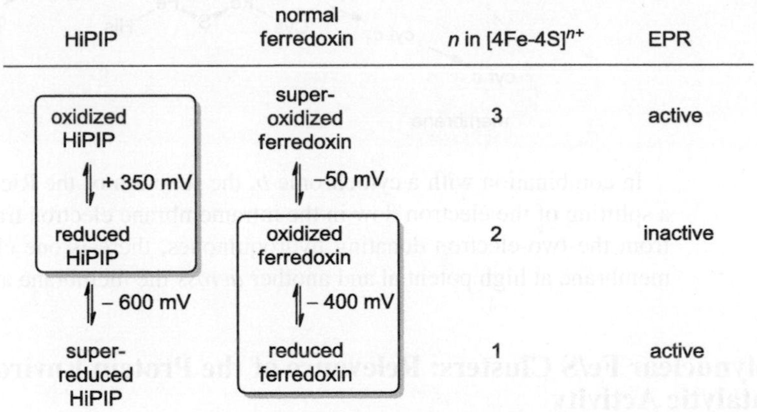

HiPIP
normal ferredoxin
n in [4Fe-4S]n+
EPR
oxidized HiPIP
super-oxidized ferredoxin
3
active
+ 350 mV
−50 mV
reduced HiPIP
oxidized ferredoxin
2
inactive
− 600 mV
− 400 mV
super-reduced HiPIP
reduced ferredoxin
1
active

(7.6)

In accordance with the explanation given for the other two classes of "inorganic" electron-transfer proteins (i.e. the cytochromes (Section 6.1) and the blue copper proteins (Section 10.1)), the geometries of the Fe/S cluster and of the whole protein change very little during electron transfer. For example, the reduction of the $[4Fe\text{-}4S]^{3+}$ form in HiPIP causes only a small elongation of Fe–S bonds, an expansion and a slightly more pronounced distortion of the cluster. In non-HIPIP ferredoxins, the diamagnetic form also seems to exhibit a somewhat higher deviation from the ideally tetrahedral arrangement of the iron atoms; a geometrically more distorted diamagnetic cluster would conform with the entatic state concept (insertion Section 2.3.1). Apparently small changes in the protein framework, such as a larger number of hydrophobic amino acid residues around the HiPIP cluster and a thus diminished access of water, determine the redox potential and stability of individual oxidation states. The expansion of the clusters upon reduction is due to the electron being transferred to nonbonding or antibonding cluster molecular orbitals; in any case, the smaller metal cations are closer to the cluster center than the larger sulfide anions (compare (7.3d)).

Experimental and theoretical studies on proteins, modified proteins and model complexes have shown that the electronic structure of [4Fe-4S] clusters is very sensitive to minute geometrical changes in bond lengths, bond angles and torsion angles. Both redox potentials and stabilities and spin–spin coupling patterns depend on the conformation, implicating for example the accessibility of spin states higher than $S = 1/2$ for odd electron forms. Like the central diamagnetic (2+) state, the neighboring paramagnetic states of the $[4Fe\text{-}4S]^{n+}$ system also contain two antiferromagnetically coupled iron dimers – one mixed-valent pair (II,III) and one homo-valent pair – each with predominantly parallel internal spin–spin coupling. In contrast to the $[2Fe\text{-}2S]^{+}$ systems, with their valence-localized state, the higher degree of delocalization (resonance) found in the $[4Fe\text{-}4S]^{n+}$ clusters can be attributed to the structurally determined orthogonality of metal orbitals that interact via superexchanging sulfide bridges. Following theoretical analyses, this favors parallel spin–spin interactions with higher resonance energies according to Hund's rule [16]. The electron delocalization as such is less susceptible to perturbation from external asymmetries induced from the protein. The participation of the cysteinate ligands in the accomodation of additional electrons is evident from a strengthening of hydrogen bond interactions X–H$\cdots{}^{-}$S(Cys) [17], suggesting an increase of effective negative charge at the cysteinate sulfur centers.

The amino acid sequence does not only determine whether a HiPIP center or the "normal" ferredoxin form of the [4Fe-4S] system occurs, it is also responsible for deciding whether a 4Fe or a 2Fe ferredoxin is formed from a cysteine-containing protein and enzymatically introduced iron (Chapter 8) and sulfide (via sulfide reductase and thiosulfate sulfur transferase). Some representative amino acid positions for the invariant cysteines in Fe/S cluster-forming proteins are summarized in (7.7) [4,15,18].

Fe/S center	typical amino acid sequence
[Rd]	: - Cys - X_2 - Cys - X_n - Cys - X_2 - Cys -
[2Fe-2S]	: - Cys - X_4 - Cys - X_2 - Cys - X_{29} - Cys -
[3Fe-3S]	: - Cys - $X_{5,7}$ - Cys - X_n - Cys -
[4Fe-4S] or "normal Fd"	: - Cys - X_2 - Cys - X_2 - Cys - X_n - Cys -
HiPIP	: - Cys - X_2 - Cys - X_{16} - Cys - X_{13} - Cys -
nonredox active (endonuclease III)	: - Cys - X_6 - Cys - X_2 - Cys - X_5 - Cys -
nitrogenase Fe protein (dimer)	: - Cys - X_{34} - Cys - - Cys - X_{34} - Cys -

(7.7)

Different amino acid sequences are obviously responsible for the formation of more special kinds of Fe/S proteins, which, in addition to [4Fe-4S] moieties, contain [3Fe-4S] centers (3Fe and 7Fe ferredoxins [19]). After suitable modification, these [3Fe-4S] centers may be converted to [4Fe-4S] centers (7.8) [11]. Structurally, they can be derived from [4Fe-4S] analogues by removal of a labile, *non-cysteinate-coordinated* iron atom from the

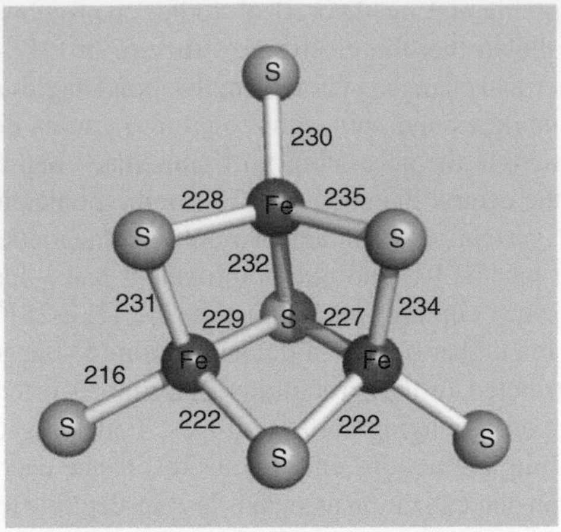

Figure 7.3
Structure and some geometrical data (bond lengths in pm) of the [3Fe-4S] cluster from *Desulfovibrio gigas* ferredoxin (according to [20]).

distorted cuboidal arrangement (Figure 7.3) [20]; another [3Fe-4S] form with *"linearly"* arranged metal centers has been detected, albeit under unphysiological (basic) conditions (7.3c), (7.8).

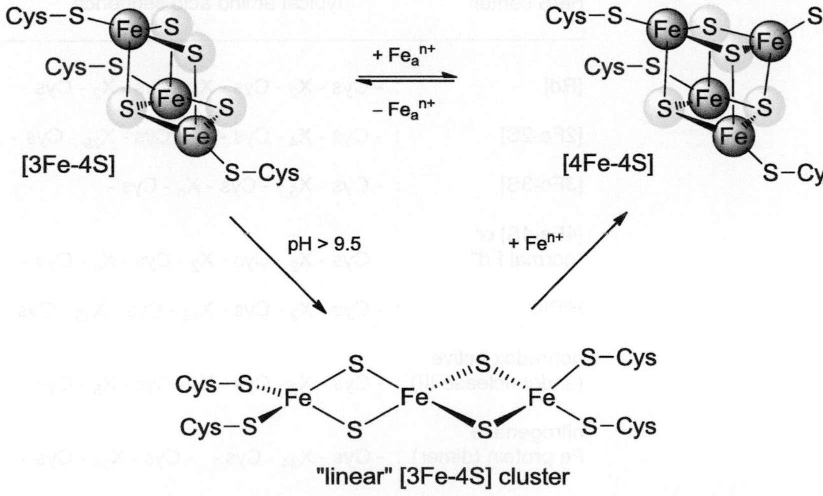

(7.8)

In addition to their occurrence in ferredoxins from microorganisms (compare Table 7.2), the [3Fe-4S] systems have been found in equilibrium with a labile [4Fe-4S] form (7.8) as a component of the enzyme aconitase (aconitate hydratase/isomerase), for example in mitochondria. This Fe/S enzyme catalyzes the equilibrium (7.9) within the Calvin cycle in what is clearly a nonredox reaction. Mechanism (7.10) has been suggested for aconitase [21].

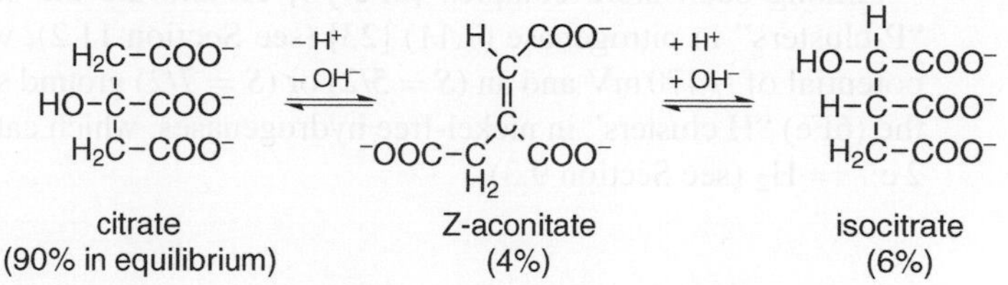

(7.9)

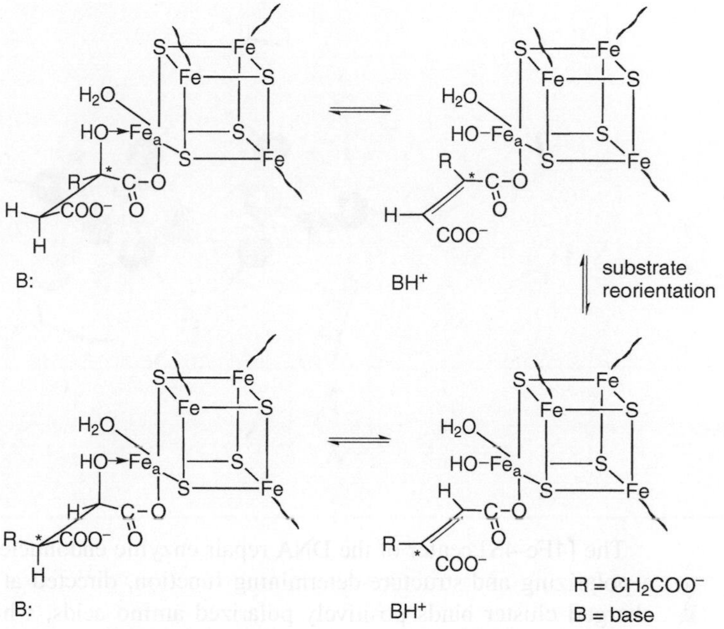

(7.10)

In the active state, the labile iron center, Fe_a, of the [4Fe-4S] form of aconitase is not coordinated by cysteinate but by water molecules. After substitution of H_2O and coordination of the chelating substrate (five-membered chelate ring, coordination number 5 or 6 at Fe_a), a sequence (7.10) of (HO)–C* bond cleavage/C–H deprotonation, rotation of the *non*-chelating Z-aconitate ligand and hydroxide–C(olefin) bond formation/C*(olefin) protonation leads to a rapid equilibration (7.9).

The electronic structure of the [3Fe-4S] centers in the oxidized form shows *one* unpaired electron ($S = 1/2$), corresponding to a formulation with three almost equally antiferromagnetically coupled high-spin Fe^{III} centers. The reduced form may be described as a combination of $1Fe^{II}$ and $2Fe^{III}$, featuring an ($S = 2$) ground state. This situation results from antiparallel spin–spin interactions between a high-spin Fe^{III} center ($S = 5/2$) and an Fe^{III}/Fe^{II} mixed-valent pair with $S = 9/2$ (i.e. exhibiting parallel internal spin–spin coupling). In some cases, excited spin states are readily accessible; according to their “open”, coordinatively unsaturated structure, the [3Fe-4S] centers are predestined for chemical-catalytic activity, including iron sensor functions.

The natural inventory of Fe/S centers is not exhausted with the 3Fe clusters; investigations of biochemical material [22–24] and model studies suggest that a variety of other possible structures with higher numbers of iron atoms (six, eight) or with nonthiolate ligands can be found in the system $Fe^{II/III}/S^{-II}$/protein.

Among such more complex [xFe-yS] centers are the sulfide-bridged double-cubane "P-clusters" in nitrogenase (7.11) [23] (see Section 11.2), with their very negative redox potential of −470 mV and an ($S = 5/2$) or ($S = 7/2$) ground state in the oxidized form, and the (6Fe) "H clusters" in nickel-free hydrogenases, which catalyze the equilibrium $2\,H^+ + 2\,e^- \rightleftharpoons H_2$ (see Section 9.3).

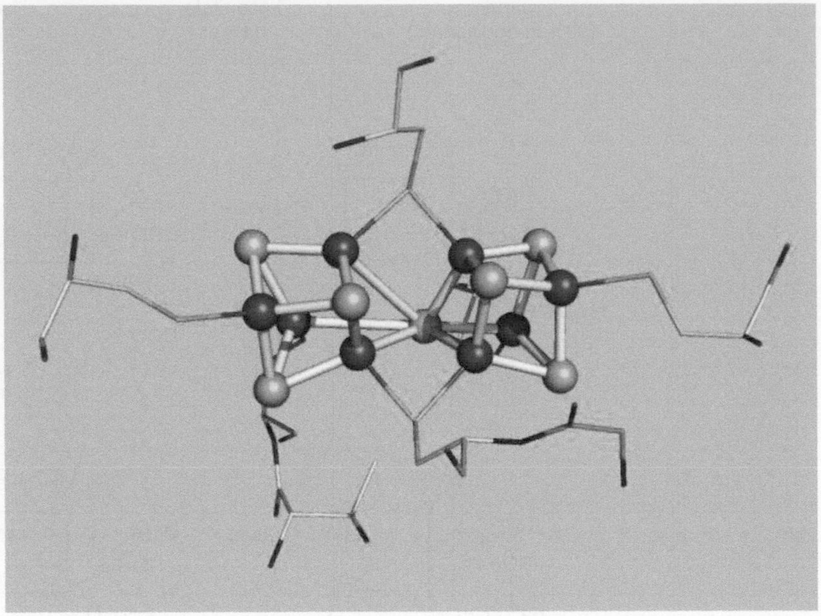

(7.11)

The [4Fe-4S] center of the DNA repair enzyme endonuclease III seems to have "only" a polarizing and structure-determining function, directed at recognition. The negatively charged cluster binds positively polarized amino acids, which in turn interact with the negatively charged phosphate backbone of the DNA [4].

7.5 Model Systems for Fe/S Proteins

Several model complexes for Fe/S centers in proteins can be prepared surprisingly easily via "spontaneous self-assembly" reactions [25]. For example, [4Fe-4S] cluster ions are obtained from four equivalents of thiol, hydrogen sulfide and iron(III) and eight equivalents of base under reducing conditions in polar aprotic solvents such as dimethylsulfoxide (DMSO) (7.12).

$$6\,RSH + 4\,NaHS + 4\,FeCl_3 + 10\,NaOR \longrightarrow Na_2[Fe_4S_4(RS)_4] + 10\,ROH + 12\,NaCl + RSSR \tag{7.12}$$

Whereas the stable [4Fe-4S] systems are typically formed with sterically unhindered thiols as model ligands for cysteine, conformational constraints imposed by preferentially chelate-forming dithiols such as *o*-xylene-α,α'-dithiol (Figure 7.1) lead to models of [2Fe-2S] dimers or, in the absence of sulfide, models of rubredoxin (Figures 7.1 and 7.4).

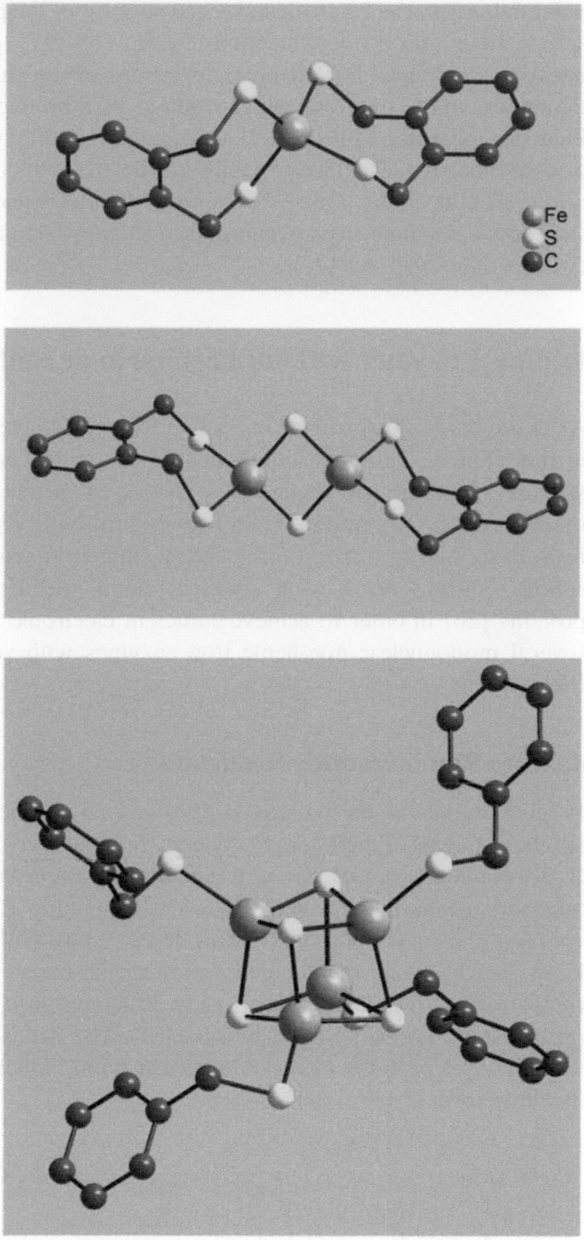

Figure 7.4
Molecular structures of model complex anions. Reprinted with permission from [25] © 1977, American Chemical Society.

Modeling of [3Fe-4S] centers and of unsymmetrically (3 + 1)-substituted clusters requires more effort, particularly the synthesis of specially designed polythiolate ligands.

As expected, most model systems show lower reaction rates for electron transfer than the natural Fe/S proteins because of the more pronounced geometrical changes during the process; in addition, the models exhibit significantly lower and thus unphysiological redox potentials. This can be attributed to the absence of the protein environment, where amine–sulfide hydrogen bond interactions N—H··· ⁻S [17], electrostatic effects in the peptide backbone and small but effective distortions of the cluster may be important. In view of the rather simple structure and function of Fe/S proteins, some "artificial proteins" of this kind (i.e. peptides with short chain lengths) can be synthesized based on known amino acid sequences (7.7). The possibility of heteroatom substitution in the clusters has been observed both in the course of chemical synthesis and in important biological materials such as Ni-containing hydrogenases (see Section 9.3) and Mo- or V-containing nitrogenases (see Sections 11.2 and 11.3).

7.6 Iron-containing Enzymes without Porphyrin or Sulfide Ligands

After the heme-iron proteins in Section 5.2 and Chapter 6 and the Fe/S proteins described in this chapter, the most important iron-containing enzymes, which apparently manage without such "special" ligands, are discussed in this section. The O_2-transporting and, therefore, non-enzymatic protein hemerythrin, with its dinuclear iron center, has already been described in Section 5.3; several other iron proteins containing neither heme nor sulfide similarly resort to an (indirect) metal–metal interaction in polynuclear metalloproteins [26] in order to achieve sufficient electronic flexibility. However, there are also several mononuclear non-heme iron enzymes with very important biological functions (Section 7.6.4).

7.6.1 Iron-containing Ribonucleotide Reductase

In most organisms, particularly higher ones, an iron-containing form of RR is responsible for the catalysis of the biologically essential reaction (7.13) [27] even though a coenzyme B_{12}-dependent form of RR is found in *Rhizobium* bacteria and lactobacilli. Anaerobic microorganisms feature an Fe/S cluster-containing RR, and still another dimanganese-containing form of bacterial RR was described by Willing, Follmann and Auling [28]. All known RRs contain metal centers and organic radicals; they catalyze the deoxygenation (reduction) of the ribose ring to yield 2′-deoxyribose in nucleotides and thus promote a crucial step in *de novo* DNA biosynthesis. The required electrons are made available by dithiols (e.g. peptides of the thioredoxin type), which can themselves be oxidized to a disulfide form (7.13).

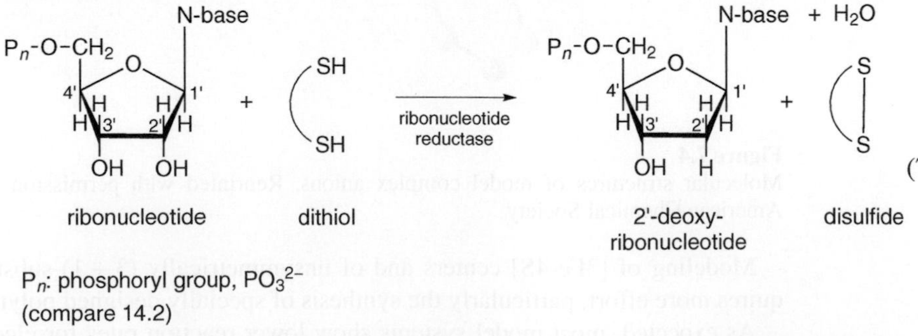

(7.13)

The iron-containing RRs are quite large proteins with molecular masses exceeding 200 kDa; this complexity is necessitated *inter alia* by the requirements for an exact control of the reaction (feedback, allosteric regulation). The most thoroughly studied enzyme of this type is the RR isolated from *E. coli*, which consists of two different dimeric subunits ($\alpha_2\beta_2$) with molecular masses of about 170 (α protein) and 87 kDa (β protein). According to x-ray crystal structure determinations [29], the active center is located at the interface between the two different subunits. Whereas the larger protein contains thiol groups as assumed direct electron donors, the smaller protein features one tyrosyl radical and one diiron site per subunit. The iron-containing RR was the first enzyme for which a stable free radical in the protein was established as an essential component; previously, it had only been assumed that the function of the coenzyme B_{12}-dependent RR was based on radical reactivity (Section 3.3.1). As with the related hemerythrin protein (Section 5.3), numerous spectroscopic and magnetic measurements had already provided information concerning the approximate structure of the diiron center and its magnetic interaction with the tyrosyl radical before details became available from a crystal structure analysis of the smaller protein subunit [29].

In the oxidized state, the trivalent high-spin iron centers in each subunit are bridged by one μ-oxo and one μ-η^1:η^1-glutamate ligand and are otherwise coordinated rather unsymmetrically (7.14): one metal center binds to two η^1-glutamate groups, one histidine ligand (N_δ coordination) and a water molecule, while the other coordinates to one H_2O molecule, one histidine ligand (N_δ) and a chelating (η^2-)aspartate in a less regular octahedral configuration. The μ-oxo/μ-carboxylato-bridged structure of the dimetal center, as also found in oxyhemerythrin, is quite characteristic and is assumed for some dimanganese-containing enzymes as well. Studies on synthetic models [30] suggest that this arrangement has a high tendency of formation ("self-assembly" process) and that it favors an efficient metal–metal interaction at a distance of about 0.33 nm (RR) and thus a certain degree of electronic flexibility.

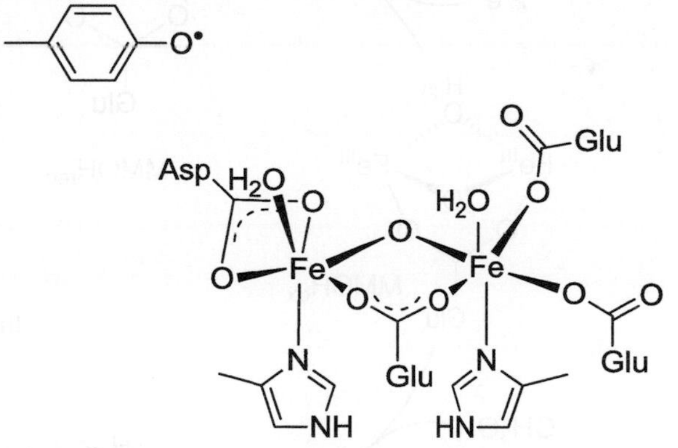

(7.14)

The presence of labile water ligands at the iron centers facilitates interactions with the tyrosyl radical situated at a distance of 0.53 nm from the metal. A magnetic interaction between antiferromagnetically coupled high-spin Fe^{III} centers and the reactive radical situated approximately along the Fe–Fe axis has been established.

The reduced Fe^{II}/Fe^{II} form of the iron dimer presumably features two μ-η^1:η^1-glutamate ligands without an oxo bridge. This state may interact with O_2 [31] to create the neutral tyrosyl radical. Additional conceivable functions of the diiron moiety are to stabilize, to protect or even to activate the tyrosyl radical in a controlled way. However, according to the results from the structure analysis [29], the situation of the tyrosyl radical at least 1 nm away from the protein surface prohibits a direct interaction (e.g. in the form of a hydrogen abstraction) between the tyrosyl radical and the substrate. A possible role of the tyrosyl radical and the iron dimer is to participate in electron-transfer processes involving a tryptophan situated at the protein surface, which also has the potential to form a reactive radical [32]. Dioxygen species such as O_2 and O_2^{2-}, further activated by the iron centers, can contribute to the oxidative formation of the tyrosyl radical. Starting from the tryptophyl radical, the deoxygenation of ribose can take place via hydrogen abstraction, loss of OH^- from a radical intermediate and reduction of the resulting radical cation species through enzymatic sulfhydryl groups; their reduction, in turn, proceeds through the external dithiol (7.13). The deoxygenation by RR enzymes may thus be viewed as the *reverse reaction of the O-insertion*, which is catalyzed by P-450 heme enzymes or by the non-heme methane monooxygenases (MMOs) described in Section 7.6.2.

7.6.2 Soluble Methane Monooxygenase

This enzyme, a multiprotein complex with a molecular mass of about 300 kDa, is employed by methano*trophic* microorganisms which use CH_4 as source of carbon and energy, the latter provided through the first step (7.15) of methane oxidation. (Metalloenzymes of methano*genic* (i.e. CH_4-producing) organisms are discussed elsewhere; compare Sections 3.3.1 and 9.5 as well as Figure 1.2.)

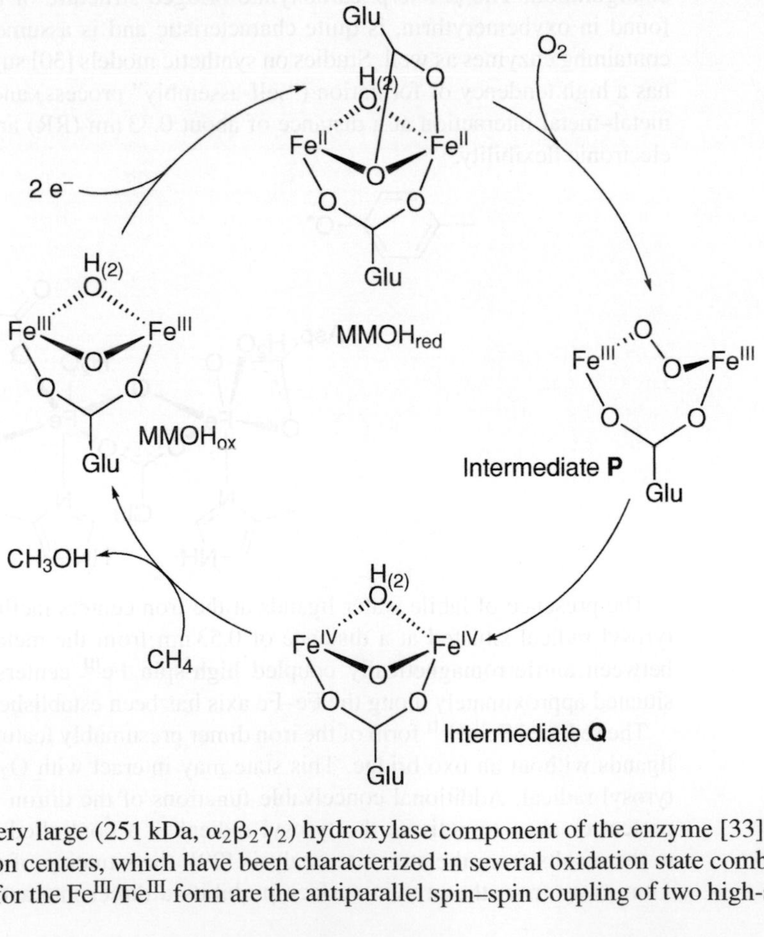

(7.15)

The very large (251 kDa, $\alpha_2\beta_2\gamma_2$) hydroxylase component of the enzyme [33] contains two diiron centers, which have been characterized in several oxidation state combinations. Typical for the Fe^{III}/Fe^{III} form are the antiparallel spin–spin coupling of two high-spin Fe^{III}

centers and the absence of an intense charge-transfer absorption in the visible spectrum. In accordance with spectroscopic data, a μ-hydroxo bridge has been found between the two coordinatively not fully saturated metal centers, in addition to a bidentate glutamate bridge (μ-η^1:η^1) and monodentate nonbridging glutamate and histidine ligands [33]. Only the high-spin Fe^{II}/Fe^{II} form is active with respect to dioxygen activation; presumably, the diiron site is the catalytic center. Other metal ions, organic cofactors and stable radicals have not been found.

The mechanism of reaction (7.15) apparently involves the formation of an oxygenated dimetal center with oxoferryl(IV) groups (see Sections 6.3 and 6.4) [34], which then effect monooxygen insertion into the C–H bond of CH_4; other hydrocarbons are also oxygenated with a certain stereospecificity. The reactivity and assumed catalytic mechanism found for MMO thus shows many parallels to the P-450 system (Section 6.2).

7.6.3 Purple Acid Phosphatases (Fe/Fe and Fe/Zn)

Polyphosphate- and phosphate ester-cleaving enzymes typically require nonredoxactive divalent metal ions such as Mg^{2+} (Section 14.1) or Zn^{2+}, as for example in alkaline phosphatase (Section 12.3) [35,36]. However, there are also iron-containing phosphatases, which are distinguished by their strong color and their activity optimum in the acidic region; such enzymes may be obtained from a variety of plants, microorganisms and animals (uteroferrin, bovine spleen phosphatase), have molecular masses of about 40 kDa and contain dimetal centers. Detailed studies of the diiron species have shown easily distinguishable colors and absorption spectra for the inactive purple Fe^{III}/Fe^{III} state and for the enzymatically active pink Fe^{II}/Fe^{III} mixed-valent form, the latter exhibiting weak antiferromagnetic spin–spin coupling.

The iron centers are bridged by a μ-hydroxo ligand and by a μ-η^1:η^1-carboxylate. A tyrosinate ligand binds to the iron(III) center in either of the two states, the coordination of this π electron-rich ligand to an oxidized metal center being responsible for intense low-energy ligand-to-metal charge transfer (LMCT) absorptions in the visible spectrum. The metals are probably coordinated by one histidine ligand each. A plausible mechanism consists in the attack of hydroxide activated by binding to Fe^{II} (compare (12.3) [37] on a Fe^{III}-coordinated phosphate derivative; variable metal–metal distances would favor such a differentiated reaction mechanism.

The native form of a phosphatase from kidney beans contains zinc(II) and iron(III) in a similarly bridged arrangement to that found in Fe^{II}/Fe^{III} phosphatases [38]. This again implies a sharing of functions between divalent and trivalent metal centers, through water activation by the (+II) ion and phosphate binding by the trivalent ion.

7.6.4 Mononuclear Non-heme Iron Enzymes

This group of enzymes with only one, apparently simple iron center includes mainly monooxygenases, dioxygenases and peroxidases, which are important in the metabolism of fatty acids and aromatic compounds and in the synthesis of amino acids, neurotransmitters, β-lactams and leukotrienes.

Iron-containing and tetrahydropterin-dependent monooxygenases (hydroxylases) catalyze the transformation of L-phenylalanine to L-tyrosine, the tetrahydropterin (compare (3.12) or (11.8)) being oxidized to its dihydro form (see (11.8)) and dioxygen becoming reduced to one molecule of water (*mono*oxygenation [39]). Obviously, both the high-spin

Fe^{III} and Fe^{II} centers and the pterin redox system (11.8) cooperate in oxygen activation and electron transfer.

Non-heme iron-containing dioxygenases are quite common, participating *inter alia* in the oxidative cleavage of aromatic compounds (compare 7.16). In their oxidized state, catechol 1,2-dioxygenases and related enzymes contain high-spin iron(III) with tyrosinate ligands [26], which give rise to an LMCT transition in the visible spectrum. According to mechanistic hypotheses, the binding of π electron-rich catecholate to the π electron-deficient and Lewis acidic high-spin Fe^{III} center in "intradiol-cleaving" enzymes results in a ligand-to-metal electron transfer and corresponding weakening of the intradiol (O)C-C(O)-bond of the thus formed o-semiquinone radical. $^3O_2$ activation may then proceed in a spin-allowed fashion via high-spin Fe^{II} (compare Section 5.2) before the final product is formed through peroxidic intermediates (7.16). There are also various extradiol-cleaving non-heme iron enzymes.

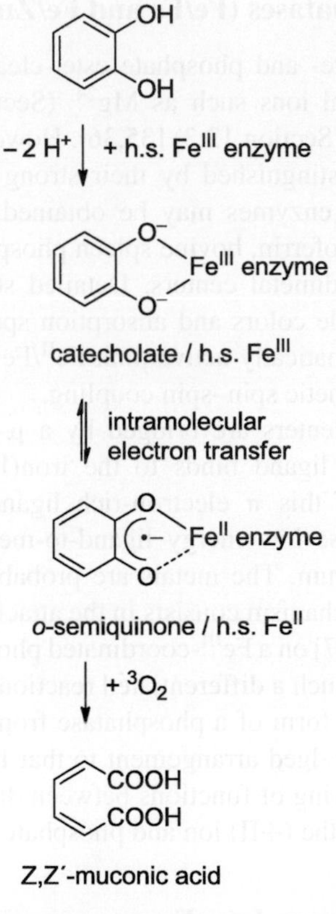

(7.16)

Lipoxygenases are among those iron-containing enzymes which, if only indirectly, catalyze reactions of dioxygen with organic substrates. In the metabolism of 1,4-diene-containing fatty acids, they stereospecifically catalyze autoxidation or "peroxidation" reactions, such as the O_2 insertion into C–H bonds as a first step in leukotriene synthesis.

Crystal structure analyses of lipogenases [40] exhibit coordinatively unsaturated high-spin iron ligated by three histidine residues and the (isoleucine) carboxy terminus. The reaction mechanism involves oxidative deprotonation of the 1,4-diene part of the coordinated substrate by Fe^{III} and $^3O_2$ binding and activation by the resulting Fe^{2+} ion [40].

No oxygenation occurs during the ring-forming reaction (7.17) to give isopenicillin N, which is catalyzed by another high-spin iron(II) enzyme, isopenicillin-N synthase; *both* oxygen atoms of O_2 are reduced to water. For the active state of the metal center, three histidine residues and a water molecule have been postulated as ligands, in addition to the coordinated reactants [41].

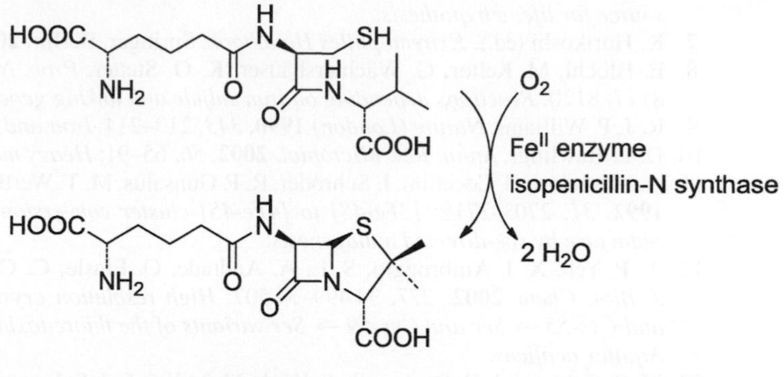

isopenicillin N

(7.17)

The unusual low-spin iron(III) center in bacterial nitrile hydratase apparently does not have a biological redox activity. This enzyme is interesting from a biotechnological point of view because it also catalyzes the formation of acrylamide from acrylonitrile in synthetic fiber production (7.18). The enzyme can use both Fe^{III} and Co^{III} [42], which are bonded to partially oxygenated cysteinato sulfur atoms (7.19).

$$R{-}CN + H_2O \xrightarrow{\text{nitrile hydratase}} R{-}\overset{\displaystyle O}{\overset{\|}{C}}{-}NH_2 \qquad (7.18)$$

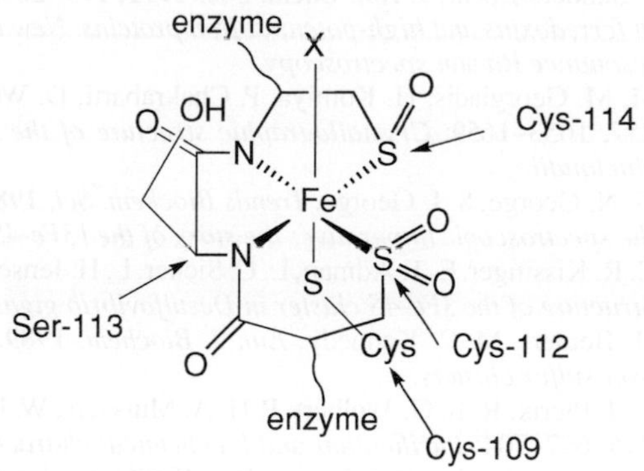

(7.19)

Iron is also redox-inactive in a particular non-nickel-containing form of hydrogenase [43].

References

1. R. Lill, *Nature (London)* **2009**, *460*, 831–838: *Function and biogenesis of iron-sulphur proteins.*
2. R. Grabowski, A. E. M. Hofmeister, W. Buckel, *Trends Biochem. Sci.* **1993**, *18*, 297–300: *Bacterial l-serine dehydratases: a new family of enzymes containing iron-sulfur clusters.*
3. T. V. O'Halloran, *Science* **1993**, *261*, 715–725: *Transition metals in control of gene expression.*
4. C.-F. Kuo, D. E. McRee, C. L. Fisher, S. F. O'Handley, R. P. Cunningham, J. A. Tainer, *Science* **1992**, *258*, 434–440: *Atomic structure of the DNA repair [4Fe-4S] enzyme endonuclease III.*

5. C. C. Correll, C. J. Batie, D. P. Ballou, M. L. Ludwig, *Science* **1992**, *258*, 1604–1610: *Phthalate dioxygenase reductase: a modular structure for electron transfer from pyridine nucleotides to [2Fe-2S].*
6. G. Wächtershäuser, *System. Appl. Microbiol.* **1988**, *10*, 207–210: *Pyrite formation, the first energy source for life: a hypothesis.*
7. K. Horikoshi (ed.), *Extremophiles Handbook*, Springer, Berlin, **2010**.
8. E. Blöchl, M. Keller, G. Wächtershäuser, K. O. Stetter, *Proc. Natl. Acad. Sci. USA* **1992**, *89*, 8117–8120: *Reactions depending on iron sulfide and linking geochemistry with biochemistry.*
9. R. J. P. Williams, *Nature (London)* **1990**, *343*, 213–214: *Iron and the origin of life.*
10. D. E. Rawlings, *Annu. Rev. Microbiol.* **2002**, *56*, 65–91: *Heavy metal mining using microbes.*
11. A. Manodori, G. Cecchini, I. Schröder, R. P. Gunsalus, M. T. Werth, M. K. Johnson, *Biochemistry* **1992**, *31*, 2703–2712: *[3Fe-4S] to [4Fe-4S] cluster conversion in Escherichia coli fumarate reductase by site-directed mutagenesis.*
12. A. P. Yeh, X. I. Ambroggio, S. L. A. Andrade, O. Einsle, C. Chatelet, J. Meyer, D. C. Rees, *J. Biol. Chem.* **2002**, *277*, 34 499–34 507: *High resolution crystal structures of the wild type and Cys-55 → Ser and Cys-59 → Ser variants of the thioredoxin-like [2Fe-2S] ferredoxin from Aquifex aeolicus.*
13. M. S. Gebhard, J. C. Deaton, S. A. Koch, M. Millar, E. I. Solomon, *J. Am. Chem. Soc.* **1990**, *112*, 2217–2231: *Single-crystal spectral studies of* $Fe(SR)_4$*- [R = 2,3,5,6-*$(Me)_4C_6H$*]: the electronic structure of the ferric tetrathiolate active site.*
14. T. A. Link, *FEBS Lett.* **1997**, *412*, 257–264: *The role of the "Rieske" iron sulfur protein in the hydroquinone oxidation (*Q_p*) site of the cytochrome* bc_1 *complex: the "proton-gated affinity change" mechanism.*
15. C. W. Carter Jr., J. Kraut, S. T. Freer, R. A. Alden, L. C. Sieker, E. Adman, L. H. Jensen, *Proc. Nat. Acad. Sci USA* **1972**, *69*, 3526–3529: *A comparison of* Fe_4S_4 *clusters in high-potential iron protein and in ferredoxin.*
16. L. Noodleman, J. G. Norman, J. H. Osborne, A. Aizman, D. A. Case, *J. Am. Chem. Soc.* **1985**, *107*, 3418–3426: *Models for ferredoxins: electronic structures of iron-sulfur clusters with one, two, and four iron atoms.*
17. G. Backes, Y. Mino, T. M. Loehr, T. E. Meyer, M. A. Cusanovich, W. V. Sweeney, E. T. Adman, J. Sanders-Loehr, *J. Am. Chem. Soc.* **1991**, *113*, 2055–2064: *The environment of* Fe_4S_4 *clusters in ferredoxins and high-potential iron proteins. New information from x-ray crystallography and resonance Raman spectroscopy.*
18. M. M. Georgiadis, H. Komiya, P. Chakrabarti, D. Woo, J. J. Kornuc, D. C. Rees, *Science* **1992**, *257*, 1653–1659: *Crystallographic structure of the nitrogenase iron protein from Azotobacter vinelandii.*
19. G. N. George, S. J. George, *Trends Biochem. Sci.* **1988**, *13*, 369–370: *X-ray crystallography and the spectroscopic imperative: the story of the [3Fe-4S] clusters.*
20. C. R. Kissinger, E. T. Adman, L. C. Sieker, L. H. Jensen, *J. Am. Chem. Soc.* **1988**, *110*, 8721–8723: *Structure of the 3Fe-4S cluster in Desulfovibrio gigas ferredoxin II.*
21. H. Beinert, M. C. Kennedy, *Eur. J. Biochem.* **1989**, *186*, 5–15: *Engineering of protein bound iron-sulfur clusters.*
22. A. J. Pierik, R. B. G. Wolbert, P. H. A. Mutsaers, W. R. Hagen, C. Veeger, *Eur. J. Biochem.* **1992**, *206*, 697–704: *Purification and biochemical characterization of a putative [6Fe-6S] prismane-cluster-containing protein from Desulfovibrio vulgaris (Hildenborough).*
23. J. Kim, D. C. Rees, *Science* **1992**, *257*, 1677–1682: *Structural models of the metal centers in the nitrogenase molybdenum-iron protein.*
24. J. Fritsch, P. Scheerer, S. Frielingsdorf, S. Kroschinsky, B. Friedrich, O. Lenz, C. M. T. Spahn, *Nature (London)* **2011**, *479*, 249–252: *The crystal structure of an oxygen-tolerant hydrogenase uncovers a novel iron-sulphur centre.*
25. R. H. Holm, *Acc. Chem. Res.* **1977**, *10*, 427–434: *Synthetic approaches to the active sites of iron-sulfur proteins.*
26. E. I. Solomon, T. C. Brunold, M. I. Davis, J. N. Kemsley, S.-K. Lee, N. Lehnert, F. Neese, A. J. Skulan, J. Zhou, *Chem. Rev.* **2000**, *100*, 235–349: *Geometric and electronic structure/function correlations in non-heme iron enzymes.*
27. P. Reichard, *Science* **1993**, *260*, 1773–1777: *From RNA to DNA, why so many ribonucleotide reductases?*

28. A. Willing, H. Follmann, G. Auling, *Eur. J. Biochem.* **1988**, *170*, 603–611: *Ribonucleotide reductase of Brevibacterium ammoniagenes is a manganese enzyme.*
29. P. Nordlund, B.-M. Sjöberg, H. Eklund, *Nature (London)* **1990**, *345*, 593–598: *Three-dimensional structure of the free radical protein of ribonucleotide reductase.*
30. D. M. Kurtz, *Chem. Rev.* **1990**, *90*, 585–606: *Oxo- and hydroxo-bridged diiron complexes: a chemical perspective on a biological unit.*
31. J. M. Bollinger Jr., D. E. Edmondson, B. H. Huynh, J. Filley, J. R. Norton, J. Stubbe, *Science* **1991**, *253*, 292–298: *Mechanism of assembly of the tyrosyl radical-dinuclear iron cluster cofactor of ribonucleotide reductase.*
32. R. C. Prince, G. N. George, *Trends Biochem. Sci.* **1990**, *15*, 170–172: *Tryptophan radicals.*
33. A. C. Rosenzweig, S. J. Lippard, *Acc. Chem. Res.* **1994**, *27*, 229–236: *Determining the structure of a hydroxylase enzyme that catalyzes the conversion of methane to methanol in methanotropic bacteria.*
34. S.-K. Lee, B. G. Fox, W. A. Froland, J. D. Lipscomb, E. Münck, *J. Am. Chem. Soc.* **1993**, *115*, 6450–6451: *A transient intermediate of the methane monooxygenase catalytic cycle containing an* $Fe^{IV}Fe^{IV}$ *cluster.*
35. N. Sträter, T. Klabunde, P. Tucker, H. Witzel, B. Krebs, *Science* **1995**, *268*, 1489–1492: *Crystal structure of a purple acid phosphatase containing a dinuclear Fe(III)-Zn(II) active site.*
36. J. B. Vincent, M. W. Crowder, B. A. Averill, *Trends Biochem. Sci.* **1992**, *17*, 105–110: *Hydrolysis of phosphate monoesters: a biological problem with multiple chemical solutions.*
37. E. G. Mueller, M. W. Crowder, B. A. Averill, J. R. Knowles, *J. Am. Chem. Soc.* **1993**, *115*, 2974–2975: *Purple acid phosphatase: a diiron enzyme that catalyzes a direct phopho group transfer to water.*
38. J. L. Beck, J. de Jersey, B. Zerner, M. P. Hendrich, P. G. Debrunner, *J. Am. Chem. Soc.* **1988**, *110*, 3317–3318: *Properties of the Fe(II)-Fe(III) derivative of red kidney bean purple phosphatase. Evidence for a binuclear Zn-Fe center in the native enzyme.*
39. T. A. Dix, S. J. Benkovic, *Acc. Chem. Res.* **1988**, *21*, 101–107: *Mechanism of oxygen activation by pteridine-dependent monooxygenases.*
40. N. C. Gilbert, S. G. Bartlett, M. T. Waight, D. B. Neau, W. E. Boeglin, A. R. Barash, M. E. Newcomer, *Science* **2011**, *331*, 217–219: *The structure of human 5-lipoxygenase.*
41. L. J. Ming, L. Que, A. Kriauciunas, C. A. Frolik, V. J. Chen, *Inorg. Chem.* **1990**, *29*, 1111–1112: *Coordination chemistry of the metal binding site of isopenicillin N synthase.*
42. P. K. Mascharak, *Coord. Chem. Rev.* **2002**, *225*, 201–214: *Structural and functional models of nitrile hydratase.*
43. S. Dey, P. K. Das, A. Dey, *Coord. Chem. Rev.* **2013**, *257*, 42–63: *Mononuclear iron hydrogenase.*

8 Uptake, Transport and Storage of an Essential Element, as Exemplified by Iron

Descriptions of the structures and physiological functions of metal centers in enzymes and proteins have always played an important role in bioinorganic chemistry. Complex mechanisms operate between the abundance of an element in an organism (compare Section 2.1) and its specific function (e.g. in an enzyme). These necessarily selective and controlled processes for the uptake, transport, storage and directed transfer of an element (e.g. to the designated apoprotein) must occur under temporally and spatially well-defined conditions; the term "trafficking" is thus often used. The intricate aspect of the space and time dependence of "bioinorganic processes" in actual organisms has been repeatedly pointed out [1].

Metallome

Just as the protein content of a certain cell type is called the "proteome", so the "metallome" is the sum of individual metal species in a cell [2]. This includes free as well as complexed metal ions and their ensembles with biomolecules. Metallomics comprises all metal-assisted biological functions and can be divided into qualitative (speciation), quantitative and comparative metallomics, the latter of which involves changes in the metallome due to changes in the outer conditions. Thus the following information is obtained in a metallome:

- how a metal is distributed inside a cell of a certain type;
- its coordination environment;
- its concentration [3].

While the problems addressed by metallomics include:

- structural–functional analysis of metalloproteins and models thereof;
- regulation of the uptake, accumulation and metabolism of metals in biological systems;
- the state of the organism (health) with regard to the availability of required elements [4].

Bioinorganic Chemistry: Inorganic Elements in the Chemistry of Life – An Introduction and Guide, Second Edition.
Written and Translated by Wolfgang Kaim, Brigitte Schwederski and Axel Klein.

The compounds responsible for the transport and storage of iron, the physiologically most abundant and most versatile transition metal, have been rather thoroughly studied. Less factual information is available for the other elements, but comparable mechanisms to those of iron transport and storage operate in them as well.

8.1 The Problem of Iron Mobilization: Oxidation States, Solubility and Medical Relevance

Iron is an essential trace element for almost all organisms, with the exception of lactobacilli, participating in such diverse biological processes as photosynthesis and respiration, N_2 fixation, methanogenesis, H_2 metabolism, oxygen transport and DNA biosynthesis. Its distribution in an adult human has been summarized in Table 5.1. Complex regulatory mechanisms serve in the uptake, transport and storage of iron [5,6]. These processes may be divided into several single steps:

- active or passive resorption during food ingestion, involving for example dissolution, redox reactions or complexation;
- selective transport of the iron ions through membranes into cells;
- processes within the cells, such as incorporation into a protein;
- elimination from the metabolism through either excretion or temporary storage.

An excess of free iron, in particular high-spin Fe^{2+}, is dangerous for any organism as radicals can be generated in the presence of dioxygen or peroxide (compare (4.6) or (5.2)) according to equations (8.1) and (8.2) [7].

$$Fe^{II} + {}^{3}O_2 \rightarrow Fe^{III} + O_2^{\bullet -} \quad (8.1)$$

$$Fe^{II} + H_2O_2 \rightarrow Fe^{III}(OH^-) + OH^{\bullet} \quad (8.2)$$

Potentially pathogenic microorganisms need a continuous supply of iron as a growth-determining factor for their reproduction. Accordingly, the availability of iron to invading bacteria in a multicellular organism plays an important role for infectious diseases such as tuberculosis, leprosy and cholera, in which a decrease in the iron content of human blood plasma is symptomatic [8]. Since microorganisms cannot activate tightly bound iron in thc blood serum, they have to utilize "free" iron species; effectively iron-scavenging and membrane-diffusive complexing ligands can thus exhibit antibiotic properties. The iron of heme-type porphyrin complexes is very tightly bound and can be liberated only inside of cells through enzymatic action.

The metabolism of iron in the human organism has general medical relevance. Of the recommended 10–20 mg iron in the daily food supply (see Table 2.3), on average only about 10–15% is resorbed. Both iron deficiency as caused by malnutrition (→ anemia [9]) and iron overload due to gene defects (thalassemia), as well as excessive iron intake as a suspected long-term risk factor for coronary heart diseases, can cause severe pathological symptoms. Blood infusions result in an acute iron enrichment in the body, since a human being can only excrete about 1–3 mg iron per day. Complexing agents such as polychelating deferrioxamine (see (2.1d), (8.10)) can bind iron even if it is coordinated by the transport protein transferrin (Section 8.4.1) or by storage proteins; however, they cannot release iron from heme. The resulting ferrioxamine complex can be excreted from the body via the kidneys. Complex formation, or more exactly, substitution (recomplexation), is rather slow, making high and continually supplied doses of the drug necessary for a successful therapy.

Careful studies of naturally occurring iron-complexing ligands, the "siderophores", and of their synthetic analoga, are therefore of great interest in medicine, where such molecules are used not only in primary or secondary hemochromatosis (iron poisoning) but also as antibiotics or drug delivery agents [10,11]. The aim is to find or develop chelating systems that can be applied orally and which are therapeutically active in small doses without being degraded in the gastrointestinal tract, the circulatory system or the liver.

Although iron is the most abundant transition metal present in the biosphere, it is not readily bioavailable. In an aqueous aerobic environment, Fe^{3+} is the predominant iron species, but in contrast to substitutionally labile Fe^{2+}, Fe^{3+} is insoluble at pH 7 if strong complexing agents are absent (5.1). According to the solubility product of 10^{-38} M^4 for $Fe(OH)_3$, the theoretical concentration of Fe^{3+} would only be about 10^{-17} M; however, the formation of aqua/hydroxo complexes contributes to a significantly higher if still very small solubility. Thus, even those microorganisms which have survived the development of an oxidizing atmosphere will make use of special low-molecular-weight compounds in order to accumulate iron(III) from their environment; if necessary even from solid phases, such as iron oxide-containing particles [12]. These special soluble chelating ligands have an extremely high affinity for iron(III) and are referred to as "siderophores" ("iron carriers"). They provide iron for further transport, storage and incorporation into proteins and act via reduction/protonation processes, at membrane receptors for example [13].

Plants also need iron for growth, particularly for chlorophyll biosynthesis. Iron-efficient plants are capable of extracting iron in sufficient amounts even from lime-rich soil with high pH values and thus very low concentrations of soluble iron. Iron-complexing phytosiderophores have been isolated from the roots of such plants. Both, the relatively simple iron transport systems of microorganisms and the more complex mechanisms of the iron metabolism in higher organisms will be discussed in this chapter.

8.2 Siderophores: Iron Uptake by Microorganisms

Many different siderophores are known, isolated from bacteria, yeast and fungi. All are naturally occurring chelate ligands with low molecular masses of between 500 and 1000 Da and a high affinity for iron(III). Their biosynthesis is regulated by the supply of iron and a DNA-binding regulatory protein, FUR (**Fe** **u**ptake **r**egulation), which is activated by Fe^{2+} and plays an essential role in the feedback mechanism [5]. All siderophores form very stable chelate complexes with approximately octahedrally coordinated high-spin Fe^{3+} [13,14]. A quantitative measure for the "stability" of the complex formed between siderophore and iron(III) is the complex formation constant, K_f:

$$Fe^{3+} + Sid^{n-} \rightleftharpoons FeSid^{(3-n)} \qquad (8.3)$$

$$K_f = \frac{[FeSid^{(3-n)}]}{[Fe^{3+}][Sid^{n-}]} \qquad (8.4)$$

where [ ]: molar concentrations; Fe^{3+}: hydrated iron(III); and Sid^{n-}: anionic siderophore ligand.

Although generally very large, the constant K_f varies over a wide range; from 10^{23} M^{-1} for the aerobactin complex to about 10^{49} M^{-1} for the Fe^{3+}/enterobactin complex (Table 8.1), which is sufficient to solubilize Fe^{3+} from its highly insoluble oxides or hydroxides. The constants for the corresponding Fe^{2+} complexes are much smaller, due to the smaller charge and larger ionic radius of that ion (see Table 2.6); thus, a release mechanism for iron can be effected by reduction coupled to proton transfer. According to the Nernst equation for the concentration dependence of the redox potential, Scheme (8.5) shows the correlation between the stability constant, K_f, the redox potential of the Fe^{2+}/Fe^{3+} transition and the acid dissociation constant, K_a, of the siderophore ligand. Keeping K_a constant, an increasing dissociative stability of the Fe^{3+} complex thus correlates directly with a more negative redox potential (Table 8.1).

$$(Fe^{III}Sid) + 3\,H^+ + e^- \rightleftharpoons Fe^{2+} + H_3Sid$$

$$E = E_0 + 0.059\,V \cdot \log \frac{[(Fe^{III}Sid)][H^+]^3}{[Fe^{2+}][H_3Sid]}$$

$$= E_0 + 0.059\,V \cdot \log \frac{[Fe^{3+}]}{[Fe^{2+}]} \cdot K_a \cdot K_f \qquad (8.5)$$

$$\text{with} \quad 3\,H^+ + Sid^{3-} \overset{K_a}{\rightleftharpoons} H_3Sid$$

$$\text{and} \quad K_a = \frac{[H_3Sid]}{[H^+]^3[Sid^{3-}]}$$

The majority of effective siderophores can be divided into three groups: α-hydroxycarboxylates, hydroxamates and catecholates (8.6). In all cases, the ligands are able to form unstrained and unsaturated five-membered-ring chelate systems with negatively charged oxygen atoms as coordination centers. Each functional group is bidentate. Three of these ligands make up an ideal hexacoordinate Fe^{3+}-binding site. This situation results in a high stability of corresponding complexes with the "hard", highly charged Lewis acid Fe^{3+}.

Table 8.1 Stability constants and redox potentials for iron complexes of naturally occurring siderophores (from [13–16]).

Siderophore ligand	$\log K_f$ (Fe^{3+} complex)	E_0 (mV) (at potential limiting high pH)
catecholate		
enterobactin	~49	−990
bacillibactin	47.6	
hydroxamate		
coprogen	29.35	−447
deferrioxamine B	30.6	−468
ferrichrome A	32.0	−440
rhodotorulic acid	62.2 (3 : 2 stoichiometry!)	−419
α-hydroxycarboxylate		
mugineic acid	32.5	−102
rhizoferrin	25.3	−82
mixed		
aerobactin (2 hydroxamate, 1 α-hydroxycarboxylate)	27.6	−336
mycobactin (2 hydroxamate, 1 oxazoline)	26.6	

Ferrichromes (8.7) and ferrioxamines (2.1), as well as complexes based on rhodotorulic acid (see (8.12)) and aerobactin (see (8.13)), are typical representatives of hydroxamate siderophore complexes, while rhizoferrin (8.8) is a hydroxycarboxylate siderophore.

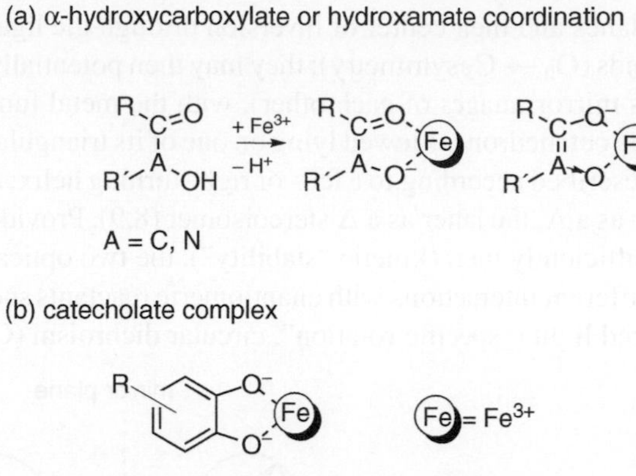

(8.6)

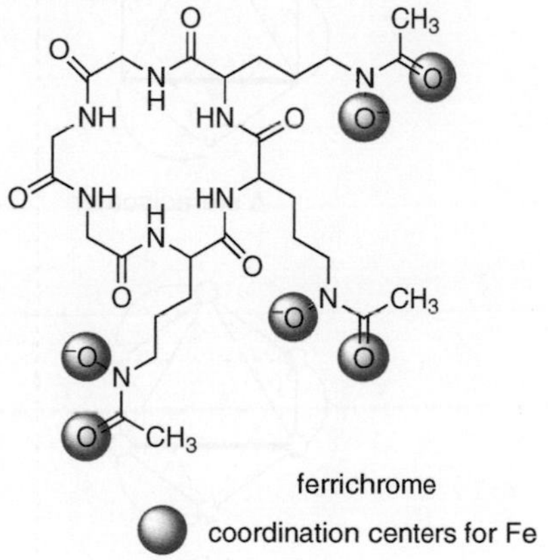

(8.7)

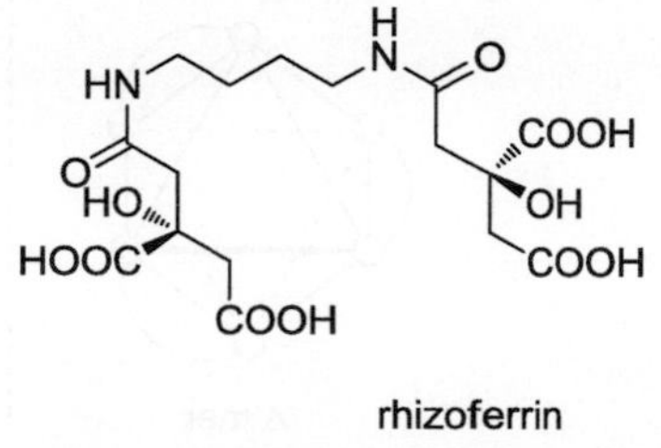

(8.8)

Ferrichromes contain cyclic hexapeptides based on glycine and N-hydroxy-l-ornithine, in which the three hydroxamine groups are derivatized with acetyl functions. They can be synthesized by fungi and also by bacteria. *In vivo*, ferrichrome adopts a Λ configuration around the iron atom, while the enantiomeric Δ form is much less effective with regard to bacterial iron transport.

A prominent member of the ferrioxamine ligand family is deferrioxamine B (2.1). This "linear" molecule features three non-equivalent hydroxamate groups as part of its chain. It is synthesized *in vivo* by *Streptomyces* species and is used in the prevention of iron poisoning, after blood transfusions for example (see Section 8.1) [11].

Optical Isomerism in Octahedral Complexes

Even strictly "octahedral" metal complexes with 90° bond angles may lose mirror planes and their center of inversion through the ligation of three identical chelate ligands ($O_h \rightarrow C_3$ symmetry); they may then potentially occur in enantiomeric forms (i.e. as mirror images of each other), with the metal functioning as the chirality center. If the octahedron is viewed lying on one of its triangular faces, this isomerism can also be described according to a left- or right-turning helix; the former arrangement is referred to as a Λ, the latter as a Δ stereoisomer (8.9). Provided that the metal-to-ligand bond is sufficiently inert (kinetic "stability"), the two optical isomers can be distinguished by different interactions with enantiomeric reactants such as biological receptors or polarized light ("specific rotation", circular dichroism (CD), in absorption spectroscopy).

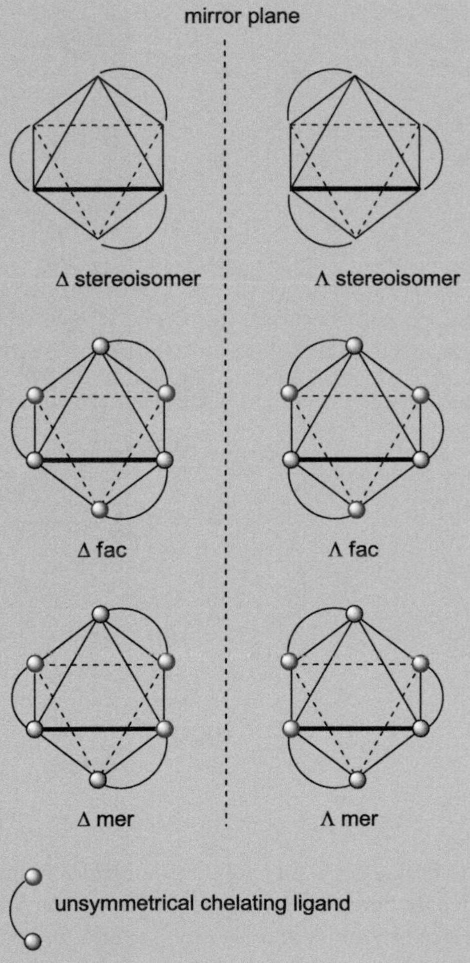

(8.9)

Another kind of (positional) isomerism, the *fac/mer* alternative, can occur in tris-chelated complexes if the coordinating sites in the chelate ring are not equivalent, as is the case with the hydroxamates. In the *fac*-arrangement (derived from "facial"), equivalent coordinating sites from different chelate rings are situated on one of the triangular faces of the octahedron, whereas in the *mer*-arrangement (derived from "meridional"), such equivalent sites are found lying on a "meridian" of the octahedron.

The combination of three different, individually unsymmetrical chelate arrangements, as it occurs in the siderophore complexes, either in a chain (2.1) or in macrocyclic fashion, leads to an increase in the number of possible isomers [17]. Example (8.10) illustrates all possible combinations for Δ-ferrioxamine; for the specific nomenclature, the reader is referred to [14]. As a result of spatial restrictions due to the chain length or to ring conformations, only a few isomers are actually realized; ideally, one form is highly preferred [18].

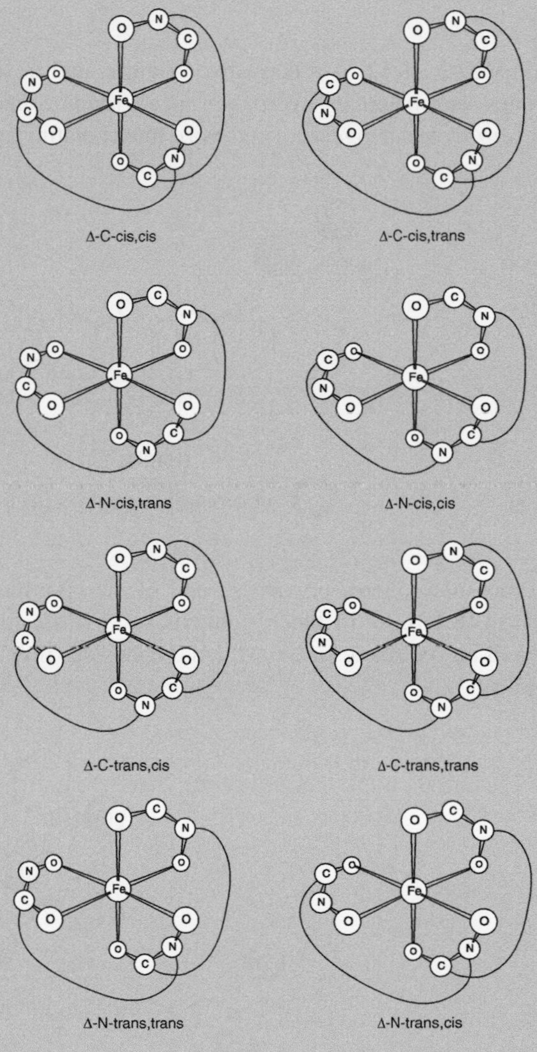

(8.10)

Rhodotorulic acid (8.11) is a dipeptide of N-hydroxy-l-ornithine and is synthesized by the yeast organism *Rhodotorula pilimanea*. In contrast to most other siderophores, which form 1 : 1 complexes with iron, rhodotorulic acid features a 3 : 2 stoichiometry. In addition to iron(III), chromium(III) may also be tightly bound.

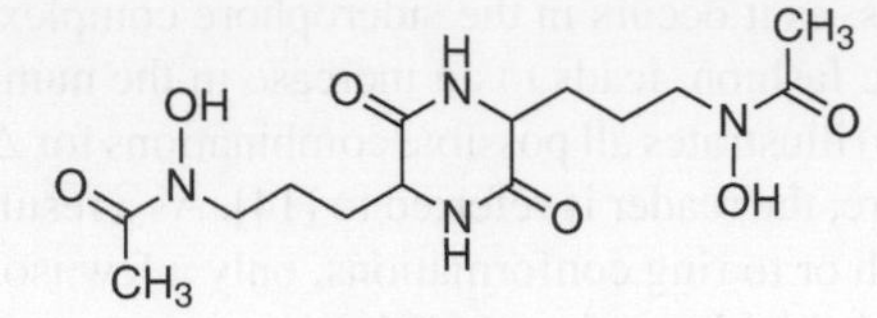

rhodotorulic acid

(8.11)

Aerobactin (8.12) is a derivative of citric acid in which the peripheral carboxylic acid moieties are replaced by hydroxamic acid groups. Some strains of *E. coli*, as well as bacteria like *Aerobacter aerogenes*, synthesize this chelating agent.

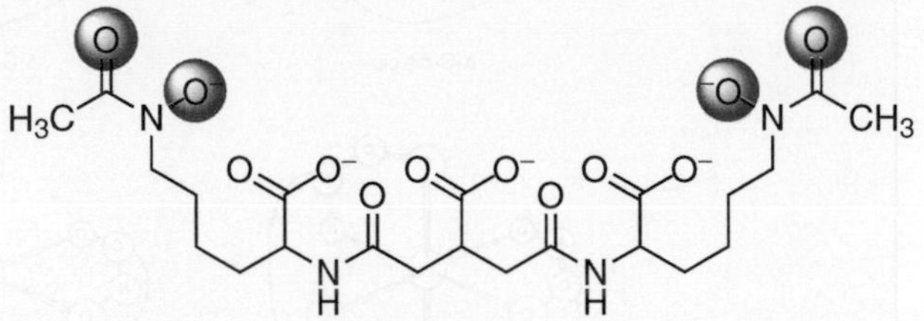

aerobactin

coordination centers for Fe

(8.12)

Some siderophores contain several of the coordination features mentioned here. Mycobactin (8.13), for instance, contains two hydroxamic acid groups, one phenolate function and an oxazoline group, which donates an (imine) nitrogen atom for coordination to the iron.

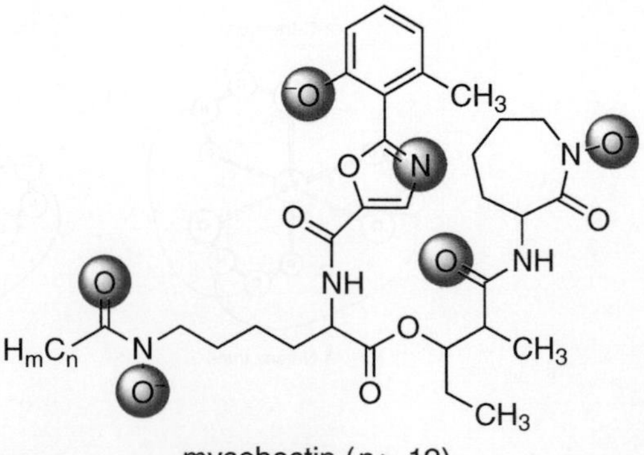

mycobactin ($n > 12$)

coordination centers for Fe

(8.13)

Like rhodotorulic acid, coprogenes (8.14) isolated from moulds contain one diketopiperazine ring.

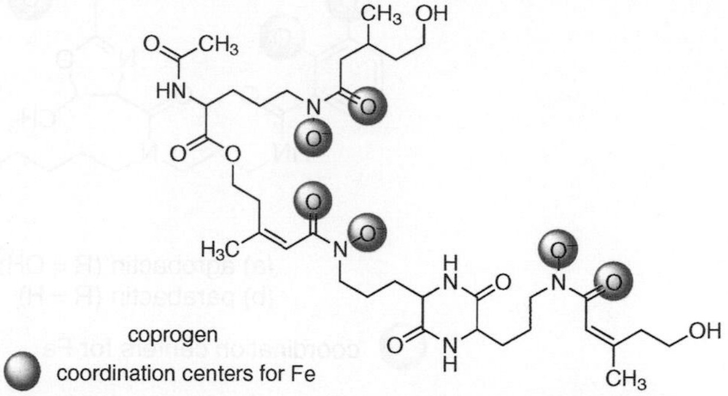

(8.14)

While hydroxamate siderophores are mostly found in higher microorganisms such as fungi and yeast, the catecholate siderophores are synthesized mainly by bacteria. Enterobactin (8.15), parabactin and agrobactin (8.16) are the most prominent representatives of the siderophores from the catecholate group; in contrast to the hydroxamates, they form negatively charged complexes with Fe^{3+}. Enterobactin (8.15), also called enterochelin, is among the strongest chelators for iron; it may be isolated from *Salmonella thyphimurium* and *E. coli*, among others [13]. Despite its extremely high affinity for iron ($K_f \approx 10^{49}$ M^{-1}), which can even result in the "extraction" of Fe^{3+} from glass, enterobactin is not suitable for a therapy of iron poisoning. One reason is that it contains a hydrolytically labile triester ring and catecholate moieties that are sensitive towards oxidation, being transformed to *o*-semiquinones and further to *o*-quinones (see Section 10.3). In addition, the free ligand is quite insoluble in aqueous solution. Finally, the iron complex of enterobactin supports the growth of higher bacteria and may thus lead to their propagation in the human body (→ infection risk). Nevertheless, chemical derivatives of enterobactin are being tested as promising chelating agents for iron therapy. Approaches include varying the backbone with its threefold symmetry (mesitylene, triamine or cyclododecane derivatives) and maintaining the catecholate binding sites, or keeping the enterobactin scaffold substituted with 8-hydroxyquinoline moieties, for example. These binding sites (clioquinol, 5-chloro-7-iodo-8-hydroxyquinoline) have been shown to remove metal ions from amyloid plaques as they are found in Alzheimer patients and thus show therapeutic potential in Alzheimer's disease [16].

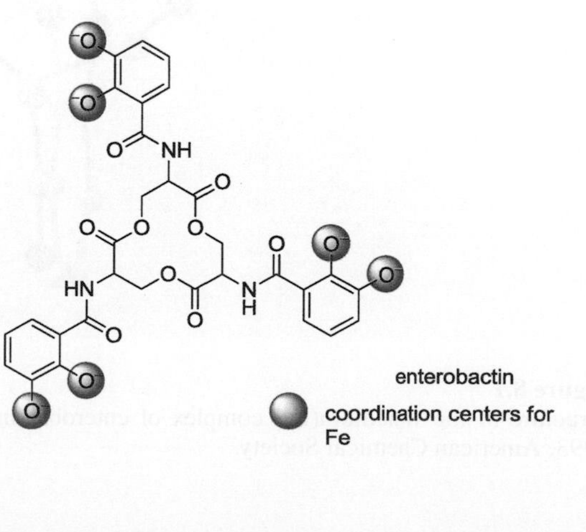

(8.15)

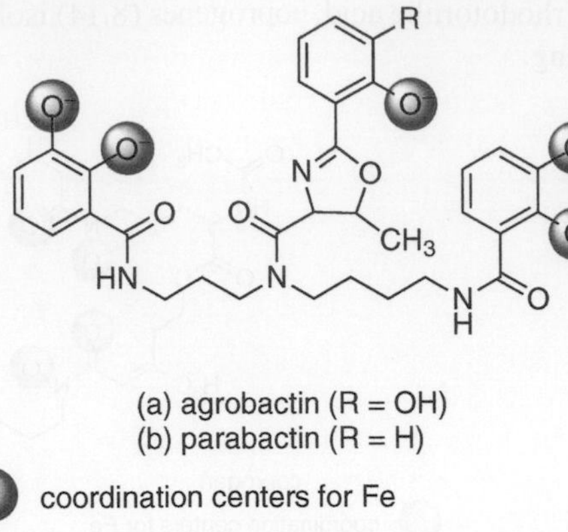

(8.16)

Although a crystal structure analysis of the Fe^{3+} complex of enterobactin has not yet been accomplished, absorption (CD) spectroscopy of an isomorphous Cr^{3+} complex and structural analysis of the V^{IV} analogue (Figure 8.1) point towards a Δ configuration [15]. Optical isomers of the naturally occurring Fe^{3+} complexes are biologically ineffective. The extraordinarily high stability of the enterobactin complex is caused *inter alia* by multiple intramolecular hydrogen bonds between the amide–NH groups and coordinating catecholate oxygen atoms. Furthermore, these amide functions are probably crucial for recognition at the membrane receptor and for the proton-induced reductive release of complexed iron.

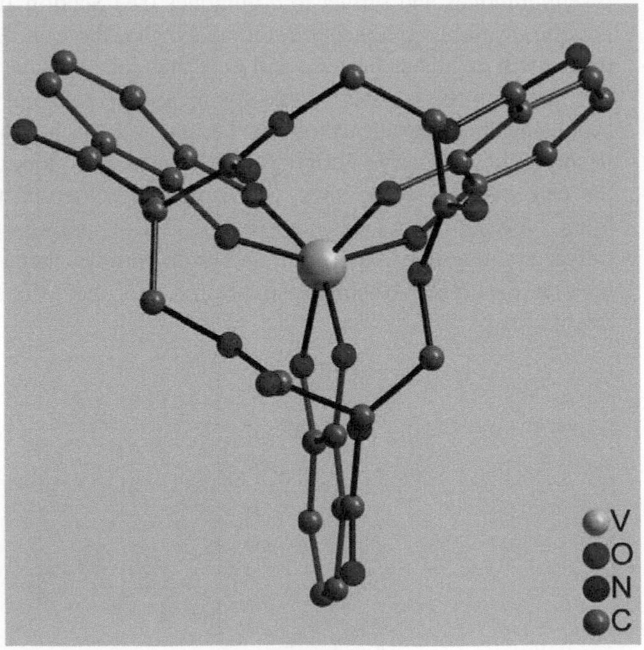

Figure 8.1
Structure of the vanadium(IV) complex of enterobactin. Reprinted with permission from [15] © 1993, American Chemical Society.

The siderophore complexes are too large to pass through the porins (proteins that cross cell membranes and allow passive diffusion of molecules), so Gram-negative bacteria employ specific outer membrane receptors. Transport to the cytosolic membrane is made possible by periplasmic binding proteins, and ATP-binding cassette proteins deliver the complex to the cytosol (compare Section 13.4). Instead of the elusive original, an enterobactin mimic (H_6-mecam (8.17)) was crystallized inside the periplasmic ferric enterobactin binding protein from *Campylobacter jejuni*. Surprisingly, the siderophore mimic forms a $[\{Fe(mecam)\}_2]^{6-}$ dimer, which bridges a pair of protein molecules [19]. In Gram-positive bacteria, a membrane-anchored binding protein serves for transport into the cytosol, where dissociation of the complex occurs. According to the principles discussed here, a mechanism of redox-facilitated release has been suggested. The relevant redox potentials, however, cannot be obtained *in vivo*, unless a decrease in pH takes place and substantial quantities of a potent Fe^{2+} chelator are present.

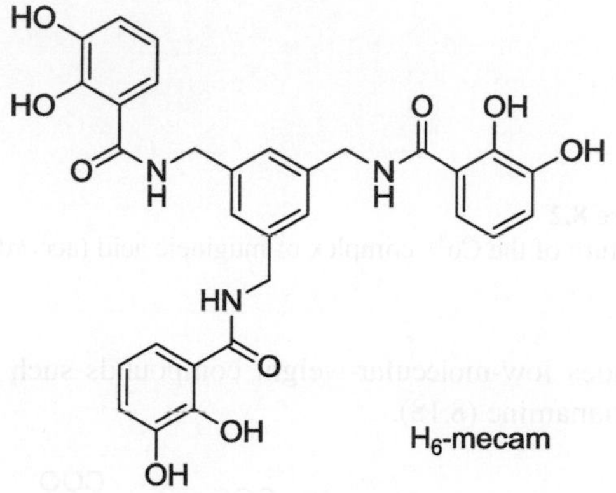

(8.17)

As the coordination chemistry of In^{3+} and of Ga^{3+} is very similar to that of Fe^{3+}, there are efforts underway to make siderophore chelating agents available for radiodiagnostics (Section 18.1.4). By varying the lipophilicity of the ligands, a relatively organ-specific distribution of the radioisotopes ^{111}In and ^{67}Ga can be achieved (see Table 18.1).

8.3 Phytosiderophores: Iron Uptake by Plants

Higher plants require iron for many components of the photosynthetic apparatus (Figure 4.7, Table 4.1) and for the biosynthesis of chlorophyll. Considering the widely varying supply of iron in the soil and the different tolerances of plants towards iron (e.g. rice is very sensitive to iron deficiency), that "mineral" component has to be extracted from the oxidic material via the roots. Symbiotic microorganisms can often synthesize siderophores, which then also mobilize iron for the host plant. Plants themselves rely on two strategies for iron uptake: reduction-based (nongraminaceous plants) and chelation-based (graminaceous plants, e.g. grasses). Acidification of the soil by ATPase activity (a drop in pH by one unit would result in a 1000fold increase in Fe^{3+} solubility) has not been confirmed. In the reduction-based strategy, Fe^{3+} is reduced to Fe^{2+} by a ferric chelate reductase (FRO2), which uses electrons from NADH [20]. The group of actual phytosiderophore ligands for the chelation of Fe^{3+}

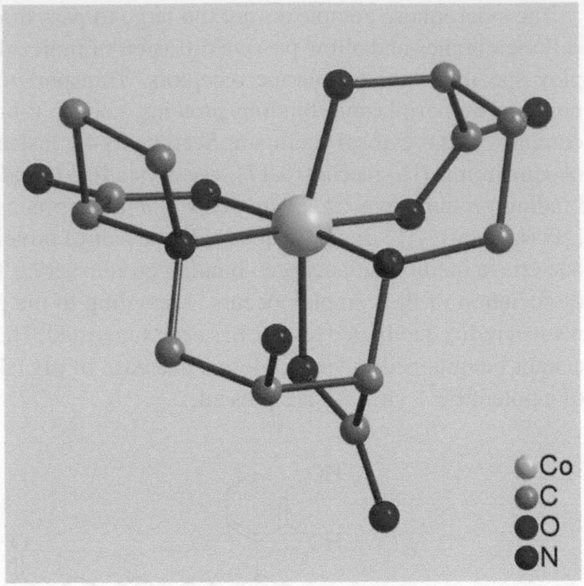

Figure 8.2
Structure of the Co^{3+} complex of mugineic acid (according to [21]).

includes low-molecular-weight compounds such as the amino acids mugineic acid and nicotianamine (8.18).

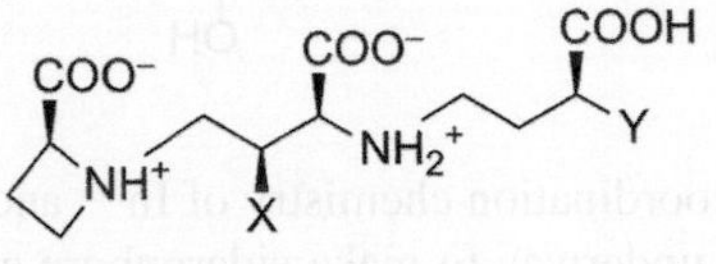

(8.18)

These compounds, which have four chirality centers in *S*-configuration, contain a four-membered azetidine ring. Figure 8.2 shows the molecular structure of the Co^{3+} complex of mugineic acid; carboxylate and amino groups of the hexadentate ligand are involved in the coordination to the metal.

8.4 Transport and Storage of Iron

In complex organisms such as human beings, the problem of the selective transport of iron to cells of very different types is of great importance. To begin with, the uptake of iron from food should be effective; for instance, the potentially chelating reductant ascorbate (vitamin C; see (3.12)) enhances a rapid resorption of iron (see (8.5)), while the nonreducing but strongly Fe^{3+}-binding phosphates may counteract such a resorption from food. In higher organisms, scavenging and transport of iron are not effected by low-molecular-weight siderophores but rather by fairly large non-heme iron proteins: the transferrins [22]. When

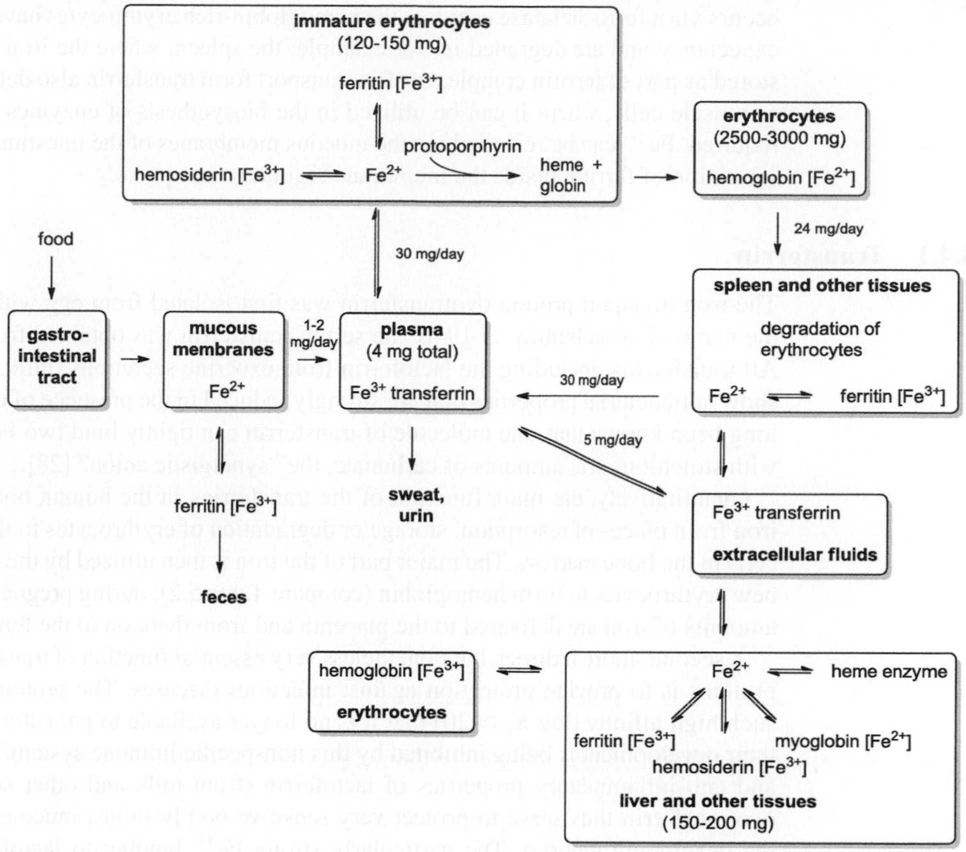

Figure 8.3
Simplified flow chart of the iron metabolism in the human body (according to [5]).

iron is set free from its transport system and released, for example within the cell, it must be either directly used or stored, due to the potentially hazardous character of free iron according to (8.1) and (8.2). Highly specialized non-heme iron proteins, particularly ferritin and the insoluble hemosiderin, serve to store this unused iron [23–26]. Storage and transport systems must function rapidly and completely reversibly under physiological conditions in order to preclude local excess or deficiency symptoms. Such transport and storage systems have not only been found in animals; the iron-storage protein phytoferritin has been detected in plants, while certain bacteria have been shown to have heme-containing bacterioferritins.

Figure 8.3 gives a simplified overview specifying the roles of transferrin and ferritin in the mammalian (human) metabolism. Iron enters the body by absorption, primarily in the duodenum through the action of enterocytes, where DMT1, a transport protein specific for Fe^{2+}, shuttles it into the cell after reduction of Fe^{3+} by a ferrireductase. Fe^{2+} has to cross the enterocyte and exit through ferriportin, a transmembrane iron transporting protein [27]. Fe^{3+} is liberated from the transferrin transport system in the blood-forming cells of the bone marrow and is then taken up by ferritin, presumably after intermediary reduction to more labile Fe^{2+}. The uptake by the ferritin storage system eventually involves reoxidation to Fe^{3+}. While iron incorporation into the porphyrin macrocycle during heme biosynthesis occurs via a ferrochelatase enzyme, the hemoglobin-rich erythrocytes have only a finite life expectancy and are degraded in, for example, the spleen, where the iron released is again stored as part of ferritin complexes. The transport form transferrin also delivers iron to liver or muscle cells, where it can be utilized in the biosynthesis of enzymes or myoglobin. If required, Fe^{2+} can be resorbed via the mucous membranes of the intestines, where the iron saturation of ferritin inside the membrane regulates the uptake.

8.4.1 Transferrin

The iron transport protein ovotransferrin was first isolated from egg white in 1900 under the name of conalbumin; in 1946, the serum transferrin was obtained from human blood. All transferrins, including the lactoferrin from exocrine secretions (milk, mucosal tissue), show antibacterial properties that are strongly reduced in the presence of excess iron. It has long been known that one molecule of transferrin can tightly bind two Fe^{3+} ions together with stoichiometric amounts of carbonate, the "synergistic anion" [28].

Quantitatively, the main function of the transferrins in the human body is to transport iron from places of resorption, storage or degradation of erythrocytes to the blood-forming cells in the bone marrow. The major part of the iron is then utilized by the precursor cells of new erythrocytes to form hemoglobin (compare Table 5.2); during pregnancy, considerable amounts of iron are delivered to the placenta and from there on to the fetus.

A second, more indirect, but nonetheless very essential function of transferrin and related proteins is to provide protection against infectious diseases. The protein binds iron with such high affinity ($\log K_f \approx 20$) that it is no longer available to parasitic microorganisms; their development is being inhibited by this nonspecific immune system. The antibacterial and anti-inflammatory properties of lactoferrin (from milk and other secretions) and of ovotransferrin thus serve to protect very sensitive bodily fluids, mucous membranes and the developing embryo. The particularly strong Fe^{3+} binding to lactoferrin presumably plays an important role in the development of cells that are directly involved in immune reactions of the body; therefore, efforts are being made to obtain human lactoferrin from appropriately modified "transgene" organisms.

The transferrins (ovotransferrin, serum transferrin and lactoferrin) are glycoproteins with molecular masses of about 80 kDa. Their single polypeptide chain is folded to form two lobes (see Figure 8.4), each of which contains one binding site for iron. The amino acid sequences for the C- and N-termini of human serum transferrin show 42% homology. The size of the carbohydrate portion of transferrins varies between different species and can even vary among different regions of the same organism. The polysaccharide chains do not contribute to iron transport; their function is to engage in specific interactions with receptors.

All transferrins can bind two high-spin Fe^{3+} ions per molecule, as well as other metal ions such as Cr^{3+}, Al^{3+} (see Section 17.6), Cu^{2+}, Mn^{2+}, Co^{3+}, Co^{2+}, Cd^{2+}, Zn^{2+}, VO^{2+}, Sc^{2+}, Ga^{3+}, Ni^{2+} and trivalent lanthanide ions. The transport of Al^{3+} by transferrins into the central nervous system (unlike Fe^{3+}, Al^{3+} cannot be remobilized by reduction) is an important aspect of aluminum toxicity (see Section 17.6) [29]. While the iron-free apoenzyme is colorless, a red-brown color indicates the coordination of Fe^{3+}. The binding of the metal ion to either of the two coordination sites in the C- and N-terminal region involves the simultaneous "synergistic" binding of a carbonate ion and the release of three protons (charge effect). The protons can be either produced from the bound water ligands (hydrolysis) or can originate from the protein. The two coordination sites for iron are not completely equivalent, the

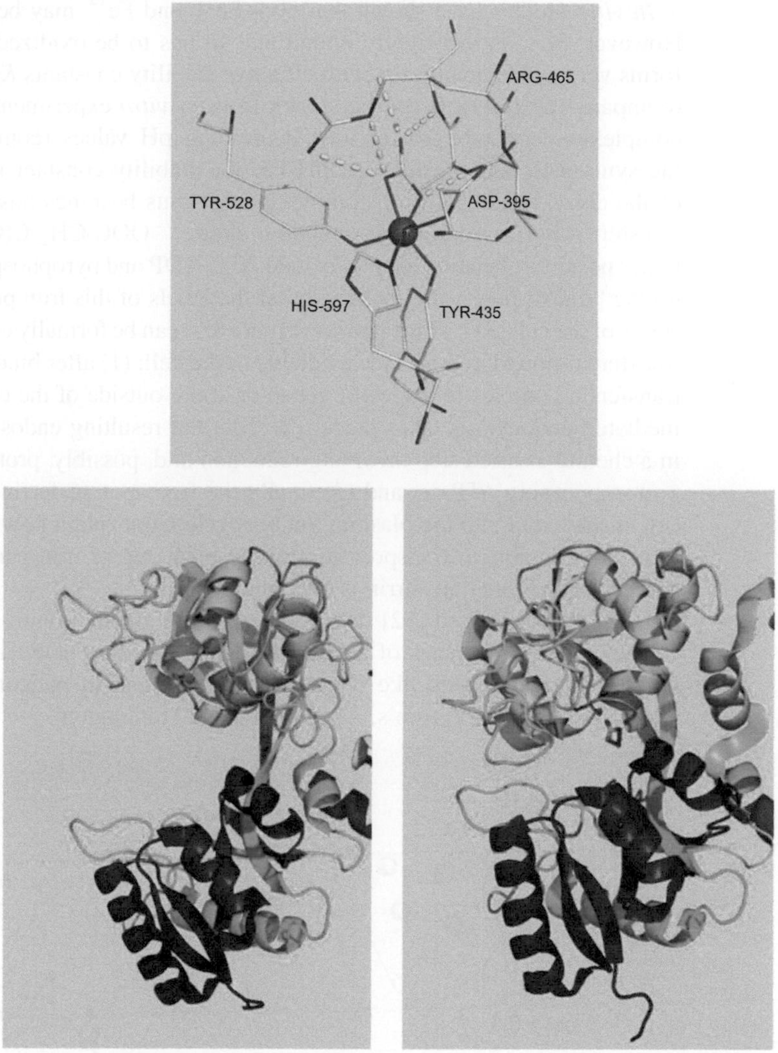

Figure 8.4
(a) Coordination environment of Fe^{3+} and $CO_3{}^{2-}$ (Fe^{3+}, red sphere; C, grey; O, red; N, blue; yellow lines indicate polar interactions/hydrogen bonds). (b,c) Reversible conformational change upon $Fe^{3+}/CO_3{}^{2-}$ binding by human lactoferrin (N-terminal region, PDB code 1LFG and 1CB6) [22b].

C-terminus showing a slightly higher affinity for Fe^{3+} ions, which results in complex formation at lower pH values. For iron-saturated lactoferrin, the following ligands have been found in the coordination sphere of each of the two approximately octahedrally configured metal centers (Figure 8.4): two tyrosinate residues, which are responsible for the intense color of the Fe^{3+} form (ligand-to-metal charge transfer, LMCT), one η^1-aspartate, one histidine ligand and one η^2-coordinated (chelating) carbonate, which links the metal to the protein via hydrogen bonds. The reversible transformation of the "open" conformation of the bilobal protein to a closed, compact structure following binding of Fe^{3+} and of the synergistic anion $CO_3{}^{2-}$ (Figure 8.4) is an essential feature of its efficient transport function ("iron shuttle").

In vitro studies have shown that both Fe^{2+} and Fe^{3+} may be taken up by transferrins. However, Fe^{2+} is less tightly bound and so has to be oxidized within the protein. Fe^{3+} forms very stable complexes with effective stability constants K_f in the range of $10^{20}\,M^{-1}$ (compare Table 8.1). It can be shown from *in vitro* experiments that the stability of the complexes decreases greatly with decreasing pH values (compare (8.5)), amplified by the synergistic anion effect. At pH 4.5, the stability constant is already lower than that of the corresponding citrate complex, and it thus becomes possible to release iron from transferrin by the addition of chelating citrate, $^{-}OOC\text{-}CH_2\text{-}C(OH)(COO^{-})\text{-}CH_2\text{-}COO^{-}$. Other possible chelating agents include ATP, ADP and pyrophosphate, presumably making up the "transit pool". It is believed that the levels of this iron pool reflect the overall iron status of the cell [31]. Three phases (Figure 8.5) can be formally distinguished in the way the transferrin-bound iron is made available to the cell: (1) after binding of the iron-containing transferrin complex to a specific receptor at the outside of the cell, a membrane receptor-mediated endocytosis takes place; (2) from the resulting endosomes, the iron is released in a chelate-assisted step involving reduction and, possibly, protonation with the help of a proton-pumping ATPase; and (3) finally the free apotransferrin is transferred back to the cell surface and into the plasma. Such a cycle can explain how the small total amount of transferrin is able to transport an average of 40 mg of iron per day when the individual capacity of human transferrin is only about 7 mg iron.

It has been reported [32] that transferrin can assemble into long protein fibrils which can release nanoparticles of an iron oxide/hydroxide (FeO(OH)) resembling the mineral lepidocrocite. Deposits like these have been found in patients with neurodegenerative diseases such as Parkinson's, Alzheimer's and Huntington's.

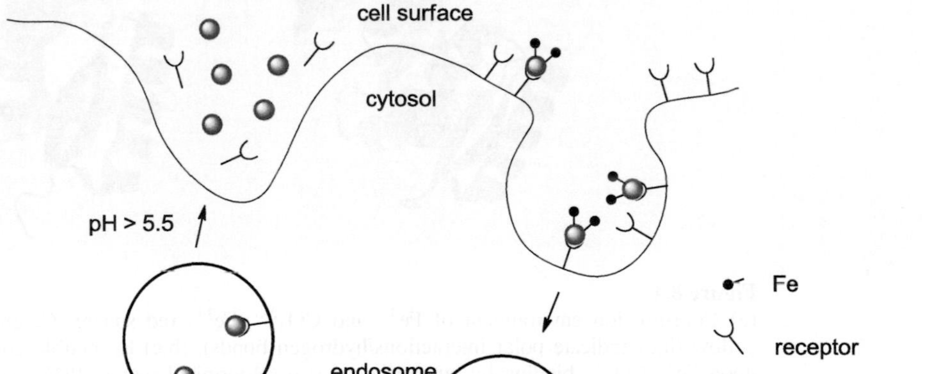

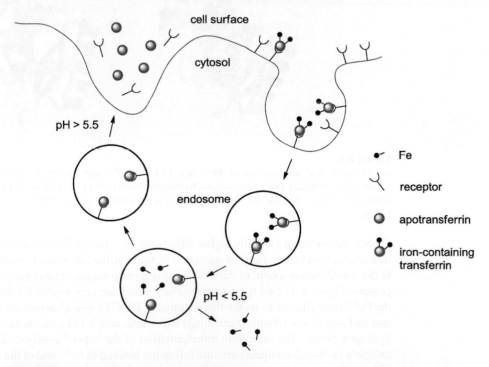

Figure 8.5
Transferrin recognition and demetallation by example of a human tumor cell (according to [30]).

8.4.2 Ferritin

Iron that has been released into the cell as described in Figure 8.5 must either be used immediately for biosynthesis or stored in a safe form. Particularly after the advent of a biogenically O_2-enriched atmosphere, there must be a system which serves (1) as a storage for iron (i.e. a mineral that is increasingly less bioavailable due to its precipitation as Fe^{3+} hydroxide/oxide) and (2) as a safekeeping site, inhibiting the uncontrolled radical-producing reactions of reduced free iron with O_2 or its metabolic products (8.1), (8.2). Two very large proteins, ferritin and hemosiderin, serve as such iron stores; the soluble and much better defined ferritin has been more thoroughly studied [23–26]. As early as 1935, Laufberger had isolated and crystallized ferritin and suggested it had a role as an iron store due to its high iron content of up to 20% Fe by weight. Ferritin is found in animals as well as in plants. Approximately 13% of the iron in the human body is present as ferritin (Table 5.2). It is mainly found in the liver, spleen and bone marrow. The very large molecule consists of an inorganic core, a nanoparticle of variable size and composition, which is surrounded and held together by a complex protein shell.

Apoferritin, the iron-free protein component of ferritin, has an average molecular mass of 440 kDa. It can be generated from iron-containing ferritin by reduction with sodium dithionite, $Na_2S_2O_4$, or ascorbate in the presence of suitable chelating agents. The water-soluble protein consists of 24 equivalent subunits, which arrange to form a hollow sphere with an outer diameter of about 12 nm (Figure 8.6). The empty inside has a diameter of approximately 7.5 nm, and in holoferritin this space is filled with inorganic material [25].

The crystal structure of apoferritin has been determined (Figure 8.6). According to this structure analysis, each subunit consists of four long α helices, forming a bundle, and a fifth short α helix arranged perpendicularly to these. Two of the long helices are connected by a loop, and two of these loops from neighboring subunits arrange to form a β sheet. A total of 24 subunits form the apoferritin molecule, with its high symmetry (rhombic dodecahedron, cubic space group $F432$, $a \approx 18.5$ nm; see Figure 8.6). The particular spatial arrangement allows for the formation of channels along the threefold and fourfold symmetry axes. These channels play an important role in the deposition and release of iron. The six channels with fourfold symmetry are hydrophobic because of 12 lining leucine residues, whereas the eight channels with threefold symmetry are rather hydrophilic due to the presence of aspartate and glutamate and, to a lesser extent, histidine and tyrosine residues.

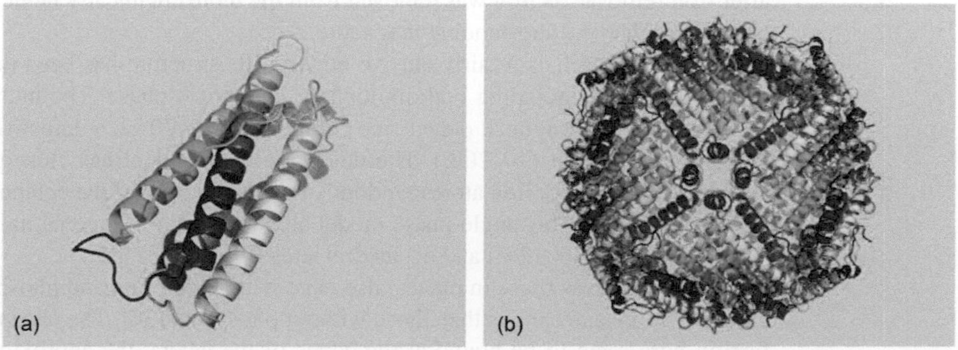

Figure 8.6
(a) Ribbon diagram of the α-helical framework of an apoferritin subunit. (b) Subunit arrangement in the apoferritin molecule (hollow sphere); view along the fourfold axis (PDB code 1FHA) [25,33].

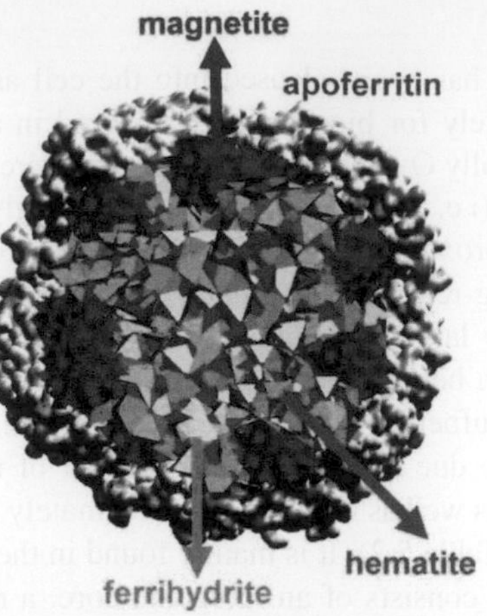

Figure 8.7
Mineral phases in a native ferritin iron core from horse spleen. Reprinted with permission from [34] © 2008, American Chemical Society.

Ferritins of mammals are largely isomorphous. However, ferritins isolated from bacteria such as *E. coli* differ a great deal, even in their amino acid sequences; bacterioferritins may contain additional heme groups with axial bis(methionine) coordination (Section 6.1), as well as miniferritins with 12 instead of 24 subunits: the so-called Dps (DNA binding proteins from starved cells) [23].

In the center of the hollow sphere formed by mammalian apoferritin there is space for the inorganic core, which can contain a maximum of 4500 iron centers in oxidic form; a typical filling involves approximately 1200 iron centers. The inorganic nucleus of ferritin represents such a large space of high electron density that it is clearly visible in the electron microscope without further modification, and the exact nature of the iron core has been the focus of many investigations, using a variety of different techniques. The native ferritin iron core has been shown to consist of a polyphasic structure with three different mineral phases: ferrihydrite, magnetite and hematite (Figure 8.7) [34]. The proportion of each phase changed with gradual iron removal, since ferrihydrite is more labile to chemical reduction and mobilization than the magnetite phase. *In vitro*, the core size did not vary significantly after iron removal, as iron was removed from the more chemically labile ferrihydrite inner core rather than the surrounding magnetite.

Ferrihydrite itself is a fairly elusive mineral. Its structure has been described by many different models, including both multiphase and single-phase. The multiphase model requires a double-hexagonal, defect-free component (50–70%), a defective and a nanosized hematite component (10–20%). The defect-free component has close-packed hexagonal layers of oxygen with iron atoms randomly occupying 50% of the octahedral sites; the unit cell is hexagonal. The single-phase model also features a hexagonal unit cell with iron in both tetrahedral (10–12%) and octahedral sites (Figure 8.8) [35].

Ferritins, such as those in plants, also exist with mixed iron and phosphate cores. These cores are more amorphous than those without phosphate [36]. The phosphate anions in the ferritin core seem to be bound at the edges of the layers, the approximate ratio being 9 Fe^{3+} centers per phosphate group. Most of the phosphate is not essential for the formation of ferritin, as is apparent from its variable amount. One generally important aspect of phosphates lies in the linkage of organic polymers such as proteins with inorganic particles

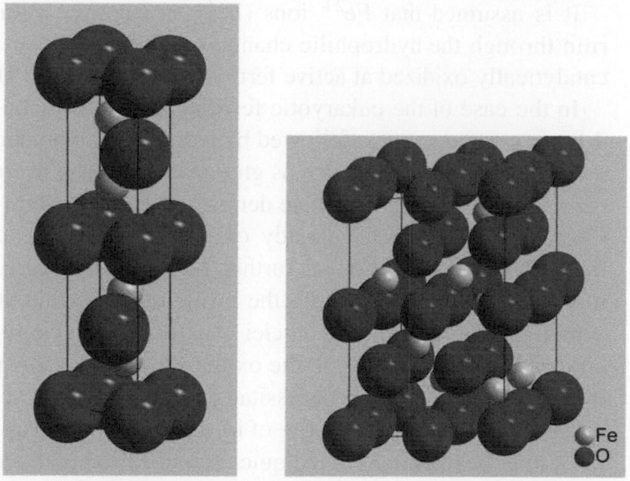

Figure 8.8
Ferrihydrite structure. Left: the double-hexagonal component of the multiphase model. Right: the hexagonal unit cell of the single-phase model. Reprinted with permission from [37] © 2007, American Association for the Advancement of Science and with permission from [38] © 1993 by the Mineralogical Society of America.

(compare Figure 15.3); formation of such a linkage can also be assumed between the inorganic layers and the protein hull of ferritin.

The magnetic behavior of the ferritin core has been studied in great detail. The results are best explained by assuming antiferromagnetic coupling between oxide-bridged iron centers (compare with ribonucleotide reductase (RR), Section 7.6.1). However, in these studies the effective magnetic moment per metal center, μ_{eff}, was determined with 3.85 Bohr magnetons and thus does not fit the ground state of free Fe^{3+} ions ($S = 5/2$), but rather a state with a spin quantum number of $S = 3/2$; a superexchange process between the iron atoms has been invoked as an explanation. The Mössbauer spectrum shows another characteristic behavior of ferritin iron [39]: while the expected six-line spectrum for the well-ordered magnetic state is obtained at low temperatures, these lines coalesce at high temperatures to form a doublet (arising from quadrupolar splitting). This effect is caused by ferromagnetic spin–spin coupling within small "nanoparticles" ($d < 20\,nm$) and their temperature-dependent orientation with respect to each other ("superparamagnetism" [40]) – a phenomenon which has parallels in various extremely finely ground metal oxides. It is thus only consequential that nanoparticles with special material properties are now being generated synthetically inside the ferritin shell [40,41].

There are two basically different possibilities for the formation of ferritin from apoferritin: if Fe^{3+} is present as polymeric oxide/hydroxide aggregate, the apoferritin shell, in equilibrium with its subunits, can assemble the ferritin framework around the already existing iron core (template effect); alternatively, the formation of ferritin can also be envisaged as a redox process in which Fe^{2+} is oxidized to Fe^{3+} in the presence of apoferritin and an electron acceptor (ultimately O_2) *during* deposition into the apoferritin. Since the biosynthesis of apoferritin precedes that of ferritin and dissociation into subunits requires fairly unphysiological conditions, the second alternative (8.19) is now well accepted.

$$2\,Fe^{2+} + O_2 + 2\,H_2O \longrightarrow \left[{}^{3+}Fe{-}O{-}O{-}Fe^{3+}\right] \longrightarrow (Fe_2O_3{\times}H_2O)_{caged\ solid} + 2\,H^+ \qquad (8.19)$$

It is assumed that Fe^{2+} ions (Fe^{3+} is inactive in this respect) can enter the apoferritin through the hydrophilic channels at a rate of about 3000 Fe/s and that they are then catalytically oxidized at active ferrioxidase centers (8.19) [24].

In the case of the eukaryotic ferritins, the first reaction intermediate characterized is a diferric peroxo species, followed by a diferric oxo product.

Carboxylate residues such as glutamate or aspartate are abundant at the inner surface of the apoferritin. Their complete derivatization in the form of esters blocks the iron loading. Carboxylate-bound and already oxidized iron centers can then serve as nuclei for the deposition and oxidation of further Fe^{2+} and thus for the growth of the iron core. The nucleation channels lead into the inside of apoferritin at the fourfold symmetry axes, so a total of 16 oxo-iron(III) nuclei can form and give rise to the highly ordered mineral core [24]. Ultimately, O_2 is the oxidizing agent for this process, because, after all, it was its bioproduction which necessitated such a soluble, vesicular iron store and protection system. However, not all 100% of iron has to be stored as Fe^{3+} in ferritin; small amounts of Fe^{2+} may be important to the quick mobilization of the mineral. In such a case, the proton flux inevitable for the oxidation reaction (8.19) would also be smaller.

According to the typical condensation polymerization pattern of aqua and hydroxo complexes (8.20) [42–44], the iron oxide/hydroxide forms more or less well-ordered "quasicrystalline" clusters, which grow until the inside of the apoferritin is filled.

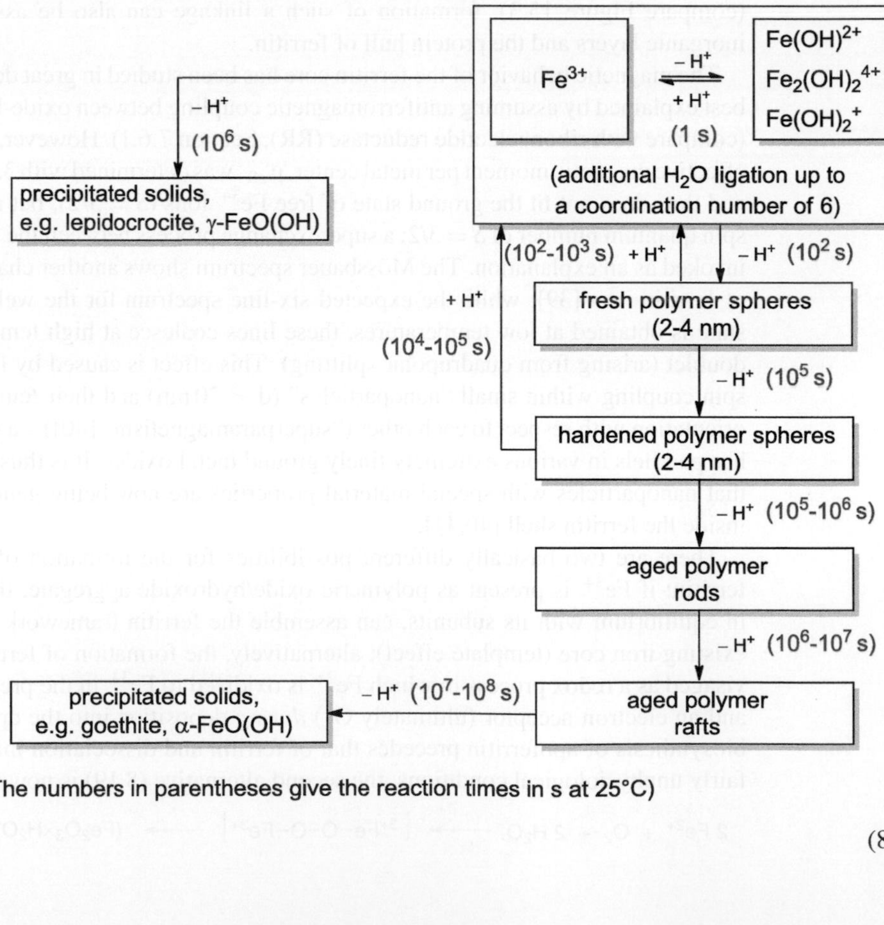

(The numbers in parentheses give the reaction times in s at 25°C)

(8.20)

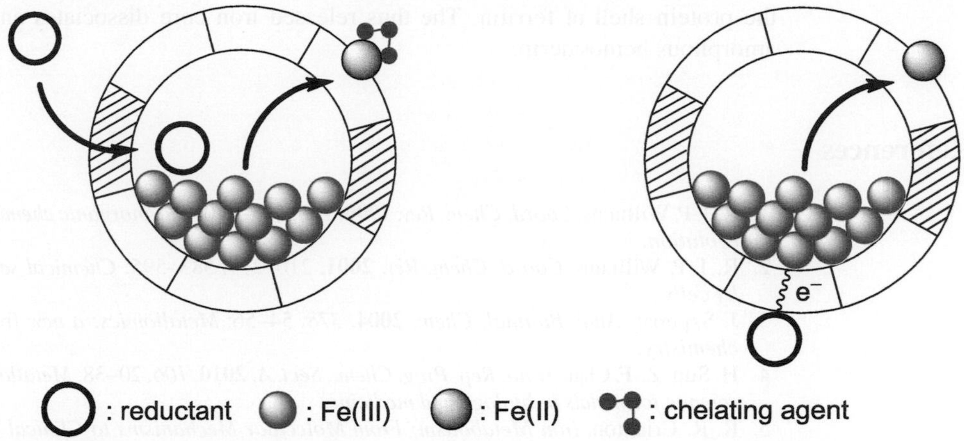

Figure 8.9
Schematic representation of different alternatives for the reductive mobilization of iron from ferritin (according to [45]; see text for details).

A probable alternative form of physiological iron mobilization from ferritin involves electron transfer through the protein shell. It has been demonstrated by *in vitro* experiments that ferritin-bound iron can be removed from the core as Fe^{2+} using nonphysiological reductants such as sodium dithionite. Furthermore, it has been shown that a dynamic equilibrium exists between Fe^{3+} ions inside and outside of ferritin.

The different possibilities for iron mobilization are summarized in Figure 8.9. Representation (a) shows the hydrophilic channels with threefold symmetry for the transport of Fe^{2+} *out of* the ferritin and the hydrophobic channels with fourfold symmetry (shaded) for the *import* of small reductants. Fe^{2+} chelators can shift the equilibrium and thus simplify the release of iron. Another possibility (b) is an electron transfer through the protein shell, with Fe^{3+} reduced and transported out through the hydrophilic channels as Fe^{2+}.

A thorough understanding of the mechanisms of release of iron from ferritin is of great importance in the study of neurodegeneration, since it is known that iron accumulates in the brains of patients with neurodegenerative diseases and that the ferritin core composition varies between healthy specimens and patients with Alzheimer's disease [34]. It is still unknown where the electron transfer takes place (i.e. directly or through the protein), how the required water reaches the surface of the mineral core and whether chelating molecules bind Fe^{2+} outside the protein or within channels [24].

8.4.3 Hemosiderin

In addition to ferritin, hemosiderin is another iron storage form found in organisms, particularly during iron overload. Hemosiderin was first isolated from horse spleen in 1929. While the structures of the iron cores of ferritin and hemosiderin are similar, the protein component of hemosiderin is largely unknown. With approximately 35% Fe, the iron/protein ratio is even higher in the extremely large hemosiderin (typically 4 MDa) than in ferritin, and it is assumed that this insoluble species is formed via lysosomal decomposition of ferritin. According to this hypothesis, proteases in the lysosome contribute to the degradation of the protein shell of ferritin. The thus released iron core dissociates and reforms as the amorphous hemosiderin.

References

1. R. J. P. Williams, *Coord. Chem. Rev.* **1990**, *100*, 573–610: *Bio-inorganic chemistry: its conceptual evolution.*
2. R. J. P. Williams, *Coord. Chem. Rev.* **2001**, 216–217, 583–595: *Chemical selection of elements by cells.*
3. J. Szpunar, *Anal. Bioanal. Chem.* **2004**, *378*, 54–56: *Metallomics: a new frontier in analytical chemistry.*
4. H. Sun, Z.-F. Chai, *Annu. Rep. Prog. Chem., Sect. A.* **2010**, *106*, 20–38: *Metallomics: an integrated science for metals in biology and medicine.*
5. R. R. Crichton: *Iron Metabolism: From Molecular Mechanisms to Clinical Consequences*, 3rd Edition, John Wiley & Sons, New York, **2009**.
6. *Biochim. Biophys. Acta* **2010**, *1800*, 687–898: special issue: *Ferritin: structures, properties and applications.*
7. B. Halliwell, J. M. C. Gutteridge, *Trends Biochem. Sci.* **1986**, *11*, 372–375: *Iron and free radical reactions: two aspects of antioxidant protection.*
8. E. D. Letendre, *Trends Biochem. Sci.* **1985**, *10*, 166–168: *The importance of iron in the pathogenesis of infection and neoplasia.*
9. N. S. Scrimshaw, *Sci. Am.* **1991**, *265*(4), 46–52: *Iron deficiency.*
10. M. J. Miller, F. Malouin, *Acc. Chem. Res.* **1993**, *26*, 241–249: *Microbial iron chelators as drug delivery agents: the rational design and synthesis of siderophore-drug conjugates.*
11. P. V. Bernhardt, *Dalton Trans.* **2007**, 3214–3220: *Coordination chemistry and biology of chelators for the treatment of iron overload disorders.*
12. W. Schneider, *Chimia* **1988**, *42*, 9–20: *Iron hydrolysis and the biochemistry of iron – the interplay of hydroxide and biogenic ligands.*
13. S. Dhungana, A. L. Crumbliss, *Geomicrobiol. J.* **2007**, *22*, 87–98: *Coordination chemistry and redox processes in siderophore-mediated iron transport.*
14. K. N. Raymond, G. Müller, B. F. Matzanke, *Top. Curr. Chem.* **1984**, *123*, 49–102: *Complexation of iron by siderophores. A review of their solution and structural chemistry and biological functions.*
15. T. B. Karpishin, T. M. Dewey, K. N. Raymond, *J. Am. Chem. Soc.* **1993**, *115*, 1842–1851: *The vanadium(IV) enterobactin complex: structural, spectroscopic, and electrochemical characterization.*
16. A. Moulinet d'Hardemare, G. Gellon, C. Philouze, G. Serratrice, *Inorg. Chem.* **2012**, *51*, 12 142–12 151: *Oxinobactin and sulfoxinobactin, abiotic siderophore analogues to enterobactin involving 8-hydroxyquinoline subunits: thermodynamic and structural studies.*
17. H. Bickel, G. E. Hall, W. Keller-Schierlein, V. Prelog, E. Vischer, A. Wettstein, *Helv. Chim. Acta* **1960**, *43*, 2129–2138: *Über die Konstitution von Ferrioxamin B.*
18. F. Vögtle: *Supramolecular Chemistry*, John Wiley & Sons, New York, **1993**.
19. A. Mueller, A. J. Wilkinson, K. S. Wilson, A.-K. Duhme-Klair, *Angew. Chem. Int. Ed. Engl.* **2006**, *45*, 5132–5136: *An* $[\{Fe(mecam)\}_2]_6$*- bridge in the crystal structure of a ferric enterobactin binding protein.*
20. C. M. Palmer, M. L. Guerinot, *Nat. Chem. Biol.* **2009**, *5*, 333–340: *Facing the challenges of Cu, Fe and Zn homeostasis in plants.*
21. Y. Mino, T. Ishida, N. Ota, M. Inoue, K. Nomoto, T. Takemoto, H. Tanaka, Y. Sugiura, *J. Am. Chem. Soc.* **1983**, *105*, 4671–4676: *Mugineic acid-iron(III) complex and its structurally analoguous cobalt(III) complex: characterization and implication for absorption and transport of iron in gramineous plants.*
22. (a) K. Gkouvatsos, G. Papanikolaou, K. Pantopoulos, *Biochim. Biophys. Acta* **2012**, *1820*, 188–202: *Regulation of iron transport and the role of transferrin*; (b) K. Mizutani, M. Toyoda, B. Mikami, *Biochim. Biophys. Acta* **2012**, *1820*, 203–211: *X-ray structures of transferrins and related proteins.*

23. X. Liu, E. C. Theil, *Acc. Chem. Res.* **2005**, *38*, 167–175: *Ferritins: dynamic management of biological iron and oxygen chemistry.*
24. E. C. Theil, R. K. Behera, T. Tosha, *Coord. Chem. Rev.* **2013**, *257*, 579–586: *Ferritins for chemistry and for life.*
25. R. R. Crichton, J.-P. Declercq, *Biochim. Biophys. Acta* **2010**, *1800*, 706–718: *X-ray structures of ferritins and related proteins.*
26. T. G. St. Pierre, J. Webb, S. Mann, *Ferritin and hemosiderin: structural and magnetic studies of the iron core*, in S. Mann, J. Webb, R. J. P. Williams (eds.): *Biomineralization*, VCH, Weinheim, **1989**.
27. M. D. Garrick, *Genes and Nutrition* **2011**, *6*, 45–54: *Human iron transporters.*
28. B. F. Anderson, H. M. Baker, G. E. Norris D. W. Rice, E. N. Baker, *J. Mol. Biol.* **1989**, *209*, 711–734: *Structure of human lactoferrin: crystallographic structure analysis and refinement at 2.8 Å resolution.*
29. S. J. A. Fatemi, F. H. A. Kadir, D. J. Williamson, G. R. Moore, *Adv. Inorg. Chem.* **1991**, *36*, 409–448: *The uptake, storage, and mobilization of iron and aluminum in biology.*
30. L. Zecca, M. B. Youdim, P. Riederer, J. R. Connor, R. R. Crichton, *Nat. Rev. Neurosci.* **2004**, *5*, 863–873: *Iron, brain aging and neurodegenerative disorders.*
31. F. Petrat, H. de Groot, R. Sustmann, U. Rauen, *Biol. Chem.* **2002**, *383*, 489–502: *The chelatable iron pool in living cells: a methodically defined quantity.*
32. S. Ghosh, A. Mukherjee, P. J. Sadler, S. Verma, *Angew. Chem. Int. Ed. Engl.*, **2008**, *47*, 2217–2221: *Periodic iron nanomineralization in human serum transferring fibrils.*
33. T. Tosha, M. R. Hasan, E. C. Theil, *Proc. Natl. Acad. Sci. USA* **2008**, *105*, 18 182–18 187: *The ferritin Fe2 site at the diiron catalytic center controls the reaction with O_2 in the rapid mineralization pathway.*
34. N. Galvez, B. Fernandez, P. Sanchez, R. Cuesta, M. Creolin, M. Clemente-Leon, S. Trasobares, M. Lopez-Haro, J. J. Calvino, O. Stephan, J. M. Dominguez-Vera, *J. Am. Chem. Soc.* **2008**, *130*, 8062–8068: *Comparative structural and chemical studies of ferritin cores with gradual removal of their iron contents.*
35. F. M. Michel, H.-A. Hosein, D. B. Hausner, S. Debnath, J. B. Parise, D. R. Strongin, *Biochim. Biophys. Acta* **2010**, *1800*, 871–885: *Reactivity of ferritin and the structure of ferritin-derived ferrihydrite.*
36. R. K. Watts, R. J. Hilton, D. M. Graff, *Biochim. Biophys. Acta.* **2010**, *1800*, 745–759: *Oxido-reduction is not the only mechanism allowing ions to traverse the ferritin protein shell.*
37. F. M. Michel, L. Ehm, S. M. Antao, P. L. Lee, P. J. Chupas, G. Liu, D. R. Strongin, M. A. A. Schoonen, B. L. Phillips, J. B. Parise, *Science* **2007**, *316*, 1726–1729: *The structure of ferrihydrite, a nanocrystalline material.*
38. V. A. Drits, B. A. Sakharov, A. L. Salyn, A. Manceau, *Clay Miner.* **1993**, *28*, 185–207: *Structural model for ferrihydrite.*
39. T. G. St. Pierre, K.-S. Kim, J. Webb, S. Mann, D. P. E. Dickson, *Inorg. Chem.* **1990**, *29*, 1870–1874: *Biomineralization of iron: Mössbauer spectroscopy and electron microscopy of ferritin cores from the chiton Acanthopleura hirtosa and the limpet Patella laticostata.*
40. R. Dagani, *Chem. Eng. News* **1992**, November 23, 18–24: *Nanostructured materials promise the advance range of technologies.*
41. F. Meldrum, B. R. Heywood, S. Mann, *Science* **1992**, *257*, 522–523: *Magnetoferritin: in vitro synthesis of a novel magnetic protein.*
42. K. S. Hagen, *Angew. Chem. Int. Ed. Engl.* **1992**, *31*, 1010–1012: *Synthetic models for iron–oxygen aggregation and biomineralization.*
43. C. M. Flynn, *Chem. Rev.* **1984**, *84*, 31–41: *Hydrolysis of inorganic iron(III) salts.*
44. M. Henry, J. P. Jolivet, J. Livage, *Struct. Bonding (Berlin)* **1992**, *77*, 153–206: *Aqueous chemistry of metal cations: hydrolysis, condensation and complexation.*
45. P. M. Harrison, G. C. Ford, D. W. Rice, J. M. A. Smith, A. Treffry, J. L. White, *Front. Bioinorg. Chem.* **1986**, *2*, 268–277: *The three-dimensional structure of apoferritin: a framework controlling ferritin's iron storage and release.*

9 Nickel-containing Enzymes: The Remarkable Career of a Long-overlooked Biometal

9.1 Overview

For a long time, nickel was the only element of the "late" 3d transition metals for which a biological role could not be definitely established [1]. This changed only in 1975, when B. Zerner discovered that urease is a nickel enzyme (see Chapter 1 and Section 9.2). The role of nickel in methanogenesis was discovered a few years later by serendipity, when the group of R. Thauer carried out growth experiments on methanogenic bacteria (Figure 1.2) [2,3]. This led to the findings that coenzyme F_{430} is a nickel porphinoid (Section 9.5), that some hydrogenases are nickel enzymes (Section 9.3) and that carbon monoxide dehydrogenase (CODH) from methanogens contains nickel (Section 9.4).

The reasons for this "oversight" were manifold: nickel ions do not exhibit a very characteristic light absorption (color) in the presence of physiologically relevant ligands. Mössbauer effects are not easily accessible for nickel isotopes, and even paramagnetic Ni^{I} (d^9) or Ni^{III} (d^7) cannot always be unambiguously detected by electron paramagnetic resonance (EPR) spectroscopy due to the lack of metal isotope hyperfine coupling (the natural abundance of ^{61}Ni with $I = 3/2$ is only 1.25%). In addition, it has now been shown that Ni is often only one of several components of complex enzymes, which may otherwise contain several coenzymes as well as additional inorganic material. For instance, nickel atoms in enzymes remained undetected for a long time due to their frequent association with Fe/S clusters. However, by applying more sensitive detection methods in atomic absorption or emission spectroscopy (AAS or AES), magnetic measuring (SQUID susceptometry) or EPR spectroscopy using ^{61}Ni-enriched material, some nickel-containing enzymes of plants and microorganisms have now been established and partly characterized (Table 9.1). Nickel is not particularly rare in the lithosphere or in sea water, where it is soluble as Ni^{2+}. In view of its very low potential requirement as an ultratrace element, no real deficiency symptoms have been reported for human beings [4]. Ni^{2+}-specific antibodies are responsible for the not uncommon "nickel allergy" [5]. The bioinorganic chemistry of nickel is unique mainly in view of the organometallic species that occur in a number of biological processes catalyzed by nickel enzymes (compare cobalt in Chapter 3). Many of these "organometallic" transformations, occurring exclusively in archaea and bacteria, involve the consumption or production of CH_4 (from CO_2), and this biological "C_1" chemistry has a large impact on the global C cycle (see insertion Chapter 15) [6]. For instance, the development of

Bioinorganic Chemistry: Inorganic Elements in the Chemistry of Life – An Introduction and Guide, Second Edition.
Written and Translated by Wolfgang Kaim, Brigitte Schwederski and Axel Klein.

Table 9.1 Ni-dependent enzymes and their abundance in organisms.

Ni-dependent proteins	Organisms
urease (Section 9.2)	archaea, bacteria, eukaryotes
[NiFe] hydrogenases (Section 9.3)	archaea, bacteria
carbon monoxide dehydrogenase (CODH) (Section 9.4)	archaea, bacteria
acetyl-*CoA* synthase/decarbonylase (ACS) (Section 9.4)	archaea, bacteria
methyl-coenzyme M reductase (MCR) (including the F_{430} cofactor) (Section 9.5)	archaea
superoxide dismutase (Ni-SOD) (Section 9.6)	bacteria

oxygen-producing bacteria to overcome the dominating methane-producing bacteria has been discussed in terms of a lower nickel demand and a lowered volcanic nickel supply around 2.7 billion years ago [7], as a consequence of which the earth's atmosphere changed dramatically in the period between 2.5 and 2.3 billion years ago (see Chapter 5).

According to current knowledge [1,3,7–12], nickel is present in six essentially different kinds of enzymes (Table 9.1). Additionally, a tetrapyrrole complex with as yet unspecified function, the nickel-containing tunichlorin, has been isolated from a tunicate species [13]. With respect to biosynthetic utilization, nickel and the chemically related cobalt are largely confined to evolutionary "old" organisms [11].

9.2 Urease

Urease enzymes, which may be isolated for example from bacteria or plant products such as jack beans (*Canavalia ensiformis*), have an interesting history [14,15]. Contrary to the opinion of R. Willstätter, they were the first enzyme to be prepared in pure crystalline form (J. Sumner, 1926; see Figure 1.1). However, their nickel content wasn't determined until about 50 years later [14,15].

The "classical" urease (urea amidohydrolase) catalyzes the degradation of urea to carbon dioxide and ammonia (9.1).

$$H_2N{-}CO{-}NH_2 + H_2O \xrightarrow{\text{urease}} [H_2N{-}COO^- + NH_4^+] \longrightarrow 2\,NH_3 + CO_2 \tag{9.1}$$

Urea is a very stable molecule, which normally hydrolyzes very slowly to give isocyanic acid and ammonia, the half-life value of the uncatalyzed reaction (9.2) being 3.6 years at 38 °C.

$$H_2N{-}CO{-}NH_2 + H_2O \longrightarrow NH_3 + H_2O + H{-}N{=}C{=}O \tag{9.2}$$

The catalytic activity of the enzyme increases the rate of complete hydrolysis by a factor of about 10^{14}. Such an astonishing acceleration can only be explained by a change in the reaction mechanism (see Figure 2.5). While the uncatalyzed reaction involves a direct elimination of ammonia, the enzyme presumably catalyzes a hydrolysis reaction with carbamate, $H_2N{-}COO^-$, as the first intermediate. A metal-to-substrate binding would facilitate this latter reaction mechanism. Additional support for substrate binding to the metal comes from the fact that phosphate derivatives binding strongly to nickel inhibit the activity of the enzyme.

The crystal structures of three bacterial ureases, from *Klebsiella aerogenes* (PDB codes 2KAU for free enzyme, 1FWJ and 1FWE for acetohydroxamic acid (AHA)-inhibited variants) [15,16], *Bacillus pasteurii* (PDB code 6UBP) [17] and *Helicobacter pylori* (PDB codes 1E9Z for free enzyme (see Figure 1.1), 1E9Y complexed with AHA) [18], indicate the conservation of the main structural features of the reaction center (9.3), in which two Ni^{II} atoms are separated by 3.5 Å and bridged by a carbamylated lysine residue. One Ni^{II} atom is five-coordinate (NiN_2O_3), while the other is described as "pseudotetrahedral" (NiN_2O_2), with a weakly bound ligand [15–20]. This ligand may be a water molecule, primarily a terminal ligand to the five-coordinate Ni^{II} atom.

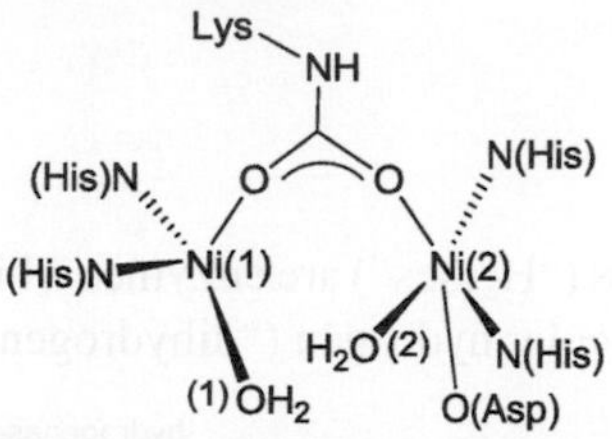

(9.3)

The holoenzyme of jack bean urease (Figure 1.1) consists of six equivalent subunits (91 kDa); each contains two dimetal sites (PDB code 3LA4) [15,20]. Extended x-ray absorption fine structure (EXAFS) measurements as well as magnetic and absorption spectroscopic studies are in agreement with the crystal structure results. Magnetic analysis points to an equilibrium between high-spin ($S = 1$) and low-spin ($S = 0$) forms [19], which is typical for five-coordinate or distorted six-coordinate Ni^{II} due to small and variable energy differences between d_{z^2} and $d_{x^2-y^2}$ orbitals (compare Figure 2.7).

The reaction mechanism shown in (9.4) was proposed in agreement with model studies and the previously mentioned experimental data [15,19,20]. It involves an electrophilic attack of one of the nickel centers on the carbonyl oxygen atom and a nucleophilic attack of a nickel hydroxo species on the carbonyl carbon center (push–pull mechanism; see also Section 12.2).

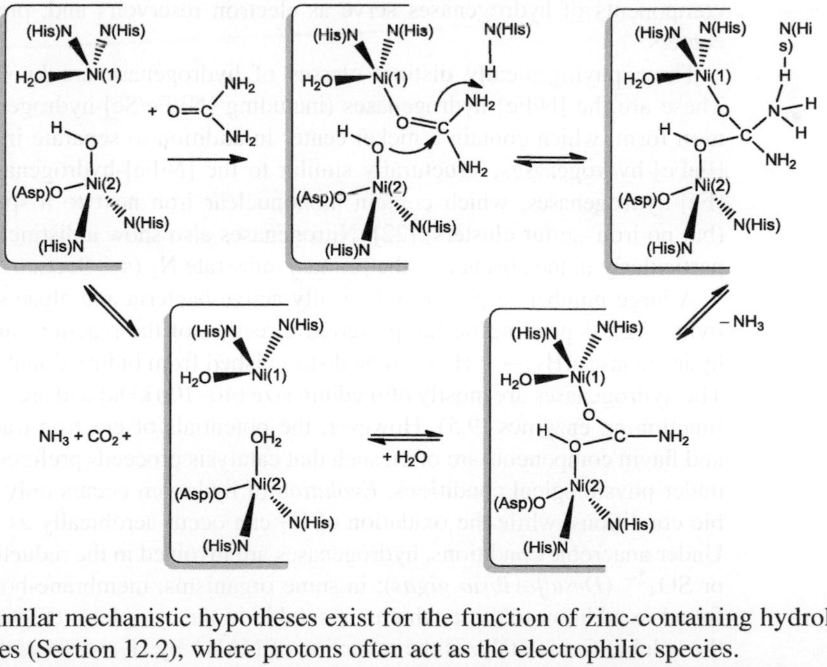

(9.4)

Similar mechanistic hypotheses exist for the function of zinc-containing hydrolytic enzymes (Section 12.2), where protons often act as the electrophilic species.

In the 1980s it was found that the Gram-negative bacterium *Helicobacter pylori* is largely responsible for gastric ulcers and stomach cancer. While this was not generally accepted until about 2000, the Nobel Prize in Medicine was awarded in 2005 to B. J. Marshall and J. R. Warren for their contributions to this finding. Probably more than 50% of the world's population harbors *H. pylori* in the upper gastrointestinal tract, and the bacterium owes its ability to survive the harsh conditions there (pH ~1) to a high concentration of the enzyme urease. This enables the bacterium to produce large amounts of ammonia (NH_3) from urea, which buffers the strong acidity.

9.3 Hydrogenases

Hydrogenases ("H_2ases") are enzymes which catalyze the reversible two-electron oxidation (9.5) of molecular hydrogen ("dihydrogen").

$$H_2 \xrightleftharpoons{\text{hydrogenase}} 2\,H^+(aq) + 2\,e^- \tag{9.5}$$

This reaction plays a major role in the course of dinitrogen "fixation" (Section 11.2), in microbial phosphorylation and in the fermentation of biological substances to methane, among other things. Anaerobic and some aerobic microorganisms contain hydrogenase enzymes; H_2 can either serve as an energy source instead of NADH or occur as the product of reductive processes [10,21,22]. Nature's concept of hydrogen conversion or the reverse process of hydrogen generation is based on heterolytic splitting of dihydrogen ($H_2 \rightleftharpoons H^+ + H^-$) at catalytic centers. The acidity of H_2, which is extremely low ($pK_a = 35$), is dramatically increased on binding to a metal. Thus, for reaction (9.5), a two-electron transfer is feasible under physiological conditions at about −0.3 V [23]. The simple one-electron reduction of a proton to give a hydrogen atom would require much more negative potentials of less than −2 V and thus completely unphysiological conditions. The inorganic components of hydrogenases serve as electron reservoirs and, presumably, as catalytic centers.

Three phylogenically distinct classes of hydrogenase are known today [10,23–27]. These are the [NiFe]-hydrogenases (including [Ni/Fe/Se]-hydrogenases), the most common form, which contain a nickel center in addition to separate iron–sulfur clusters, the [FeFe]-hydrogenases, structurally similar to the [NiFe]-hydrogenases, and the so-called [Fe]-hydrogenases, which contain mononuclear iron next to a special organic cofactor (but no iron–sulfur clusters) [22]. Nitrogenases also show a distinct hydrogenase activity, particularly in the absence of the primary substrate N_2 (see Section 11.2).

A large number of photosynthetically active bacteria and algae show hydrogenase activity [22]; depending on the preferred direction of the reaction, unidirectional "uptake" hydrogenases ($H_2 \rightarrow 2\,H^+$) can be distinguished from bidirectional "reversible" enzymes. The hydrogenases are mostly of medium size (40–100 kDa) and are, in principle, reversibly functioning enzymes (9.5). However, the potentials of electron-transferring Fe/S cluster and flavin components are often such that catalysis proceeds preferentially in one direction under physiological conditions. *Evolution* of hydrogen occurs only under strictly anaerobic conditions, while the oxidation of H_2 can occur aerobically as well as anaerobically. Under anaerobic conditions, hydrogenases are involved in the reduction of CO_2, NO_3^-, O_2 or SO_4^{2-} (*Desulfovibrio gigas*); in some organisms, membrane-bound hydrogenases are found in addition to those which are soluble in the cytoplasm (Figure 9.1). In these cases, the soluble form catalyzes the reduction of NAD^+ by H_2, whereas the electrons generated via the membrane-localized oxidation of dihydrogen are inserted into the respiratory chain and serve in the production of high-energy phosphates [22].

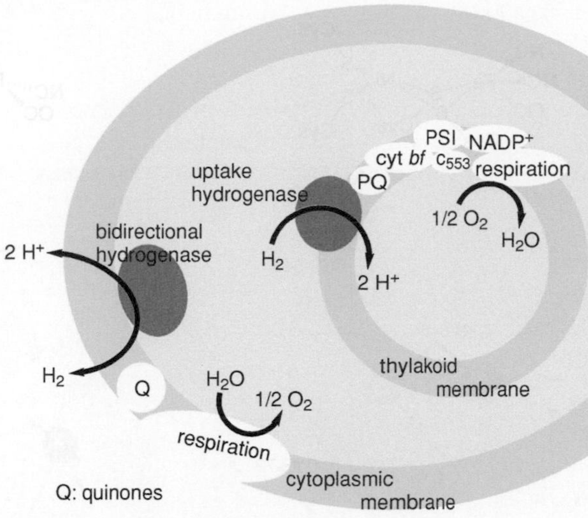

Figure 9.1
Arrangement of the protein complexes (compare Chapters 4 and 6), including both hydrogenases, in the cyanobacterium *Anacystis nidulans* (now called *Synechococcus elongatus*). Reprinted with permission from [22] © 2007, American Chemical Society.

Hydrogenases catalyze the degradation of potentially dangerous free dihydrogen, H_2, as inadvertently produced during reduction reactions; on the other hand, the hydrogenase-catalyzed "Knallgas" reaction (9.6) protects strictly anaerobic life processes by scavenging and deactivation of O_2, for example during N_2 fixation (Section 11.2).

$$2\,H_2 + O_2 \rightleftharpoons 2\,H_2O \qquad (9.6)$$

Hydrogenases are currently being thoroughly studied for their potential use in the controlled microbial degradation of organic material and in the production of CH_4 and H_2 as energy carriers [25,26]. However, the enzymes have proven to be very sensitive and labile, which has prevented large-scale application so far.

Both the [FeFe]- and the [NiFe]-hydrogenases contain bimetallic active centers featuring the biologically very unusual CO and CN^- ligands (Figure 9.2) [27]. The [NiFeSe]-enzymes incorporate a selenocysteine coordinated to the Ni in place of one of the terminal cysteine residues of the standard [NiFe]-enzymes. For [NiFe]-hydrogenases, a number of different redox states through which the system shuttles during catalytic activity have been found, including the oxidation states Ni^{III} and Ni^{I}. Further states, the functions of which are not yet clear, have also been detected, mainly by EXAFS, EPR and Fourier transform infrared (FTIR) spectroscopy. Due to the presence of the CO and CN^- ligands and the ease of detection of the corresponding $C{\equiv}O$ or $C{\equiv}N$ stretching vibrations, the technique of FTIR spectroelectrochemistry (*in situ* electrolysis + spectroscopy) [28] has turned out to be very useful [24].

Although much structural and spectroscopic information concerning these various states has been obtained, the mechanism of catalysis by [NiFe]-hydrogenases and the nature of the participating species and oxidation states are still controversial [22–24,29]. Considering that CO, CN^- and acetylene (ethyne) reversibly inhibit hydrogenases, there is a great deal of evidence that H_2 is split in the active center, while hydrophobic tunnels help to pass H_2 quickly through the protein to the Ni site [24]. Additionally, the electrons are transported through the enzyme molecule by a special relay system comprising Fe/S clusters (Figure 9.3).

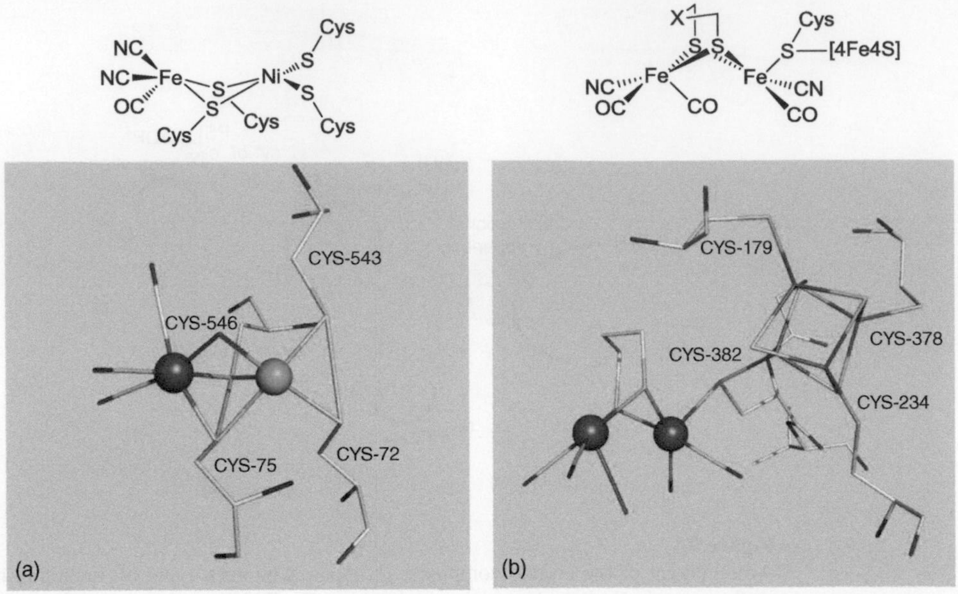

Figure 9.2
Schematic representation of the active centers of (a) *Desulfovibrio gigas* [NiFe]-hydrogenase (PDB code 1YRQ) and (b) *Desulfovibrio desulfuricans* [FeFe]-hydrogenase (PDB code 1HFE) [24]. The identity of the bridgehead atom X in the *D. desulfuricans* structure has been debated; most likely it is X = N.

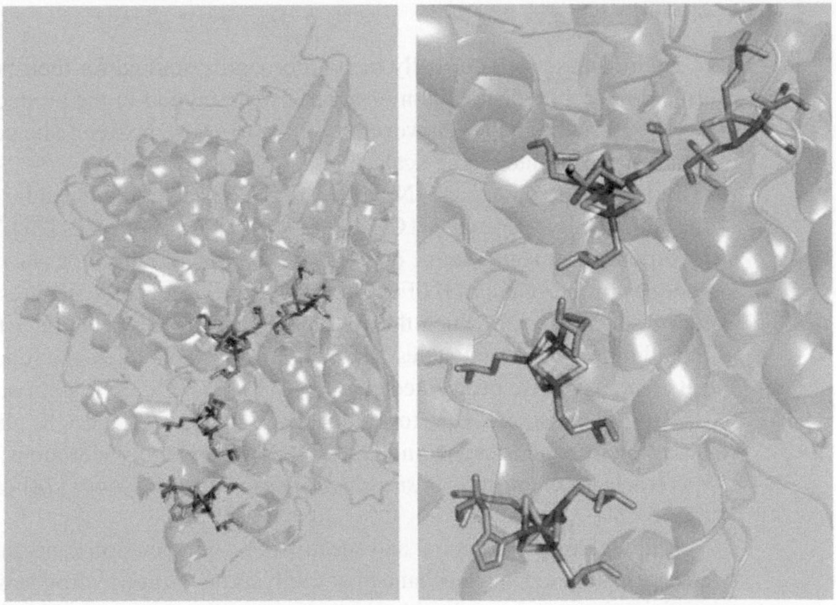

Figure 9.3
Structure of the [NiFe]-hydrogenase of *Desulfovibrio vulgaris* at 1.35 Å resolution, representing the reduced active enzyme. The crystal was first treated with H_2, then with CO, and was subsequently photolysed in order to cleave the CO and replace it with H_2 (PDB code 1UBU).

Any mechanism for hydrogenase catalysis must also take into account that these enzymes also accelerate the H/D exchange with water, according to (9.7).

$$H_2 + D_2O \xrightleftharpoons{\text{hydrogenase}} HD + HDO \tag{9.7}$$

Corresponding models postulate the heterolytic cleavage of H_2 (as already outlined) where the hydridic hydrogen atom (H^-) remains at the metal while the protic component (H^+) binds to a metal-coordinated sulfide, to hydroperoxo centers or, as the reaction proceeds, to other basic sites within the protein [23].

The nature of the assumed bond between hydrogen and the metal (Fe or Ni) [21,30] is of particular interest. More than Fe–H species, the hydrides of nickel are well known from organometallic chemistry. For the primary approach and binding of H_2 to the active site of the enzyme, the possibility of side-on (η^2) coordination of dihydrogen to Ni (or Fe) is discussed, since this binding mode has now been found in many complexes [21,31].

9.4 CO Dehydrogenase = CO Oxidoreductase = Acetyl-*CoA* Synthase

Many methanogenic and acetogenic (i.e. methane and acetic acid-producing) bacteria contain a "CODH" enzyme which catalyzes the oxidation (9.8) of CO to CO_2. In biochemistry, oxidation is often equated with dehydrogenation; however, CO does not contain any hydrogen atoms and the term "CO oxidoreductase" instead of "CO dehydrogenase" would thus be more appropriate.

$$CO + H_2O \rightleftharpoons CO_2 + 2\,H^+ + 2\,e^- \tag{9.8}$$

The reaction is enzymatically reversible and may therefore serve as an alternative method of CO_2 fixation (assimilation) by photosynthetic bacteria. The other biological function of the enzyme is to catalyze the reversible formation of acetyl-*CoA* (see (3.15)) in combination with *CoA* itself and a methyl source (9.9). Corrinoid proteins with a CH_3–[Co] functionality (see Section 3.3.2), a methyl transferase and an Fe/S-containing disulfide reductase also contribute to this reaction [1,9,32–35].

$$H_3C\text{–}[Co] + CO + HSCoA \rightleftharpoons CH_3C(O)S\text{–}CoA + H^+ + [Co]^- \tag{9.9}$$

The resulting acetyl-*SCoA* ("activated acetic acid") can be carboxylated to pyruvate $CH_3C(O)COO^-$ in autotrophic bacteria. In methanogenic bacteria, the further degradation of acetic acid to CO_2 and CH_4 presumably proceeds via CO as an intermediate. Acetyl-*CoA* synthase (ACS), together with CODH, is a key player in the metabolism of anaerobic microorganisms via the Wood–Ljungdahl pathway (Figure 9.4) and a major component of the global carbon cycle [33,36] (see also insertion Section 15.3.2).

It has been shown that the CODH of all anaerobic bacteria contains nickel, while aerobic species require molybdopterin (see Section 11.1). Several crystal structures have been obtained which show that the enzyme is a mushroom-shaped homodimer containing metal clusters [32–35,37]. The cluster "C" is a [3Fe-4S] cluster bridged to a heterodinuclear [NiFe] site, the catalytic center for CO oxidation (Figure 9.5). Clusters B and D are $[4Fe\text{-}4S]^{2+/1+}$ clusters that transfer electrons between the C cluster and external redox proteins.

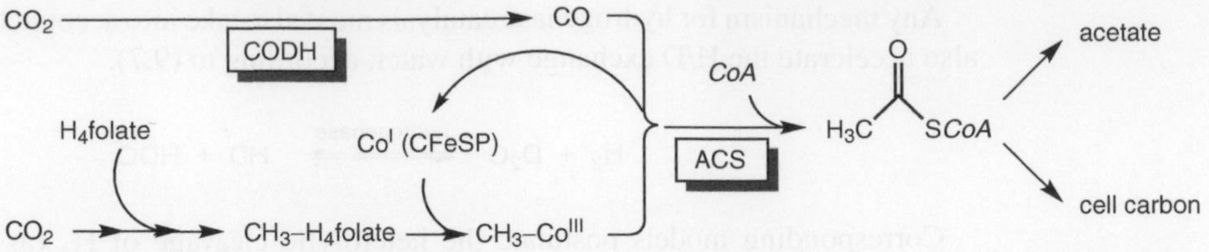

Figure 9.4
Diagram of the Wood–Ljungdahl pathway used in the carbon metabolism of anaerobic organisms. Adapted with permission from [35] © 2004, American Chemical Society.

From this structure and further spectroscopic evidence, the mechanism for the CO oxidation can be described as depicted in Figure 9.6. H_2O is bound to the C cluster in the bridging Fe··· Ni position, losing a proton (C_{red1}). When CO is bound (presumably to Ni), the Fe-bound OH^- is able to perform a nucleophilic attack on CO. The Ni–CO(OH) intermediate (not shown) is deprotonated and CO_2 is formed, while two electrons are transferred on to the C cluster ($C_{red1} \rightarrow C_{red2}$). The two electrons are subsequently transferred to the B cluster, then the D cluster, and finally to external electron acceptors.

The ACS active site (the "A cluster") consists of a [4Fe-4S] cluster, which is CysS-bridged to a proximal Ni site (Ni_p), which is thiolate-bridged to the so-called "distal Ni ion" (Ni_d), which has a square-planar thiolato- and carboxamido-type N_2S_2 coordination environment (Figure 9.7) [33,35]. There is a similarity between this arrangement and the hydrogenases. The coordination sphere of Ni_p is completed by an exogenous ligand L of so far unknown identity.

There are two main lines of proposed mechanisms for ACS. One describes all steps involving only diamagnetic intermediates and thus needs to postulate a biologically rather unlikely Ni^0 state [9,35,38], while the other comprises paramagnetic species [33,34]

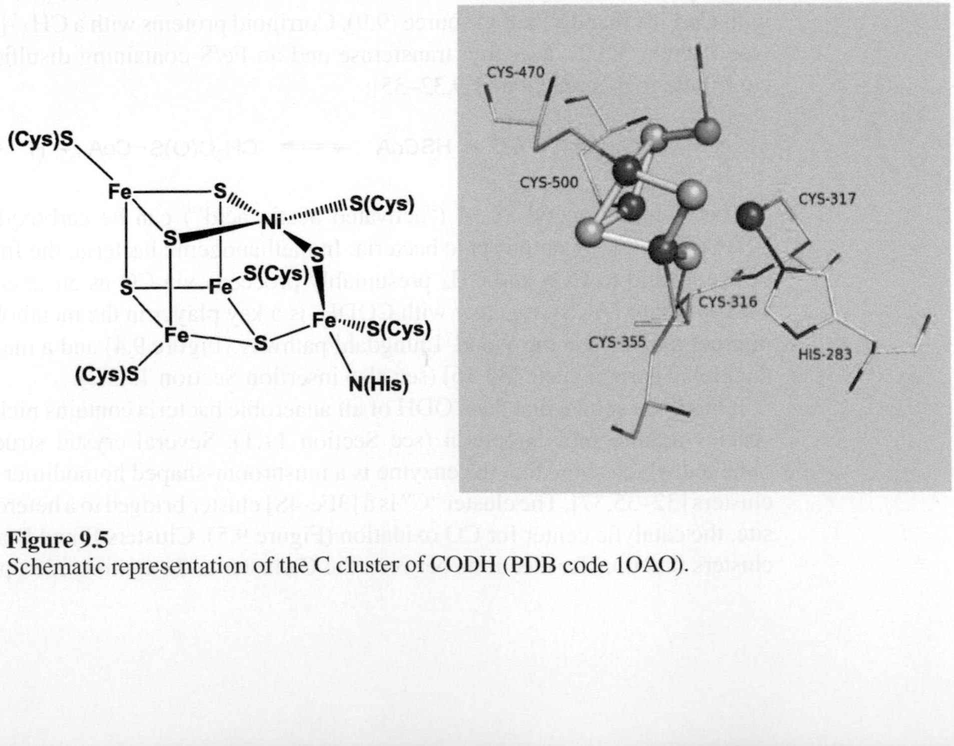

Figure 9.5
Schematic representation of the C cluster of CODH (PDB code 1OAO).

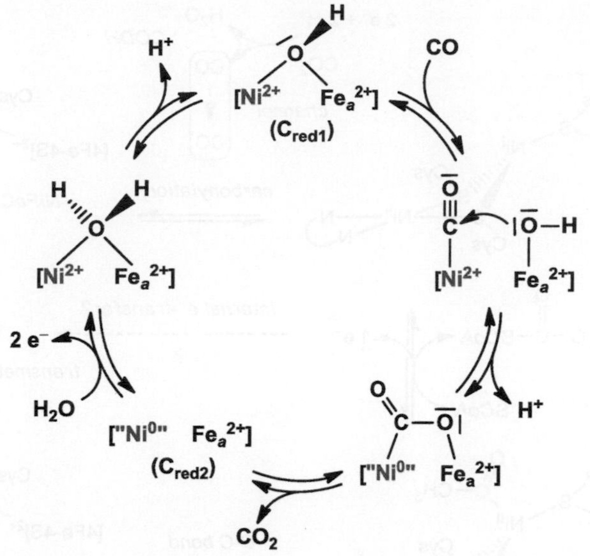

Figure 9.6
Assumed catalytic mechanism for CODH [32]. "Ni^0" is a consequence of the electron count. The oxidation state of nickel in this step remains unclear.

(Figure 9.8). What is generally agreed is that the crystal structures of CODH/ACS reveal the presence of a hydrophobic channel connecting the CODH active site (C cluster) to the ACS A cluster. This gas channel opens at the metal proximal to the [4Fe-4S] cluster, suggesting that Ni_p is the site of CO binding. In its isolated form, CODH/ACS is oxidized and EPR-silent, consistent with a $[4Fe\text{-}4S]^{2+}$ cluster linked to two $S = 0$ Ni^{2+} ions. However, this species does not exhibit ACS activity and requires the addition of an electron to become catalytically active. It was previously assumed that the electron came from the $[4Fe\text{-}4S]^{2+}$ cluster, but quantum chemical calculations of EPR and Mössbauer parameters suggest a $[4Fe\text{-}4S]^{2+} \cdots Ni_p^{I} \cdots Ni_d^{II}$ arrangement [33]. The addition of CO to this species results in the formation of the "NiFeC" state, which contains the diagnostic EPR signal (g = 2.08, 2.07 and 2.03). No other species in the ACS pathway has been successfully characterized, and there is little further agreement concerning the mechanism [33,35]. In summary, the ACS catalytic cycle appears to involve metal-centered catalysis via bioorganometallic intermediates that resemble the Monsanto process for industrial acetate production [34,35]. The insertion of CO into a metal–alkyl bond is well known from the organometallic chemistry of nickel. Model complexes with Ni^{II}–CH_3, Ni^{II}–C(O)CH_3,

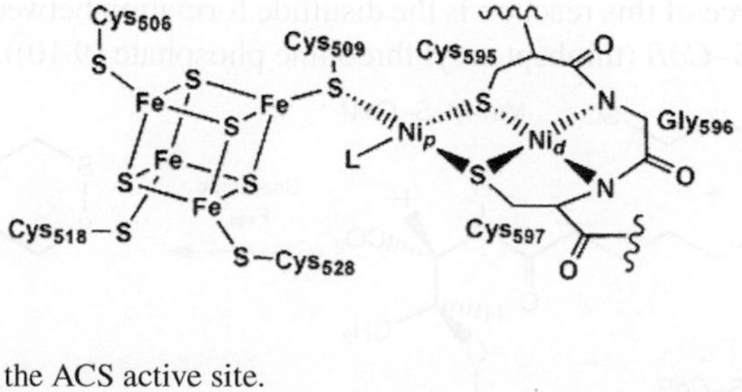

Figure 9.7
Schematic drawing of the ACS active site.

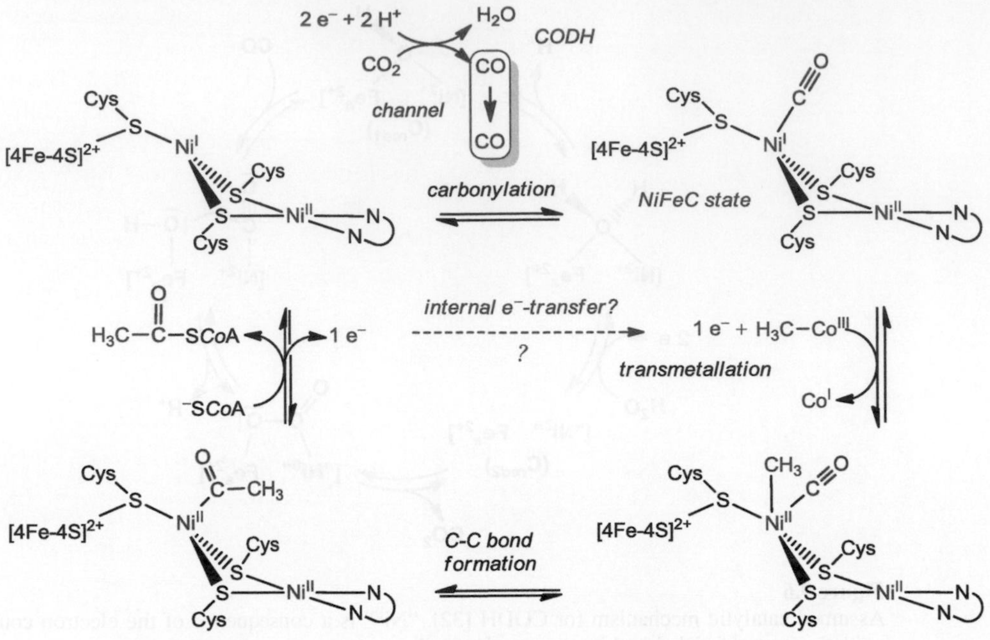

Figure 9.8
Proposed mechanism for ACS. Adapted with permission from [33] © 2006, American Chemical Society. The depicted radical mechanism for the transmetallation reaction requires a one-electron input, and since no external electron transfer has been found for ACS, the electron may be redonated internally.

Ni^{II}–H and Ni^{I}–CO functions have been obtained with tetradentate "tripod" ligands and the reaction sequence Ni–CH_3 → Ni–C(O)CH_3 → CH_3COSR' can be established [39]. The use of a low-valent metal center as a nucleophile is rare in biochemistry and is reminiscent of the reactions of cobalamin-dependent methyltransferases such as methionine synthase (Chapter 3) [8–10].

9.5 Methyl-coenzyme M Reductase (Including the F_{430} Cofactor)

Methyl-coenzyme M reductase (MCR) serves methanogenic bacteria in the eventual formation and liberation of methane by catalyzing the reduction of methyl-*CoM* (2-methylthioethanesulfonate) [1,3,8,33,40–42].

Reaction (9.10) is the last step in the energy-producing synthesis of CH_4 from CO_2 by autotrophic archaebacteria such as *Methanobacterium thermoautotrophicum*. The driving force of this reaction is the disulfide formation between *CoM* and the additional component HS–*CoB* (thioheptanoyl threonine phosphate (9.10)) [40,42].

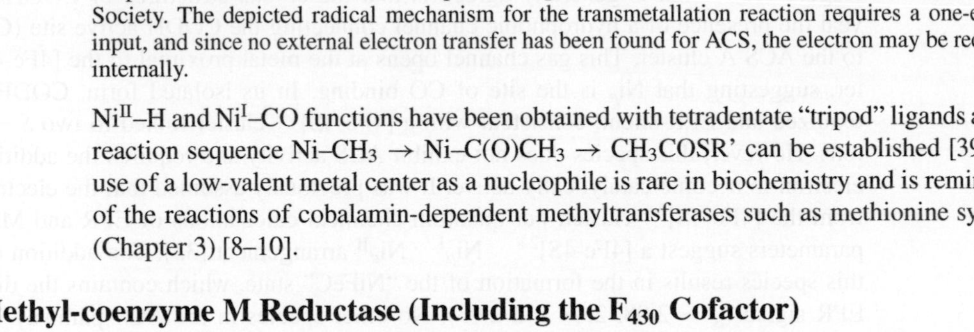

(9.10)

Several coenzymes are essential for the entire CO_2 to CH_4 conversion process (9.11), which requires a total of eight electrons. Methanofuran serves in the uptake of CO_2 and for the transformation of the carboxylic function during the first $2e^-$ reduction. The resulting formyl group, -CHO, is transferred to tetrahydromethanopterin, which functionally resembles tetrahydrofolic acid (3.12) of eukaryotes. After two more two-electron-transfer steps (formyl → hydroxymethyl → methyl), the resulting methyl group is transferred to *CoM*, which releases methane in a reaction catalyzed by MCR (9.10). The electrons required for the various reduction steps are obtained through the oxidation of molecular hydrogen, which is catalyzed by various hydrogenases.

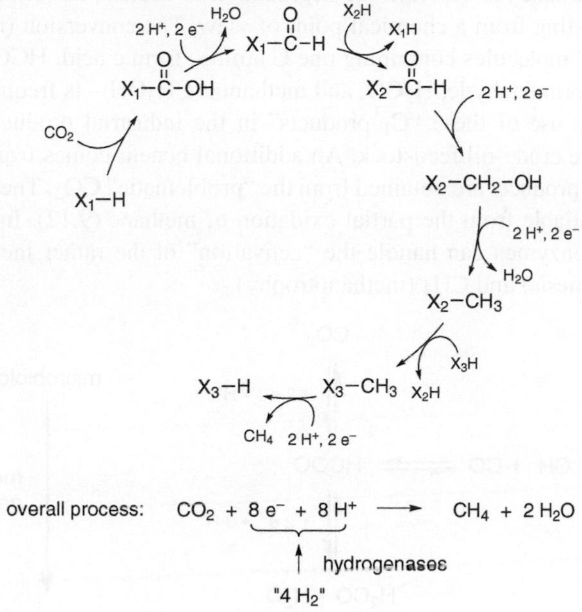

(9.11)

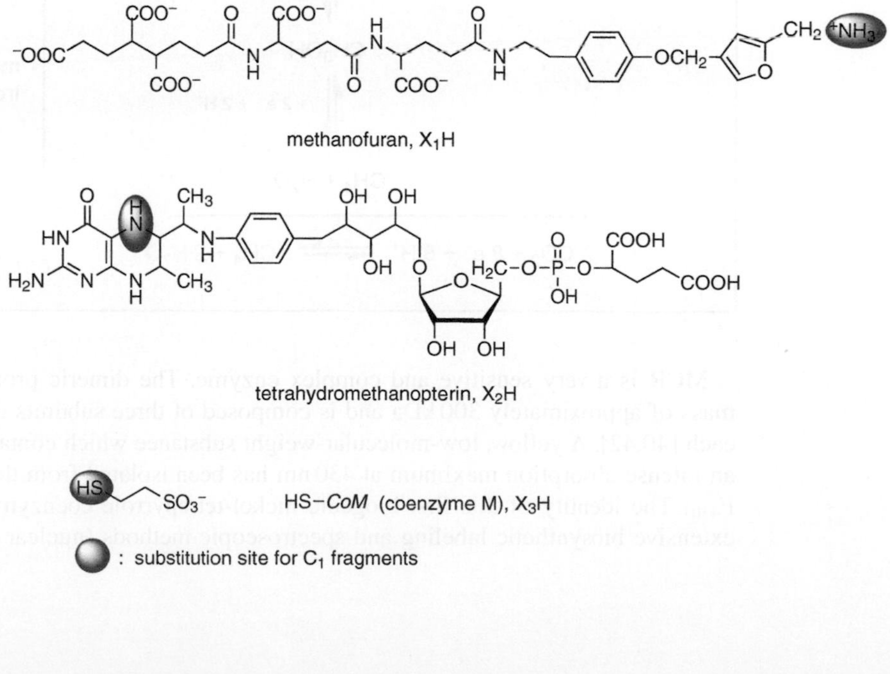

Natural and Artificial (Industrial) C_1 Chemistry

Methanogenic bacteria participate in the anaerobic microbial degradation of the organic components of sludge (to methane) in sewage plants. By volume, this is one of the largest biotechnological processes. In addition, microbial methane production takes place in sediments (→ natural gas), during the cultivation of rice and in the digestive system of ruminants. The atmospheric trace component CH_4 has received much attention lately due to its contribution to the "greenhouse effect". While most biogenic methane results from the degradation of acetate, its formation from CO_2 is more interesting from a chemical point of view. The conversion (reduction) of CO_2 to "organic" molecules containing one C atom – formic acid, HCOOH (or formiate, $HCOO^-$), formaldehyde, H_2CO, and methanol, CH_3OH – is frequently discussed in terms of the use of these "C_1 products" in the industrial production of chemicals replacing the crude-oil feedstock. An additional benefit comes from the fact that the "useful" C_1 products are obtained from the "problematic" CO_2. The same C_1 products are also available from the partial oxidation of methane (9.12). Importantly, nickel-containing enzymes can handle the "activation" of the rather inert molecules CO_2 (methanogenesis) and CH_4 (methanotrophy).

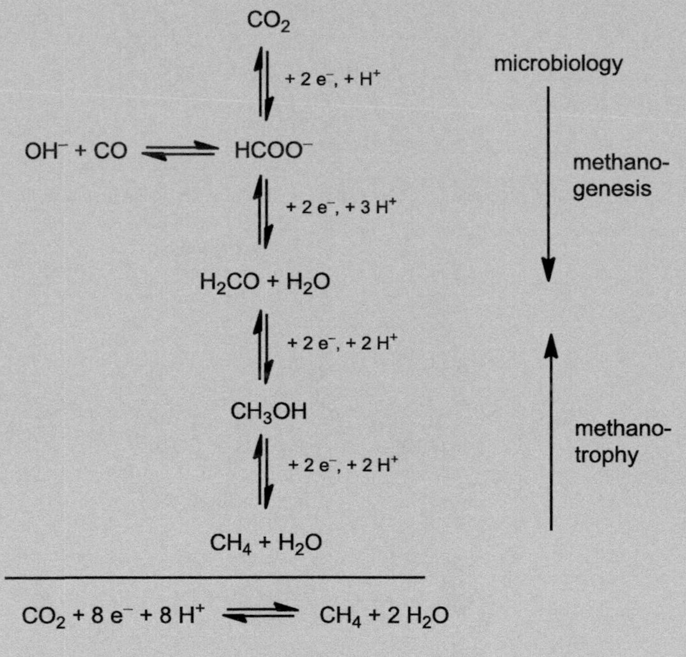

(9.12)

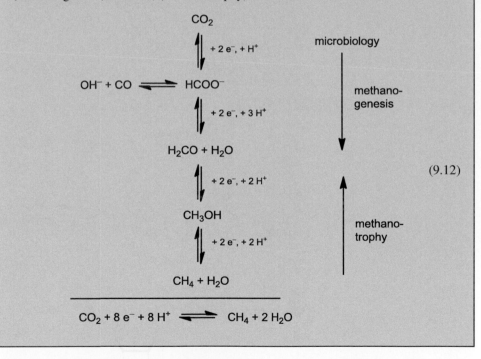

MCR is a very sensitive and complex enzyme. The dimeric protein has a molecular mass of approximately 300 kDa and is composed of three subunits of 68, 47 and 38 kDa each [40,42]. A yellow, low-molecular-weight substance which contains nickel and shows an intense absorption maximum at 430 nm has been isolated from the enzyme: the factor F_{430}. The identity of this first biogenic nickel-tetrapyrrole coenzyme was elucidated by extensive biosynthetic labeling and spectroscopic methods (nuclear magnetic resonance,

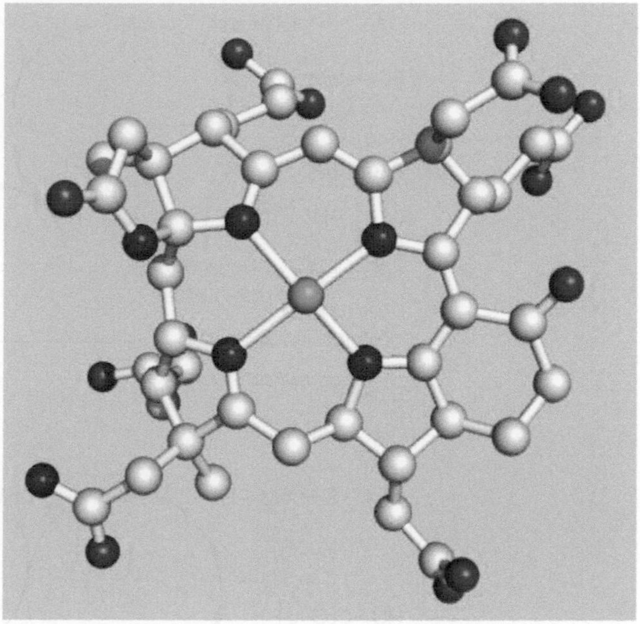

Figure 9.9
Structure of the F_{430} cofactor (PDB code 1HBN). Green, Ni; grey, C; red, O; blue, N [42].

NMR), as well as crystal structure determination [42]. It features a highly hydrogenated porphin system (Figure 9.9) with anellated lactam and cyclohexanone rings in which the π chromophore only extends over three of the four nitrogen atoms (2.9). The underlying ring structure has been named "hydrocorphin" in order to highlight the relationships with both porphyrins and the not-cyclically-conjugated corrins (F_{430} as "missing link" in the evolution of tetrapyrroles [43]).

The partially saturated character of the macrocycle and the anellation of the additional saturated rings allow a marked structural flexibility, especially with regard to a folding of the tetrapyrrole ring in the direction of an S_4-distortion (saddle-like, see Figure 2.6). As an important consequence, both the spin crossover to low-spin Ni^{II} ($S = 0$) and the transition to the d^9-configurated Ni^I state are facilitated for the otherwise not axially activated high-spin nickel(II) center ($S = 1$). A low-spin d^8 configuration (i.e. spin-pairing) only results after strong distortion of the octahedral ligand field; the degeneracy of the e_g orbitals, which would disfavor spin pairing, has to be effectively lifted (Figure 2.7). As expected for a $Ni^I = d^9$ species, the EPR spectrum of reduced F_{430} indicates a half-filled $d_{x^2-y^2}$ orbital with spin delocalization to the equatorial nitrogen atoms and a nucleophilic electron pair in the d_{z^2} orbital [44,45].

Analysis of the mechanistic function of nickel in the enzyme-bound cofactor F_{430} is far from complete, and two different mechanisms are under discussion (Figure 9.10). Catalytic mechanism I (Figure 9.10, top) postulates as the key step a nucleophilic attack of the Ni^I (reduced state) of F_{430} at the methyl group of CH_3–S*CoM*. The thus generated CH_3–Ni^{III} state (so-called MCR_{Me}) has not yet been observed within the catalytic cycle, although an artificially generated CH_3–Ni^{III} species (from Ni^I + MeI) can be established by x-ray diffraction (XRD) and x-ray absorption spectroscopy (XAS) [46]. It is assumed that it

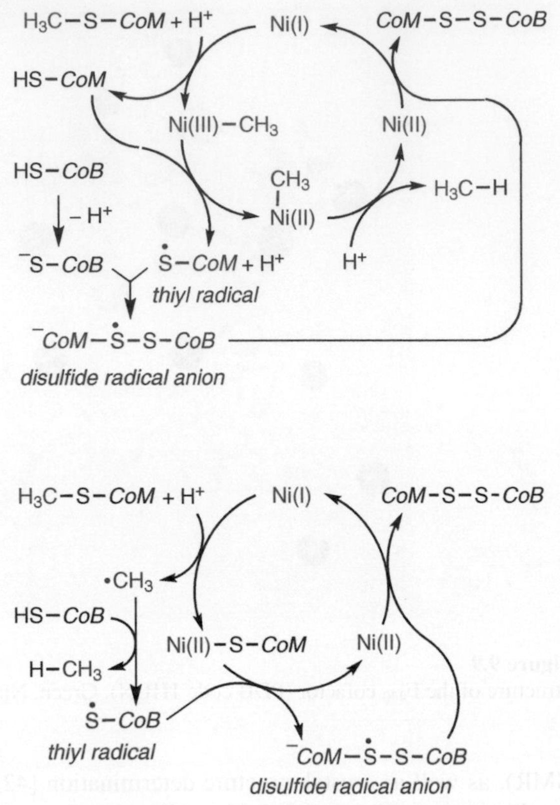

Figure 9.10
Two postulated mechanisms for MCR. Adapted with permission from [40] © 2005, Royal Society of Chemistry.

subsequently withdraws an electron from *CoM* thiolate to form CH_3–Ni^{II} and a *CoM* thiyl radical. Correspondingly, CH_3–Ni^{II} becomes protonated, producing Ni^{II} and CH_4. The *CoM* thiyl radical reacts with HS–*CoB* to give a heterodisulfide anion radical, and finally the excess electron returns to nickel, converting Ni^{II} to Ni^{I}. Mechanism II (Figure 9.10, bottom) includes in its key step an attack of Ni^{I} at the thioether sulfur of CH_3–S*CoM*, generating a free methyl radical (also highly reactive and thus not directly detected), which is converted to CH_4 by withdrawing a hydrogen atom from the sulfhydryl group of HS–*CoB*. The thiyl radical form of *CoB* thus produced combines with *CoM* thiolate to form the heterodisulfide radical anion, whereas the excess electron is returned to the nickel.

Bioorganometallics II: The Organometallic Chemistry of Cobalt and Nickel

Organometallic compounds are defined as containing a direct metal–C(ligand) bond, as in complexes of the neutral molecule CO and "organic" ligands such as methyl and phenyl. The latter are best described as carbanions (e.g. $^{-}CH_3$ or $^{-}C_6H_5$). In aqueous solution, a species such as $^{-}CH_3$ (systematically called "methanide"; isoelectronic to

NH_3) is a very strong base and will immediately yield CH_4. However, when bound to a relatively electronegative transition metal, the "methanide" loses much of its basic character, and corresponding complexes can be stable towards hydrolysis, such as the Co–CH_2R bonds in alkylcobalamins (Chapter 3). In a similar way, complexes containing the strongly basic hydride ligand ($^{-}$H) can be discussed. Of the series of 3d metals (Sc to Zn), cobalt and its neighbor nickel are the only two with the ability to form such relatively unreactive M–C bonds, and thus, from the viewpoint of an organometallic chemist, "nature's choice" for nickel and cobalt is not unexpected. In organometallic catalysis, the occurrence of organo-nickel or -cobalt complexes containing M–CO, M–CN or M–H functions (cf. hydrogenases), of complexes with rather stable M–CH_3 bonds (cf. MeB_{12}, MCR) and of complexes which can catalyze the insertion of CO into a metal–alkyl bond (M–CH_3 + CO → M–C(O)CH_3; cf. ACS) is well established.

It can be concluded that noble metals of the platinum group (Ru, Os, Rh, Ir, Pd, Pt) and gold should also form unreactive M–C bonds. This is indeed so, and explains why these elements find various applications in organometallic catalysis. Organometallic complexes of these metals do not play a role in biochemistry, due to their low natural abundance. However, since many organometallic complexes of these elements are largely unreactive towards hydrolysis under physiological conditions, such complexes have gained significance for their use in medical diagnosis and treatment, for example as anticancer drugs (see Section 19.5). (See Chapter 3 for Bioorganometallics I.)

9.6 Superoxide Dismutase

The reactive oxygen species (ROS) superoxide ($O_2^{\bullet-}$) is an inevitable cytotoxic byproduct of aerobic metabolism, and, if not eliminated, can lead to a variety of health disorders (see also Chapters 5 and 10). A number of metalloenzymes known as superoxide dismutases (SODs) are able to catalyze the disproportionation of metastable $O_2^{\bullet-}$ into hydrogen peroxide (H_2O_2) and O_2 through alternate oxidation and reduction of the catalytic metal centers. These SODs have been extensively characterized and employ metal ions of Fe, Mn or Cu/Zn to catalyze the disproportionation reaction. Since the most commonly discussed SOD is of the Cu/Zn type, the role and occurrence of SODs is further discussed in Section 10.5, and only the specific properties of the Ni-containing SOD are outlined in this chapter.

Ni-SOD has been characterized relatively recently and not much is known regarding the mechanism [1,11,12,47,48]. However, the overall protein fold, spectroscopy, ligands and coordination geometry in Ni-SOD are quite distinct from those of other SODs. For example, Ni-SOD in its reduced Ni^{II} form (Ni-SOD_{red}) has a square-planar N_2S_2 geometry arising from the coordination of a primary amine-N of His1, carboxamido-N from Cys2 and two thiolato-S donor atoms from Cys2 and Cys6 (Figure 9.11). Oxidation of Ni-SOD_{red} to the Ni^{III} oxidation state produces a structural change about the metal center; the N-terminal histidine group coordinates to Ni^{III} in the axial position.

The presence of carboxamido-N and primary amine-N provides an unusual set of donors and raises questions with regard to the properties they impart on the nickel center, with few other such examples known in biology [12,48]. Another striking feature of Ni-SOD is the presence of two coordinated cysteine thiolates, which are responsible for the modulation

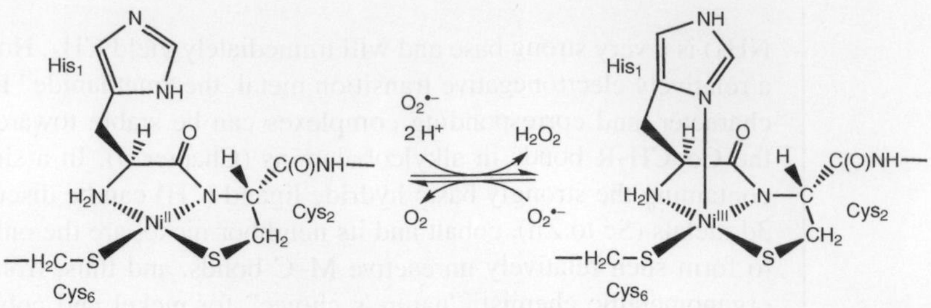

Figure 9.11
Proposed mechanism for Ni-SOD.

of the $Ni^{II/III}$ redox couple to a value suitable for the physiological function (see also Section 9.7). In order to be able to reduce and oxidize $O_2^{\bullet -}$, the $Ni^{II/III}$ redox couple of the active center has to lie in between those of $O_2^{\bullet -}/O_2$ (−0.16 V) and $O_2^{\bullet -}/H_2O_2$ (0.89 V vs normal hydrogen electrode, NHE). Additional questions arise as to how superoxide coordinates, concerning the source of protons in the oxidative half-reaction, and how these tie into the overall Ni-SOD catalytic mechanism.

9.7 Model Compounds

Since the most striking aspect of the Ni-containing enzymes is their oxidation states, ranging from Ni^0 to Ni^{III}, it is not surprising that many model compounds have been designed to mimic this behavior. In organometallic compounds, nickel frequently occurs in low oxidation states such as +I or 0, which are important for catalytic hydrogenations (cf. hydrogenases), desulfurization (cf. MCR) and carbonylation reactions (cf. CODH/ACS). On the other hand, Ni^{III} has long been considered a rather unusual oxidation state of the metal. However, a large number of Ni^{III} complexes with chelating amide ligands, deprotonated peptides, oximes and thiols have been synthesized [48–51], stimulated not least by the detection of trivalent nickel in oxidized if not necessarily active enzymes (cf. MCR or Ni-SOD). An important aspect in modeling Ni^{III} sites is that the coordination geometry should be restricted (e.g. by chelating ligands), since the electronic configurations d^8 (Ni^{II}) and d^7 (Ni^{III}) show markedly different geometrical preferences. Applying such strategies and using highly electron-rich thiolate (RS^-), sulfide (S^{2-}) or amide (R_2N^-) donor ligands (σ and π donation), it has been possible to lower the potential for the $Ni^{II/III}$ couple to such an extent that Ni^{III} species become stable even under physiological conditions and reach the relatively low potentials of the nickel-containing enzymes [48–51].

Further Reading

A. Sigel, H. Sigel, R. K. O. Sigel (eds.) *Metal Ions in Life Science, Vol. 2: Nickel and its Surprising Impact in Nature*, John Wiley & Sons, Hoboken, **2007**.

References

1. R. K. Thauer, *Science* **2001**, *293*, 1264–1265: *Nickel to the fore*.
2. P. Schönheit, J. Moll, R. K. Thauer, *Arch. Microbiol.* **1979**, *123*, 105–107: *Nickel, cobalt, and molybdenum requirement for growth of Methanobacterium thermoautotrophicum*.
3. R. K. Thauer, *Microbiol.* **1998**, *144*, 2377–2406: *Biochemistry of methanogenesis: a tribute to Marjory Stephenson*.
4. B. T. A. Muyssen, K. V. Brix, D. K. DeForest, C. R. Janssen, *Environ. Rev.* **2004**, *12*, 113–131: *Nickel essentiality and homeostasis in aquatic organisms*.
5. F. O. Nestle, H. Speidel, M. O. Speidel, *Nature (London)* **2002**, *419*, 132–133: *High nickel release from 1- and 2-euro coins*.
6. R. K. Thauer, *Angew. Chem. Int. Ed. Engl.* **2010**, *49*, 6712–6713: *Functionalization of methane in anaerobic microorganisms*.
7. (a) K. O. Konhauser, E. Pecoits, S. V. Lalonde, D. Papineau, E. G. Nisbet, M. E. Barley, N. T. Arndt, K. Zahnle, B. S. Kamber, *Nature (London)* **2009**, *458*, 750–754: *Oceanic nickel depletion and a methanogene famine before the great oxidation event*; (b) M. A. Saito, *Nature (London)* **2009**, *458*, 714–715: *Less nickel for more oxygen*.
8. B. Jaun, R. K. Thauer, *Nickel-alkyl bond formation in the active site of methyl-coenzyme M reductase*, in A. Sigel, H. Sigel, R. K. O. Sigel (eds.) *Metal Ions in Life Science, Vol. 6: Metal–Carbon Bonds in Enzymes and Cofactors*, RSC Publishing, Cambridge, **2009**.
9. P. A. Lindahl, *Nickel-carbon bonds in acetyl-coenzyme A synthases/carbon monoxide dehydrogenases*, in A. Sigel, H. Sigel, R. K. O. Sigel (eds.) *Metal Ions in Life Science, Vol. 6: Metal–Carbon Bonds in Enzymes and Cofactors*, RSC Publishing, Cambridge, **2009**.
10. (a) J. C. Fontecilla-Camps, *Structure and function of [NiFe]-hydrogenases*, in A. Sigel, H. Sigel, R. K. O. Sigel (eds.) *Metal Ions in Life Science, Vol. 6: Metal–Carbon Bonds in Enzymes and Cofactors*, RSC Publishing, Cambridge, **2009**; (b) J. C. Fontecilla-Camps, P. Amara, C. Cavazza, Y. Nicolet, A. Volbeda, *Nature (London)* **2009**, *460*, 814–822: *Structure-function relationships of anaerobic gas-processing metalloenzymes*.
11. Y. Zhang, V. N. Gladyshev, *Chem. Rev.* **2009**, *109*, 4828–4861: *Comparative genomics of trace elements: emerging dynamic view of trace elements utilization and function*.
12. R. H. Holm, P. Kennepohl, E. I. Solomon, *Chem. Rev.* **1996**, *96*, 2239–2314: *Structural and functional aspects of metal sites in biology*.
13. F.-P. Montforts, M. Glasenapp-Breitling, *Naturally occurring cyclic tetrapyrroles*, in W. Herz, H. Falk, G. W. Kirby (eds.), *Progress in the Chemistry of Organic Natural Products, Vol. 84*, Springer, Wien, **2002**.
14. (a) P. A. Karplus, M. A. Pearson, R. P. Hausinger, *Acc. Chem. Res.* **1997**, *30*, 330–337: *70 years of crystalline urease: what have we learned?*; (b) B. Zerner, *Bioorg. Chem.* **1991**, *19*, 116–131: *Recent advances in the chemistry of an old enzyme, urease*.
15. A. Balasubramanian, K. Ponnuraj, *J. Mol. Biol.* **2010**, *400*, 274–283: *Crystal structure of the first plant urease from Jack Bean: 83 years of journey from its first crystal to molecular structure*.
16. E. Jabri, M. B. Carr, R. P. Hausinger, P. A. Karplus, *Science* **1995**, *268*, 998–1004: *The crystal structure of urease from Klebsiella aerogenes*.
17. B. Zambelli, F. Musiani, S. Benini, S. Ciurli, *Acc. Chem. Res.* **2011**, *44*, 520–530: *Chemistry of Ni^{2+} in urease: sensing, trafficking, and catalysis*.
18. (a) D. Witkowska, M. Rowinski-Zyrek, G. Valensin, H. Kozlowski, *Coord. Chem. Rev.* **2012**, *256*, 133–148: *Specific poly-histidyl and poly-cysteil protein sites involved in Ni^{2+} homeostasis in Helicobacter pylori. Impact of Bi^{3+} ions on Ni^{2+} binding to proteins. Structural and thermodynamic aspects*; (b) N.-C. Ha, S.-T. Oh, J. Y. Sung, K. A. Cha, M. H. Lee, B.-H. Oh, *Nature Struct. Biol.* **2001**, *8*, 505–509: *Supramolecular assembly and acid resistance of Helicobacter pylori urease*.
19. (a) G. Schenk, N. Mitic, L. Gahan, D. L. Olus, R. P. McGeary, L. W. Guddat, *Acc. Chem. Res.* **2012**, *45*, 1593–1603: *Binuclear metallohydrolases: complex mechanistic strategies for a simple chemical reaction*; (b) D. F. Wilcox, *Chem. Rev.* **1996**, *96*, 2435–2458: *Binuclear metallohydrolases*.
20. E. L. Carter, N. Flugga, J. L. Boer, S. B. Mulrooney, R. P. Hausinger, *Metallomics* **2009**, 1, 207–221: *Interplay of metal ions and urease*.

21. J. C. Gordon, G. J. Kubas, *Organometallics* **2010**, *29*, 4682–4701: *Perspectives on how nature employs the principles of organometallic chemistry in dihydrogen activation in hydrogenases.*
22. P. M. Vignais, B. Billoud, *Chem. Rev.* **2007**, *107*, 4206–4272: *Occurrence, classification and biological function of hydrogenases: an overview.*
23. A. L. De Lacey, V. M. Fernandez, M. Rousset, R. Cammack, *Chem. Rev.* **2007**, *107*, 4304–4330: *Activation and inactivation of hydrogenase function and the catalytic cycle: spectroelectrochemical studies.*
24. J. C. Fontecilla-Camps, A. Volbeda, C. Cavazza, Y. Nicolet, *Chem. Rev.* **2007**, *107*, 4273–4303: *Structure/function relationships of [NiFe]- and [FeFe]-hydrogenases.*
25. M. L. Ghirardi, M. C. Posewitz, P.-C. Maness, A. Dubini, J. Yu, M. Seiber, *Annu. Rev. Plant Biol.* **2007**, 71–91: *Hydrogenases and hydrogen photoproduction in oxygenic photosynthetic organisms.*
26. (a) F. R. Armstrong, N. A. Belsey, J. A. Cracknell, G. Goldet, A. Parkin, E. Reisner, K. A. Vincent, A. F. Wait, *Chem. Soc. Rev.* **2009**, *38*, 36–51: *Dynamic electrochemical investigation of hydrogen oxidation production by enzymes and implications for future technology*; (b) J. A. Cracknell, K. A. Vincent, F. A. Armstrong, *Chem. Rev.* **2008**, *108*, 2439–2461: *Enzymes as working or inspirational electrocatalysts for fuel cells and electrolysis.*
27. K. D. Swanson, B. R. Duffus, T. E. Beard, J. W. Peters, J. B. Broderick, *Eur. J. Inorg. Chem.* **2011**, *7*, 935–947: *Cyanide and carbon monoxide ligand formation in hydrogenase biosynthesis.*
28. S. P. Best, S. J. Borg, K. A. Vincent, *Infrared spectroelectrochemistry*, in W. Kaim, A. Klein (eds.), *Spectroelectrochemistry*, RSC Publishing, Cambridge, **2008**.
29. P. Jayapal, M. Sundararajan, I. H. Hillier, N. A. Burton, *Phys. Chem. Chem. Phys.* **2006**, *8*, 4086–4094: *How are the ready and unready states of nickel–iron hydrogenase activated by H_2? A density functional theory study.*
30. W. Keim, *Angew. Chem. Int. Ed. Engl.* **1990**, *112*, 235–244: *Nickel: an element with wide application in industrial homogeneous catalysis.*
31. G. J. Kubas, *Chem. Rev.* **2007**, *107*, 4152–4205: *Fundamentals of H_2 binding and reactivity on transition metals underlying hydrogenase function and H_2 production and storage.*
32. P. A. Lindahl, *Angew. Chem. Int. Ed. Engl.* **2008**, *47*, 4054–4056: *Implications of a carboxylate-bound C-cluster structure of carbon monoxide dehydrogenase.*
33. S. W. Ragsdale, *Chem. Rev.* **2006**, *106*, 3317–3337: *Metals and their scaffolds to promote difficult enzymatic reactions.*
34. T. C. Harrop, P. K. Mascharak, *Coord. Chem. Rev.* **2005**, *249*, 3007–3024: *Structural and spectroscopic models of the A-cluster of acetyl coenzyme A synthase/carbon monoxide dehydrogenase: nature's Monsanto acetic acid catalyst.*
35. E. L. Hegg, *Acc. Chem. Res.* **2004**, *37*, 775–783: *Unraveling the structure and mechanism of acetyl-coenzyme A synthase.*
36. S. W. Ragsdale, *Biofactors* **1997**, *6*, 3–11: *The eastern and western branches of the Wood/Ljungdahl pathway: how the east and west were won.*
37. J. H. Jeoung, H. Dobbek, *Science* **2007**, *318*, 1461–1464: *Carbon dioxide activation at the Ni,Fe-cluster of anaerobic carbon monoxide dehydrogenase.*
38. P. A. Lindahl, *J. Biol. Inorg. Chem.* **2004**, *9*, 516–524: *Acetyl-coenzyme A synthase: the case for a $Ni_p{}^0$-based mechanism of catalysis.*
39. P. Stavroupoulus, M. C. Muetterties, M. Carrie, R. H. Holm, *J. Am. Chem. Soc.* **1991**, *113*, 8485–8492: *Structural and reaction chemistry of nickel complexes in relation to carbon monoxide dehydrogenase: a reaction system simulating acetyl-coenzyme A synthase activity.*
40. U. Ermler, *Dalton Trans.* **2005**, 3451–3458: *On the mechanism of methyl-coenzyme M reductase.*
41. S. Shima, R. K. Thauer, *Curr. Opin. Microbiol.* **2005**, *8*, 643–648: *Methyl-coenzyme M reductase and the anaerobic oxidation of methane in methanotrophic archaea.*
42. (a) W. Grabarse, F. Mahlert, E. C. Duin, M. Goubeaud, S. Shima, R. K. Thauer, V. Lamzin, U. Ermler, *J. Mol. Biol.* **2001**, *309*, 315–330: *On the mechanism of biological methane formation: structural evidence for conformational changes in methyl-coenzyme M reductase upon substrate binding*; (b) U. Ermler, W. Grabarse, S. Shima, M. Goubeaud, R. K. Thauer, *Science* **1997**, *278*, 1457–1462: *Crystal structure of methyl–coenzyme M reductase: the key enzyme of biological methane formation.*
43. A. Eschenmoser, *Angew. Chem. Int. Ed. Engl.* **1988**, *27*, 5–39: *Vitamin B_{12}: experiments concerning the origin of its molecular structure.*

44. (a) S. Ebner, B. Jaun, M. Goenrich, R. K. Thauer, J. Harmer, *J. Am. Chem. Soc.* **2010**, *132*, 567–575: *Binding of coenzyme B induces a major conformational change in the active site of methyl-coenzyme M reductase*; (b) D. Hinderberger, R. P. Piskorski, M. Goenrich, R. K. Thauer, A. Schweiger, J. Harmer, B. Jaun, *Angew. Chem. Int. Ed. Engl.* **2006**, *45*, 3602–3607: *A nickel-alkyl bond in an inactivated state of the enzyme catalyzing methane formation.*
45. C. Holliger, A. J. Pierik, E. J. Reijerse, W. R. Hagen, *J. Am. Chem. Soc.* **1993**, *115*, 5651–5656: *A spectroelectrochemical study of factor F_{430} nickel(II/I) from methanogenic bacteria in aqueous solution.*
46. P. E. Cedervall, M. Dey, X. Li, R. Sarangi, B. Hedman, S. W. Ragsdale, C. M. Wilmot, *J. Am. Chem. Soc.* **2011**, *133*, 5626–5628: *Structural analysis of a Ni-methyl species in methyl-coenzyme M reductase from Methanothermobacter marburgensis.*
47. J. Wuerges, J.-W. Lee, Y.-I. Yim, H.-S. Yim, S.-O. Kang, K. Djinovic Carugo, *Proc. Nat. Acad. Sci. USA* **2004**, *101*, 8569–8574: *Crystal structure of nickel-containing superoxide dismutase reveals another type of active site.*
48. D. P. Barondeau, C. J. Kassmann, C. K. Bruns, J. A. Tainer, E. D. Getzoff, *Biochemistry* **2004**, *43*, 8038–8047: *Nickel superoxide dismutase structure and mechanism.*
49. E. M. Gale, A. K. Patra, T. C. Harrop, *Inorg. Chem.* **2009**, *48*, 5620–5622. *Versatile methodology toward NiN_2S_2 complexes as nickel superoxide dismutase models: structure and proton affinity.*
50. J. Shearer, L. M. Long, *Inorg. Chem.* **2006**, *45*, 2358–2360. *A nickel superoxide dismutase maquette that reproduces the spectroscopic and functional properties of the metalloenzyme.*
51. I. Zilbermann, E. Maimon, H. Cohen, D. Meyerstein, *Chem. Rev.* **2005**, *105*, 2609–2625: *Redox chemistry of nickel complexes in aqueous solutions.*

10 Copper-containing Proteins: An Alternative to Biological Iron

For many iron-containing proteins, there are "parallel" copper-dependent analogues with comparable functions (Table 10.1). The correspondence between the reversibly O_2-binding proteins hemerythrin (Fe; see Section 5.3) and hemocyanin (Cu; see Section 10.2) has already been mentioned; both metals are also featured in electron-transfer proteins of photosynthesis and respiration, in the metabolism of dioxygen (e.g. in oxidases or oxygenases) and in the deactivation of toxic intermediates of O_2 reduction (see also Section 16.8).

Despite obvious functional similarities, however, iron and copper also show some general differences in their physiological appearance and function:

- Unlike iron in heme, biological copper does not occur in the form of tetrapyrrole coordination compounds. It is the imine-nitrogen atom in the imidazole ring of histidine which is able to form strong and, most importantly, kinetically inert bonds to copper in both relevant oxidation states, (+I) and (+II). The metal can thus be retained in the protein without having to be bound by special macrocycles.
- As a general rule, the redox potentials for $Cu^{I/II}$ transitions are higher than those for $Fe^{II/III}$ redox pairs, with both physiological and nonphysiological ligands. Copper proteins such as ceruloplasmin are thus able to catalyze the oxidation of Fe^{II} to Fe^{III} (ferroxidase reactivity).
- In neutral aqueous solution and in sea water, the *oxidized* form Cu^{2+} is more soluble than Cu^{+}, which forms insoluble compounds with halide and sulfide; in contrast, the oxidized form is *less* soluble in the $Fe^{II/III}$ system (5.1). In view of the biogenic O_2 production during early evolution, this difference also had geochemical implications in terms of increasing iron precipitation and copper mobilization: copper is a "modern" bioelement [1].
- Due to its later appearance and bioavailability in evolution, copper is often found in the extracellular space, whereas iron occurs mainly within cells (see also Section 13.1) [2].

Since human beings do not require copper proteins for their stoichiometric oxygen transport, the total amount of about 150 mg copper in the body of an adult is relatively small. Nevertheless, there is only a little tolerance for deviations, mainly because of the essential role of this trace element [3] in superoxide deactivation and in the respiratory chain (see Figure 6.1). In this context, some copper-related pathological disorders [4,5] should be mentioned:

- Wilson's disease involves a hereditary dysfunction of the primary copper storage capability of the body, in the protein ceruloplasmin (Table 10.2). The metal ion is then

Bioinorganic Chemistry: Inorganic Elements in the Chemistry of Life – An Introduction and Guide, Second Edition.
Written and Translated by Wolfgang Kaim, Brigitte Schwederski and Axel Klein.

Table 10.1 Correspondence of iron and copper proteins.

Function	Fe protein	Cu protein
O_2 transport	hemoglobin (*h*) hemerythrin (*nh*)	hemocyanin
oxygenation	cytochrome P-450 (*h*) methane monooxygenase (*nh*) catechol dioxygenase (*nh*)	tyrosinase quercetinase (dioxygenase)
oxidase activity	peroxidases (*h*) peroxidases (*nh*)	amine oxidases laccase
electron transfer	cytochromes (*h*)	blue Cu proteins
antioxidative function	peroxidases (*h*) bacterial superoxide dismutases (*nh*)	superoxide dismutase (Cu, Zn) from erythrocytes
NO_2^- reduction	heme-containing nitrite reductase (*h*)	Cu-containing nitrite reductase

h, heme system; *nh*, non-heme system.

accumulated in liver and brain, leading to dementia, liver failure and ultimately death. A therapy of this disease, and of acute copper poisoning, requires the administration of Cu-specific chelate ligands such as D-penicillamine (see (2.1) and [6]). This ligand contains both S(thiolate) and N(amine) coordination sites, in order to guarantee specificity for copper(I/II), and a hydrophilic carboxylic function, which makes the resulting complex excretable.

- Acute copper deficiency may occur particularly in newborn infants, since the complex metal transport and storage mechanisms involving serum albumin, ceruloplasmin and metallothionein (see Section 17.3) stabilize only several months after birth. Due to the essential role of copper in respiration (cytochrome *c* oxidase; see Section 10.4), this deficiency can cause an insufficient oxygen utilization in the brain and thus permanent damage. Infants are also very sensitive to excessive supply of copper; they usually have a high saturation concentration in the liver directly after birth. Corresponding childhood cirrhosis has thus been described for particularly copper-exposed parts of the world (India, Tyrol, parts of Germany).
- Menke's "kinky hair" syndrome is based on a hereditary dysfunction of intracellular copper transport [4,7]. The resulting copper deficiency symptoms in infants include severe disturbances in mental and physical development accompanied by the occurrence of kinky hair; an effective therapy must rely on intravenously administered copper compounds. The occurrence of sparse, kinky hair due to disorders in the copper metabolism illustrates that this element participates in the formation of connective tissue (collagen, keratin; see Section 10.3). The faulty gene responsible for Menke's syndrome has been localized on the X chromosome and cloned. The corresponding ATPase transport protein (see Section 13.4) features six Cys-X_2-Cys motifs as presumable copper binding sites [7].
- Defects (mutations) in the copper-dependent superoxide dismutase (SOD) are responsible for the neurodegenerative (paralytic) disorder known as Lou Gehrig's disease or familial amylotrophic lateral sclerosis (ALS) [8]. Copper has also received much attention for its involvement in other neurodegenerative diseases, such as Parkinson's and Alzheimer's diseases and prion-based disorders [5,9]. The observed oxidative stress and protein misfolding have been connected to disturbances of metal homeostasis.

Copper and molybdenum (see Section 11.2) are both metals with affinity for N *and* S donor atoms and thus behave as antagonists; cattle raised on Mo-rich and Cu-poor soil may

Table 10.2 Some representative copper proteins.

Function and typical proteins	Molecular mass (kDa)	Cu type(s)	Occurrence, reactivity
electron transfer ($Cu^{I} \rightleftharpoons Cu^{II} + e^-$)			
plastocyanin	10.5	1 type 1 ($E = 0.3$–0.4 V)	participation in plant photosynthesis (see Figure 4.6)
azurin	15	1 type 1 ($E = 0.2$–0.4 V)	participation in bacterial photosynthesis
"blue" oxidases ($O_2 \rightarrow 2\,H_2O$)			
laccase	60–140	1 type 1 ($E = 0.4$–0.8 V) 1 trimer	oxidation of polyphenols and polyamines in plants
ascorbate oxidase	2×75	2 type 1 ($E = 0.4$ V) 2 trimers	oxidation of ascorbate to dehydroascorbate in plants (3.12)
ceruloplasmin	130	2 type 1 ($E = 0.4$ V) 1 trimer	Cu transport and storage Fe mobilization and oxidation oxidase and antioxidation function in human and animal serum
"non-blue" oxidases ($O_2 \rightarrow 4\,H_2O_2$)			
galactose oxidase	68	1 type 2	alcohol oxidation in fungi (10.9)
amine oxidases	>70	1 type 2	degradation of amines to carbonyl compounds (10.10) cross-linking of collagen
monooxygenases ($O_2 \rightarrow H_2O$ + substrate O)			
dopamine β-monooxygenase	4×70	8 type 2	side-chain oxidation of dopamine to norepinephrine in the adrenal cortex (10.6)
tyrosinase	42	2 type 3	ortho-hydroxylation of phenols and subsequent oxidation to *o*-quinones in skin, fruit pulp etc. (10.8)
dioxygenases ($O_2 \rightarrow 2$ substrate-O)			
quercetinase	110	2 type 2	oxidative cleavage of quercetin in fungi
terminal oxidase ($O_2 \rightarrow 2\,H_2O$)			
cytochrome *c* oxidase	>100	Cu_A Cu_B	end point of the respiratory chain (see Figures 6.1 and 6.2)
superoxide degradation ($2\,O_2^{\bullet -} \rightarrow O_2 + O_2^{2-}$)			
Cu,Zn-superoxide dismutase	2×16	2 type 2	$O_2^{\bullet -}$ disproportionation e.g. in erythrocytes
dioxygen transport			
hemocyanin	n × 50 (mollusks) n × 75 (arthropods)	2 type 3 per n	O_2 transport in hemolymph of molluscs and arthropods (n = 8)
functions in the nitrogen cycle			
nitrite reductase	3×36	3 type 1 3 "type 2"	NO_2^- reduction (dissimilatory)
N_2O reductase	2×70	2 Cu_A 4 Cu_Z	reduction of N_2O to N_2 (10.12) in the nitrogen cycle

Table 10.3 Characteristics of "classical" copper centers in proteins.

Generalized coordination geometry	Function, structure, characteristics
type 1 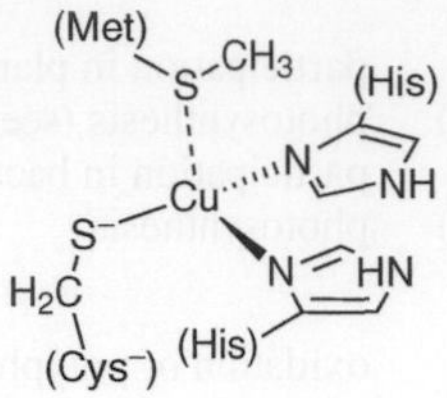	**type 1:** "blue" copper centers function: reversible electron transfer $Cu^{II} + e^- \rightleftarrows Cu^I$ structure: strongly distorted, (3 + 1) coordination absorption of the copper(II) form at about 600 nm, molar extinction coefficient $\varepsilon > 2000\ M^{-1}\ cm^{-1}$, LMCT transition $S(Cys) \rightarrow Cu^{II}$ EPR/ENDOR of the oxidized form: small $^{63,65}Cu$ hyperfine coupling and *g* anisotropy, interaction of the electron spin with **-S-CH$_2$-**; $Cu^{II} \rightarrow S(Cys)$ spin delocalization
type 2 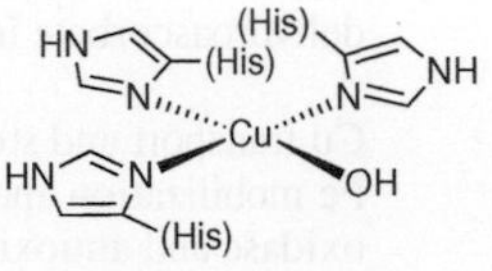	**type 2:** normal, "non-blue" copper function: O_2 activation from the Cu^I state in cooperation with organic coenzymes structure: essentially planar with weak additional coordination (Jahn–Teller effect for Cu^{II}), typically weak absorptions of Cu^{II}, $\varepsilon < 1000\ M^{-1}\ cm^{-1}$, ligand–field transitions (d → d) normal Cu^{II} EPR
type 3 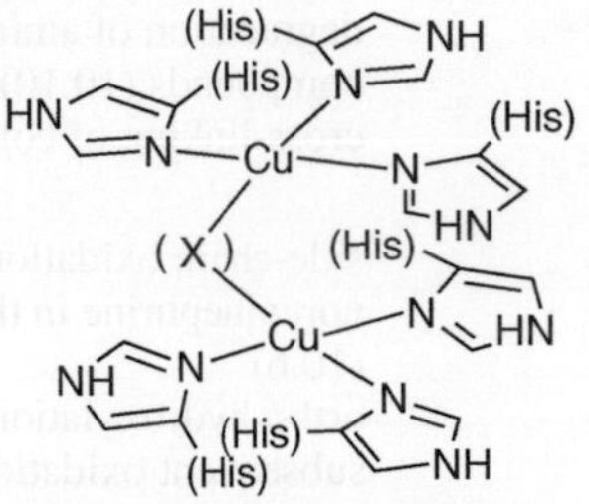	**type 3:** copper dimers function: O_2 uptake from the Cu^I–Cu^I state structure: (bridged) dimer, Cu–Cu distance about 360 pm after O_2 uptake, intense absorptions around 350 and 600 nm, $\varepsilon = 20\,000$ and $1000\ M^{-1}\ cm^{-1}$, LMCT transitions $O_2^{2-} \rightarrow Cu^{II}$ EPR-inactive Cu^{II} form (antiferromagnetically coupled d^9 centers)

hence develop severe copper deficiencies [10]. The resulting disorders can be counteracted by supplementing copper compounds to the animals' feed. Presumably, copper is not available for the metabolism because of its tight binding to thiomolybdates, $MoO_nS_{4-n}{}^{2-}$, which are formed from molybdenum and sulfur-containing substances in the digestive tract of ruminants (see (11.4)) [11]. Similar "secondary" copper deficiencies can occur in the presence of excess Fe, Zn or Cd.

From a structural and spectroscopic point of view, several "types" of biological copper center can be distinguished in proteins, according to a generally accepted convention (Table 10.3) [1,12]. In addition to the original type 1 – type 3 centers, there are Cu centers with approximate type 1 – type 3 ("type 1", "type2", or "type 3" in the following) structures, combinations such as type 2/type3 trimers and special copper centers such as Cu_A, Cu_B, Cu_M, Cu_H and Cu_Z (see Sections 10.3 and 10.4) [1,13] which do not fit into the classical scheme. The Cu^I centers in non-enzymatic sensor, transport and storage proteins are predominantly coordinated by cysteinate residues (see also the metallothioneins, Section 17.3).

10.1 Type 1: "Blue" Copper Proteins

The type 1 copper centers have received their additional attribute because of the *intensely* blue color of the corresponding Cu^{II} proteins, which have been given appropriate names such as "azurin" and "plastocyanin" (Table 10.3; κυαηοζ: dark blue). This color is

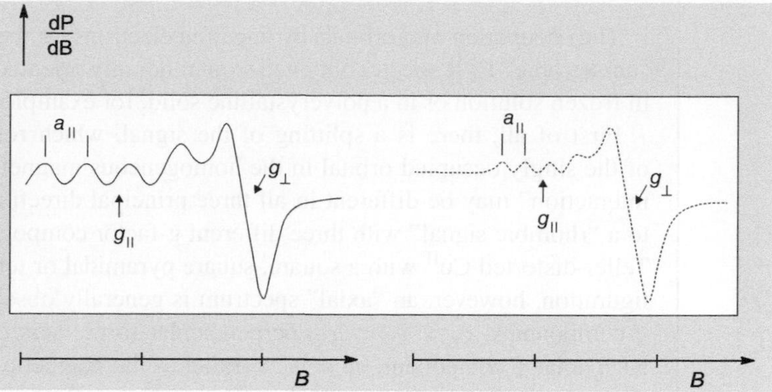

Figure 10.1
Typical anisotropic EPR signals of a normal Cu^{II} complex (—) and of a "blue" copper(II) protein (···); (first derivative spectra).

distinctive because metal centers are optically so "diluted" in metalloproteins that only intense absorptions in the visible region resulting from symmetry-allowed electronic transitions give rise to conspicuous colors. In contrast, the comparatively pale blue color of normal Cu^{2+}, as in crystalline copper(II)sulfate pentahydrate, is the result of "forbidden" electronic transitions between d orbitals of different symmetries ("ligand–field" transitions) (2.12); in these cases, the molar extinction coefficients ε are less than $100\,M^{-1}\,cm^{-1}$. The copper(II) centers of "blue" copper proteins, however, show much higher ε values of about $3000\,M^{-1}\,cm^{-1}$. In the electron paramagnetic resonance (EPR) spectra of type 1 copper(II) sites, the interaction of the unpaired electron (Cu^{II} has a d^9 configuration) with the magnetically not very different isotopes ^{63}Cu and ^{65}Cu (nuclear spin $I = 3/2$ for both isotopes) leads to a markedly decreased hyperfine splitting $a_{||}$ (Figure 10.1) when compared to that of normal Cu^{II} centers [12].

Electron Paramagnetic Resonance II

Unlike the Co^{II} state of cobalamins, with low-spin d^7 configuration (3.4) and the unpaired electron located in the d_{z^2} orbital, normal Cu^{II} complexes with their d^9 configuration have the $d_{x^2-y^2}$ orbital occupied by the single electron. In d^9 systems, the octahedral configuration is unstable because of the ambiguity (10.1) that results from incomplete occupation of a degenerate orbital (e_g) (2.12). The system circumvents this ambiguous situation (10.1) through geometrical distortion; that is, a lowering of the symmetry removes the orbital degeneracy: first-order Jahn–Teller effect (compare Figure 2.7).

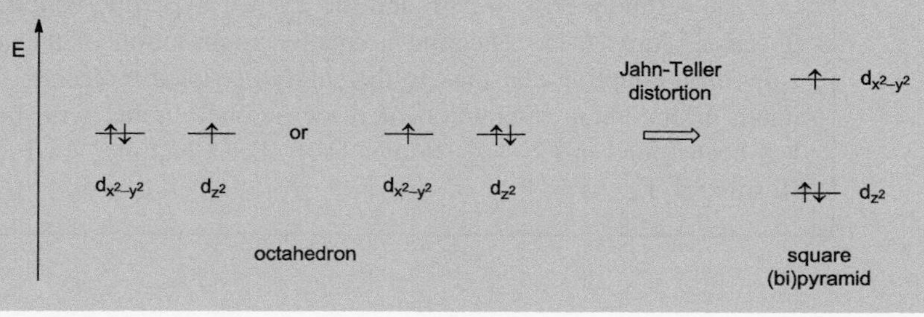

(10.1)

The occupation of d orbitals by unpaired electrons can be deduced from the typical "anisotropic" EPR spectra obtained from randomly oriented, immobilized complexes in frozen solution or in a polycrystalline solid, for example.

First of all, there is a splitting of the signal, which reflects the local symmetry of the singly occupied orbital in the homogeneous magnetic field. This "anisotropic interaction" may be different in all three principal directions of space and will lead to a "rhombic signal" with three different g-factor components. In the case of Jahn–Teller-distorted Cu^{II} with a square, square pyramidal or tetragonal bipyramidal configuration, however, an "axial" spectrum is generally observed, with two coinciding g components, $g_x = g_y = g_\perp$, perpendicular to the axis of the magnetic field, and a separate g component, $g_z = g_\parallel$, parallel to the magnetic field (Figure 10.1). For a d^9 configuration with $(d_{x^2-y^2})^1$ there is $g_\parallel > g_\perp \approx 2.01$, and furthermore, the interaction of the unpaired electron with the nuclear spin $I = 3/2$ of the copper isotopes $^{63,65}Cu$ is markedly greater in the axial direction than for the components perpendicular to the magnetic field: $a_\parallel > a_\perp$. Accordingly, the low-field g component is split into four lines (four nuclear spin orientations, $M = +3/2, +1/2, -1/2, -3/2$, relative to the magnetic field); only transitions with $\Delta S = \pm 1$ and $\Delta I = 0$ are allowed (10.2).

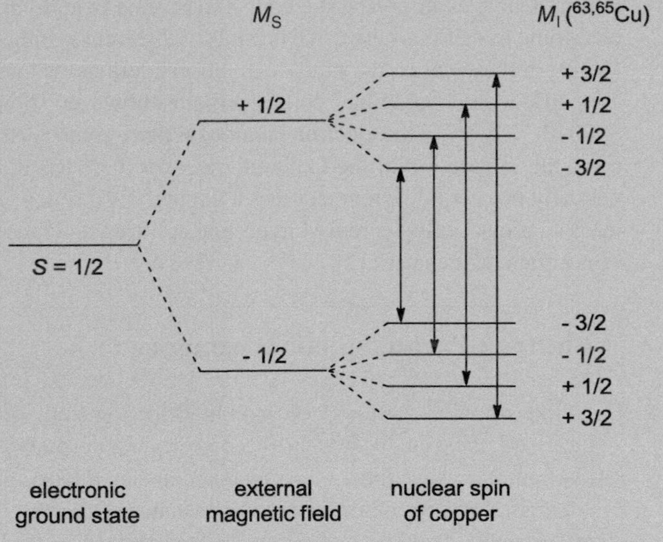

(10.2)

In blue copper proteins, with their distorted metal configuration even in the Cu^{II} state and with a sizable covalent character of the Cu-thiolate bond, both the g anisotropy (i.e. the difference $g_\parallel - g_\perp$), and the $^{63,65}Cu$ hyperfine coupling are markedly reduced (Figure 10.1), reflecting a smaller contribution of the metal, with its large spin–orbit coupling constant, to the singly occupied molecular orbital. On the other hand, a significant interaction of the cysteinate ligand with the unpaired electron has been found in ENDOR studies [14]. (See Chapter 3 for Electron Paramagnetic Resonance I.)

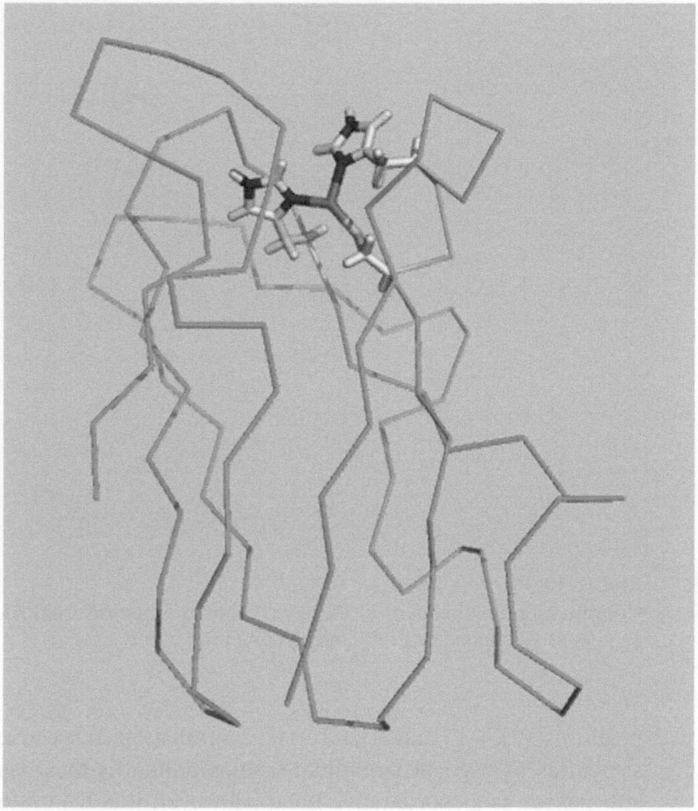

Figure 10.2
Structure of the Cu^{II} form of plastocyanin from poplar leaves (*Populus nigra*; α carbon representation of the polypeptide chain with some selected amino acid side chains (PDB code 1PLC)) [15].

Crystal structure analyses of several "blue" copper proteins (Figures 10.2 and 10.3) [15–17] have shown that the metal centers feature very irregular "distorted" coordination (10.3).

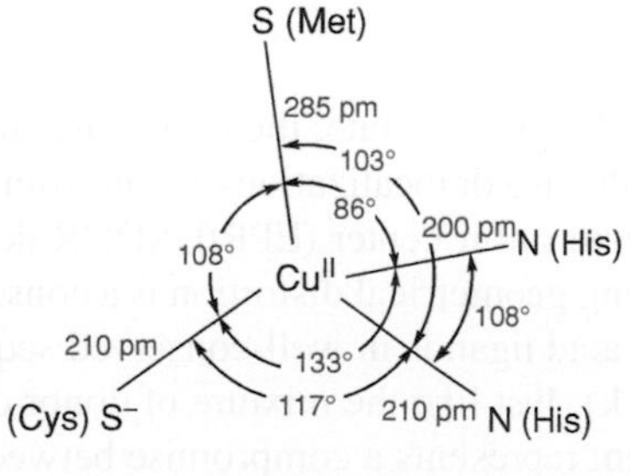

(10.3)

Two histidine residues and one hydrogen bond-forming cysteinate ligand are strongly bound in an approximately trigonal planar arrangement, while weakly bound (10.3) methionine (or glutamine as in stellacyanin) and, in some instances, a very weakly coordinating oxygen atom from a peptide bond complete the coordination environment in the axial

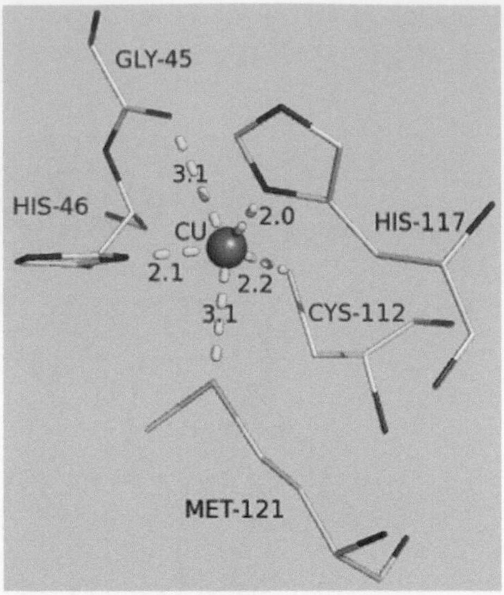

Figure 10.3
Coordination environment of the copper center in azurin from *Alcaligenes denitrificans* (α carbon centers as filled circles; PDB code 2AZA) [16].

positions ("3 + 1" and "3 + 1 + 1" coordination, respectively). The description of these structures as trigonal pyramidal or bipyramidal is thus based on the evaluation of what still passes as a coordinative bond. However, it is the cysteinate ligand in particular that is responsible for the unusual behavior of type 1 copper centers [12]. As in the case of oxidized rubredoxin (7.4), the intense absorption of the oxidized form is attributed to a ligand-to-metal charge transfer (LMCT) transition, that is, an electronic charge is transferred from the π and σ electron-rich thiolate ligand to the electron-poor Cu^{II} center via light excitation (10.4).

$$(R\text{-}CH_2\text{-}S^-)Cu^{II}(His)_2(Met) \xrightarrow[LMCT]{h\nu} (R\text{-}CH_2\text{-}S^{\bullet})Cu^{I}(His)_2(Met)^* \qquad (10.4)$$

Even in the ground state, the cysteinate sulfur atom donates charge to the metal center, which results in a delocalization of spin from the metal (smaller EPR coupling constant) to the cysteinate sulfur center (EPR/ENDOR detectable coupling with $^-$**S**-C**H**$_2$-) [14].

The strong geometrical distortion is a consequence of the incorporation of the coordinating amino acid ligands in well-conserved sequences His-X_k-Cys-X_n-His-X_m-Met (n, m = 2–4; large k). Just like the mixture of donor centers (two N, two S), this strongly distorted arrangement represents a compromise between $Cu^I = d^{10}$, with its preferred tetrahedral or trigonal coordination through "soft" (e.g. sulfur) ligands, and $Cu^{II} = d^9$, with preferentially square planar or square pyramidal geometry and N ligand coordination. Presumably, the irregular high-energy arrangement at the metal (10.3) largely resembles the transition-state geometry between the tetrahedral and the square planar equilibrium configurations of the two involved oxidation states, resulting in a higher rate of electron transfer (e.g. within the photosynthetic apparatus; Figures 4.6 and 4.7) [17]. Similar "intermediate geometries", anisotropy of the actual electron-transfer process, and electron delocalization between the metal and the (porphyrin or sulfur) ligands have been found for the two other major classes of protein with inorganic electron-transfer centers, namely the iron/sulfur and the cytochrome proteins (Sections 6.1 and 7.2–7.4).

Good model compounds for the "blue", type 1 copper proteins were unknown for quite some time as the typical spectroscopic properties can only be mimicked using special ligands with well-controlled thiolate coordination centers [18]; simple thiolates would immediately reduce Cu^{II}. "Masking" of blue copper centers is possible through isomorphous metal exchange with Hg^{2+}, which prefers a similar coordination but is neither EPR nor charge-transfer active [19].

As for the redox pair $Fe^{II/III}$, the potentials for the $Cu^{I/II}$ transitions are generally higher in bioinorganic systems than in simple model complexes. The potential range for proteins with type 1 copper centers runs from 0.18 V (stellacyanin, with Gln instead of Met as axial ligand) to 0.68 V (rusticyanin). The special stabilization of the lower oxidation state by ligands such as Met and the Cu^{II}-*destabilizing* deviation from the square planar or square pyramidal configuration are responsible for the rather high $Cu^{I/II}$ potentials.

10.2 Type 2 and Type 3 Copper Centers in O_2-activating Proteins: Oxygen Transport and Oxygenation

The type 3 dinuclear copper centers feature an EPR-inactive oxidized state due to antiferromagnetic coupling of the Cu^{2+} ions (see (4.11)). These centers are always found in connection with the activation of $^3O_2$; the O_2-transporting hemocyanin (*Hc*) and several O_2-dependent oxidases (electron transfer and catalysis) and monooxygenases (transfer of one O from O_2) are featured in this group (see Table 10.3).

It is remarkable that triplet dioxygen is bound very rapidly by the diamagnetic Cu^I centers of the deoxy forms. A diamagnetic ground state due to antiparallel spin–spin coupling of Cu^{II} centers is also observed for the oxy forms. Low-lying electronically excited (triplet) configurations of the copper(I) centers, such as $3d^9 4p^1$ and $3d^9 4s^1$, may be important for O_2 binding by the deoxy form; such configurations have also been implicated for synthetic $3d^{10} \cdots 3d^{10}$ systems with potentially available 4s and 4p levels [20].

Structural information is available for the metal-containing center of the highly associated hemocyanin (*Hc*), the O_2-transferring protein of certain mollusks and arthropods (e.g. spiders) [21]. Two coordinatively unsaturated Cu^I centers, each anchored in the protein by one weakly and two strongly bound histidine residues, are situated at a distance of approximately 350 pm in the (colorless) deoxy form (10.5) of each subunit.

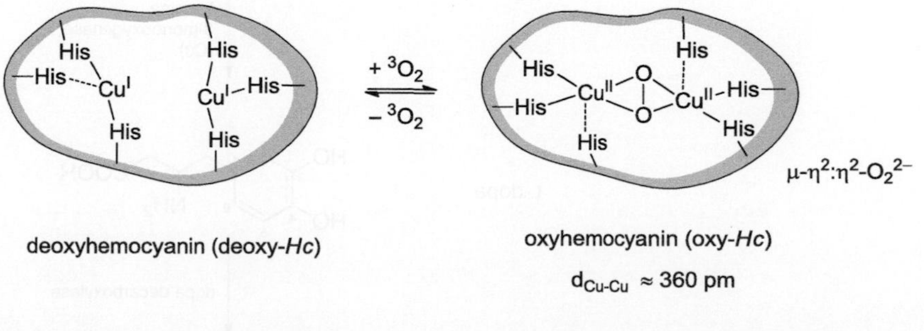

(10.5)

Based on model compounds and spectroscopic studies of the oxy form, it had been assumed that it contains either *cis*-μ-η^1:η^1 or μ-η^2:η^2 coordinated "dioxygen" in the peroxide oxidation state. The latter alternative was only seriously considered after corresponding model compounds (see (5.5)) had been prepared [22]; this alternative does not require an additional ligand and is in better agreement with the strongly weakened O–O bond and the little changed Cu–Cu distance after O_2 binding [21].

The assignment of the O_2 oxidation state is based on resonance Raman measurements of the O–O vibrational frequency (5.3) and on $(O_2^{2-}) \rightarrow (Cu^{II})$ LMCT absorptions, both in comparison with model compounds. While the reaction of O_2 with Cu^I is not totally unexpected, the reversibility of the dioxygen binding by the protein remains astonishing, given the distinct changes in oxidation states and the multiple coordination. In realistic model systems, such reversibility has been observed when protein-modeling N-polychelate ligands such as trispyrazolylborates (see (4.14) and (5.5)) are used. The high association of the *Hc* subunits to form protein complexes of more than 1 MDa has to be viewed in the context of the (limited) degree of cooperativity achieved by this O_2 transport system (see Section 5.2). Nevertheless, even the O_2 transport in giant squid species, which may weigh up to 150 kg and can accelerate to 30 km/h, is based on the hemocyanin system.

The group of monooxygenase (hydroxylase) enzymes includes not only iron-containing heme enzymes like cytochrome P-450 systems but also, with a more special selectivity, copper-containing enzymes such as tyrosinase [23] and dopamine β-monooxygenase. The latter plays a role in the biological oxidation of phenylalanine via the anti-Parkinson drug DOPA to (nor)epinephrine (10.6) and thus in the important biosynthesis of hormones and neurotransmitters. The O_2-dependent *ortho*-hydroxylation of phenols by tyrosinase to produce catechols with possible subsequential oxidation to *o*-quinones has a quite general biochemical relevance. Numerous natural products, including vitamins such as ascorbic acid and hormones, contain unsaturated rings which are substituted with vicinal polyhydroxo or polyalkoxy groups.

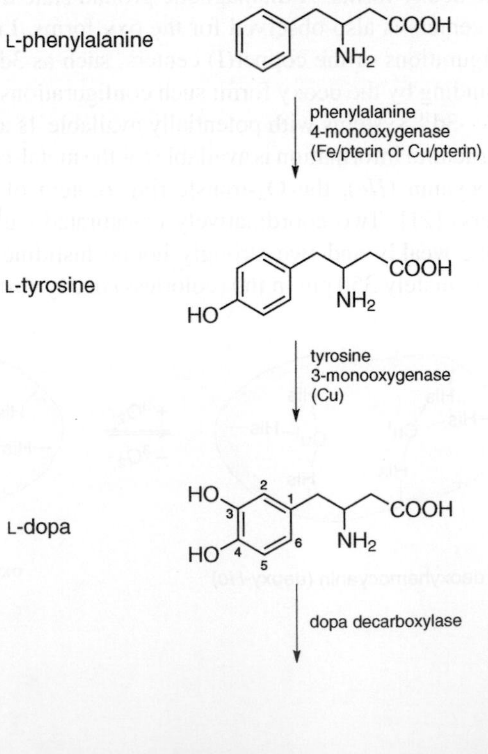

(10.6)

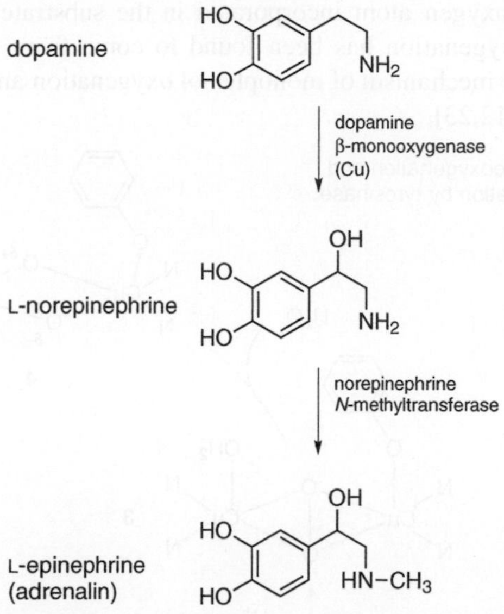

(10.6 *continued*)

The copper enzyme-catalyzed oxidative transformation of catechol derivatives to light-absorbing *o*-quinones is clearly visible after their polymerization to melanins, the color of which ranges from red to dark brown (10.7). The melanin pigments of skin, hair and feathers and the pigments of rotten (i.e. air-oxidized) fruit (enzymatic browning) are all polymeric *o*-quinone derivatives [24]. Copper enzyme-containing banana fruit pulp tissue can therefore be utilized in an electrode membrane as a dopamine-sensitive biosensor, the "bananatrode" [25]. In contrast, single Cu^{I} centers act as binding sites for the common plant hormone ethylene, $H_2C{=}CH_2$, which is believed to coordinate in π (η^2 side-on) fashion [26].

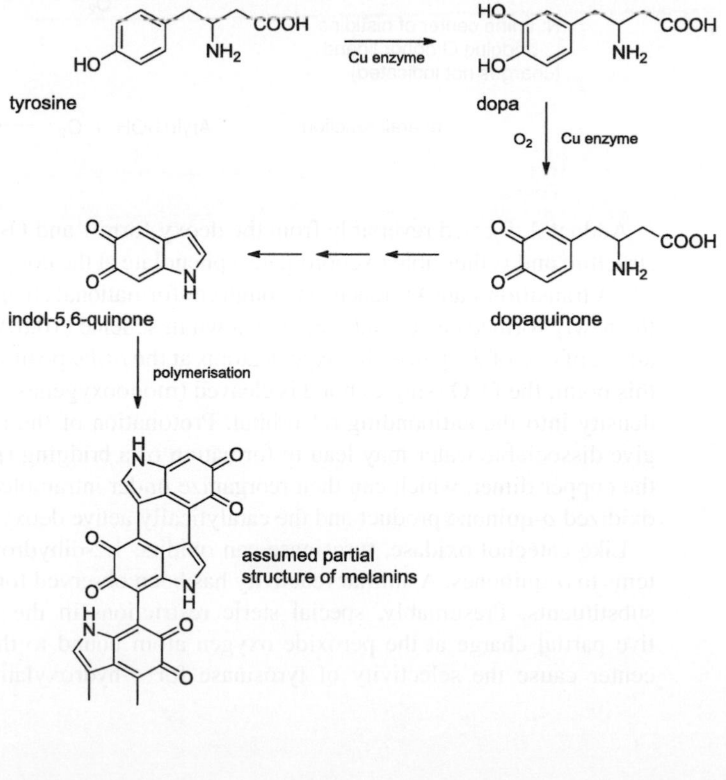

(10.7)

The oxygen atom incorporated in the substrates following copper enzyme-catalyzed monooxygenation has been found to come from O_2, as shown by isotopic labeling; a possible mechanism of monophenol oxygenation and oxidation to *o*-quinone is depicted in (10.8) [12,23].

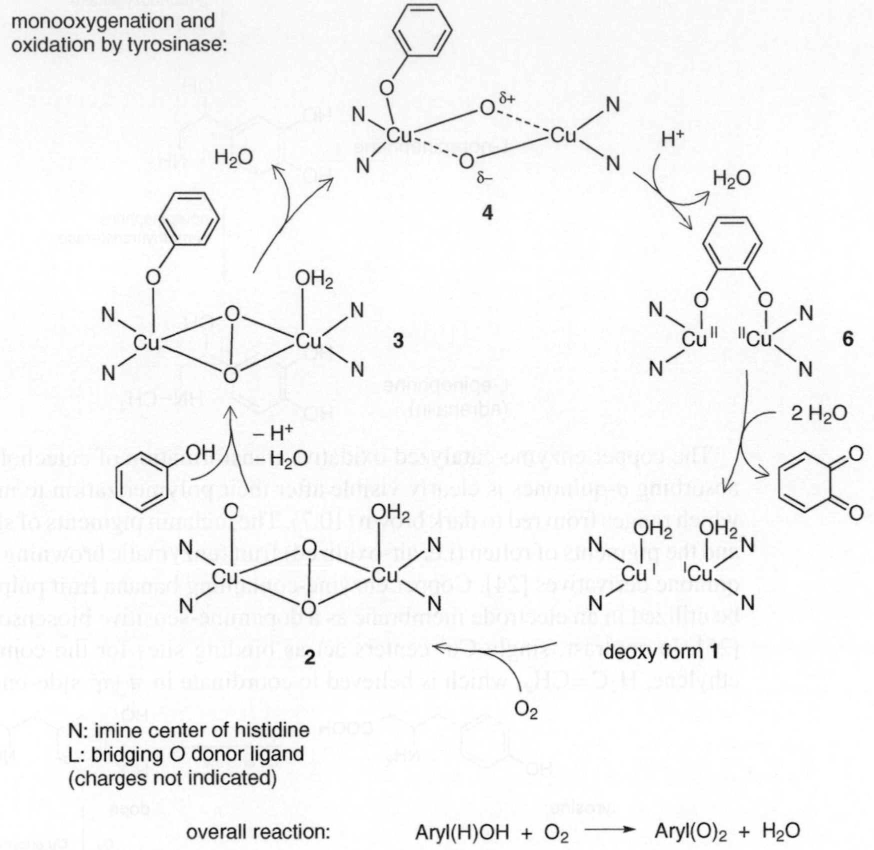

(10.8)

Adduct 2, formed reversibly from the deoxy form 1 and O_2 features a μ-η^2:η^2 peroxide structure and is thus able to coordinate a phenolate at the copper center via its oxygen atom (**3**). A transition state **4** is reached through conformational changes, and a chelate complex of the newly formed catecholate (**5**, not shown in scheme (10.8)) is created after electrophilic attack of one of the peroxidic oxygen atoms at the *ortho* position of the aromatic system. At this point, the O–O (single) bond is cleaved (monooxygenase activity) by shifting electron density into the antibonding σ^* orbital. Protonation of the remaining hydroxo ligand to give dissociable water may lead to formation of a bridging (μ-)catecholato complex **6** of the copper dimer, which can then reorganize under intramolecular electron transfer to the oxidized *o*-quinone product and the catalytically active deoxy state of the enzyme (10.8).

Like catechol oxidase, tyrosinase can oxidize 1,2-dihydroxy-substituted aromatic systems to *o*-quinones. A similar reactivity has been observed for aromatics with 1,2-diamino substituents. Presumably, special steric restrictions in the transition state and a positive partial charge at the peroxide oxygen atom bound to the lower coordinated copper center cause the selectivity of tyrosinase for *o*-hydroxylation of phenols. Not only is this reaction essential for the synthesis of active substances and for the formation of melanin but *o*-polyphenols can be further transformed under ring opening (see (7.16)) with the help of non-heme iron-containing catechol dioxygenases (Section 7.6.4), a reaction that is important for the microbial degradation of aromatic compounds in the environment.

Dopamine β-monooxygenase (DβM) catalyzes the specific monooxygenation in an aliphatic side chain (10.6). As in the related, well-characterized peptidylglycine-α-hydroxylating monooxygenase (PHM) [27], it has two uncoupled copper centers, separated by a distance of more than 1 nm. The O_2-binding Cu_M center is coordinated by two histidine residues, a weakly coordinating methionine and one or two H_2O molecules. The initial species formed after oxygen binding to Cu^I is believed to be a Cu^{II}-superoxo species with end-on coordinated $O_2^{\bullet -}$. Electron transfer from the distal Cu_H center, followed by proton transfer, results in the formation of a Cu^{II}-hydroperoxo intermediate, which breaks up under loss of H_2O following further electron/proton transfer to form the reactive $[Cu{=}O]^+$ species, which may be held responsible for the C–H activation in a rebound mechanism.

10.3 Copper Proteins as Oxidases/Reductases

In addition to Cu-dependent monooxygenases, there are several copper-containing "blue" (i.e. type 1 containing) and "non-blue" oxidases (without type 1 Cu) which convert both atoms of dioxygen to H_2O (blue oxidases) or H_2O_2 (non-blue oxidases). Laccase, ascorbate oxidase and the multifunctional protein ceruloplasmin are blue oxidases, while galactose oxidase and amine oxidases belong to the group of non-blue oxidase enzymes. Most of these oxidase enzymes have a complicated structure and often contain several types of copper centers. The large protein ceruloplasmin, for example, shows ferroxidase and antioxidation activity (see Section 10.5), even though its primary function in the cytoplasm most likely lies in the transport and storage of copper and in the regulation and mobilization of iron. The necessity for this mutual control between Cu and Fe can be understood by considering the correspondence mentioned in Table 10.1; disorders of the copper storage mechanism may thus lead to anemia as a secondary iron deficiency symptom.

Structural data for laccase and ascorbate oxidase [28] show that "type 2" and "type 3" copper centers are in such close proximity to each other (Figure 10.4) that, in effect, a copper trimer with new properties may be formulated [29]. This arrangement seems to favor a four-electron reduction of O_2 to two H_2O; in return, four oxidation equivalents become available for the polyphenolic substrates.

The crystal structure analysis of ascorbate oxidase from zucchini peels shows a copper trimer and a separate type 1 Cu center separated by a distance of more than 1200 pm (Figure 10.4). In the trimer, two metal centers, each coordinated by three histidine residues, are bridged by dissociable hydroxide (inactive oxidized form), while the third, coordinatively unsaturated "type 2" copper center, is bound to two histidine ligands and one OH^-. This structural arrangement illustrates the function of such enzymes: the transformation of the $4e^-$ oxidation potential of O_2 (trimer as O_2 coordination site) [28] into separate $1e^-$-oxidation equivalents, which are then passed on vectorially to a substrate binding site through the type 1 copper center. In the fully reduced Cu^I state, the bridging hydroxide ligand is dissociated, lowering the coordination number, while the Cu–Cu

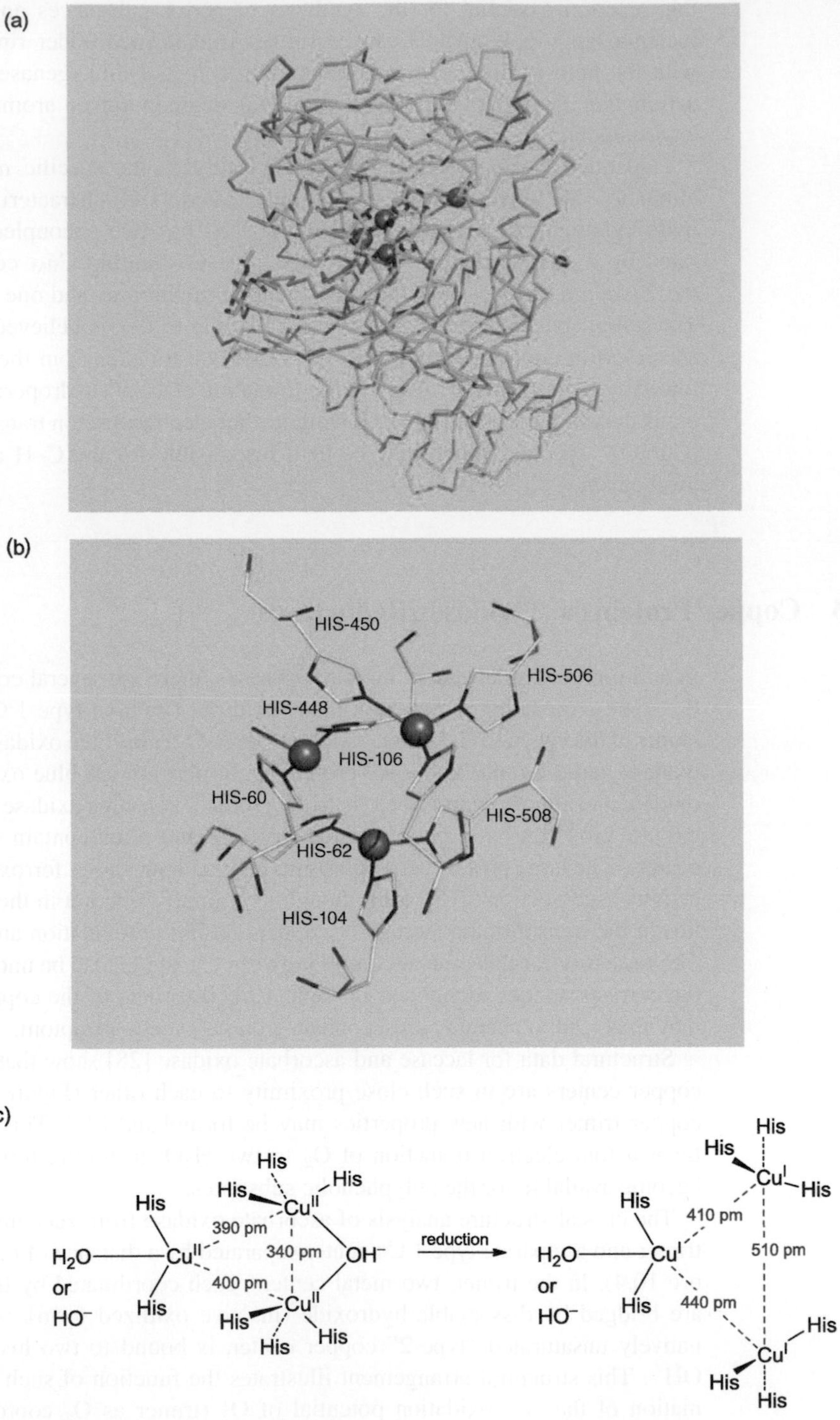

Figure 10.4
(a) Protein folding of the inactive (oxidized) form of ascorbate oxidase from zucchini with copper centers (golden spheres). (b) Active-site structure of Cu_3 region. (c) Dimensions and ligands in the oxidized and reduced forms of the tricopper site (PDB code 1ASO) [28].

distances experience a marked increase (Figure 10.4b). In the oxygenated form, a hydroperoxide ion, HO_2^-, is end-on coordinated to one of the oxidized type 3 copper centers [28].

Detailed inspection of the amino acid sequences reveals an increasingly complex development of related copper proteins in the series plastocyanin (1 Cu), ascorbate oxidase (3 Cu) and ceruloplasmin (6 Cu).

The reactivity of non-blue copper-dependent oxidases such as the stereospecific galactose oxidase (10.9) and amine oxidases (10.10) with mainly histidine-coordinated type 2 copper is based on the interaction of the single metal center ($Cu^{I,II}$) with organic redox cofactors; the overall result is a *two-electron reactivity*, which is required for the transformation $O_2 \rightarrow H_2O_2$. Earlier suggestions involving a Cu^I/Cu^{III} transition have not been substantiated.

$$\mathrm{RR'CHOH} + O_2 \xrightarrow{\text{galactose oxidase}} \mathrm{RR'C{=}O} + H_2O_2 \tag{10.9}$$

$$\mathrm{RCH_2NH_2} + O_2 + H_2O \xrightarrow{\text{amine oxidase}} \mathrm{RCHO} + H_2O_2 + NH_3 \tag{10.10}$$

Galactose oxidase (68 kDa), as isolated from the fungus *Dactylium dendroides* [30], contains a single copper(II) center in a square pyramidal arrangement with two histidines, one special tyrosinate/tyrosyl ligand and the substrate binding site in the equatorial plane and a weakly coordinating tyrosine in the axial position. The equatorial tyrosine is substituted ("covalently linked") with the sulfur atom of a cysteine residue in *o*-position to the phenolic oxygen atom, and furthermore, it exhibits a π–π interaction (stacking) with an aromatic tryptophane side chain. The equatorial ligand is assumed to undergo a redox-change tyrosyl radical/tyrosinate anion (see Sections 4.3 and 7.6.1) [30], which, in combination with the Cu^I/Cu^{II} transition, serves to effect catalysis of the two-electron process (10.9).

Organic redox cofactors are also present in copper-containing amine oxidases; the amino acid-derived *o,p*-quinonoid system 6-hydroxydopa quinone ("topaquinone", PAQ) has been established (Figure 10.5) [31–33]. An intraenzymatic electron transfer between Cu^{II}/coenzyme catecholate and Cu^I/coenzyme semiquinone forms has been suggested (10.11) [32,33]. Amine oxidases have a number of significant physiological functions; among other things, they are important in the metabolism and crosslinking of connective tissue such as collagen.

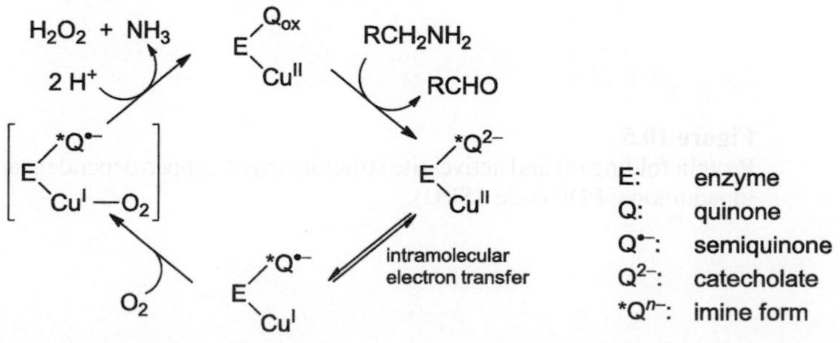

(10.11)

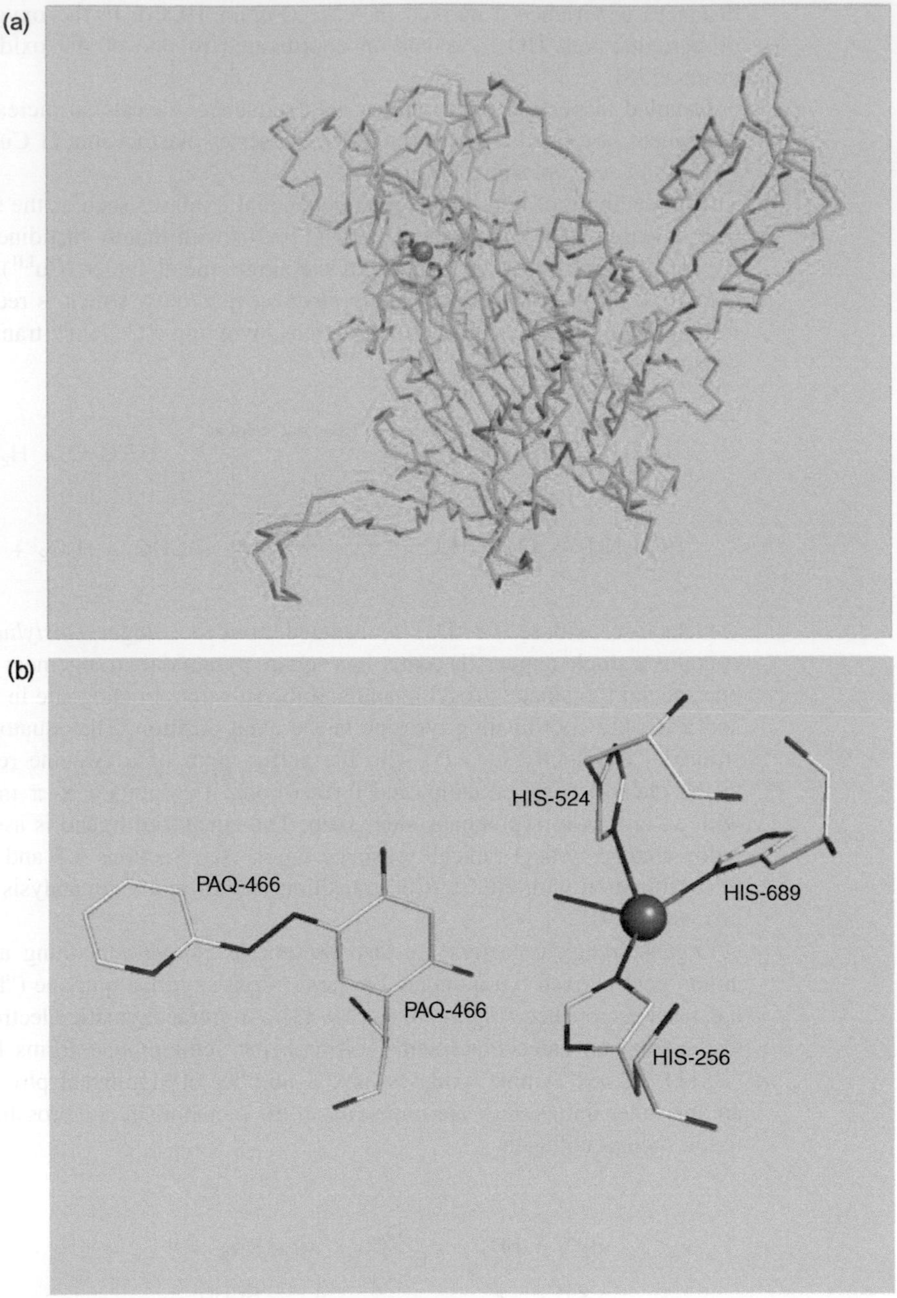

Figure 10.5
Protein folding (a) and active-site structure (b) of copper-dependent amine oxidase from *E. coli* (PAQ: topaquinone; PDP code 1SPU).

Copper-containing redox enzymes with unusual spectroscopic properties of their metal centers have been isolated from bacteria of importance to the global nitrogen cycle (see Figure 11.1) [34], including nitrite- and nitrous oxide (N_2O)-reducing bacteria and microorganisms that oxidize ammonia to hydroxylamine (NH_2OH) (10.12).

$$N_2O + 2\,e^- + 2\,H^+ \xrightarrow{N_2O\text{ reductase}} N_2 + H_2O \qquad (10.12)$$

The copper centers of environmentally relevant (O_3 depletion, greenhouse gas) N_2O reductase from *Pseudomonas stutzeri* [13] include "Cu_A", a dinuclear, doubly cysteinate-bridged arrangement with EPR-detectable delocalized Cu^I/Cu^{II} ("$Cu^{1.5}/Cu^{1.5}$") mixed valence. This long-wavelength-absorbing and electron-transferring Cu_A site was first observed in cytochrome *c* oxidase, which is introduced in Chapter 11. The catalytically active "Cu_Z" center is a tetracopper cluster [4Cu:2S] with seven histidine ligands and two sulfide ions in a low-symmetry arrangement (Figure 10.6), which leaves space for the binding of linear N=N=O [13].

A nitrosyl complex structure Cu^I–NO was postulated as intermediate [35,36] in the denitrification by copper-containing bacterial nitrite reductases to N_2O and N_2 (there

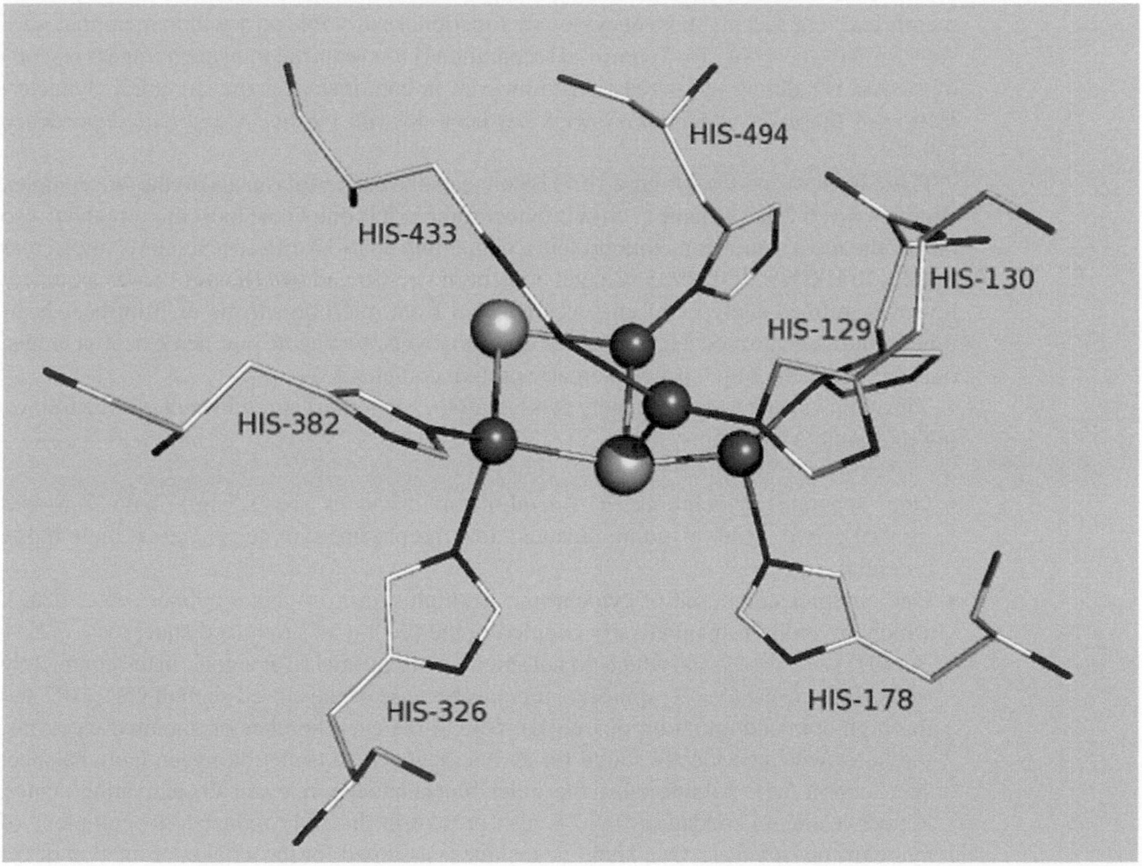

Figure 10.6
Cu_Z active site from *P. stutzeri* N_2O reductase (copper: brown; sulfur: yellow; PDB code 3SBR) [13].

are also polyheme-containing forms, see Section 6.5) [34,36]. Only following this suggestion have simple nitrosyl complexes of Cu^{I} been synthesized and structurally characterized [37]. The crystal structure analysis of nitrite reductase from *Achromobacter cycloclastes* shows a trimeric enzyme with 3 × 2 Cu centers [38], type 1 and "type 2" copper centers being approximately 1250 pm apart. Due to their unusual tetrahedral coordination geometry, the individual pseudo type 2 copper centers are assumed to be the binding and reduction sites for NO_2^- [38]; they are located at the bottom of solvent-filled channels and each such center is bound to histidine ligands of different protein subunits.

10.4 Cytochrome *c* Oxidase

Cytochrome *c* oxidase, the "Atmungsferment" (respiratory enzyme) of Otto Warburg (Nobel Prize in Medicine 1931), is a membrane protein complex and as such the site of the last phosphorylation in the respiratory chain (see Figures 6.1 and 6.2). It serves in the transformation of O_2 and H^+ to water and, as a terminal O_2-*consuming* system, represents the counterpart to the dioxygen-*producing* manganese-containing centers in the photosynthetic membrane. The fact that both enzymes are functional and stable only within membranes results from the necessity for a controlled separation (i.e. a vectorial transmembrane transport) of e^- and H^+ during redox reactions; however, in both instances the structural characterization of these very complex systems has been delayed by their membrane dependence [39,40].

Cytochrome *c* oxidase (Figure 10.7) is connected to the periplasm and to the bc_1-complex (Figures 6.1, 6.2 and scheme (7.5)) via cytochrome *c*. It is one of the most important but also one of the most complex metalloproteins, containing up to 13 different subunits (molecular mass > 100 kDa). Until 1985, a metal content of two Cu and two (heme) Fe was assumed; however, refined analyses of enzymes isolated from microorganisms or from beef heart mitochondria confirmed a total of three copper centers, two heme *a* or heme a_3 iron atoms, one Zn^{2+} and one Mg^{2+} per monomeric protein complex [39].

The main subunit (approximately 60 kDa) of the monomeric protein contains the following inorganic components:

- One separate cytochrome *a* (metal-metal distances > 1.5 nm) with low-spin iron(III), two axial histidine ligands, low porphyrin symmetry and a high redox potential.
- One complex composed of cytochrome a_3 (high-spin iron) and a copper center (Cu_B), which are antiferromagnetically coupled in the (Fe^{III}, Cu^{II}) oxidized state, $(S = 5/2) - (S = 1/2) \rightarrow (S = 2)$, and which do not show a conventional EPR signal. In the completely reduced form, the $(S = 2)$ spin state remains because of the closed shell of $Cu^{I} = d^{10}$ and the high-spin configuration of iron(II). Due to the even number of unpaired electrons, the latter would be ideally suited for an interaction with triplet dioxygen (spin balance; see Section 5.2). Additionally, the other biochemically relevant O_2-activating center, copper in the +I oxidation state, is also present in the fully reduced "a_3 complex" of cytochrome *c* oxidase. One histidine residue is assumed for the axial coordination of cyt a_3, with the sixth coordination site remaining empty for the binding of O_2; Cu_B faces

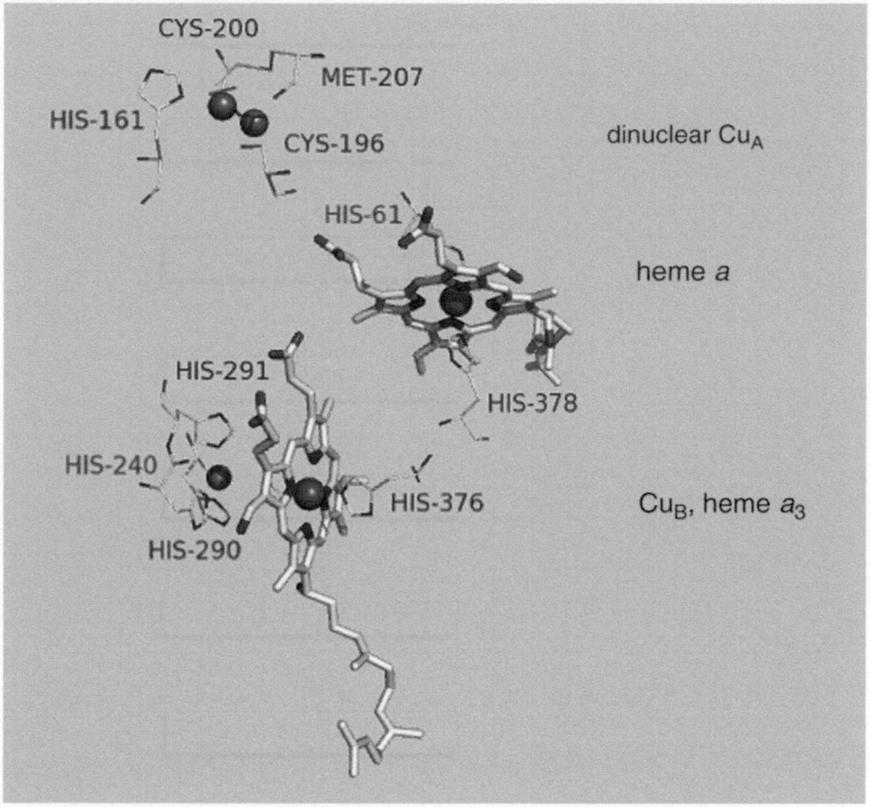

Figure 10.7
Schematic representation of redox centers in the dimeric multiprotein complex cytochrome *c* oxidase of the mitochondrial membrane from bovine heart (copper: brown spheres; iron: red; PDB code 1OCO; see also Figure 6.2).

this free position at a distance of less than 0.5 nm and is ligated by three histidine groups (Figure 10.5).

Outside of the actual membrane region, subunit II (26 kDa) contains a "Cu_A" dinuclear copper center with unusual spectroscopic characteristics in the oxidized state, namely a weak long-wavelength absorption, a very small *g* anisotropy and $^{63,65}Cu$ hyperfine coupling in the EPR spectrum. This form of Cu_A represents a delocalized mixed-valent dimer $Cu^{1.5}/Cu^{1.5}$, as supported by the $^{63,65}Cu$ EPR hyperfine splitting $a_{\parallel}$ into seven lines [41].

Cytochrome *a* and Cu_A have mainly electron-transfer functions at physiologically high potentials; like the four-electron oxidation of two H_2O molecules (4.9), the mechanism of the four-proton/four-electron reduction of O_2 can be formulated as a stepwise process (10.13) [40].

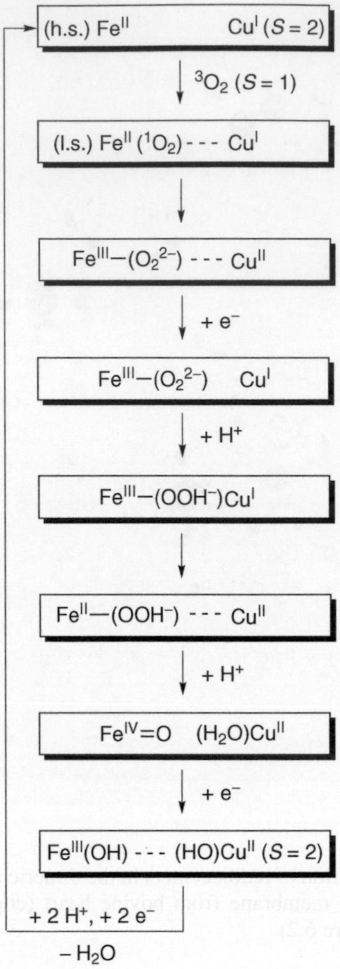

overall reaction:

$$O_2 + 4\,e^- + 4\,H^+ \longrightarrow 2\,H_2O \quad \text{or}$$

$$O_2 + 4\text{ cyt } c^{2+} + 8\,H^+_{inside} \longrightarrow 2\,H_2O + 4\text{ cyt } c^{3+} + 4\,H^+_{outside} \qquad (10.13)$$

In the active (reduced) form (reduction via Cu_A and *cyt a*), the a_3 complex contains high-spin iron(II). After coordination of O_2 which has reached this site through diffusion via a channel, the oxidation of and electronic coupling between iron and copper centers effects its rapid inner-sphere reduction to a (hydro)peroxo ligand bridging Fe^{III} and Cu^{II}. Uptake of protons and electrons (vectorial e^-/H^+ transport through the membrane, → phosphorylation) can lead to cleavage of the O–O single bond [42] and to the known oxoferryl-heme system with an ($S = 1$) state for $[Fe^{IV}{=}O]^{2+}$ (see Section 6.3). Further proton and electron addition can cause oxoferryl reduction; a copper-associated oxidizable tyrosine is also assumed to participate in the mechanism [42]. Both hydroxo metal centers of this Fe^{III}/Cu^{II} resting state can then be stepwise reduced back to the active form under dissociation of water. The four-electron reduction of O_2 proceeds so rapidly that most mechanistic studies have to involve low-temperature quenching techniques or time-resolved spectroscopy [40], including the crucial aspect of proton translocation.

In contrast to water oxidase, which is a $^3O_2$-*releasing* manganese enzyme (Section 4.3), the critical states of the O_2-*consuming* enzyme cytochrome *c* oxidase feature an *even* number of unpaired electrons, which favors the effective binding and ensuing transformation of O_2. Gaseous carbon monoxide competes with O_2 for binding by cytochrome *c* oxidase, as in non-enzymatic hemoglobin, which transfers O_2 reversibly at the beginning of the respiratiory process. Toxicologically more important, however, is the effective inhibition of the small amount of enzymatic cytochrome *c* oxidase by soluble cyanide, which thereby blocks the terminal step of respiration by irreversible coordination to the essential metal centers.

10.5 Cu,Zn- and Other Superoxide Dismutases: Substrate-specific Antioxidants

SODs catalyze the disproportionation ("dismutation") of toxic $O_2^{\bullet-}$ to O_2 and H_2O_2. This reaction is also catalyzed by many non-enzymatic transition metal species, albeit in a less controlled fashion; as a rule, "free" transition metals ions are physiologically undesirable. The hydrogen peroxide formed in this process can further disproportionate to yield O_2 and H_2O (see (6.11)) in reactions catalyzed by catalases, or it can be utilized via peroxidase enzymes (see Sections 6.3, 11.4 and 16.8). In addition to the Cu,Zn-containing SOD from the cytoplasm of eucaryotes, there are other forms which contain nickel (see Section 9.6), iron (bacterial and plant SODs) or manganese (SODs from mitochondria, bacteria [43]). Iron- and manganese-containing SODs have been structurally characterized [44]; the M^{III} centers are coordinated in a trigonal bipyramidal fashion by three histidine residues, one η^1-aspartate and one solvent ligand.

Considering the metastability and reactivity of the radical anion $O_2^{\bullet-}$, there seems to be no need for a special catalyst specificity of SODs; in essence, oxygen-affine metal cations are required that can easily change their oxidation number by one unit. In fact, O_2-activating centers such as the hemes, type 3 copper proteins, non-heme iron dimers, copper/cofactor complexes and cytochrome *c* oxidase contain *two* interacting redox centers (2 metals or 1 metal + 1 π system) in order to circumvent just that one-electron reduction of O_2 to $O_2^{\bullet-}$; however, these measures cannot completely prevent the production of small quantities of superoxide $O_2^{\bullet-}$ or its conjugated acid $HO_2^{\bullet}$ (5.2) through "leakage" of one-electron equivalents [43]. Under physiological circumstances, the dismutation of $O_2^{\bullet-}$ must proceed very rapidly (i.e. at the diffusion limit) in order to prevent uncontrolled oxidations by this radical anion or its follow-up products in reactions with transition metal ions (see (3.11) and (4.6)). A major requirement for SODs is therefore their resistance to the aggressive substrate, $O_2^{\bullet-}$, *and* to the products O_2^{2-} and O_2.

The relatively small (2×16 kDa) Cu,Zn-SOD from erythrocytes, previously also known as "erythrocuprein", is structurally well characterized [45]; each subunit contains one copper and one zinc ion, bridged by a deprotonated, resonance-stabilized imidazolate ring (10.14) of a histidine side chain. The other amino acid ligands are three His (Cu) and two His and one Asp^- (Zn), respectively. Compared to a regular tetrahedron, the geometry is more severely distorted at the copper site than at the zinc center (Figure 10.8). An

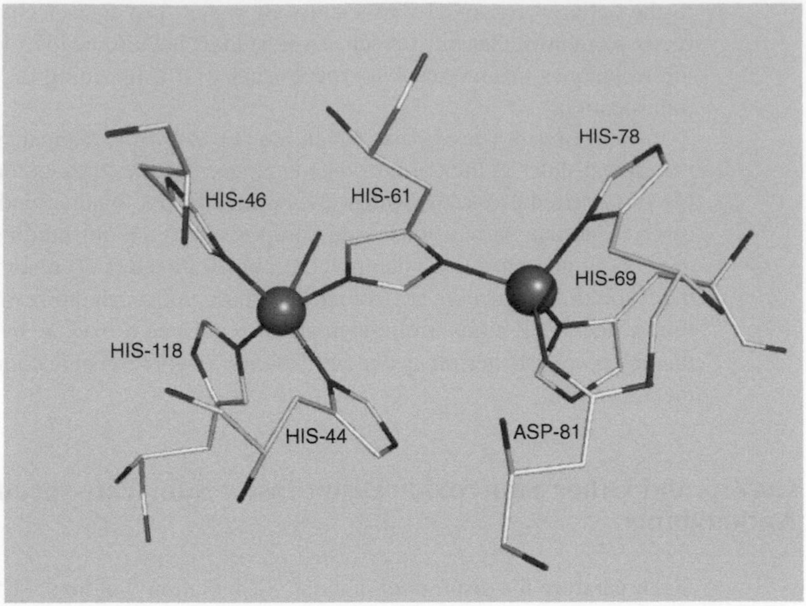

Figure 10.8
Structure of the dimetal center in the Cu,Zn-SOD of bovine erythrocytes (copper: brown sphere; zinc: grey; PDB code 2SOD) [45].

additional coordination site for $O_2^{\bullet-}$ is thus created at the catalytically active copper center (→ square pyramidal arrangement; see Figure 10.8), which may be temporarily occupied by labile H_2O.

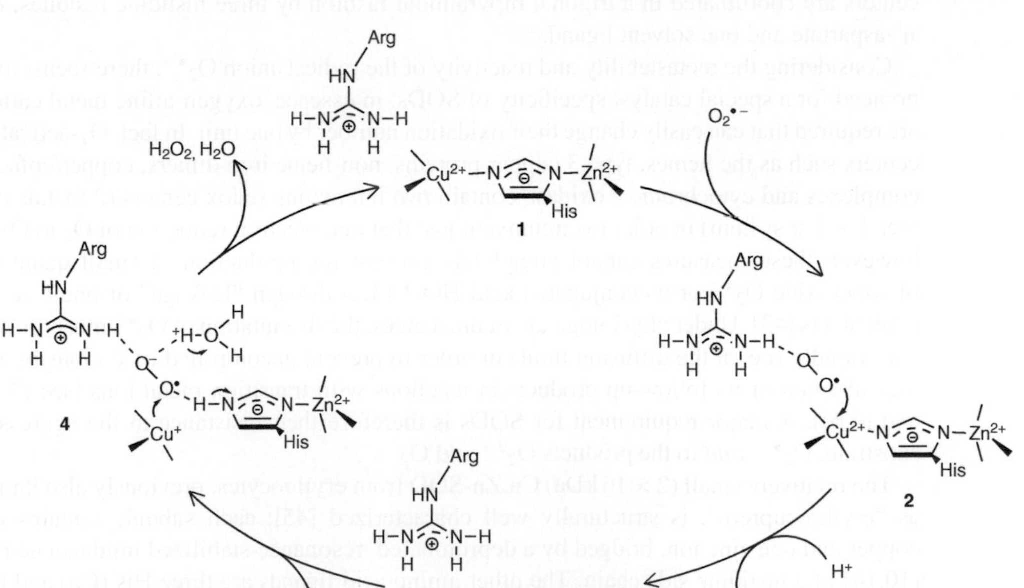

(10.14)

The detailed mechanism of the dismutation reaction involves as its essential part the ability of the redox-active metal center (here: Cu) to oxidize metastable superoxide in one oxidation state and to reduce it in the other. According to (10.14) and (10.15), it is assumed that $O_2^{\bullet-}$ is oxidized to O_2 by the Cu^{II} species **1** (SODs must *not* be oxidatively attacked by O_2!); the now reduced copper(I) center can then be replaced at the imidazole ring by a proton, resulting in a normal, geometrically relaxed zinc complex **3** with histidine. Next, the coordinatively unsaturated copper(I), which is still anchored in the protein, can be oxidized by hydrogen-bonded superoxide anion (**4**). The thus formed basic (hydro)peroxide (**5.2**) is then protonated (e.g. by the imidazole ring of still Zn-coordinated histidine) and transformed into H_2O_2. The driving force for this reaction lies in the affinity of copper(II) for the imidazole nitrogen of histidine. Removal of the Zn^{2+} ion seems to cause only a small decrease in the enzymatic activity; the role of this metal ion appears to be the introduction of structural stabilization. On the other hand, the high concentration of Cu,Zn-SOD in erythrocytes has led to the hypothesis that it might be primarily a metal storage protein and that its SOD activity is only a secondary function.

hypothetical mechanism:

$$Zn-N \overset{\ominus}{\frown} N-Cu^{II} + O_2^{\bullet-} \longrightarrow Zn-N \overset{\ominus}{\frown} N-Cu^{I} + O_2$$

$$Zn-N \overset{\ominus}{\frown} N-Cu^{I} + H^+ \longrightarrow Zn-N \overset{\frown}{} NH + Cu^{I}$$

$$Cu^{I} + O_2^{\bullet-} \longrightarrow Cu^{II} -- O_2^{2-}$$

$$Cu^{II} -- O_2^{2-} + H^+ + Zn-N \overset{\frown}{} NH \longrightarrow Zn-N \overset{\ominus}{\frown} N-Cu^{II} + H_2O_2$$

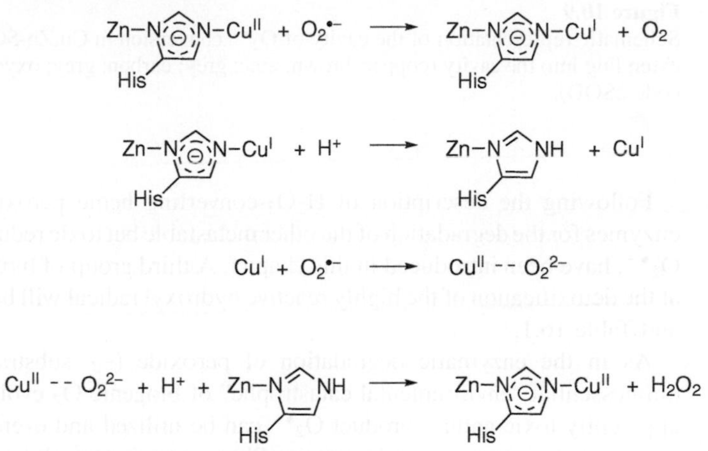

(10.15)

Zn: three-coordinate Zn^{II} center (2 His, 1 Asp$^-$)
Cu: three-coordinate copper center (3 His)

overall reaction:

$$\underset{(-0.5)}{2\ O_2^{\bullet-}} + 2\ H^+ \xrightarrow{\text{SOD}} \underset{(-I)}{H_2O_2} + \underset{(0)}{O_2} \quad \text{(oxygen oxidation states)}$$

The very rapid, almost diffusion-controlled reaction of the enzyme with $O_2^{\bullet-}$ (i.e. the successful conversion with nearly every collision of the reactants) is strongly assisted by electrostatic interactions, which lead the small monoanion $O_2^{\bullet-}$ into the protein through a funnel-shaped, approximately 1–2 nm deep channel. Near the catalytic site, it can be additionally positioned by the positively charged guanidinium group of an arginine residue, which also offers the possibility of hydrogen bonding ((10.14) **2**, **4** and Figure 10.9) [46]. Directed mutations can influence the field gradient inside the channel and thus even further increase the SOD activity [46]. Small halide (F$^-$) and pseudohalide ligands such as cyanide, CN$^-$, azide, N$_3^-$, and thiocyanate, NCS$^-$, can compete with $O_2^{\bullet-}$ for the binding to SOD.

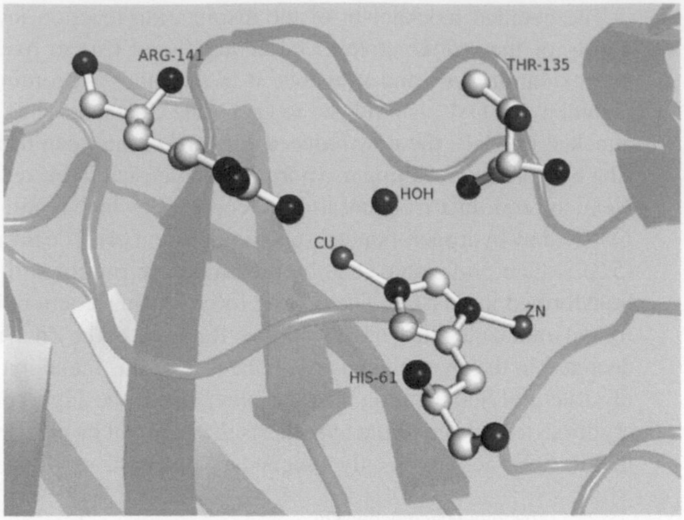

Figure 10.9
Schematic representation of the cavity of $O_2^{\bullet-}$ conversion in Cu,Zn-SOD, with one water molecule extending into the cavity (copper: brown; zinc: grey; carbon: grey; oxygen: red; nitrogen: blue; PDB code 2SOD).

Following the description of H_2O_2-converting heme peroxidases in Section 6.3, the enzymes for the degradation of the other metastable but toxic reduction product of dioxygen, $O_2^{\bullet-}$, have been introduced in this chapter. A third group of biological antioxidants aimed at the detoxification of the highly reactive hydroxyl radical will be presented in Section 16.8 and Table 16.1.

As in the enzymatic degradation of peroxide (→ substrate oxidation) and in the quintessential "environmental catastrophe" of biogenic O_2 evolution (→ respiration), the apparently toxic natural product $O_2^{\bullet-}$ can be utilized and even deliberately produced by organisms for certain special purposes. Phagocytes (neutrophils), which are essential for the immune system of higher organisms, produce large amounts of superoxide and follow-up products such as H_2O_2 and ClO^- with the help of a "respiratory burst oxidase" in order to kill invading microorganisms. However, this potent defense system can malfunction, giving then rise to autoimmune diseases such as rheumatoid arthritis. In such cases, SOD can be administered as an antiinflammatory drug. A SOD therapy is also indicated after exposure to ionizing radiation, which mainly produces oxygen-containing radicals (see Section 18.1); connections between SOD activity and the rate of aging and the role of defect mutants for neurodegenerative diseases such as amylotrophic lateral sclerosis [8] are under investigation.

In summary, an overview (Figure 10.10) [1] of the most important activities of biological copper (apart from regulatory, transport and storage functions) reveals its close association with dioxygen, its reaction products such as N/O compounds and its metabolites, and with inorganic and organic radicals [47]. So far, only copper(I) and, as part of the $Cu^{I/II}$ redox system, copper(II) have been established as biological copper oxidation states: a clear contrast with the wider oxidation state range of iron from Fe^{I} to at least Fe^{IV}. From a chemists' point of view, this reflects the nobler character of Cu vs. Fe.

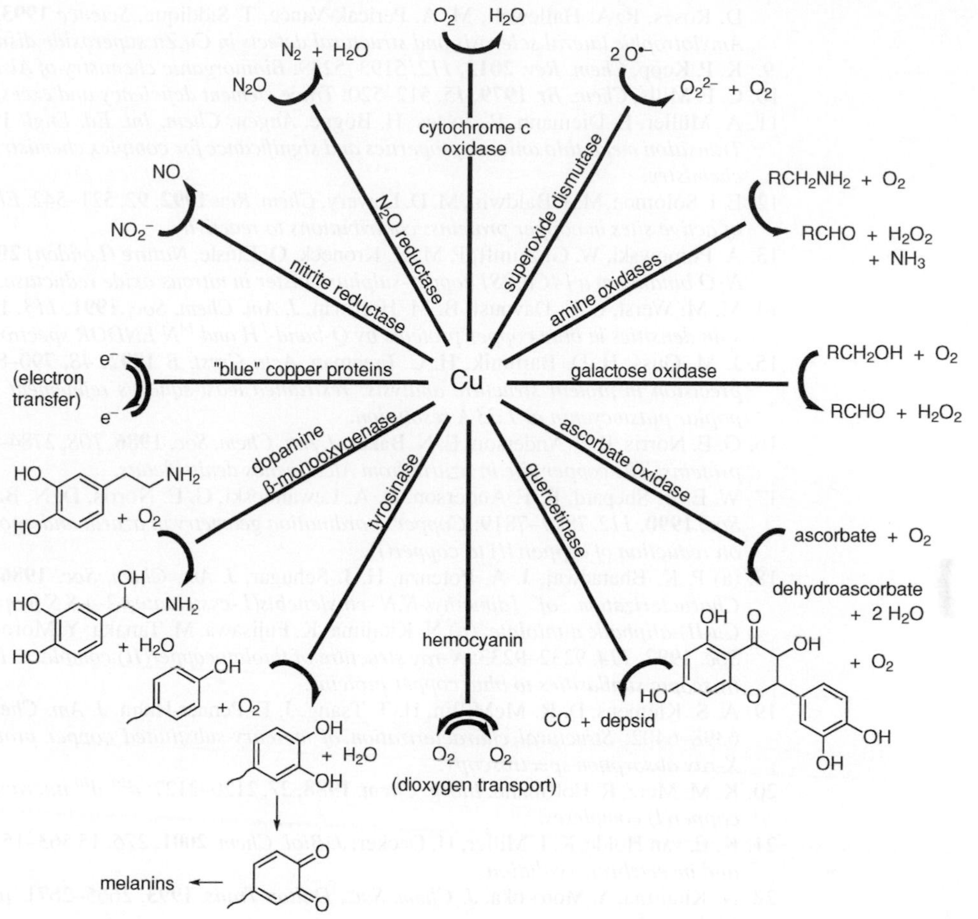

Figure 10.10
Essential metabolic functions of copper proteins [1].

References

1. W. Kaim, J. Rall, *Angew. Chem. Int. Ed. Engl.* **1996**, *35*, 43–60: *Copper – a "modern" bioelement*.
2. R. J. P. Williams, *The copper and zinc triads in biology*, in A. J. Welsh, S. K. Chapman (eds.), *The Chemistry of Copper and Zinc Triads*, Royal Society of Chemistry, Cambridge, **1993**.
3. M. C. Linder, C. A. Goode: *Biochemistry of Copper*, Plenum Press, New York, **1991**.
4. H. Koduma, C. Fujisawa, W. Bhadhprasit, *Curr. Drug Metab.* **2012**, *13*, 237–250: *Inherited copper transport disorders: biochemical mechanisms, diagnosis, and treatment*.
5. H. Kozlowski, M. Luczkowski, M. Remelli, D. Valensin, *Coord. Chem. Rev.* **2012**, *256*, 2129–2141: *Copper, zinc and iron in neurodegenerative diseases (Alzheimer's, Parkinson's and prion diseases)*.
6. P. Delangle, E. Mintz, *Dalton Trans.* **2012**, *41*, 6359–6370: *Chelation therapy in Wilson's disease: from D-penicillamine to the design of selective bioinspired intracellular Cu(I) chelators*.
7. K. Davies, *Nature (London)* **1993**, *361*, 98–98: *Cloning the Menkes disease gene*.
8. H.-X. Deng, A. Hentati, J. A. Tainer, Z. Iqbal, A. Cayabyab, W.-Y. Hung, E. D. Getzoff, P. Hu, B. Herzfeldt, R. P. Roos, C. Warner, G. Deng, E. Soriano, C. Smyth, H. E. Parge, A. Ahmed, A.

D. Roses, R. A. Hallewell, M. A. Pericak-Vance, T. Siddique, *Science* **1993**, *261*, 1047–1051: *Amylotrophic lateral sclerosis and structural defects in Cu,Zn superoxide dismutase.*

9. K. P. Kepp, *Chem. Rev.* **2012**, *112*, 5193–5239: *Bioinorganic chemistry of Alzheimer's disease.*
10. C. F. Mills, *Chem. Br.* **1979**, *15*, 512–520: *Trace element deficiency and excess in animals.*
11. A. Müller, E. Diemann, R. Jostes, H. Bögge, *Angew. Chem. Int. Ed. Engl.* **1981**, *20*, 934–955: *Transition metal thio anions: properties and significance for complex chemistry and bioinorganic chemistry.*
12. E. I. Solomon, M. J. Baldwin, M. D. Lowery, *Chem. Rev.* **1992**, *92*, 521–542: *Electronic structures of active sites in copper proteins: contributions to reactivity.*
13. A. Pomowski, W. G. Zumft, P. M. H. Kroneck, O. Einsle, *Nature (London)* **2011**, *477*, 234–238: *N_2O binding at a [4Cu:2S] copper-sulphur cluster in nitrous oxide reductase.*
14. M. M. Werst, C. E. Davoust, B. M. Hoffman, *J. Am. Chem. Soc.* **1991**, *113*, 1533–1538: *Ligand spin densities in blue copper proteins by Q-band-^1H and ^{14}N ENDOR spectroscopy.*
15. J. M. Guss, H. D. Bartunik, H. C. Freeman, *Acta Cryst. B* **1992**, *48*, 790–811: *Accuracy and precision in protein structure analysis: restrained least-squares refinement of the structure of poplar plastocyanin at 1.33 Å resolution.*
16. G. E. Norris, B. F. Anderson, E. N. Baker, *J. Am. Chem. Soc.* **1986**, *108*, 2784–2785: *Blue copper proteins. The copper site in azurin from Alcaligenes denitrificans.*
17. W. E. B. Shepard, B. F. Anderson, D. A. Lewandoski, G. E. Norris, D. N. Baker, *J. Am. Chem. Soc.* **1990**, *112*, 7817–7819: *Copper coordination geometry in azurin undergoes minimal change on reduction of copper(II) to copper(I).*
18. (a) P. K. Bharadwaj, J. A. Potenza, H. J. Schugar, *J. Am. Chem. Soc.* **1986**, *108*, 1351–1352: *Characterization of [dimethyl-N,N′-ethylenebis(L-cysteinato)(2–)-S,S′]copper(II), a stable Cu(II)-aliphatic dithiolate*; (b) N. Kitajima, K. Fujisawa, M. Tanaka, Y. Moro-oka, *J. Am. Chem. Soc.* **1992**, *114*, 9232–9233: *X-ray structure of thiolatocopper(II) complexes bearing close spectroscopic similarities to blue copper proteins.*
19. A. S. Klemens, D. R. McMillin, H. T. Tsang, J. E. Penner-Hahn, *J. Am. Chem. Soc.* **1989**, *111*, 6398–6402: *Structural characterization of mercury-substituted copper proteins. Results from X-ray absorption spectroscopy.*
20. K. M. Merz, R. Hoffmann, *Inorg. Chem.* **1988**, *27*, 2120–2127: *d^{10}-d^{10} interactions: multinuclear copper(I) complexes.*
21. K. E. van Holde, K. I. Miller, H. Decker, *J. Biol. Chem.* **2001**, *276*, 15 563–15 566: *Hemocyanins and invertebrate evolution.*
22. N. Kitajima, Y. Moro-oka, *J. Chem. Soc., Dalton Trans.* **1993**, 2665–2671: *μ-η^2;η^2-peroxide in biological systems.*
23. H. Decker, T. Schweikardt, F. Tuczek, *Angew. Chem. Int. Ed. Engl.* **2006**, *45*, 4546–4550: *The first crystal structure of tyrosinase: all questions answered?*
24. M. G. Peter, *Angew. Chem. Int. Ed. Engl.* **1989**, *28*, 555–570: *Chemical modification of biopolymers with quinones and "quinone" methides.*
25. J. S. Sidwell, G. A. Rechnitz, *Biotechnol. Lett.* **1985**, *7*, 419–422: *"Bananatrode" – an electrochemical biosensor for dopamine.*
26. B. M. Binder, C. Chang, G. E. Schaller, *Annu. Plant Rev.* **2012**, *44*, 117–145: *Perception of ethylene by plants – ethylene receptors.*
27. S. T. Prigge, B. Eipper, R. Mains, L. M. Amzel, *Science* **2004**, *304*, 864–867: *Dioxygen binds end-on to mononuclear copper in a precatalytic enzyme complex.*
28. A. Messerschmidt, H. Luecke, R. Huber, *J. Mol. Biol.* **1993**, *230*, 997–1014: *X-ray structures and mechanistic implications of three functional derivatives of ascorbate oxidase from zucchini.*
29. J. L. Cole, P. A. Clark, E. I. Solomon, *J. Am. Chem. Soc.* **1990**, *112*, 9534–9548: *Spectroscopic and chemical studies of the laccase trinuclear copper active site: geometric and electronic structure.*
30. N. Ito, S. E. V. Phillips, C. Stevens, Z. B. Ogel, M. J. McPherson, J. N. Keen, K. D. S. Yadav, P. F. Knowles, *Faraday Discuss.* **1992**, *93*, 75–84: *Three dimensional structure of galactose oxidase: an enzyme with a built-in secondary cofactor.*
31. J. Duine, *Eur. J. Biochem.* **1991**, *200*, 271–284: *Quinoproteins: enzymes containing the quinonoid cofactor pyrroloquinoline quinone, topaquinone or tryptophan-tryptophan quinine.*
32. D. M. Dooley, M. A. McGuirl, D. E. Brown, P. N. Turowski, W. S. McIntire, P. F. Knowles, *Nature (London)* **1991**, *349*, 262–264: *A Cu(I)-semiquinone state in substrate-reduced amine oxidases.*

33. M. R. Parsons, M. A. Convery, C. M. Wilmot, K. D. S. Yadav, V. Blakeley, A. S. Corner, S. E. V. Phillips, M. J. McPherson, P. F. Knowles, *Structure* **1995**, *3*, 1171–1184: *Crystal structure of a quinoenzyme: copper amine oxidase of Escherichia coli at 2 Å resolution.*
34. P. M. H. Kroneck, J. Beuerle, W. Schumacher, *Metal-dependent conversion of inorganic nitrogen and sulfur compounds*, in H. Sigel (ed.), *Metal Ions in Biological Systems, Vol. 28*, Marcel Dekker, New York, **1992**.
35. C. L. Hulse, B. A. Averill, J. M. Tiedje, *J. Am. Chem. Soc.* **1989**, *111*, 2322–2323: *Evidence for a copper-nitrosyl intermediate in denitrification by the copper-containing nitrite reductase of Achromobacter cycloclastes.*
36. T. Brittain, R. Blackmore, C. Greenwood, A. J. Thomson, *Eur. J. Biochem.* **1992**, *209*, 793–802: *Bacterial nitrite-reducing enzymes.*
37. S. M. Carrier, C. E. Ruggiero, W. B. Tolman, *J. Am. Chem. Soc.* **1992**, *114*, 4407–4408: *Synthesis and structural characterization of a mononuclear copper nitrosyl complex.*
38. J. W. Godden, S. Turley, D. C. Teller, E. T. Adman, M. Y. Liu, W. J. Payne, J. LeGall, *Science* **1991**, *253*, 438–442: *The 2.3 Ångstrom x-ray structure of nitrite reductase from Achromobacter cycloclastes.*
39. G. C. M. Steffens, T. Soulimane, G. Wolff, G. Buse, *Eur. J. Biochem.* **1993**, *213*, 1149–1157: *Stoichiometry and redox behaviour of metals in cytochrome-c oxidase.*
40. G. T. Babcock, M. Wikström, *Nature (London)* **1992**, *356*, 301–309: *Oxygen activation and the conservation of energy in cell respiration.*
41. W. E. Antholine, D. H. W. Kastrau, G. C. M. Steffens, G. Buse, W. G. Zumft, P. M. H. Kroneck, *Eur. J. Biochem.* **1992**, *209*, 875–881: *A comparative EPR investigation of the multicopper proteins nitrous-oxide reductase and cytochrome c oxidase.*
42. M. R. A. Blomberg, P. E. M. Siegbahn, M. Wikström, *Inorg. Chem.* **2003**, *42*, 5231–5243: *Metal-bridging mechanism for O-O bond cleavage in cytochrome c oxidase.*
43. I. Fridovich, *J. Biol. Chem.* **1989**, *264*, 7761–7764: *Superoxide dismutases.*
44. G. E. O. Borgstahl, H. E. Parge, M. J. Hickey, W. F. Beyer, Jr., R. A. Hallewell, J. A. Tainer, *Cell* **1992**, *71* 107–118: *The structure of human mitochondrial manganese superoxide dismutase reveals a novel tetrameric interface of two 4-helix bundles.*
45. J. A. Tainer, E. D. Getzoff, J. S. Richardson, D. C. Richardson, *Nature (London)* **1983**, *306*, 284–287: *Structure and mechanism of copper,zinc superoxide dismutase.*
46. E. D. Getzoff, D. E. Cabelli, C. L. Fisher, H. E. Parge, M. S. Viezzoli, L. Banci, R. A. Hallewell, *Nature (London)* **1992**, *358*, 347–351: *Faster superoxide dismutase mutants designed by enhancing electrostatic guidance.*
47. W. Kaim, *Dalton Trans.* **2003**, 761–768: *The chemistry and biochemistry of the copper-radical interaction.*

11 Biological Functions of the "Early" Transition Metals: Molybdenum, Tungsten, Vanadium and Chromium

In contrast to "late" transition metals such as cobalt, nickel and copper, the metals from the left half of the transition-element periods (see Figure 1.4) favor high oxidation states, high coordination numbers and "hard", in particular negatively charged, oxygen donor coordination centers under physiological conditions. The resulting oxo and hydroxo complexes are thus often negatively charged, which is important with regard to physiological uptake and mobilization mechanisms. Whereas scandium and titanium at the beginning of the first (3d) row of the transition metals do not have any biochemical relevance, vanadium, chromium and the heavier homologues molybdenum and tungsten exhibit quite differentiated physiological functions. Molybdenum is the biologically most important element in this series; its chemistry and enzymatic function [1] in oxygen transfer and nitrogen fixation will therefore be described in more detail in the first section of this chapter.

11.1 Oxygen Transfer through Tungsten- and Molybdenum-containing Enzymes

11.1.1 Overview

The essential character of molybdenum for life processes has been known for some time from dietetics and the agricultural sciences. From a chemical point of view, however, this distinction is quite peculiar: according to current knowledge, molybdenum is the only element from the second (4d) period of the transition metals with a biological function. One explanation certainly lies in its bioavailability: Although molybdenum is quite rare in the earth's crust, like all other heavy metals from this part of the periodic table, it is quite soluble in (sea) water at pH 7 (approximately 100 mM) in its most stable, hexavalent form as molybdate(VI), MoO_4^{2-}. This ion shows a close resemblance to the biologically important sulfur-transporting sulfate ion, SO_4^{2-}. In contrast to molybdate(VI), the oxometalates, MO_n^{m-}, of the heavier homologue tungsten, and especially of those metals to the left in the periodic system, such as niobium, tantalum, zirconium and hafnium, are nearly insoluble at pH 7 due to aggregation; on the other hand, the 4d and 5d elements to the right of Mo and W in the periodic system, technetium (see Section 18.1.5), rhenium and the platinum metals, are obviously too rare to have any biological significance.

Bioinorganic Chemistry: Inorganic Elements in the Chemistry of Life – An Introduction and Guide, Second Edition. Written and Translated by Wolfgang Kaim, Brigitte Schwederski and Axel Klein.

Table 11.1 Some molybdenum-containing hydroxylases and correspondingly catalyzed reactions.

Enzyme	Molecular mass (kDa)	Prosthetic groups	Typical function
xanthine oxidase	275 (dimer)	2 Mo, 4 Fe_2S_2, 2 FAD	oxidation of xanthine to uric acid in liver and kidney (11.11), (11.12)
nitrate reductase	228 (dimer)	2 Mo, 2 cyt *b*, 2 FAD	nitrate/nitrite transformation in plants and microorganisms (11.2): $NO_3^- + 2H^+ + 2e^- \rightleftharpoons NO^- + H_2O$
aldehyde oxidase	280 (dimer)	2 Mo, 4 Fe_2S_2, 2 FAD	oxidation of aldehydes, heterocycles, amines and sulfides in liver
sulfite oxidase	110 (dimer)	2 Mo, 2 cyt *b*	sulfite/sulfate transformation in liver (sulfite detoxification, (11.5)): $SO_3^{2-} + H_2O \rightleftharpoons SO_4^{2-} + 2e^- + 2H^+$
arsenite oxidase	85	Mo, Fe_nS_n	transformation of thiolate-blocking AsO_2^- by microorganisms: $AsO_2^- + 2H_2O \rightleftharpoons AsO_4^{3-} + 2e^- + 4H^+$
formate dehydrogenase (Mo)	>100	Mo, Fe_nS_n, Se	CO_2 reduction by microorganisms: $HCOO^- \rightleftharpoons CO_2 + 2e^- + H^+$
formate dehydrogenase (W)	340	W, Fe_nS_n, Se	CO_2 reduction by microorganisms: $HCOO^- \rightleftharpoons CO_2 + 2e^- + H^+$

In 1983, a formate ($HCOO^-$) dehydrogenase isolated from *Clostridium thermoaceticum* was described which, in addition to iron, sulfur and selenium (see Sections 7.4 and 16.8), contained tungsten [2] instead of molybdenum as its activating component (see Table 11.1) [3]. Since then, several other tungsten-incorporating microorganisms have been discovered [4], particularly in thermophilic ($\approx 65\,°C$) [5] and hyperthermophilic archaebacteria ($\approx 100\,°C$) [6]. The latter are found near the hydrothermal vents which exist on the sea floor along the middle oceanic ridges; these organisms (extremophiles; see Section 7.1) exhibit a chemolithotrophic metabolism involving sulfide oxidation and CO_2 reduction in order to lead a sunlight-independent life. Analogous tungsten and molybdenum enzymes are very similar with regard to composition and oxygen transfer function; however, in agreement with common inorganic-chemical experience, the tungsten-containing enzymes require higher temperatures and a more negative redox potential for transitions between corresponding hexavalent, pentavalent and tetravalent forms. A non-redox-active tungsten enzyme is acetylene hydratase, which catalyzes the reaction $C_2H_2 + H_2O \rightarrow H_3CC(O)H$ [7].

The chromate(VI) ion, CrO_4^{2-}, which is analogous to MoO_4^{2-} and also quite soluble at pH 7, behaves as a strong oxidant and is thus unstable under physiological conditions, in agreement with the rule that the heavier homologues in a transition metal group favor higher oxidation states. In fact, CrO_4^{2-} is mutagenic and carcinogenic, as will be pointed out in Section 17.8.

In addition to being bioavailable, a metal such as molybdenum or tungsten must perform a useful function in order to count as an essential element. The molybdoenzymes constitute another class of hydroxylase (oxidase), besides iron- and copper-containing proteins (see Table 10.1). Additionally, an "FeMo cofactor" plays an essential role in the main form of the dinitrogen (N_2)-fixating nitrogenase enzyme (see Section 11.2).

11.1.2 Oxotransferase Enzymes Containing the Molybdopterin or Tungstopterin Cofactor

For oxidation states lower than +VI, the chemistry of molybdenum in aqueous solution is characterized by aggregation phenomena; that is, by the formation of "clusters". As is shown in (11.1), dimeric and trimeric ionic systems are formed with the metals connected by hydroxo or oxo bridges and coordinatively saturated by water ligands (EXAFS measurements, [8]).

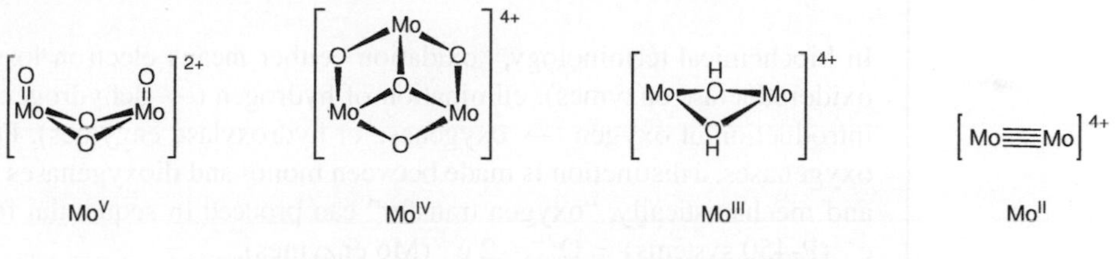

(11.1)

Such aggregation can be suppressed in the presence of a protein acting as a multidentate and protecting chelate system or under utilization of special cofactor ligands, with the result that – as with many other metals – a rather "unusual" biological coordination chemistry of these oxidation states develops. The physiologically relevant oxidation states of molybdenum lie between +IV and +VI, and the corresponding redox potentials of about −0.3 V are in the physiologically acceptable range, in contrast to those of homologous chromium complexes. In these oxidation states, molybdenum shows comparable affinity towards negatively charged O *and* S ligands such as oxide, sulfide, thiolates and hydroxide; nitrogen ligands are also bound quite well. One of the most important biological functions of molybdenum in enzymes is the catalysis of controlled oxygen transfer to or from a two-electron substrate, which involves spatially separated one-electron-transferring components such as cytochromes, Fe/S centers or flavins. A coupling of electron transfer and oxide exchange leads to the formal transfer of an oxygen *atom* from the metal center to the substrate or vice versa (oxotransferase activity $LMo^{VI}O_2 + X \rightleftharpoons LMo^{IV}O + XO$; see (11.2) [9]). The transferred oxygen does *not* originate from O_2 (oxygenation), as is often the case with Fe- and Cu-containing hydroxylases (oxygenases) (Table 10.1), and the result is a temporal and spatial separation of the electron transfer and the actual oxygen translocation. Regeneration of the reduced enzyme may proceed via O_2 as the eventual oxidant, a process which can yield peroxides or superoxide, as is well known for example from xanthine oxidase.

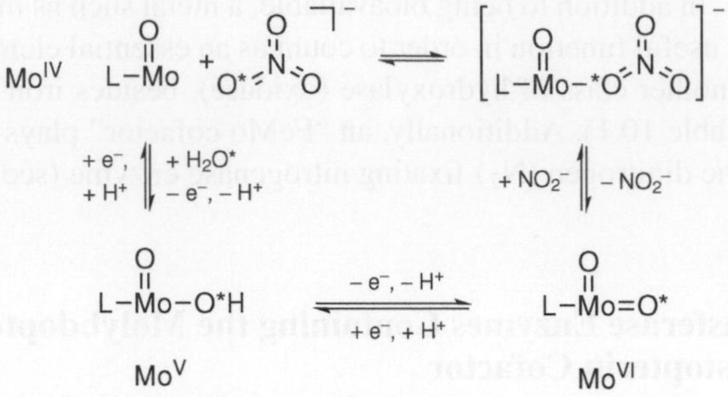

(11.2)

L: ligands in the coordination sphere of molybdenum

O*: ^{18}O labelling

> ## "Oxidation"
>
> In biochemical terminology, "oxidation" either means electron loss (→ oxidase or oxidoreductase enzymes), elimination of hydrogen (→ dehydrogenase enzymes) or introduction of oxygen (→ oxygenase or hydroxylase enzymes). For O_2-dependent oxygenases, a distinction is made between mono- and dioxygenases (see Table 10.2), and mechanistically, "oxygen transfer" can proceed in sequential form: $O = O^{\bullet-} - e^-$ (P-450 systems) $= O^{2-} - 2\,e^-$ (Mo enzymes).

Several molybdenum-dependent hydroxylases are known (Table 11.1) [10], the best characterized being xanthine oxidase, sulfite oxidase and nitrate reductase (see (11.2)). Other Mo-containing enzymes also have very important biochemical functions: Aldehyde oxidases participate in the alcohol metabolism, various nitrogen heterocycles get C–H oxygenated in α position to N, and other varieties of Mo enzyme catalyze the "oxygen atom" transfer in amine/amine oxide, arsenite/arsenate (Section 16.4) and sulfide/sulfoxide systems. The latter reaction plays an important role in the conversion of D-biotin-5-oxide to the actual coenzymatic biotin (vitamin H), as well as in the transformation of dimethylsulfoxide (DMSO) to dimethylsulfide, which is highly important for the global sulfur cycle and climate [11]. Formate and formylmethanofuran dehydrogenases containing molybdenum or tungsten are essential for the C_1 metabolism ($CO_2 \rightarrow \rightarrow \rightarrow CH_4$ (2.5)) of microorganisms. In general terms, molybdoenzymes catalyze the reactions depicted in (11.3).

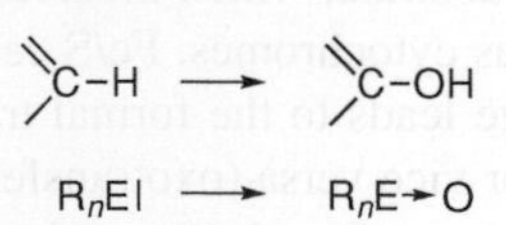

(11.3)

E = N, S, As

Malfunctions of molybdoenzymes in higher organisms are known, and are manifested for example in problems of the purine metabolism (→ gout, dysfunction of xanthine oxidase) or in neurological disorders (sulfite oxidase dysfunction [12]). The Cu/MoS antagonism known from cattle breeding has already been mentioned briefly in Chapter 10. Tetrathiomolybdates(VI), which are formed in the complex stomach of ruminants from molybdate and electron-rich sulfide, not only exhibit an intense color (low-energy ligand-to-metal charge transfer (LMCT) transition $S^{-II} \rightarrow Mo^{+VI}$) but also act as efficient chelating ligands for positively charged, albeit π electron-rich ("soft"), metal ions such as Cu^+ (11.4). The thus caused secondary deficiency of copper (which is necessary for the functioning of collagenases; Section 10.3) can be responsible for a weakening of connective tissue in animals that graze on molybdenum-rich soil.

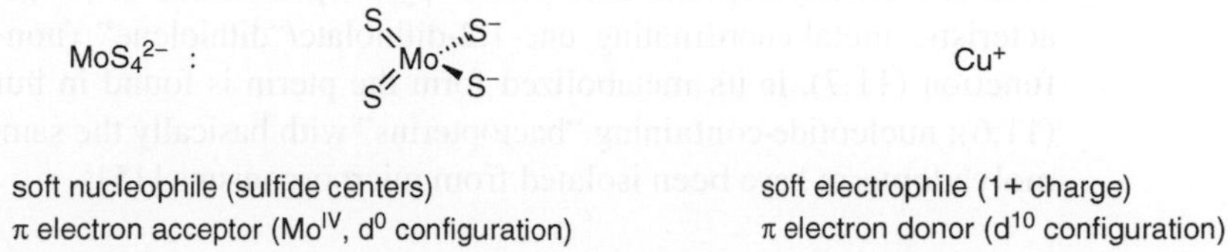

(11.4)

In biological oxidations that are not directly dependent on O_2, the oxidation equivalents are made available in one-electron steps via electron-transfer proteins. Therefore, almost all molybdoenzymes contain electron-transfer components such as cytochromes, Fe/S centers or flavins (Table 11.1). Scheme (11.5) shows a simple functional catalytic cycle for the oxidation of sulfite to sulfate, where d^1-configurated Mo^V is observed as an electron paramagnetic resonance (EPR)-detectable intermediate. Oxide ligands either get protonated after reduction due to their strongly increased basicity or will be replaced by hydroxide from the surrounding water after complete oxygen atom transfer.

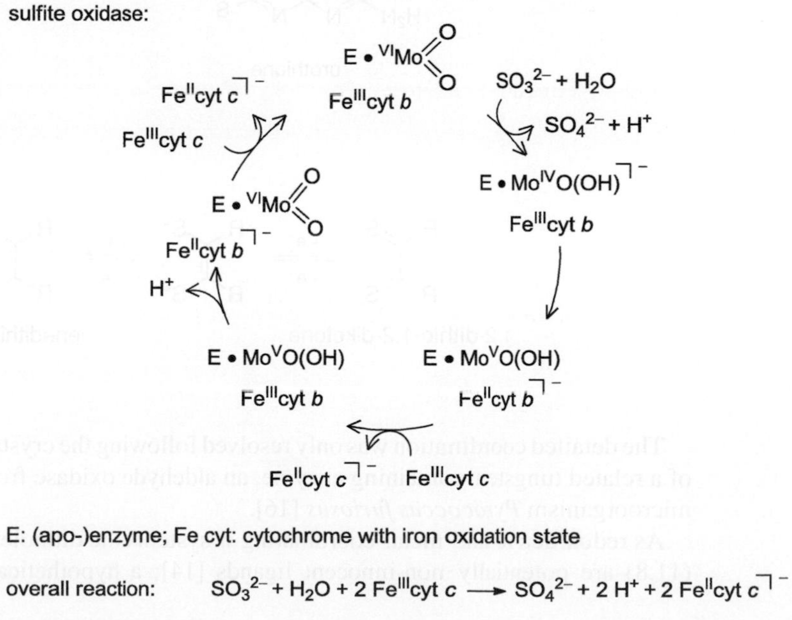

(11.5)

Initially, EPR studies of Mo^{V} intermediates were used to determine the coordination environment of the metal. The 1966 experiment of Bray and Meriwether [13] has become famous: a Frisian cow was injected with ^{95}Mo-enriched molybdate in order to allow a better EPR detection of the coupling between the unpaired electron and the nuclear spin ($I = 5/2$) of ^{95}Mo in the Mo^{V} form of xanthine oxidase as isolated from the cow's milk. From such experiments and from the observation of three accessible metal oxidation states, the participation of Mo^{VI}, Mo^{V} and Mo^{IV} in the enzymatic reactions was deduced.

All molybdenum- or tungsten-containing hydroxylases contain a Mo or W cofactor, consisting of molybdenum (or tungsten) and a special molybdopterin/tungstopterin ligand. This is a tetrahydropterin derivative, "pyranopterin" (11.6) [10], which has a very characteristic metal-coordinating ene-1,2-dithiolate/"dithiolene" (non-innocent [14]) chelate function (11.7). In its metabolized form the pterin is found in human urine as urothione (11.6); nucleotide-containing "bactopterins" with basically the same chromophor as that in molybdopterin have been isolated from microorganisms [15].

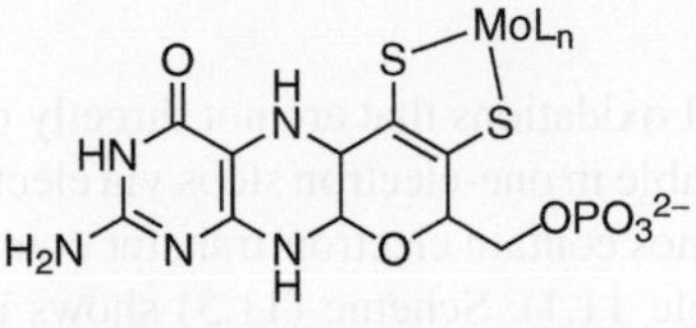

Mo-cofactor ("Moco"); L_n: ligands

(11.6)

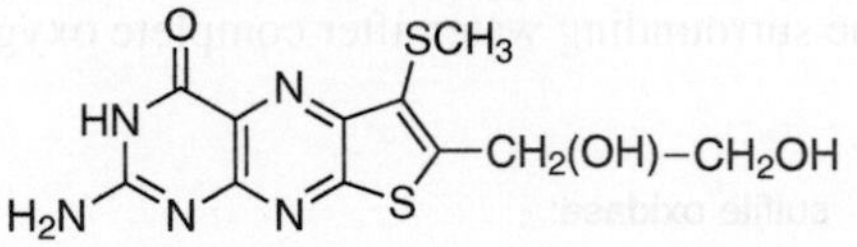

urothione

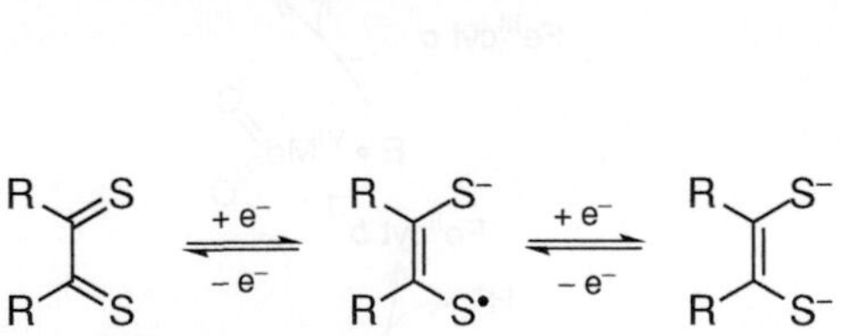

(11.7)

1,2-dithio-1,2-diketone enedithiolate(2–)

The detailed coordination was only resolved following the crystal structure determination of a related tungsten-containing enzyme, an aldehyde oxidase from the hyperthermophilic microorganism *Pyrococcus furiosus* [16].

As redox-active and metal-coordinating π systems the dithiolenes (11.7) and the pterins (11.8) are potentially non-innocent ligands [14]; a hypothetical electron-transfer chain

(11.9) consisting of several components (substrate, metal and other prosthetic groups) can thus be formulated.

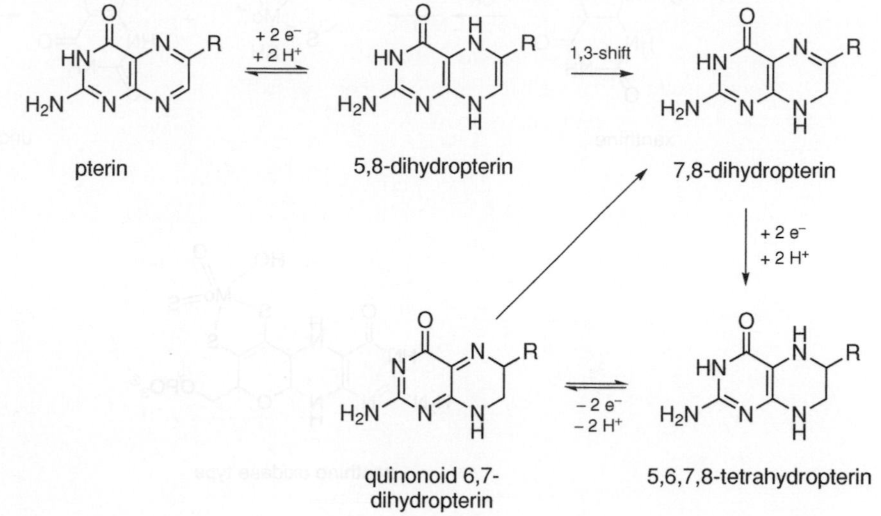

(11.8)

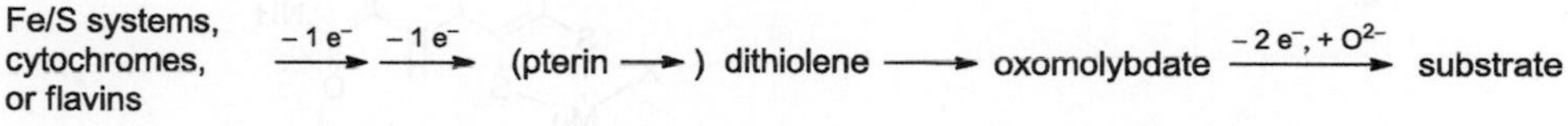

(11.9)

Although tetrahydropterins have been shown to reduce sulfur-coordinating molybdenum(VI) while being oxidized to the enzymatically reducible quinonoid dihydropterin [17], there has been no unambiguous evidence for the type of dihydro isomer *in vivo* or of a redox reactivity of Mo-coordinated molybdopterin in the enzymatic process.

Based on the coordination situation of the Mo^{VI} form, three families of molybdoenzyme are distinguished [1,10]. The sulfite oxidase family is characterized by one S,S′-coordinated pyranopterin and one cysteinate ligand and by oxo functions at the metal, which may be converted on reduction (11.10). In the Mo^{VI} state of enzymes of the xanthine oxidase family (11.11), one sulfide ligand is bound at an Mo–S distance of 215 pm (11.12). Finally, the DMSO reductase family of molybdoenzymes involves two pyranopterin ligands, an oxo function and a further monodentate ligand such as cysteinate, selenocysteinate, serin or aspartate (11.13).

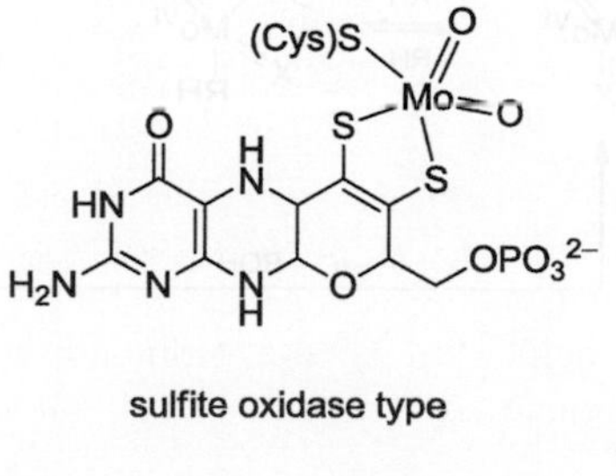

(11.10)

sulfite oxidase type

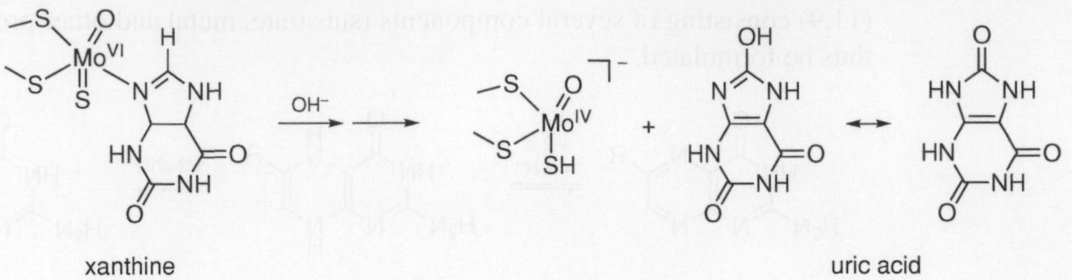

(11.11)

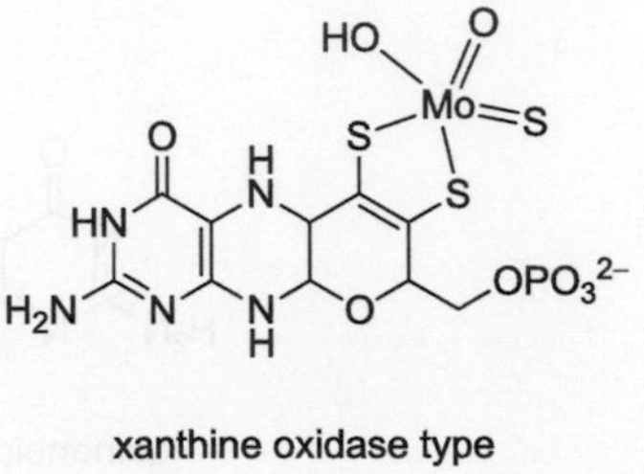

(11.12)

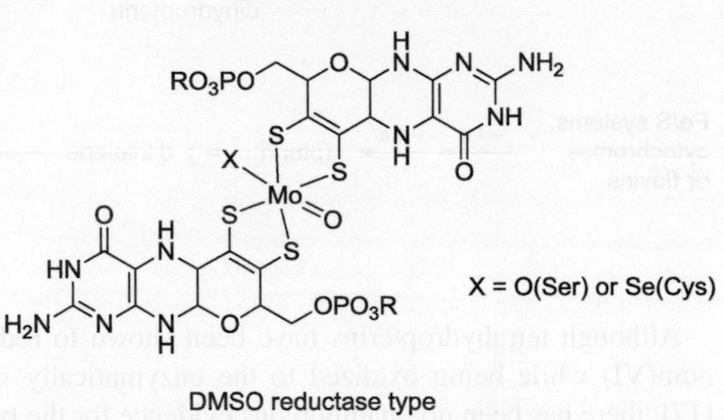

(11.13)

Mo^{V} species are frequently observed by EPR spectroscopy during the stepwise reoxidation of Mo^{IV} after completed oxygen transfer to the substrate; the occurrence of several EPR signals with partially detectable proton "superhyperfine structure" can be rationalized according to scheme (11.14).

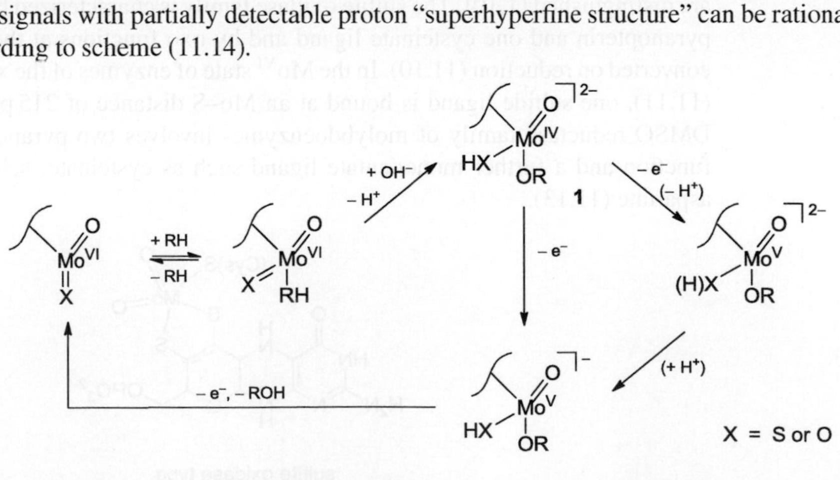

(11.14)

The activated complex **1** of tetravalent molybdenum with one oxo, one hydroxyl or sulfhydryl and one indirectly coordinated substrate ligand, R, can be one-electron oxidized under optional deprotonation. Loss of the second (and last) d electron causes release of the oxidized substrate and the resting state of the enzyme is obtained, which can again react with substrate and base to **1** in an intramolecular electron-transfer process. The mechanistic hypothesis (11.14) is supported by studies of model complexes such as (11.15) [18].

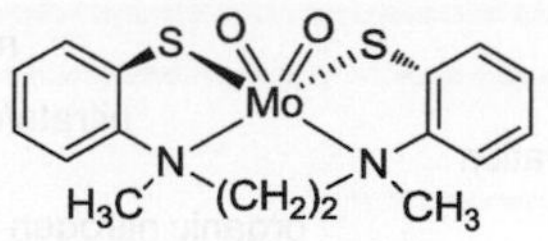

(11.15)

Similar model complexes have been used in the stepwise simulation of an enzyme-analogous cycle for the energetically much favored oxygen transfer from DMSO to the physiologically not relevant triphenylphosphine (11.16) [9].

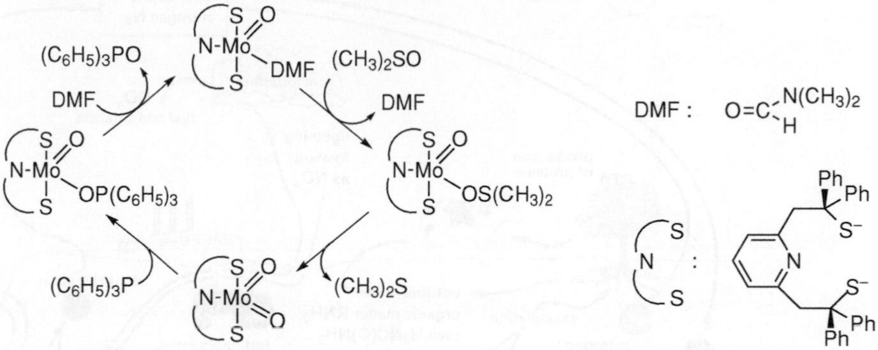

(11.16)

11.2 Metalloenzymes in the Biological Nitrogen Cycle: Molybdenum-dependent Nitrogen Fixation

The importance of the inorganic-biological nitrogen cycle (Figure 11.1) to life on earth can hardly be overestimated. Only the technical nitrogen fixation realized in ammonia synthesis (according to F. Haber and C. Bosch) guarantees to supplement the often growth-limiting nitrogen content of the soil and thus guarantee a food production adequate for the growing human population. Ammonia continues to be one of the leading products of the chemical industry. Other large-scale chemicals such as ammonium nitrate, urea and nitric acid are follow-up products of the technical "fixation" of nitrogen obtained from air. Accordingly, the overall turnover of products from synthetic nitrogen fixation tends to exceed that of the biological process [19,20]. A further source are physical-atmospherical reactions, which transform N_2 during thunderstorm discharges, for example.

The drawbacks of excessive fertilizer use lie in the pollution of soil or ground and surface water with ammonium and nitrate; in addition, the toxic gases NO and NO_2 ("NO_x"), as well as the greenhouse gas N_2O (see Section 10.3), are formed in increasing concentrations through agricultural activity and in combustion processes (catalyzed ozone formation; "summer smog").

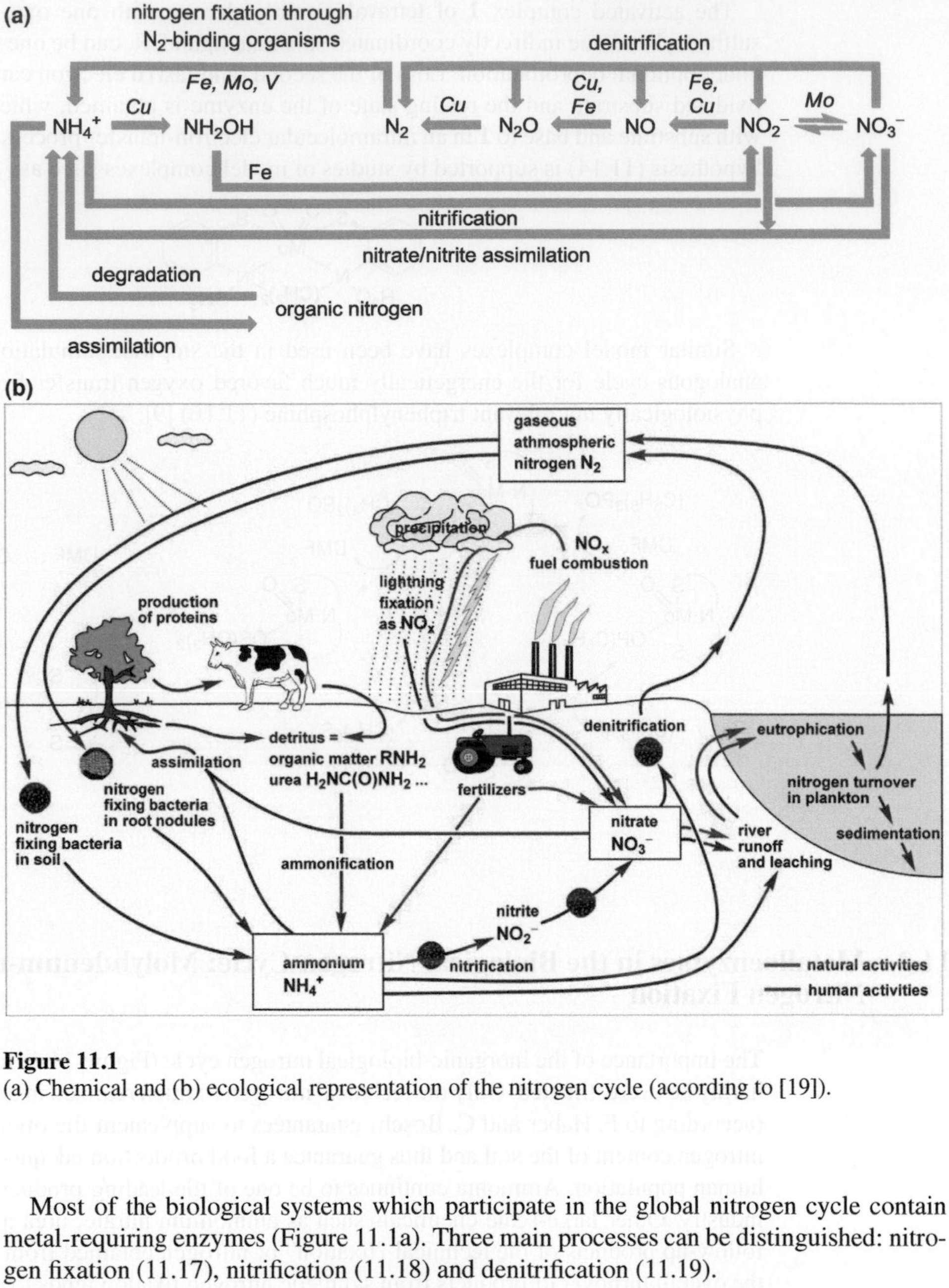

Figure 11.1
(a) Chemical and (b) ecological representation of the nitrogen cycle (according to [19]).

Most of the biological systems which participate in the global nitrogen cycle contain metal-requiring enzymes (Figure 11.1a). Three main processes can be distinguished: nitrogen fixation (11.17), nitrification (11.18) and denitrification (11.19).

$$N_2 + 3\,H_2 \xrightarrow[\text{metal oxide catalyst}]{\text{nitrogen fixation (technical): >400 °C, >100 bar}} 2\,NH_3 \quad \text{(gas phase)}$$

$$N_2 + 10\,H^+ + 8\,e^- \xrightarrow[\text{nitrogenases}]{\text{nitrogen fixation (biological)}} 2\,NH_4^+ + H_2 \qquad (11.17)$$

$$NH_4^+ + 2\,O_2 \xrightarrow{\text{nitrification}} NO_3^- + H_2O + 2\,H^+ \qquad (11.18)$$

$$2NO_3^- + 12\,H^+ + \underbrace{10\,e^-}_{\substack{\text{(from "biomass",}\\ \text{i.e. reduced carbon}\\ \text{compounds)}}} \xrightarrow{\text{denitrification}} N_2 + 6\,H_2O \qquad (11.19)$$

The oxidative process of nitrification yields nitrate: the oxidation state in which nitrogen can be assimilated by most higher plants. The crucial challenge in this reaction as performed by *Nitrosomas* bacteria is to avoid the formation of the stable dinitrogen molecule, N_2, the zerovalent state. Aerobic conditions and sufficient buffer capacity for the resulting protons in soil or mineral material (atmospheric weathering) are indispensable. Some metalloenzymes for certain stages of *de*nitrification, a process that is gaining importance for ecological reasons [21], have already been introduced: the molybdenum-containing nitrate(N^{+V}) reductase (→ nitrite), the copper or heme-iron containing nitrite(N^{+III}) reductases (see Sections 6.5 and 10.3) and the copper-containing dinitrogen monoxide(N^{+I}) reductase (10.12). Denitrification requires rather anaerobic conditions and organic substances as reductants. The potential physiological importance of nitrogen monoxide as a free radical, $NO^\bullet$, or as a metal-coordinated NO^+ intermediate (Sections 6.5 and 10.3) is only just beginning to be realized (Section 6.5). The final product of denitrification is the extremely stable, inert and volatile dinitrogen molecule. Its recycling (i.e. its reuse in the biological cycle) proceeds via the energetically and mechanistically challenging process of nitrogen fixation, as based on nitrogenase (or better "dinitrogenase") enzymes.

This process, which is comparable only to photosynthesis in its overall biological importance, is exclusively confined to procaryotic "diazotrophic" organisms, such as free bacteria of the *Azotobacter* strain or *Rhizobium* bacteria, which exist in symbiosis with the root system of leguminous plants. The restriction on dinitrogen fixation to a relatively few specialized organisms and their symbionts becomes obvious with the appearance of certain "pioneer plants" in the repopulation of, for example, nutrient-poor soil left behind by receding glaciers. After nonvolatile ("fixated") inorganic nitrogen compounds have been made available via (11.17), the element is incorporated as an amino group into organic carrier molecules such as glutaric acid, to be used further in the biosynthesis of proteins and nucleobases.

Due to the thermodynamic stability of the dinitrogen molecule, its reduction requires a large amount of energy in the form of several ATP equivalents (with Mg^{2+} as the hydrolysis catalyst; Section 14.1) and (at least) six electrons per N_2 at a physiologically quite negative potential of less than −0.3 V. In addition, the extremely low reactivity of the dinitrogen molecule requires efficient catalysts: the dinitrogenase enzymes [22,23]. Due to both thermodynamic and kinetic difficulties, the enzymatic nitrogen fixation proceeds rather slowly, with turnover numbers on the order of one per second; equation (11.17) summarizes an eight-electron process coupled with ATP hydrolysis.

Although a large number of stable synthetic complexes of N_2 are now known [24], the first of these substances, a complex (11.20) of the 4d transition metal ruthenium, has been reported only in 1965. In contrast, metal carbonyl complexes of the carbon monoxide ligand, CO, which is isoelectronic with N_2, have been known for more than 100 years. Coordination of centrosymmetrical N_2 (mostly end-on) requires a twofold attack: in addition to the normal coordinative σ bond involving the free electron pair at one nitrogen center and the electrophilic metal ion, the transition metal center should provide electron density for the relatively low-lying unoccupied molecular orbitals of the triple-bond system in N≡N via π interactions without rotational symmetry (π back bonding, push–pull mechanism; (11.20); compare also (5.7)). Thus, N_2 complexation and fixation requires d_π electron-rich metal centers, as found for example in many organometallic compounds [24].

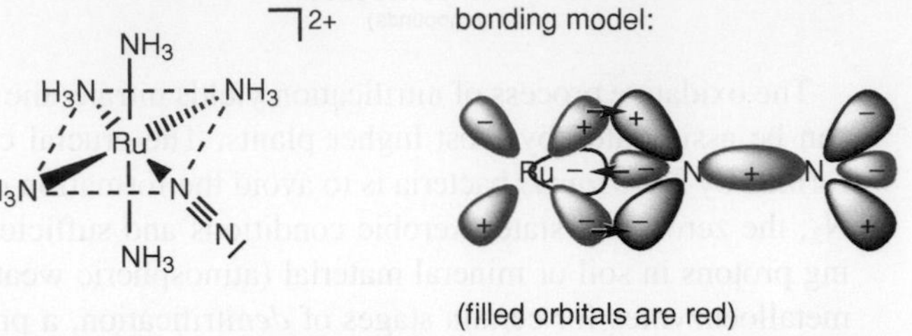

(11.20)

The low redox potentials and the necessarily high reactivity of nitrogenase enzymes further require the absence of competing (i.e. related but better-coordinating) molecules such as dioxygen, O_2. Therefore, N_2-assimilating microorganisms are either obligatory anaerobes or posses complex protection mechanisms for the exclusion of O_2 from the active site of nitrogenase enzymes. These mechanisms include iron-containing proteins, which serve as O_2 sensors. Nitrogenase activity is also inhibited by the isoelectronic substrates carbon monoxide, $^-C{\equiv}O^+$, and $N{\equiv}O^+$; characteristic reaction products are obtained with several other small molecules containing multiple bonds (Table 11.2).

Table 11.2 Reduction (hydrogenation) reactions catalyzed by nitrogenises.

Substrate	Products	Number of electrons required
|N≡N|	$2\ NH_3 + H_2$	8 e⁻
H–C≡C–H	C_2H_4 or Z-$C_2H_2D_2$ (from C_2D_2)[a]	2 e⁻
H–C≡N|	$CH_4 + NH_3$ (CH_3NH_2)	6 e⁻ (4 e⁻)
$H_3C{-}N^+{\equiv}C\|^-$	$CH_3NH_2 + CH_4$	6 e⁻
$\langle N{=}N^+{=}N\rangle^-$	$N_2H_4 + NH_3$ ($N_2 + NH_3$)	6 e⁻ (2 e⁻)
$\langle N{=}N^+{=}O\rangle$	$N_2 + H_2O$	2 e⁻
$\overset{CH_2}{HC{=}CH}$	1/3 $\overset{CH_2}{H_2C{-}CH_2}$ + 2/3 $H_3C{-}CH{=}CH_2$	2 e⁻
$2\ H^+$	H_2	2 e⁻

[a]Partially four-electron reduction to C_2H_6 in vanadium-dependent nitrogenase.

In the case of conventional (i.e. molybdenum-containing) nitrogenase, both the exclusive *Z*(*cis*)-hydrogenation of acetylene to the ethylene state and the cleavage of the isocyanide triple bond are remarkable. Furthermore, nitrogenases possess an intrinsic hydrogenase activity, which leads to the obligatory production of dihydrogen, H_2, during the biological N_2 fixation reaction (11.21); no H_2 evolution was observed during the reduction of the unphysiological substrates listed in Table 11.2.

$$N_2 + 8\,H^+ + 8\,e^- \longrightarrow H_2 + 2\,NH_3 \underset{pK_a = 9.2}{\overset{2\,H^+}{\rightleftharpoons}} 2\,NH_4^+ \qquad (11.21)$$

Even N_2 pressures of up to 50 bar do not preclude the formation of 25% H_2 per reduction equivalent according to (11.21); therefore, a simple displacement equilibrium cannot explain this H_2 evolution. On the other hand, dihydrogen (in equilibrium) inhibits the N_2 fixation process. It is now assumed (11.22) that the reduction of the enzyme, E, proceeds via one-electron/one-proton addition steps and involves binding of N_2 only after the addition of the third reduction equivalent [23]. In the hypothetical sequence (11.22), the immediately bound and reduced hydrogen, formulated as side-on coordinated H_2, gets replaced by N_2.

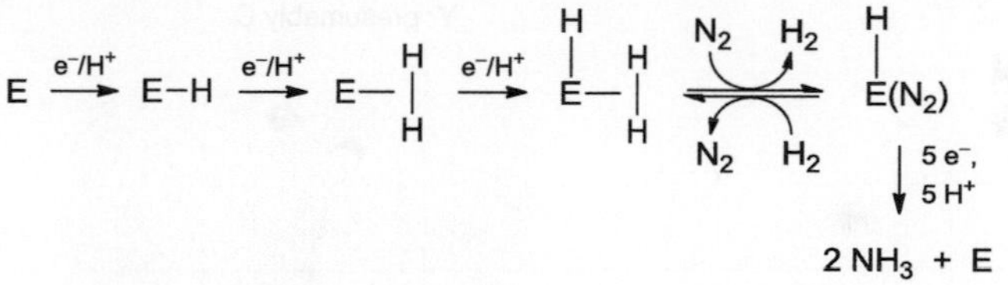

(11.22)

In most cases, the dihydrogen thus obtained is immediately reoxidized by hydrogenases (Section 9.3), yielding protons and energy.

Nitrogenase activity is also inhibited by an excess of the product, NH_4^+, by the presence of the assumed intermediate hydrazine, N_2H_4 (see (11.25)) and in the absence of essential inorganic components. Required are the already mentioned Mg^{2+} for ATP hydrolysis, sulfur in the form of sulfide or transformable sulfate, iron and, at least for the "classical" nitrogenase, molybdenum; the heavier homologue tungsten cannot serve as a substitute.

The most extensively studied molybdenum-dependent form of nitrogenase (see Figure 11.3) has an $(\alpha_2\beta_2)(\gamma_2)$ two-protein composition, with the special dimeric iron protein (γ_2), the "dinitrogenase reductase" (about 60 kDa), being essential for the function of the enzyme. Bound *between* the two γ subunits of this protein is a single [4Fe–4S] cluster, which can be reduced to a paramagnetic form at the physiologically very negative potential of −0.35 V. This "Fe protein" of nitrogenase contains two Mg^{2+}/ATP receptors, because two ATP molecules have to be hydrolyzed for each transferred electron. Binding of ATP or ADP further lowers the potential by about 0.1 V.

The second component, the actual "dinitrogenase" or "FeMo protein", is an $\alpha_2\beta_2$ tetramer (220 kDa; see Figure 11.3), which contains two very special Fe/S systems, the [8Fe-7S] P clusters between α and β subunits and two "FeMo cofactors" (FeMo co or M clusters) in the α subunits, each with the inorganic composition of $MoFe_7S_9$. The breakthrough structural elucidation of a nitrogenase from *Azotobacter vinelandii* in 1992 [25–27] has confirmed some of the assumptions made previously on the basis of physical measurements and chemical models. However, the actual structure of the $MoFe_7S_9$ cluster (11.23, PDB code 3U7Q [29a]), consisting of two subclusters, $[MoFe_3S_3]$ and $[Fe_4S_3]$, which are bridged by three sulfide ions, was not deduced by the model studies. It has long been known from Fe and Mo extended x-ray absorption fine structure (EXAFS) measurements of the protein and the anionic FeMo cofactor (1.5 kDa, extractable with N-methylformamide, $HN(CH_3)C(O)H$) that a molybdenum-modified cuboidal Fe/S cluster is present and that the tetravalent molybdenum (reduced form) is bound to some O or N donor atoms. In fact, the six-coordinate molybdenum of the M cluster is located at the corner of a metal sulfide heterocubane structure and is coordinated to the outside by a histidine and a tetraanionic chelating homocitrate ligand ((11.24): homocitric acid = (R)-2-hydroxy-1,2,4-butanetricarboxylic acid). The possibly multiple function of the homocitrate ligand (structural, catalysis support, assembly factor) is still being investigated [28]. A hexacoordinate (μ_6) atom surrounded by iron atoms at the center of the M cluster has been identified as a carbido (C^{4-}) ligand, originating from the methyl group of S-adenosylmethionine [29].

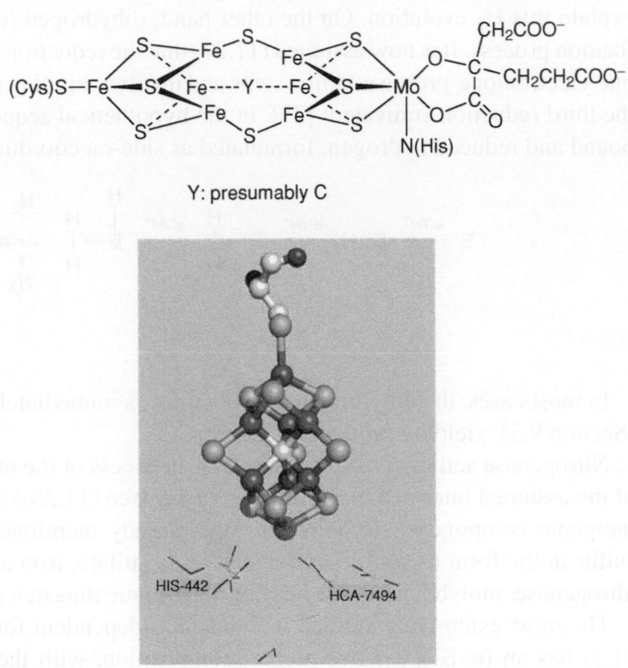

(11.23)

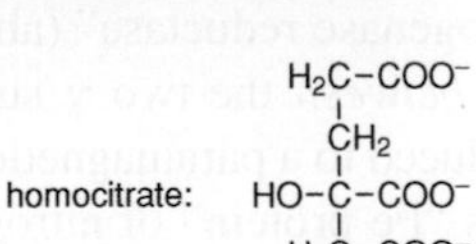

(11.24)

In the reduced state, the cofactor contains tetravalent molybdenum; however, the total electron spin of $S = 3/2$ is mostly located at the various different sulfur-coordinated iron centers, which, according to Mössbauer data, show some degree of electron delocalization. In the C_2 symmetrical (Figure 11.3) dinitrogenase protein, the M clusters are separated by about 7 nm, whereas the closest distance between M and P cluster is about 2 nm.

In the reduced Fe protein (dinitrogenase reductase), the function of the unusual [4Fe–4S] cluster is to provide an electron flow at low potential using simultaneous ATP hydrolysis as the driving force (two ATP per electron). Inside the FeMo protein, the polynuclear P clusters can be reduced to the all-Fe^{II} form, and presumably regulate the low potential transfer of electrons to the FeMo cofactors.

Concerning the actual mechanism of the metalloenzyme-catalyzed transformation of N_2 to NH_3, the leading hypotheses [30] take into account the results obtained for non-enzymatic complexes, for example of zerovalent molybdenum [31]. A multistep reaction sequence is certainly necessary for this energetically and mechanistically demanding process (11.21); according to (11.25), and in analogy to the heterogeneously catalyzed technical synthesis of ammonia, the catalyst is required to prevent the formation of "free", high-energy intermediates.

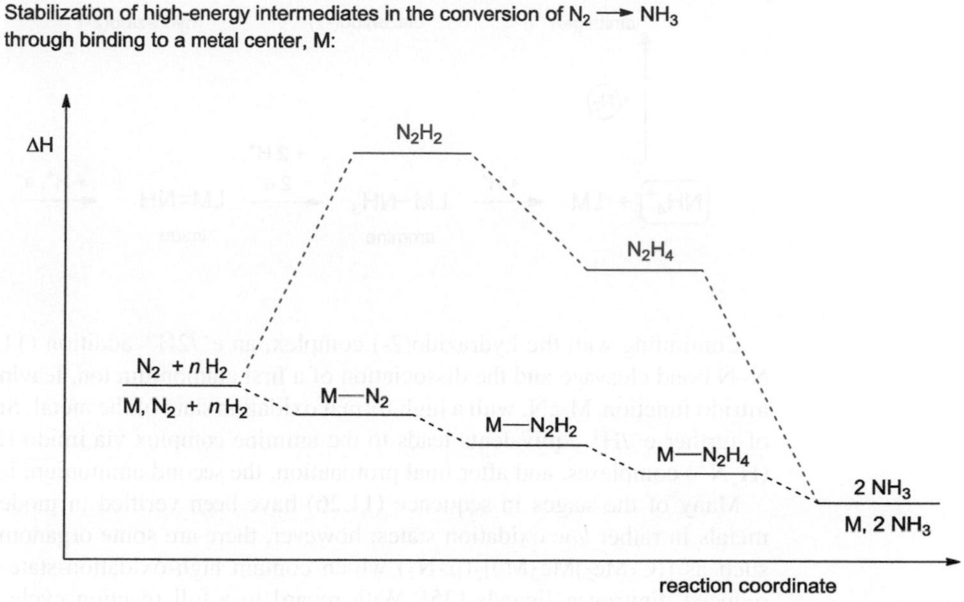

(11.25)

The coordinative saturation of the Mo center (11.22) and the existence of heteroatom-free nitrogenases suggest an iron rather than a molybdenum coordination of N_2; as with O_2 (see Table 5.1), the alternatives [32] for the coordination mode can range from end-on binding (η^1) to side-on bridging. In synthetic chemistry, the μ-η^2:η^2-coordination of N_2 is rarely observed [33], end-on bridging (μ-η^1:η^1) being much more common [24]. The cluster asymmetry caused by the molybdenum center presumably facilitates electrophilic attack during the stepwise reduction of dinitrogen.

A catalysis mechanism based on model reactions of complexes with nonbridging, end-on coordinated N_2 can be formulated as in (11.26) (see [23]). Multiple stepwise addition of e^-/H^+ before the actual dinitrogen fixation results in the displacement of H_2 (11.22), as caused by the coordination of N_2, followed by the combination of N_2 and e^-/H^+ to form a diazenido(1–) ligand, HNN^- (11.26). This ligand can react with another e^-/H^+ equivalent to yield a terminal hydrazido(2-) ligand, H_2N–N^{2-}, which can also be formulated as a neutral aminonitrene ligand, H_2N–N, after redistribution of oxidation states. In fact, hydrazine can be detected as an incompletely reduced form of dinitrogen following rapid quenching of the nitrogenase reaction sequence. Nonreductive protonation of the diazenido ligand will yield the N^{-I} species diazene, H–N=N–H, which is unstable in free form; however, metal complexes can stabilize this molecule through multiple coordination and hydrogen bond interactions [34].

Possible intermediate states in the metal-catalyzed reduction of N_2 with reference to the different nitrogen ligands (LM = metal complex with one free coordination site):

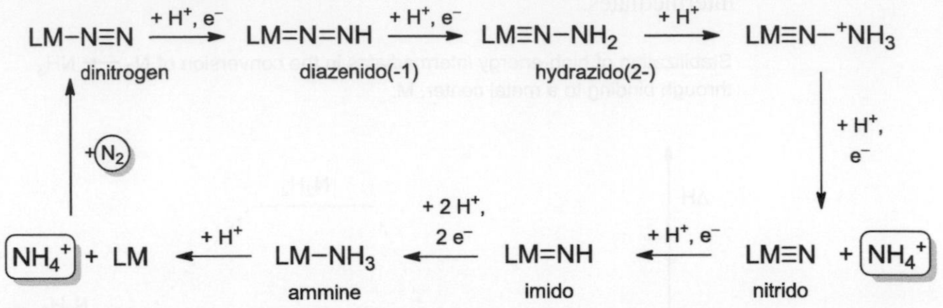

(11.26)

Continuing with the hydrazido(2-) complex, an $e^-/2H^+$ addition (11.26) will result in N–N bond cleavage and the dissociation of a first ammonium ion, leaving behind a metal-nitrido function, M≡N, with a high formal oxidation state of the metal. Successive addition of further e^-/H^+ equivalents leads to the ammine complex via imido (HN^{2-}) and amido (H_2N^-) complexes, and after final protonation, the second ammonium ion is obtained.

Many of the stages in sequence (11.26) have been verified in model complexes with metals in rather *low* oxidation states; however, there are some organometallic complexes such as $[(C_5Me_5)Me_3Mo]_2(\mu\text{-}N_2)$ which contain high-oxidation-state Mo and partially reduced dinitrogen ligands [35]. With regard to a full reaction cycle and hydrazine or ammonia production, complexes of low-valent molybdenum and tungsten have been most successful.

In addition to the structural modeling of the nitrogenase metal cluster and experiments directed at the simulation of at least parts of the biocatalytic reactivity, the genetic variations of corresponding microorganisms and the effect on proteins synthesized by the resulting mutants have contributed to the current understanding of biological nitrogen fixation. A major objective of such rational protein design [1,23] is to eventually transfer the set of at least 17 *nif* (*ni*trogen *f*ixation) genes from microorganisms to important crops. The knowledge gained concerning classical (i.e. molybdenum-dependent) nitrogenase has also turned out to be very valuable for the characterization of two other established variants, the vanadium-dependent and the heteroatom-independent nitrogenase enzymes.

11.3 Alternative Nitrogenases

Contrary to a proposition by H. Bortels from the 1930s concerning the possible role of vanadium, molybdenum was for a long time considered to be indispensable for nitrogen fixation. Only more recent advances in ultratrace elemental analysis and genetic analysis of N_2-fixating organisms have provided unambiguous evidence for the existence of *alternative* nitrogenases that can be produced in the absence of molybdenum or FeMo cofactor-synthesizing genes [23,36]. In the presence of vanadium, a V-dependent nitrogenase is

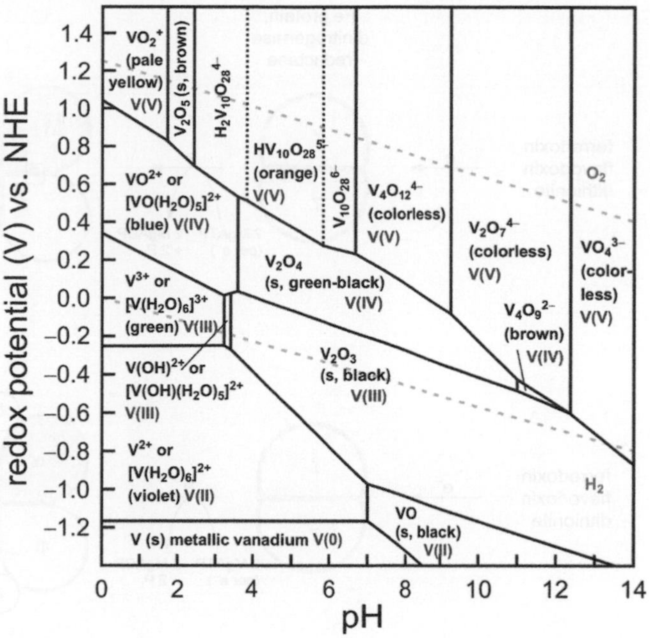

Figure 11.2
Stability or "Pourbaix" diagram of vanadium in water, showing regions of various hydrated species and solid forms as depending on redox potentials and pH values (according to [37]).

formed; if this element too is not sufficiently available, a third, "iron-only" form can be synthesized by some organisms. Where there is a sufficient supply of molybdenum, the formation of such "alternative" nitrogenases is suppressed, so under normal conditions these less effective enzymes can be tentatively regarded as "back-up" systems.

There are numerous parallels between the three types of nitrogenase, but also characteristic differences [23]. First of all, it should come as no surprise that the "second best" form of nitrogenase contains vanadium, since V and Mo are connected by a diagonal relationship in the periodic table of elements; other (bio)inorganic examples such as Mn/Ru, with regard to O_2 production (see Section 4.3), and Fe/Rh, with regard to H^+/H_2 conversion (Sections 7.4 and 9.3), can be mentioned here. The correspondence between V and Mo manifests itself in their similar chemistry in aqueous solution, among other things. The fairly complicated stability diagram in Figure 11.2 shows that – like molybdenum (see (11.1)) – vanadium has a strong tendency to form oxo/hydroxo-bridged aggregates.

Vanadium is more abundant than molybdenum in the earth's crust, but not in sea water; nevertheless, it is the most abundant first-row (3d) transition metal in that medium. The V-dependent nitrogenase has a slightly lower activity regarding N_2 reduction and exhibits small amounts of N_2H_4 among its reduction products; however, its main disadvantage is that about 50% of the reduction equivalents are used in the formation of H_2, as compared to only 25% in the case of the Mo enzyme (see (11.21)). Typically, the V-dependent nitrogenase shows rather little activity in the reduction of acetylene, yielding not only ethene, C_2H_4, but also some ethane, C_2H_6, as four-electron hydrogenation product (Figure 11.3). Remarkably, however, the V-dependent nitrogenase is more efficient than the molybdenum system at temperatures around 5 °C, a fact that may have favored the conservation of this form during

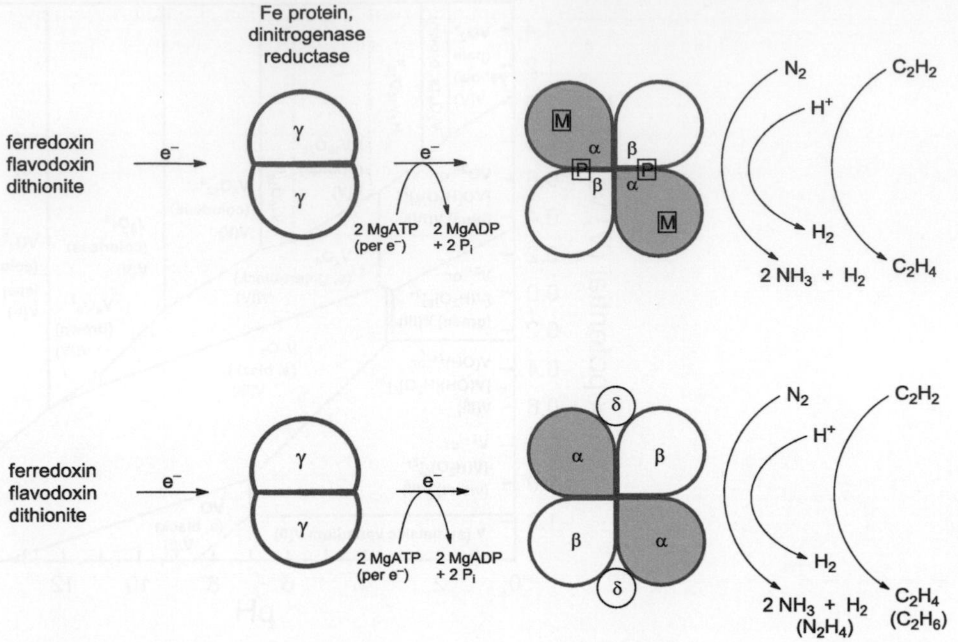

Figure 11.3
Schematic composition of nitrogenase enzymes: Mo-dependent (top) and V-dependent (bottom) forms.

evolution [23]. The most sensitive, heterometal-independent (Fe-only) nitrogenases have not yet been well characterized; their activity seems to be lower than that of the vanadium- or molybdenum-containing enzymes.

Like the molybdenum-dependent system, both alternative nitrogenases are composed of two proteins. The V-dependent enzyme in particular exhibits many common features when compared to the Mo analogue (Figure 11.3): a dimeric Fe protein with Mg^{2+}/ATP binding sites faces a larger protein, which in the case of the V system is an $\alpha_2\beta_2\delta_2$ hexamer with two small δ subunits. As well as thc basic enzyme structure, certain invariant amino acids in the protein sequence and the resulting inorganic composition of the cluster centers are also comparable for vanadium- and molybdenum-dependent nitrogenases [23] with MFe_3S_4 heterocubane structural motifs.

Characteristically, the geometrical changes at the metal cluster site are small during redox reactions of both Mo and V nitrogenases, whereas the Mo centers of oxotransferases show a marked structural difference between the +VI and +IV oxidation states. The heteroatom cofactor of vanadium-dependent nitrogenase is also extractable with N-methylformamide. Another analogy with the molybdenum system is the ($S = 3/2$) EPR signal of reduced V-dependent nitrogenase and the d^2 configuration of the heterometal (here, V^{III}). Like molybdenum, vanadium is known to form low-molecular-weight complexes in which the low-valent (≤+II) metal can bind N_2 [38]. The apparent requirement of certain heteroatoms for an efficient N_2 fixation may be rationalized by considering the pronounced capability of polythiometal centers for multi-electron "buffering" in the negative potential range. A. Müller *et al.* have therefore suggested that rhenium (which is also connected to molybdenum via a diagonal relationship) could similarly function as a heteroelement in nitrogenases; its extremely low natural abundance restricts its bioavailability, however.

11.4 Biological Vanadium Outside of Nitrogenases

While vanadium-dependent nitrogenases were established only in about 1985, the occurrence of vanadium, often in highly enriched form, in marine organisms such as tunicates and brown algae (seaweeds) and in lichen and mushrooms (toadstool) has been known for some time [39–41]. With regard to higher organisms, the inhibiting effect of vanadate(V) on phosphate-depending enzymes (see Sections 13.4 and 14.1) is well recognized, as is the possible therapeutic function of vanadyl(IV) complexes as insulin "mimics" in the glucose metabolism [42]. Considerable amounts of vanadium, particularly in the form of tetrapyrrole complexes, can be found in some kinds of biogenic crude oil (Table 2.6).

As in the molybdate/sulfate pair, the orthovanadate ion, VO_4^{3-}, resembles quite closely – but not completely [43] – the (ortho)phosphate ion, PO_4^{3-}, for example in terms of structural characteristics (tetrahedral ion with a tendency for trigonal bipyramidal coordination) and a shared preference for aggregation (see e.g. ATP (14.2)). As a transition metal, however, vanadium has empty, low-lying 3d orbitals in its maximal oxidation state of (+V); it can thus be reduced fairly easily (see Figure 11.2) to lower oxidation states such as (+IV) or (+III) using biological reductants such as glutathione (16.7). While V^{V} and V^{III} compounds can be examined by heteroatom nuclear magnetic resonance (NMR) spectroscopy (^{51}V: $I = 7/2$, 99.75% natural abundance; see also insert in Section 13.1) [40], the vanadyl(IV) ion, VO^{2+}, and its complexes are particularly well suited for EPR spectroscopical detection, because of their d^1 configuration.

For humans, the essentiality and biochemical function of the ultratrace element vanadium (uptake 10–60 mg/day) has not yet been well defined [44]. Experiments on the vanadate-supported stimulation of cardiovascular activity in insulin-treated diabetic rats were able to demonstrate that the glucose metabolism itself can be favorably influenced by the oral administration of certain vanadium compounds (see also Sections 11.5 and 19.4.4). Among effective insulin mimics are peroxovanadium(V) and vanadyl(IV) complexes, $L_nV^{IV}O$, which have the ability to cross membrane barriers if L is a lipophilic ligand [42]. On the other hand, vanadium and its aggregates inhibit enzymes such as Na^+/K^+-ATPase (see Section 13.4) and related phosphate transferases (kinases, cyclases, phosphatases, ribonucleases; see Section 14.1), even in rather low concentrations. Also, the biosynthesis of cysteine, which is an essential amino acid for metal coordination and protein conformation, is inhibited by vanadate. The iron-transport protein apotransferrin (Section 8.4.1) binds vanadium in the oxidation states III–V.

It has been known since about 1910 that certain sessile marine organisms, sea squirts (*Ascidiae*) of the tunicate group, can accumulate vanadium in their "blood cells" by a factor of above 10^7 more than the surrounding sea water. Earlier assumptions which attributed an O_2 transport function to these large amounts of reduced vanadium(+III, partly +IV) have turned out to be untenable; instead, they may possibly serve as a primitive immune system, in the template synthesis of redox-active tunichrome peptides (11.27) or as components in the anaerobic metabolism [45]. It is a matter of controversy at what pH value (neutral or strongly acidic) and in which coordination arrangement vanadium occurs in these sulfate-rich cells [45,46]. An interaction is assumed between dissolved vanadium and the labile tunichrome dipeptides, which contain potentially metal-coordinating and redox-active o-polyphenol groups and which are also very characteristic but functionally poorly defined natural products with an antimicrobial defense function in such animals.

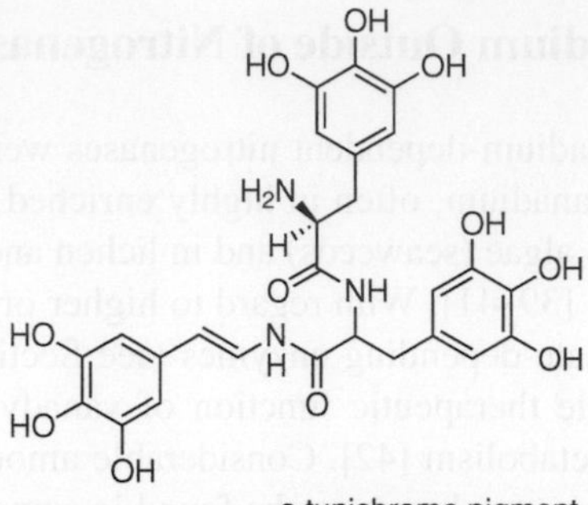

(11.27)

a tunichrome pigment

Another vanadium-containing natural product, "amavadin", can be isolated from mushrooms of the toadstool kind (genus *Amanita*). Its exact identity has been disputed for a long time [47,48]. Crystal structure analysis shows eight-coordinate vanadium (V^{V} or V^{IV}) [48] with two 2,2'-(oxyimino)dipropionate ligands (11.28). This complex is extremely stable (compare the related hydroxamate function of iron-siderophore complexes; Section 8.2), which explains the remarkable vanadium accumulation of up to 200 ppm from the surrounding soil. The function of such a vanadium accumulation is still unclear; mushrooms and fungi are generally known to synthesize efficient low-molecular-weight chelate ligands for metal ions (see Sections 17.3 and 18.1.2).

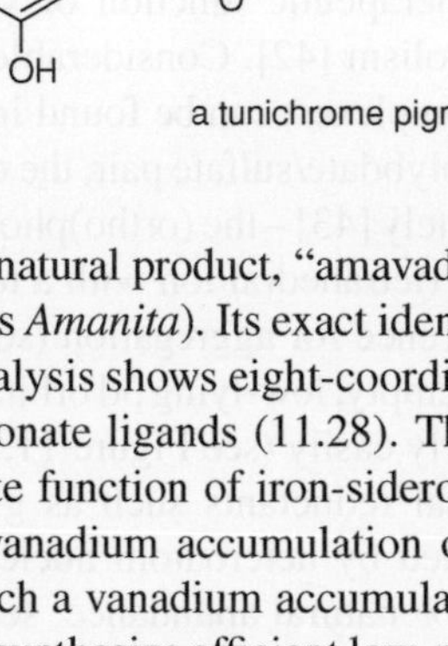

(11.28)

amavadin from toadstool (*Amanita*),
charge 1– (V^{V}) or 2– (V^{IV})

Certain red and brown algae ("seaweeds" such as *Ascophyllum nodosum*) and some species of lichen and fungi contain vanadium-dependent haloperoxidases [49–51]. These enzymes, which also occur in a more "conventional" form as hemoproteins (see Section 6.3), catalyze the halogenation of organic substrates with the help of hydrogen peroxide according to (11.29), with hypohalide acids, HalOH, as intermediates. Like many of their synthetic analogs, the "natural" halogenated hydrocarbons, which are formed in quite large amounts globally, may serve as biocides in the defense system of organisms.

$$R{-}H + H_2O_2 + Hal^- + H^+ \xrightarrow{\text{haloperoxidase}} R{-}Hal + 2\,H_2O \qquad (11.29)$$

R: organic alkyl or aryl group
Hal: Cl, Br, I

While land-living organisms like fungi and lichen feature mainly chlorinating and iodinating enzymes (cf. thyreoperoxidases; Sections 6.3 and 16.7), the bromination to yield bromoform, $CHBr_3$, for example, plays an important role for marine organisms. Hypobromite, ^-OBr, is generally assumed to be the final brominating species. The likely function of vanadium in the rather acidic bromoperoxidase enzymes (about 110 kDa) is the coordination and activation of (hydro)peroxide, which is essential for the controlled formation of reactive bromo-oxygen compounds, such as ^-OBr; oxidation of the far more abundant chloride requires higher potentials and proceeds much more slowly. However, some substrates can also be efficiently chlorinated by vanadium-containing "bromo"peroxidases [50]. In agreement with the peroxidase function, only the (+V) oxidation state of vanadium is assumed to be relevant under physiological conditions. The resting state contains $H_2VO_4^-$ (= $[V(O)_2(OH)_2]^-$) [51], which is anchored by a histidine (N_ε) to the protein. This proximal histidine is in one of the axial ligands in a trigonal bipyramidal configuration of the metal; the other O and OH ligands are engaged in extensive hydrogen bonding. Addition of H_2O_2 results in conversion to a tetragonal pyramid with side-on bonded peroxide, $VO(OH)(O_2)(His)$, which, after halide coordination, produces a hypohalide vanadate, $[VO_2(OH)(OHal)(His)]^-$ (11.30) [51]. Because of their high thermal and chemical stability, the V-dependent peroxidases are being considered as components for medical diagnostics, biotechnology and even household detergents.

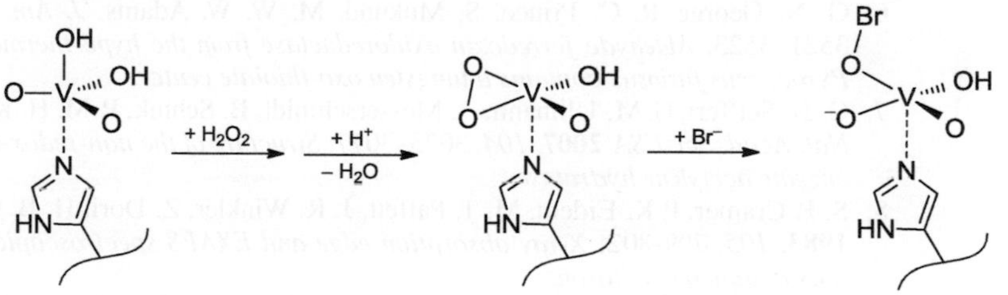

(11.30)

11.5 Chromium(III) in the Metabolism?

While chromium(VI), as chromate, CrO_4^{2-}, has been recognized as a mutagenic and even carcinogenic species (see Section 17.8), the metal has been prematurely [52,53] proposed as an essential trace element in its (+III) oxidation state, which is the most stable state in aqueous solution. Like iron(III) (see Chapter 8) or aluminum(III) (see Section 17.6), trivalent chromium can only be resorbed very slowly, due to the insolubility of the hydroxide at pH 7 and because of very slow substitution reactions. Responsible for this extraordinary sluggishness, even with regard to the exchange of coordinated water, is the stable d^3 configuration with half-filled t_{2g} orbitals in octahedral symmetry (see (2.13)); any structural deviation caused by associative or dissociative processes requires a very high activation energy.

The most discussed function of Cr^{III} concerns the participation of a chromium-containing glucose tolerance factor (GTF) in optimizing the effect of insulin [52,53]. The GTF is formulated as a typical (i.e. octahedrally configured and substitutionally inert) Cr^{III} complex, usually with two nicotinic acid ligands in *trans* position; the remaining four coordination sites can at least partially be occupied by sulfur ligands, for example from the peptide glutathione.

It has been speculated that this complex interacts with the membrane receptor for insulin (see Section 12.7). However, in microorganisms such as yeast cells, GTF does not necessarily contain chromium, which, in any case, is a ubiquitous element (occurring even in stainless steel) and thus hard to exactly specify in ultratrace quantities in biological systems (see also Section 17.6).

References

1. G. Schwarz, R. R. Mendel, M. W. Ribbe, *Nature (London)* **2009**, *460*, 839–847: *Molybdenum cofactors, enzymes and pathways*.
2. R. Hille, R. Mendel (eds.), *Coord. Chem. Rev.* **2011**, *255*, 991–1224: *Molybdenum in living systems*.
3. I. Yamamoto, T. Saiki, S. M. Liu, L. G. Ljungdahl, *J. Biol. Chem.* **1983**, *258*, 1826–1832: *Purification and properties of NADP-dependent formate dehydrogenase from Clostridium thermoaceticum, a tungsten-selenium-iron protein*.
4. R. Wagner, J. R. Andreesen, *Arch. Microbiol.* **1987**, *147*, 295–299: *Accumulation and incorporation of ^{185}W-tungsten into proteins of Clostridium acidiurici and Clostridium cylindrosporum*.
5. R. A. Schmitz, S. P. J. Albracht, R. K. Thauer, *FEBS* **1992**, *309*, 78–81: *Properties of the tungsten-substituted molybdenum formylmethanofuran dehydrogenase from Methanobacterium wolfei*.
6. G. N. George, R. C. Prince, S. Mukund, M. W. W. Adams, *J. Am. Chem. Soc.* **1992**, *114*, 3521–3523: *Aldehyde ferredoxin oxidoreductase from the hyperthermophilic archaebacterium Pyrococcus furiosus contains a tungsten oxo-thiolate center*.
7. G. B. Seiffert, G. M. Ullmann, A. Messerschmidt, B. Schink, P. M. H. Kroneck, O. Einsle, *Proc. Nat. Acad. Sci. USA* **2007**, *104*, 3073–3077: *Structure of the non-redox-active tungsten/[4Fe:4S] enzyme acetylene hydratase*.
8. S. P. Cramer, P. K. Eidem, M. T. Paffett, J. R. Winkler, Z. Dori, H. B. Gray, *J. Am. Chem. Soc.* **1983**, *105*, 799–802: *X-ray absorption edge and EXAFS spectroscopic studies of molybdenum ions in aqueous solution*.
9. R. H. Holm, *Coord. Chem. Rev.* **1990**, *100*, 183–221: *The biologically relevant oxygen atom transfer chemistry of molybdenum: from synthetic analogue systems to enzymes*.
10. R. Hille, *Trends Biochem. Sci.* **2002**, *27*, 360–367: *Molybdenum and tungsten in biology*.
11. J. H. Weiner, R. A. Rothery, D. Sambasivarao, C. A. Trieber, *Biochim. Biophys. Acta* **1992**, *1102*, 1–18: *Molecular analysis of dimethylsulfoxide reductase: a complex iron-sulfur molybdoenzyme of Escherichia coli*.
12. K. V. Rajagopalan, J. L. Johnson, *J. Biol. Chem.* **1992**, *267*, 10199–10202: *The pterin molybdenum cofactors*.
13. R. C. Bray, L. S. Meriwether, *Nature (London)* **1966**, *212*, 467–469: *Electron spin resonance of xanthine oxidase substituted with molybdenum-95*.
14. W. Kaim, B. Schwederski, *Coord. Chem. Rev.* **2010**, *254*, 1580–1588: *Non-innocent ligands in bioinorganic chemistry – an overview*.
15. B. Krüger, O. Meyer, *Biochim. Biophys. Acta* **1987**, *912*, 357–364: *Structural elements of bactopterin from Pseudomonas carboxydoflava carbon monoxide dehydrogenase*.
16. M. K. Chan, S. Mukund, A. Kletzin, M. W. W Adams, D. C. Rees, *Science* **1995**, *267*, 1463–1469: *Structure of a hyperthermophilic tungstopterin enzyme, aldehyde ferredoxin oxidoreductase*.

17. S. J. N. Burgmayer, A. Baruch, K. Kerr, K. Yoon, *J. Am. Chem. Soc.* **1989**, *111*, 4982–4984: *A model reaction for Mo(VI) reduction by molybdopterin.*
18. Z. Xiao, C. G. Young, J. H. Enemark, A. G. Wedd, *J. Am. Chem. Soc.* **1992**, *114*, 9194–9195: *A single model displaying all the important centers and processes involved in catalysis by molybdoenzymes containing* $[Mo^{VI}O_2]^{2+}$ *active sites.*
19. R. Söderlund, T. Rosswall, The nitrogen cycles, in E. Frieden (ed.), *Biochemistry of the Elements, Vol. 1, Part B*, Plenum Press, New York, **1982**.
20. J. Rockström, W. Steffen, K. Noone, Å. Persson, F. S. Chapin III, E. Lambin, T. M. Lenton, M. Scheffer, C. Folke, H. Schellnhuber, B. Nykvist, C. A. De Wit, T. Hughes, S. van der Leeuw, H. Rodhe, S. Sörlin, P. K. Snyder, R. Costanza, U. Svedin, M. Falkenmark, L. Karlberg, R. W. Corell, V. J. Fabry, J. Hansen, B. H. Walker, D. Liverman, K. Richardson, C. Crutzen, J. Foley, *Nature (London)* **2009**, *461*, 472–475: *A safe operating space for humanity.*
21. S. J. Ferguson, *Trends Biochem. Sci.* **1987**, *12*, 354–357: *Denitrification: a question of the control and organization of electron and ion transport.*
22. L. C. Seefeldt, B. M. Hoffman, D. R. Dean, *Annu. Rev. Biochem.* **2009**, *78*, 701–722: *Mechanism of Mo-dependent nitrogenase.*
23. R. R. Eady, *Chem. Rev.* **1996**, *96*, 3013–3030: *Structure-function relationships of alternative nitrogenases.*
24. R. A. Henderson, G. J. Leigh, C. J. Pickett, *Adv. Inorg. Chem. Radiochem.* **1983**, *27*, 197–292: *The chemistry of nitrogen fixation and models for the reactions of nitrogenase.*
25. M. M. Georgiadis, H. Komiya, P. Chakrabarti, D. Woo, J. J. Kornuc, D. C. Rees, *Science* **1992**, *257*, 1653–1659: *Crystallographic structure of the nitrogenase iron protein from Azotobacter vinelandii.*
26. O. Einsle, F. A. Tezcan, S. L. A. Andrade, B. Schmid, M. Yoshida, J. B. Howard, D. C. Rees, *Science* **2002**, *297*, 1696–1700: *Nitrogenase MoFe-protein at 1.16Å resolution: a central ligand in the FeMo-cofactor.*
27. J. Kim, D. C. Rees, *Nature (London)* **1992**, *360*, 553–560: *Crystallographic structure and functional implications of the nitrogenase molybdenum-iron protein from Azotobacter vinelandii.*
28. a) T. Spatzal, M. Aksoyoglu, L. Zhang, S. L. Andrade, E. Schleicher, S. Weber, D. C. Rees, O. Einsle, *Science* **2011**, *334*, 940: *Evidence for interstitial carbon in nitrogenase FeMo cofactor*; b)A. W. Fay, M. A. Blank, J. M. Yoshizawa, C. C. Lee, J. A. Wiig, Y. Hu, K. O. Hodgson, B. Hedman, M. W. Ribbe, *Dalton Trans.* **2010**, *39*, 3124–3130: *Formation of a homocitrate-free iron-molybdenum cluster on NifEN: implications for the role of homocitrate in nitrogenase assembly.*
29. J. A. Wiig, Y. Hu, C. C. Lee, M. W. Ribbe, *Science* **2012**, *337*, 1672–1675: *Radical SAM-dependent carbon insertion into the nitrogenase M-cluster.*
30. Y. Hu, M. W. Ribbe, *Acc. Chem. Res.* **2010**, *43*, 475–484: *Decoding the nitrogenase mechanism: the homologue approach.*
31. F. Studt, F. Tuczek, *Angew. Chem. Int. Ed. Engl.* **2005**, *44*, 5639–5642: *Energetics and mechanism of a room-temperature catalytic process for ammonia synthesis (Schrock cycle): comparison with biological nitrogen fixation.*
32. H. Deng, R. Hoffmann, *Angew. Chem. Int. Ed. Engl.* **1993**, *32*, 1062–1065: *How* N_2 *might be activated by the FeMo cofactor of nitrogenase.*
33. J. Ho, R. J. Drake, D. W. Stephan, *J. Am. Chem. Soc.* **1993**, *115*, 3792–3793: $[Cp_2Zr(\mu\text{-}PPh)]_2[((THF)_3Li)_2(\mu\text{-}N_2)]$*: a remarkable salt of a zirconene phosphinidene dianion and lithium dication containing side-bound dinitrogen.*
34. D. Sellmann, W. Soglowek, F. Knoch, M. Moll, *Angew. Chem. Int. Ed. Engl.* **1989**, *28*, 1271–1272: *Transition metal complexes with sulfur ligands. 48. Nitrogenase model compounds:* $[\mu\text{-}N_2H_2\{Fe(\text{"}NHS_4\text{"})\}_2]$*, the proto type for the coordination of diazene on iron-sulfur center and its stabilization via strong N-H*$\cdots$ *S hydrogen bonds.*
35. R. R. Schrock, T. E. Glassman, M. G. Vale, *J. Am. Chem. Soc.* **1991**, *113*, 725–726: *Cleavage of the N-N bond in a high-oxidation-state tungsten or molybdenum hydrazine complex and the catalytic reduction of hydrazine.*
36. J. R. Chisnell, R. Premakumar, P. E. Bishop, *J. Bacteriol.* **1988**, *170*, 27–33: *Purification of a second alternative nitrogenase from a nifHDK deletion strain of Azotobacter vinelandii.*
37. R. M. Garrels, C. L. Christ, *Solutions, Minerals, and Equilibria*, Freeman & Cooper, San Francisco, **1965**.

38. D. Rehder, C. Woitha, W. Priebsch, H. Gailus, *J. Chem. Soc., Chem. Commun.* **1992**, 364–365: *trans-[Na(thf)][V(N$_2$)$_2$(Ph$_2$PCH$_2$CH$_2$PPh$_2$)$_2$]: structural characterization of a dinitrogenvanadium complex, a functional model for vanadium nitrogenase.*
39. R. Wever, K. Kustin, *Adv. Inorg. Chem.* **1990**, *35*, 81–115: *Vanadium: a biologically relevant element.*
40. D. Rehder, *Bioinorganic Vanadium Chemistry*, Wiley-VCH, Weinheim, **2008**.
41. A. Butler, C. J. Carrano, *Coord. Chem. Rev.* **1991**, *109*, 61–105: *Coordination chemistry of vanadium in biological systems.*
42. Y. Shechter, *Diabetes* **1990**, *39*, 1–5: *Insulin-mimetic effects of vanadate: possible implications for future treatment of diabetes.*
43. M. Krauss, H. Basch, *J. Am. Chem. Soc.* **1992**, *114*, 3630–3634: *Is the vanadate anion an analogue of the transition state of RNAse A?*
44. F. H. Nielsen, *FASEB J.* **1991**, *5*, 2661–2667: *Nutritional requirements for boron, silicon, vanadium, nickel, and arsenic: current knowledge and speculation.*
45. M. J. Smith, D. Kim, B. Horenstein, K. Nakanishi, K. Kustin, *Acc. Chem. Res.* **1991**, *24*, 117–124: *Unraveling the chemistry of tunichrome.*
46. J. P. Michael, G. Pattenden, *Angew. Chem. Int. Ed. Engl.* **1993**, *32*, 1–23: *Marine metabolites and the complexation of metal ions: facts and hypotheses.*
47. E. Bayer, E. Koch, G. Anderegg, *Angew. Chem. Int. Ed. Engl.* **1987**, *26*, 545–546: *Amavadin, an example of selective vanadium binding in nature. Complexation studies and a new structure proposal.*
48. R. E. Berry, E. M. Armstrong, R. L. Beddoes, D. Collison, S. N. Ertok, M. Helliwell, C. D. Garner, *Angew. Chem. Int. Ed. Engl.* **1999**, *38*, 795–797: *The structural characterization of amavadin.*
49. J. W. P. M. van Schijndel, E. G. M. Vollenbroek, R. Wever, *Biochim. Biophys. Acta* **1993**, *1161*, 249–256: *The chloroperoxidase from the fungus Curvularia inaequalis: a novel vanadium enzyme.*
50. A. Butler, J. V. Walker, *Chem. Rev.* **1993**, *93*, 1937–1944: *Marine haloperoxidases.*
51. S. Raugei, P. Carloni, *J. Phys. Chem. B* **2006**, *110*, 3747–3758: *Structure and function of vanadium haloperoxidases.*
52. J. B. Vincent, *Dalton Trans.* **2010**, *39*, 3787–3794: *Chromium: celebrating 50 years as an essential element?*
53. J. Vincent, *The Bioinorganic Chemistry of Chromium*, John Wiley & Sons, New York, **2012**.

12 Zinc: Structural and Gene-regulatory Functions and the Enzymatic Catalysis of Hydrolysis and Condensation Reactions

12.1 Overview

With approximately 2 g per 70 kg body weight, zinc is the second most abundant transition element in the human organism, following iron; the metal also plays an important role in many other living beings [1]. Under physiological conditions, this element occurs only in the dicationic state; the Zn^{2+} ion is diamagnetic and colorless in its complexes, due to the closed-shell d^{10} configuration. While the metal ion itself is therefore not easily excited electronically, there are also no naturally occurring zinc(II) complexes with (colored) tetrapyrrole ligands, although synthetic complexes of this kind are very stable. As a consequence of all this, the ubiquitous zinc-containing proteins could only be characterized as such after advanced analytical methods had become available; meanwhile, several hundred different zinc proteins are known (see Table 12.1). These include numerous essential enzymes which catalyze the metabolic conversion (synthases, polymerases, ligases, transferases) or degradation (hydrolases) of proteins, nucleic acids, lipids, porphyrin precursors and other important bioorganic compounds. Other functions lie in the structural fixation of special rate- and/or stereoselectivity-determining protein conformations in oxidoreductases, and in the structural stabilization of insulin, of hormone/receptor complexes and of transcription-regulating factors. It is thus not surprising that zinc deficiency can lead to severe pathological effects [2] and that the heavier homologues, cadmium and mercury, are toxic at least partly because they can replace zinc in its enzymes (see Sections 17.3 and 17.5).

The wound-healing effect of zinc-containing ointments was known in the ancient world (→ Zn^{2+}-containing collagenase), and in the last few decades zinc has increasingly been used as a remedy for growth disorders due to malnutrition (→ interactions between Zn^{2+} and growth hormones) [3]. Zinc deficiency can also cause a variety of other symptoms, such as a lack of appetite, a reduced sense of taste, an enhanced disposition for inflammations and an impairment of the immune system (acquired immune deficiency syndrome (AIDS)-like symptoms). Particularly high zinc contents have been found in the tissues of fetuses and infants, as well as in the reproductive organs, especially in seminal fluid – a further indication of the essential catalytic function of zinc for metabolic processes. The daily zinc requirement, which is not always met in contemporary diets and which increases with alcohol consumption, has been estimated at between 3 mg (small children) and 25 mg

Bioinorganic Chemistry: Inorganic Elements in the Chemistry of Life – An Introduction and Guide, Second Edition.
Written and Translated by Wolfgang Kaim, Brigitte Schwederski and Axel Klein.

Table 12.1 Representative zinc-containing proteins.

Zinc protein	Molecular mass (kDa)	Ligands	Function
carboanhydrase (CA)	30	3 His 1 H_2O	hydrolysis (12.6)
carboxypeptidase (CPA)	34	2 His 1 η^2-Glu 1 H_2O	hydrolysis (12.2), (12.11)
thermolysin	35	2 His 1 η^2-Glu 1 H_2O	hydrolysis (12.2)
5-aminolevulinic acid dehydratase (ALAD)	8 × 35	8 × {3 S, 1 N/O}	condensation (12.18)
alcohol dehydrogenase (ADH)	2 × 40	2 × {2 Cys, 1 His, 1 H_2O}	oxidation of 1° or 2° alcohols via NAD^+ (12.19)
glyoxalase	2 × 23	2 × 2 His 2 × 2 Glu? 2 H_2O	reduction of α-dicarbonyl compounds by glutathione (12.21)
superoxide dismutase (SOD)	2 × 16	2 × {2 His, 1 μ-His^-, 1 Asp}	disproportionation of $O_2^{\bullet-}$ (10.15)
transcription factors	TFIIIA: 40 GALA: 17	n × {2 His, 2 Cys} 2 × 4 Cys	structural function: formation of specifically folded domains
insulin hexamer	6 × 6	2 × {3 His, n L}	structural function: stabilization of oligomeric storage forms
metallothionein	6	≤7 × 4 Cys	transport and storage protein

(pregnant women). A relatively large tolerance exists for higher doses before symptoms of zinc poisoning become manifest.

From a chemical point of view, the most obvious biologically effective function of divalent zinc is its Lewis acidity. Lewis acids are, generally speaking, electron-deficient species (frequently cations), which can, as the proton, H^+, undergo acid-base reactions (see also Section 2.2). Lewis acids are thus able to catalyze condensation and (in reverse) hydrolysis reactions at *physiological* pH by polarizing the substrates, including H_2O (12.1). Such reactions include the polymerization of RNA (condensation) and the cleavage of peptides, proteins or esters (hydrolysis) (12.2). In chemical synthesis, such reactions are generally catalyzed by strong acids or bases, but except in the case of gastric fluid (see Figure 13.2), such extreme pH conditions are physiologically not feasible. An alternative is the use of an electrophilic polarizing agent, namely a Lewis acidic metal cation with a rather high effective charge.

$$Zn^{2+} \longleftarrow {}^{\delta-}\text{substrate}^{\delta+} \qquad (12.1)$$

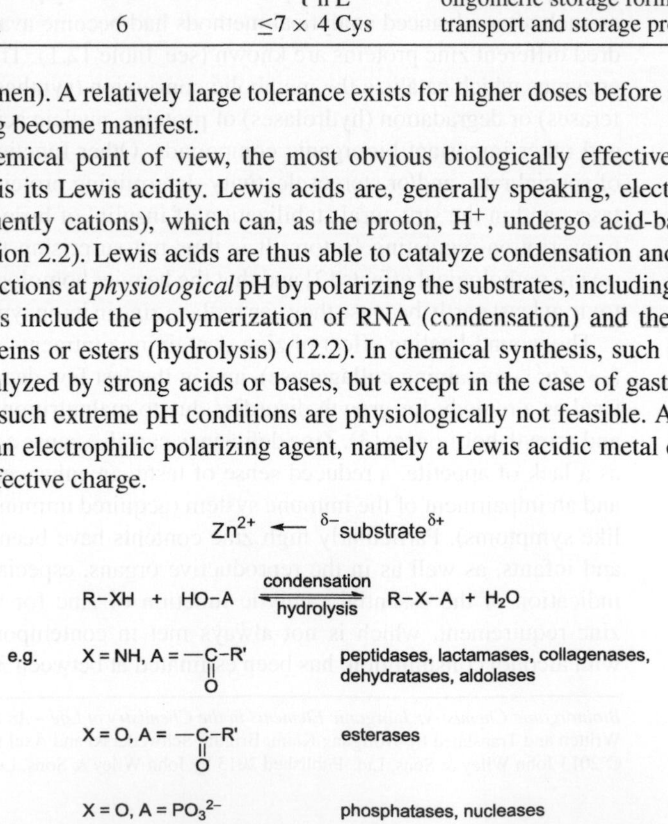

(12.2)

Whereas a direct attack by the Lewis *acidic* metal can occur at nucleophilic substrates, the Zn^{2+} ion can also undergo an "Umpolung" to a Lewis *basic* species $[Zn–OH]^+$ after reaction with H_2O, a process that is very important for enzymatic hydrolysis. The underlying deprotonation of an aqua complex (12.3) is a well-known reaction and precedes every metal hydroxide precipitation. However, the polymerization of deprotonated aqua complexes (see (8.20)) is not possible for an immobilized *monofunctional* system buried inside of a protein (see also (5.12)).

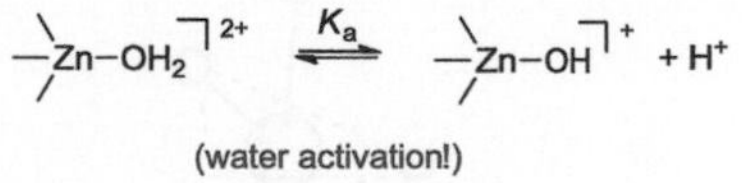

(12.3)

The pK_a value, which is still about 10 for free $[Zn(OH_2)_6]^{2+}$, can drop to less than 7 in enzymatic systems (12.3), while the ability of metal-bound hydroxide to attack electrophilic centers in hydrolyzable substrates is largely retained.

The function of substrate activation first requires a firm anchoring of the metal ion in the enzyme; like Cu^{2+}, Zn^{2+} forms kinetically inert bonds, particularly to histidine (see Section 2.3.1). Zinc(II) ions thus differ from other, in some regards quite similar, divalent ions such as Mg^{2+}, high-spin Mn^{2+}, Fe^{2+} and Co^{2+}, which also have a lower Lewis acidity. In contrast to Cu^{2+} and Ni^{2+}, however, Zn^{2+} is not redox active, which excludes unwanted electron-transfer processes; due to the d^{10} configuration (→ no ligand field effects) and the position in the periodic table, zinc(II) centers prefer rather low coordination numbers and exhibit isotropic (i.e. spatially undirected) polarization effects. The apoenzyme can therefore determine the metal coordination geometry, and during enzymatic catalysis, rather large substrates may coordinate to the metal in such a way that a distorted geometry resembling the transition state of the reaction becomes sterically possible. In fact, the entatic state concept (see insertion Section 2.3.1) had been developed in view of the first structures of zinc-containing enzymes, with their typical unsaturated metal coordination regarding amino acid ligands [1]. In contrast to the kinetically inert bonds between the metal and some protein side-chains, the activation of water (→ hydrolysis function) requires a very labile binding of this molecule during catalysis; in fact, Zn^{2+} belongs to those metal ions which exchange water very rapidly.

The ligand field stabilization phenomenon (see (2.14)), which often contributes to the adoption of an octahedral configuration, is not effective in ions with a filled (Zn^{2+}), half-filled (high-spin Mn^{2+}) or empty (Mg^{2+}) d shell. Physicochemical measurements have strongly benefitted from the possibility of an isomorphous metal exchange in zinc enzymes between Zn^{2+} and high-spin Co^{2+} (d^7); the latter does not show a marked preference for octahedral (O_h) versus tetrahedral (T_d) coordination [4] due to a rather favorable situation with completely filled bonding e_g orbitals (T_d) versus not yet filled bonding t_{2g} orbitals in the O_h situation (12.4). The open d shell of the Co^{2+} ion, however, allows several electronic transitions in the visible region of the electromagnetic spectrum. Therefore, the presumed isomorphous substitution of Zn^{2+} by Co^{2+} has often been used, for example in the determination of pK_a values of ionizable groups in the coordination sphere of the metal ion.

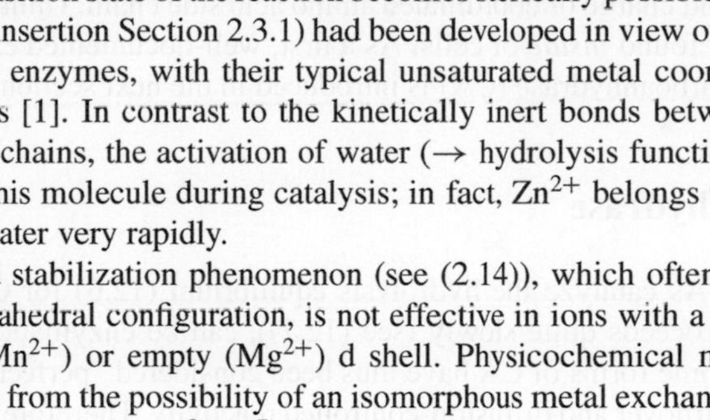

(12.4)

Starting from the typical coordination number for Zn^{II}, 4, the additional coordination of a substrate (metal catalysis as reaction of coordinated ligands) leads to structures with five-coordinate metal. Such systems do not unambiguously prefer one of the prototypical geometries – the square pyramid or the trigonal bipyramid (12.5) – and thus do not impose a metal-based restriction on the substrate specificity of the enzyme (protein → selectivity; metal → activity; see Section 2.3.1).

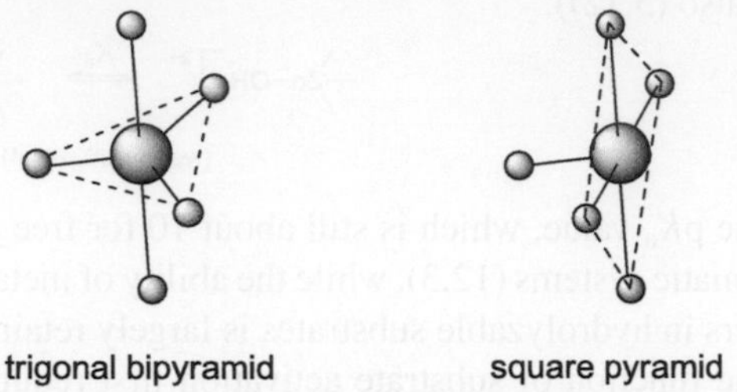

(12.5)

In addition to its polarization function in hydrolyzing or condensation-catalyzing enzymes, zinc may also have a structural, conformation-determining role. Special examples include the zinc-containing superoxide dismutase (Section 10.5), in which Zn^{II} binds and activates a crucial histidin(at)e ligand, and alcohol dehydrogenase (ADH; Section 12.5), in which one of the two different Zn^{II} centers positions and electronically activates the substrate for the redox reaction with an organic coenzyme. Protein-bound zinc is not coordinated only to histidine but also, in part or sometimes even exclusively, to negatively charged sulfur (cysteinate) or oxygen (e.g. glutamate) donor ligands. In addition to a rather variable coordination number, Zn^{2+} thus shows a useful flexibility with regard to the kind and charge of coordinated amino acid side chain. Unlike copper, most of the biological zinc is found *inside* of cells. As a first, well-documented example of zinc-containing enzymes, carboanhydrase (CA) is introduced in the next section.

12.2 Carboanhydrase

CAs catalyze the hydrolysis equilibrium (12.6) for CO_2. This reaction, which normally proceeds quite slowly (see (12.7)), can be enzymatically accelerated by a factor of 10^7; some forms of CA have thus been considered "perfectly evolved enzymes", with maximal turnover and diffusion-controlled reactivity. Therefore, it is not surprising that the details of the enzymatic catalysis of this apparently simple inorganic reaction (12.6) have been studied with great interest [5], even using quantum chemical approaches [6,7]. CAs are biologically very important enzymes which play essential roles in processes such as photosynthesis (→ efficient CO_2 uptake), respiration (→ rapid CO_2 disposal), (de)calcification (→ formation and degradation of carbonate-containing skeletons; Section 15.3.2) and pH control (→ buffering). In human erythrocytes, for example, one isozymic form of CA is the most abundant protein component after hemoglobin. CA is an important component for the necessary "bio"-recycling of CO_2 following its anthropogenically caused introduction into the atmosphere.

$$H_2O + CO_2 \rightleftharpoons HCO_3^- + H^+ \qquad (12.6)$$

Due to the quite varied sites of utilization in organisms, there are several structurally similar but differently effective and pH-dependent variants (isozymes) of the CA enzyme

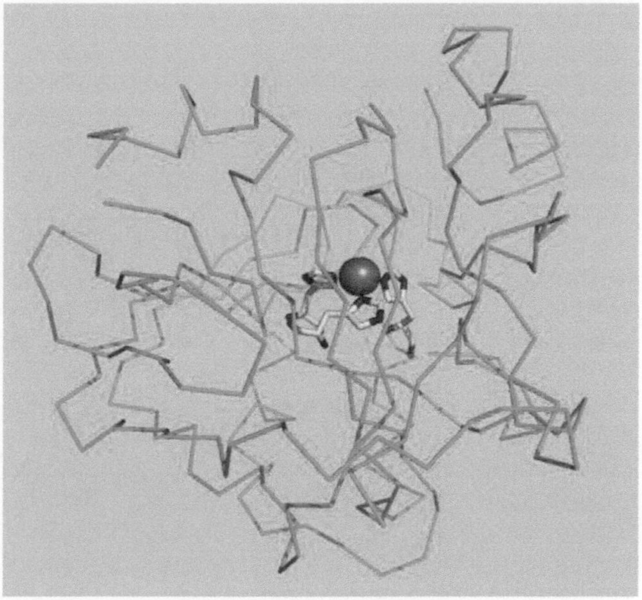

Figure 12.1
Structural representation of human CA II, showing the protein folding and the Zn^{2+} coordination to three imidazole rings of histidine side chains (PDB code 4CAC) [8].

family. Figure 12.1 shows a structural representation of human CA, form II(c). The protein is of medium size ($4 \times 4 \times 5.5$ nm), is composed of 259 amino acids and has a molecular mass of approximately 30 kDa; the dipositive zinc ion is coordinated to three neutral histidine residues and located at the bottom of a 1.6 nm deep conical cleft, which contains hydrophilic and lipophilic regions.

Structural analysis of the substrate-free enzyme shows that the fourth coordination site is occupied by a water molecule, which is connected to other amino acid side chains and water molecules via hydrogen bonds (see (12.10a)). The coordination geometry of zinc is a distorted tetrahedron; the metal can be removed by external chelators, leaving behind an inactive apoprotein. As in many other proteins, the water molecules retained by hydrogen bonds constitute an integral component for structure *and* function inside and at the surface of the enzyme (Figure 12.2). In the case of CA, the ordered network of water molecules (Figure 12.3) is of special importance because CA is a very efficient hydrolysis enzyme [9].

In addition to theoretical calculations, many detailed experimental studies, involving for example ^{18}O labeling and measurements of the effects of metal substitution, H/D-exchange, inhibitors and pH on reaction rates and equilibrium position, have suggested that the rate-determining step in CO_2 hydrolysis by CA is not the CO_2/HCO_3^- conversion but rather proton shuttling under participation of amino acid side chains and the H_2O network [6,9,10].

As can be seen from the equilibrium (12.6), the hydrolysis of CO_2 is a strongly pH-dependent process. In addition to "physically" dissolved (hydrated) carbon dioxide, the carbonic acid molecule, H_2CO_3, which is present in small amounts (12.7), must be considered. The astonishingly slow hydration of CO_2, with a half-life of about 20 seconds at 25 °C, cannot be influenced significantly by conventional acid or base catalysis. Due to

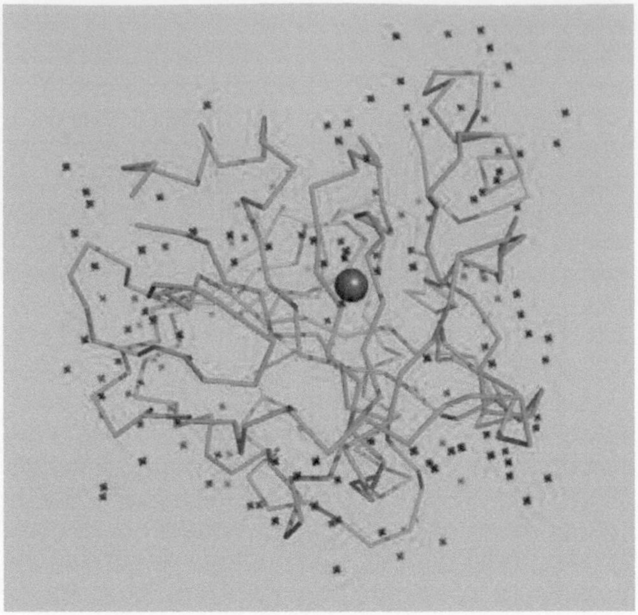

Figure 12.2
Protein backbone structure of human CA I, supplemented by water molecules (PDB code 1HCB).

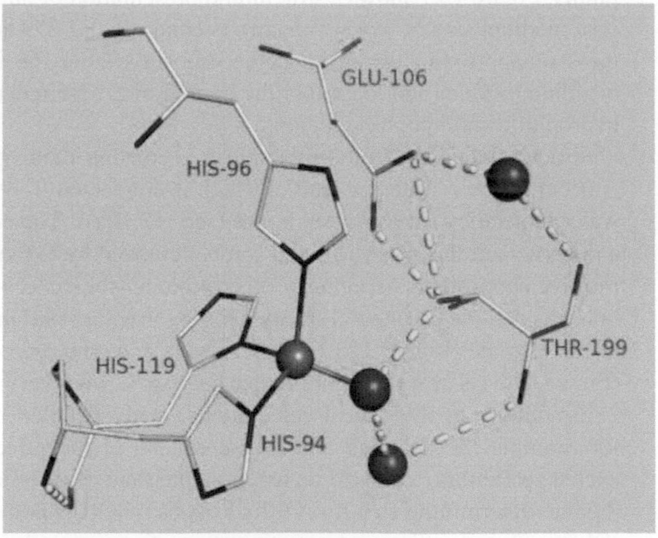

Figure 12.3
Representation of the water network and proton relay system (dashed yellow lines: hydrogen bonds) surrounding the zinc ion (blue sphere) in CA I (red spheres: H_2O molecules; PDB code 1CAC).

the symmetrical character of linear O=C=O (a molecule without a permanent dipole moment), the activation of this polarizable molecule (12.8) requires a combination of (Lewis) acid attack at the oxygen *and* (Lewis) base attack at the carbon atom ("push–pull effect", bifunctional catalysis).

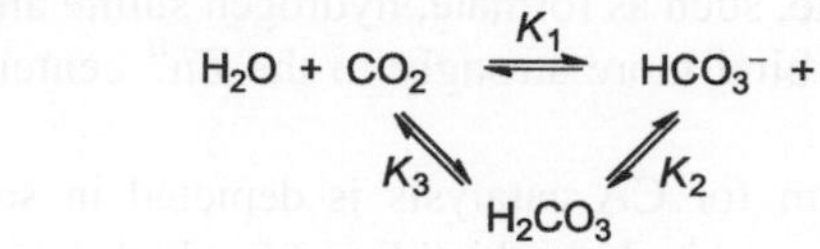

$$H_2O + CO_2 \underset{}{\overset{K_1}{\rightleftharpoons}} HCO_3^- + H^+ \qquad H_2O + CO_2 \overset{K_3}{\rightleftharpoons} H_2CO_3 \overset{K_2}{\rightleftharpoons} HCO_3^- + H^+$$

equilibrium constants:
$K_1 = 4.44 \times 10^{-7}$
$K_2 = 1.72 \times 10^{4}\,M^{-1}$
$K_3 = 2.58 \times 10^{3}\,M$

$$HO^- + CO_2 \underset{k_{-4}}{\overset{k_4}{\rightleftharpoons}} HCO_3^- \tag{12.7}$$

rate constants (uncatalyzed):
$k_4 = \sim 8500\,M^{-1}s^{-1}$
$k_{-4} = \sim 2 \times 10^{-4}\,s^{-1}$

$$\overset{\delta-}{O}=\overset{\delta+}{C}=\overset{\delta-}{O} \Longrightarrow \text{(Lewis) acid}; \quad \Uparrow \text{(Lewis) base at C} \tag{12.8}$$

Non-enzymatic catalysis of this reaction can thus be achieved to some extent by using ambivalent "amphoteric" systems such as sulfurous or hypobromous acid or monohydroxo metal complexes. These compounds feature both basic free electron pairs and an electron pair acceptor (Lewis acid) function; however, the efficiency of such small, unspecific systems and of model complexes is much smaller than that of CA.

In CA, a zinc-bound *hydroxide* group, formed via indirect deprotonation of the originally coordinated water ligand caused by histidine-64, serves as the center for nucleophilic attack at the carbon atom of CO_2. Model studies have shown that the Zn^{II}–OH entity (which does not polymerize, due to shielding by the protein) features a nucleophilicity sufficient for an attack at CO_2. While the metal center itself contributes to attract, orient and polarize the CO_2 molecule Zn···O=C=O [7], the actual productive attack at the oxygen centers of the substrate occurs from acidic protons inside the previously mentioned hydrogen bond network; the very rapid rate of hydrolysis at physiological pH is favored by the poised "proton-relay system" depicted in Figure 12.3. Proton shifts in hydrogen-bonded systems can be extremely rapid, particularly when there is a fairly rigid structure. In effect, the productive attacks on CO_2 occur from the components of water, H^+ and OH^-, and the autoprotolysis of water has thus been proposed as the rate-determining step for catalysis by CA [9]. The most simple reaction sequence, in accord with the experiments but without structural details, is depicted in (12.9), in which E: apoprotein.

$$[E{-}Zn{-}OH]^+ \xrightarrow{CO_2} [E{-}Zn(OH)CO_2]^+ \longrightarrow [E{-}Zn{-}HCO_3]^+ \xrightarrow{H_2O} [E{-}Zn{-}H_2O]^{2+} + HCO_3^- \tag{12.9}$$

(with $[E{-}Zn{-}H_2O]^{2+}$ returning to $[E{-}Zn{-}OH]^+$)

Several alternatives have been discussed for the binding of the resulting hydrogen carbonate in the transition state, with an unsymmetrical (4 + 1) coordination (3N, 2O; compare (12.10b)) at the metal emerging as the most plausible arrangement [10]. Metal complexes with hydrogen carbonate ligands are generally very soluble (cf. the water hardness equilibrium (15.2)) and labile (i.e. they tend to transform into the much more stable carbonate complexes). However, if the proton in the HCO_3^- ligand is replaced by an organic residue,

R, model complexes of the type $L^3Zn(RCO_3)$ can be isolated (L^3: a tris(pyrazolyl)borate ligand [11]. Studies on corresponding nitrate complexes suggest that the monodentate η^1 coordination of the potential bidentate chelate ligands XO_2^- (X = HCO or NO) is characteristic for L^3Zn^{II} complexes, contributing to facile dissociation and rapid catalysis. Substrates related to hydrogen carbonate, such as formate, hydrogen sulfite and sulfonamides, inhibit CA reactivity because they bind more strongly to the Zn^{II} center and thus stabilize the reaction intermediates.

A hypothetical mechanism for CA catalysis is depicted in scheme (12.10b), which illustrates intermediate states and includes histidine-64, which is connected to the zinc ion via the water network and features a pH-dependent conformation [6].

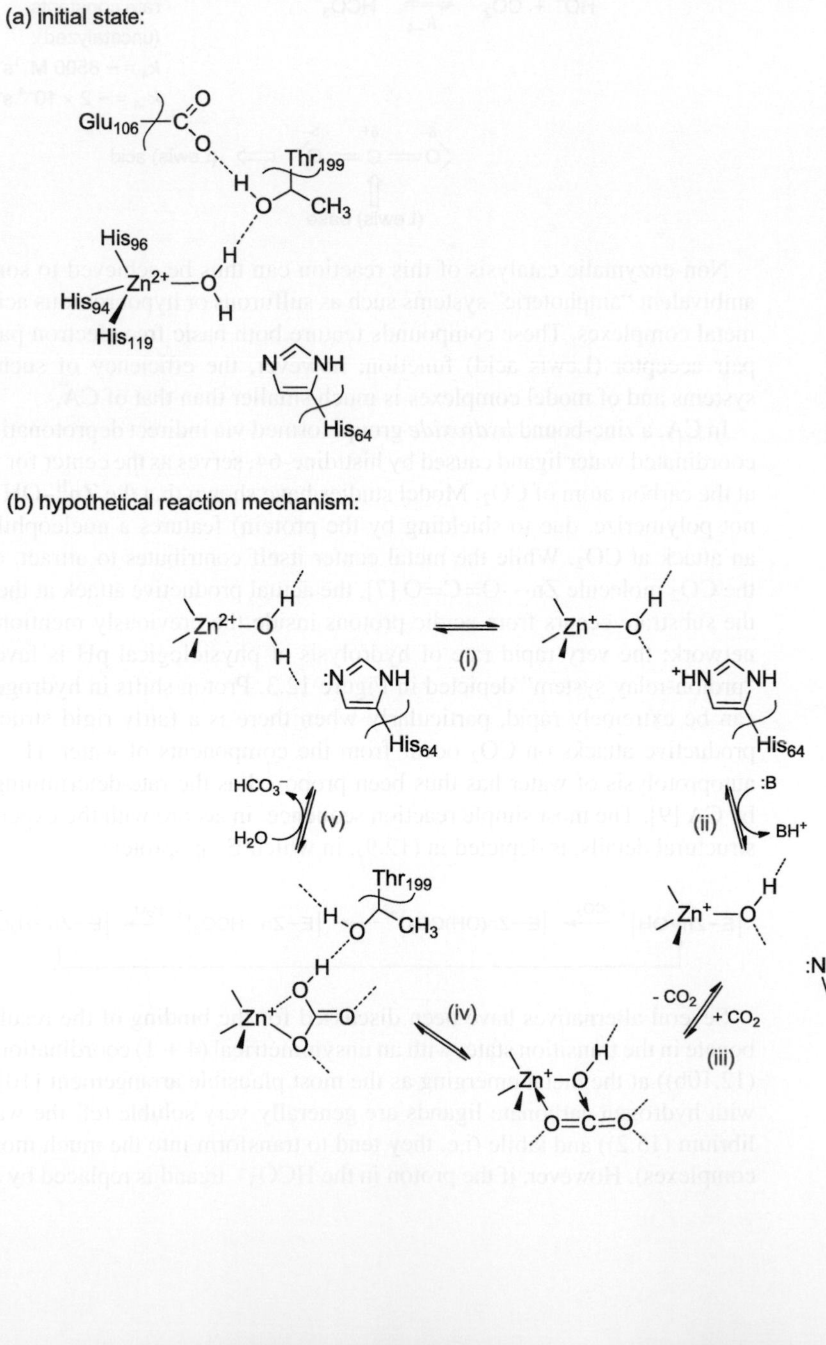

(12.10)

Starting from the aqua complex, a proton is transferred indirectly (i) to histidine-64 via interposed molecules of the water network; one proton, as a stoichiometric product of the forward reaction (12.6), is then delivered to buffer molecules situated farther away (ii). The hydroxide ligand remaining at the zinc center very rapidly reacts and forms the transition state (iii) through hydrogen bond interactions with already partly activated CO_2. Following this transition state, the system yields the product complex (iv), and in a final step (v) the labile HCO_3^- ligand is very rapidly replaced by water.

12.3 Carboxypeptidase A and Other Hydrolases

One of the most thoroughly studied peptide-hydrolyzing enzymes is carboxypeptidase A (CPA), a digestive enzyme typically isolated from the bovine pancreas. According to its specificity, it may be more precisely referred to as "peptidyl-l-amino acid hydrolase"; the enzymatic binding sites for the substrate favor a C-terminal cleavage of l-amino acids with large hydrophobic, preferably aromatic side chains such as phenylalanine (12.11). Nevertheless, CPA can also catalyze the hydrolysis of certain esters.

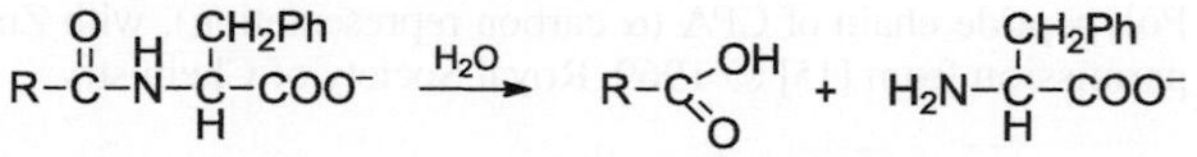

(12.11)

CPA has long been investigated with regard to its structure, its selectivity and its underlying reaction mechanism [12]. The enzyme may have diverse functions, and it can serve as a model for metal-containing or even metal-free (serine or thiol) proteases.

Crystal structure analyses are available not just for the substrate-free CPA enzyme with "natural" zinc or some other metal ions [13] but also for various enzyme/inhibitor complexes. It is assumed that the various phases of the multistage enzymatic catalysis can thus be examined in a "frozen" state. Such studies, particularly on zinc-containing proteases, have provided the experimental basis for a computer-aided "molecular modeling" of enzyme–substrate and in particular of protease enzyme–inhibitor interactions. Due to the many ways in which small peptides can influence the action of hormones, the search for inhibitors of special Zn-containing metallopeptidases such as the "angiotensin converting enzyme" (ACE) has meanwhile led to a very successful development of tailor made pharmaceuticals (drug design), for example for the treatment and control of high blood pressure [14].

CPA is similar in size to CA, with about 300 amino acids and a molecular mass of approximately 34 kDa; the catalytically essential zinc center is again located at the bottom of a cavity in the otherwise rather globular protein (Figure 12.4). The metal is coordinated to two histidine ligands and a chelating (η^2-)glutamate anion, and to one H_2O molecule; this water molecule is linked via hydrogen bonds to serine-197 and glutamate-270. As for carbonic anhydrase, several acidic and basic amino acid residues in the vicinity of the active center are important for the activity of the enzyme (see (12.14)). In addition to the acidic phenol group of tyrosine-248, these include the basic carboxylate of glutamate-270 as well as the arginine groups Arg-127 and Arg-145, each in its protonated form C(α)-CH_2-CH_2-CH_2-NH-C(=$^+NH_2$)NH_2.

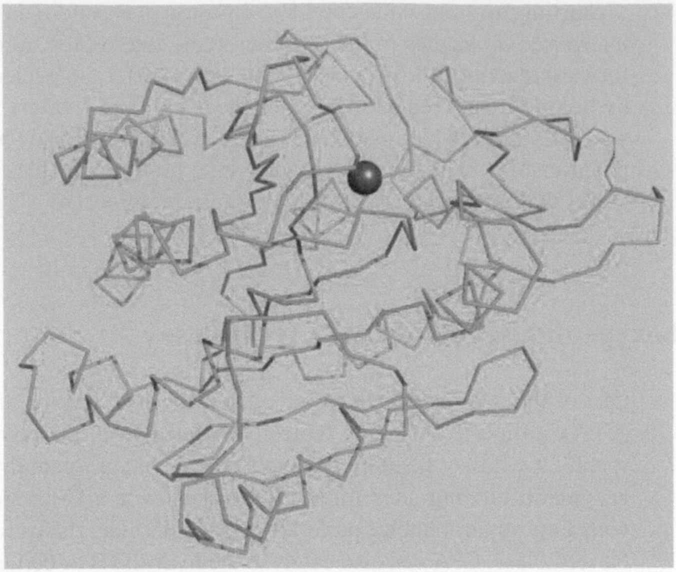

Figure 12.4
Polypeptide chain of CPA (α carbon representation), with Zn^{2+} shown as a sphere. Reprinted with permission from [15] © 1969, Royal Society of Chemistry.

Peptidase and esterase activity differ in some kinetic aspects (e.g. in the rate determining step) and in the effect of metal substitution. The peptidase function is restricted mainly to the "natural" zinc and the Co^{2+} substituted enzyme (see (12.4)), while esterase activity can be found with a variety of introduced divalent ions, the activity decreasing in the series $Mn^{2+} > Cd^{2+} > Zn^{2+} > Co^{2+} > Hg^{2+} > Pb^{2+} > Ni^{2+}$. It can thus be assumed that the reaction mechanisms for ester and peptide hydrolysis differ in some details. Also, significant structural differences have been detected after metal substitution, even though the coordination environment remained the same (2 His, η^2-Glu$^-$, H_2O) [13]. The deviation from the trigonal bipyramidal configuration (12.5) of the five-coordinate metal was smallest in the "natural" zinc system, followed by that of the Co^{2+} and Mn^{2+} complexes. With Cd^{2+}, Hg^{2+} and particularly Ni^{2+}, an approximately square pyramidal structure (12.5) results, which can also be described as an octahedral arrangement with one open coordination site.

Rapid and selective hydrolysis of peptides and proteins under physiological conditions requires sophisticated catalysis; the half-life for noncatalyzed hydrolysis of the carboxamide bond at pH 7 is about 7 years. However, both the ester and the carboxamide groups are polarizable systems, with a carbonyl (acceptor) and an alkoxy or amino (donor) component (12.12). As in the activation of CO_2, a bifunctional attack by electrophilic *and* nucleophilic reagents is thus necessary; due to the hydrolysis character of the reaction, this requires the participation of hydrogen bond-forming acidic or basic components such as polar amino acid residues and coordinated water. On the other hand, the multitude of hydrolyzable systems (autohydrolysis of the enzyme itself is also a possibility) requires a rather high substrate specificity; therefore, several other enzyme/substrate adhesion or recognition sites must be present (cooperativity, "induced fit") in addition to the enzymatic centers poised for multiple attack at the actual functional group that is to be cleaved in the substrate. These numerous stringent requirements for a rather common but also selective type of enzyme-catalyzed reaction have made CPA one of the most widely studied enzymes.

(12.12)

Two mechanistic hypotheses for CPA-catalyzed hydrolysis are presented here. Mechanism (12.13) features a direct coordination of the electrophilic metal at the carbonyl oxygen atom of the peptide or ester substrate; as a consequence, the activated carbonyl carbon center is attacked by nucleophilic glutamate-270 to form an intermediate mixed anhydride function, Glu-C(O)-O-C(O)-R [16]. Hydrolysis of that anhydride, by zinc-coordinated water or hydroxide, for example, leads to the formation of the products.

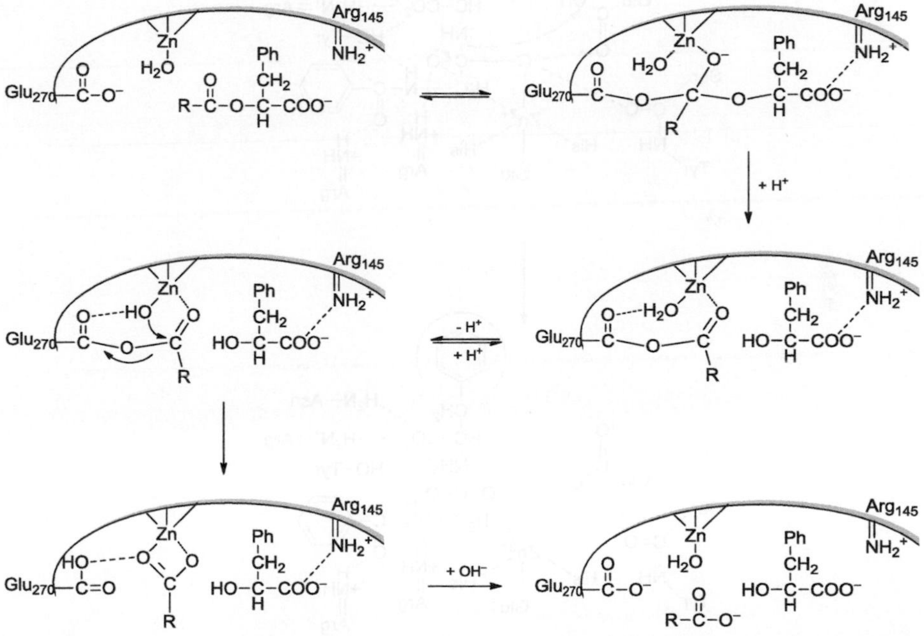

(12.13)

In fact, there is direct evidence for the occurrence of anhydrides during the hydrolysis of certain substrates by CPA. In contrast, Christiansen and Lipscomb [12] suggested a mechanism (12.14) which shows more parallels to the CA mechanism with regard to the metal function. In this scheme, the water ligand bound to five-coordinate zinc is converted to a nucleophilic, metal-modified hydroxide (12.3) through deprotonation by glutamate-270, and can then attack the carbonyl carbon center of the peptide or ester (12.14b).

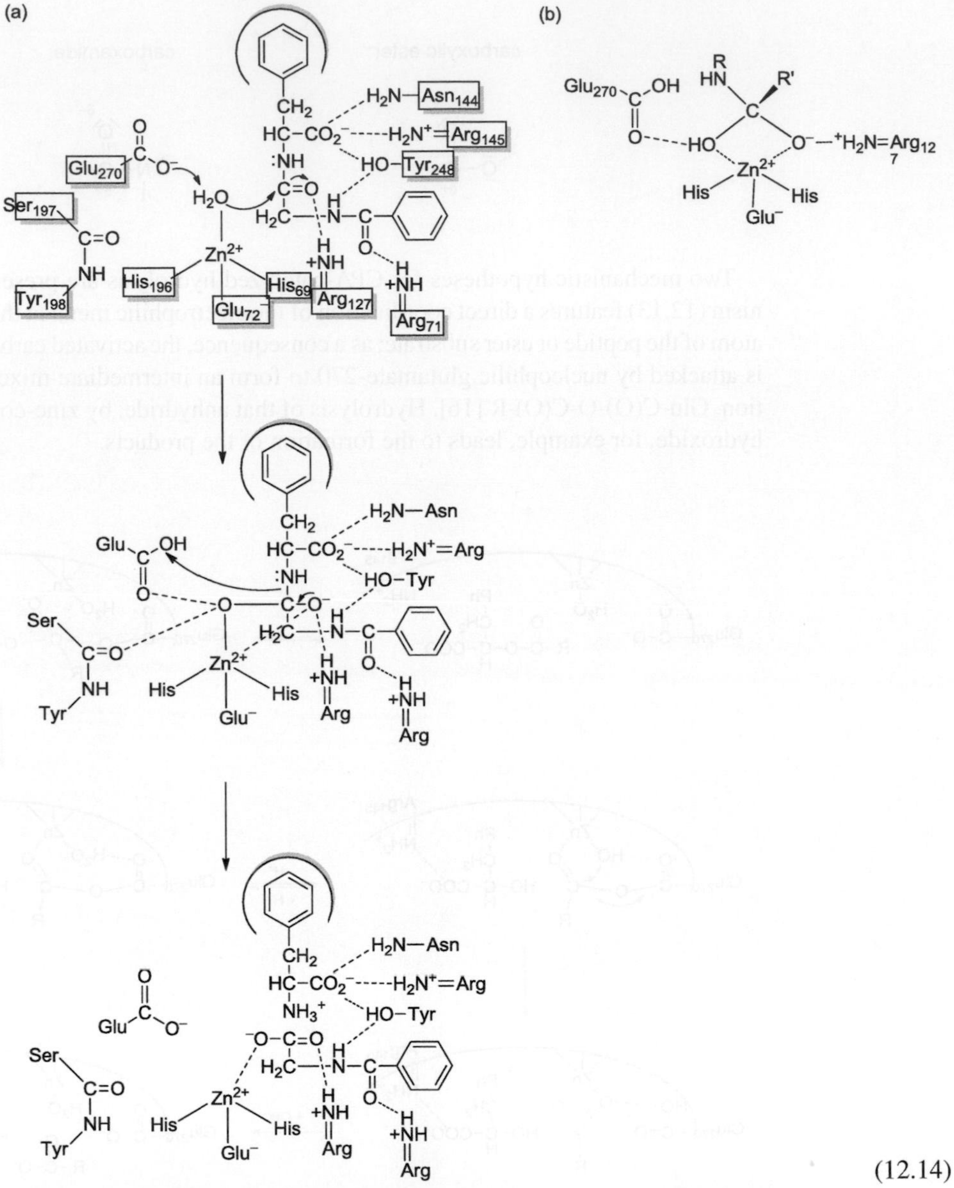

(12.14)

The electrophilic attack at the carbonyl oxygen atom occurs from the conjugated acid form of an arginine by way of hydrogen bond interactions (12.14b); hydrogen bonds also participate in the selective binding and further activation of the substrate (12.14). Additional direct polarization of the carbonyl function by the zinc ion, in combination with an attack of metal and glutamate-activated water or hydroxide at the carbonyl carbon ultimately brings about hydrolysis. In any case, the metal site, with its single positive net charge, serves to electrostatically stabilize and compensate the negative partial charges that occur during enzymatic catalysis, particularly in the transition state.

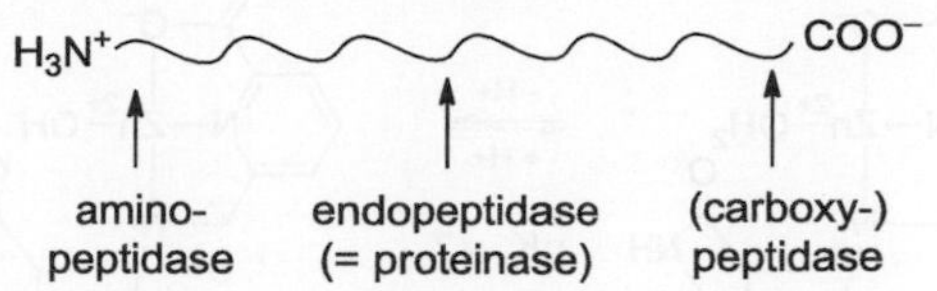

(12.15)

The structural elucidation of an *amino*peptidase [18] which attacks at the N-terminus of a polypeptide chain (12.15) has shown the existence of two neighboring zinc centers separated by a distance of 288 pm; the ligands are glutamate and aspartate groups coordinating through carbonyl or carboxylate oxygen atoms [19]. One of the two zinc centers is coordinatively unsaturated with regard to amino acid ligands, as might be expected from the foregoing.

Dinuclear zinc enzymes are also involved in β-lactamase activity, which is at the center of bacterial resistance to antibiotics and is thus a serious global health hazard [20,21].

The proteinase (endopeptidase (12.15)) thermolysin is another structurally well-documented zinc-containing protease. This enzyme, from *Bacillus thermoproteolyticus*, features four coordinated Ca^{2+} ions (see Figure 2.4 and Section 14.2) and is thus stabilized both thermally and chemically (i.e. against autohydrolysis, the degradation of its own polypeptide backbone). In contrast to peptidases and aminopeptidases, proteinases such as thermolysin, the digestive enzyme astacin [22] and the widely distributed matrix metalloproteinases (MMPs) [23] do not hydrolyze at one of the termini but within the protein. In collagenases and astacin the zinc ion features three histidine ligands [22]. Zinc-dependent tissue-dissolving MMPs such as collagenases, gelatinases and stromelysin are essential for the development of an embryo, wound healing, tumor metabolism, arthritic processes and the degradation of amyloid proteins (→ Alzheimer's syndrome) [24], amongst other things. Characteristically, MMPs are first produced as inactive "zymogen" forms, which can then be activated through the removal of a blocking cysteinate ligand from the catalytic Zn^{2+} center [23]. Essential structural characteristics of highly conserved zinc-containing proteases [22,25] include the coordination environment at the metal and the relative positions of base (glutamate), zinc and acid (arginine for CPA, histidine in the case of thermolysin) at the catalytically active site.

An interesting aspect relating to the problem of autohydrolysis is the occurrence of zinc-containing proteases as essential constituents in the toxins of poisonous snakes; they very effectively dissolve connective tissue and inhibit blood clotting. For example, five such aggressive metalloproteases have been isolated from the toxin of the western diamond rattlesnake, *Crotalus atrox*; after removal of the metal by complexation with ethylenediaminetetraacetic acid (EDTA), they completely lose their activity [26]. The even more effective (neuro)toxins of the tetanus or botulinum type have also been recognized as zinc-dependent proteases that specifically degrade synaptic membrane proteins [27,28]. On the other hand, nonmetalloproteases such as the aspartyl protease of the human immunodeficiency virus (HIV) may be inhibited by Zn^{2+} through binding of the metal ion close to the active center [29].

Using a model complex (12.16), Groves and Olson [30] were able to demonstrate that a single water ligand coordinated to zinc(II) can in fact be quite acidic ($pK_a \approx 7$; compare (12.3)). Furthermore, the hydrolysis of a carboxamide function by metal hydroxide attack at the (unoccupied) π^* orbital of the carbonyl group is accelerated by several orders of magnitude even in a small model complex (12.16), provided the groups are properly oriented.

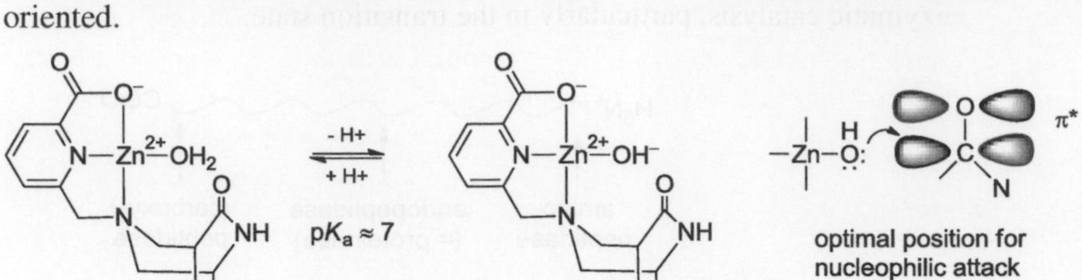

(12.16)

Among the frequently zinc-containing hydrolytic enzymes are not only proteases, peptidases, lactamases and ester-cleaving phospholipases but also phosphate ester-cleaving phosphatases and nucleases. Depending on their pH optimum, phosphatases may be classified as acidic (see Section 7.6.3) or alkaline [31]. Although phosphoryl transfer is the domain of Mg^{2+} catalysis (see Section 14.1), the alkaline phosphatases for phosphate monoester hydrolysis seem to require zinc as well. In its most active form, the enzyme isolated from *E. coli* features three neighboring metal centers: two directly catalytic Zn^{II} centers at 0.4 nm distance and one more remote (0.5–0.7 nm) Mg^{2+} ion (see Table 12.1) [31]. The special requirements for phosphate hydrolysis with a five-coordinate phosphorus atom in the transition state are discussed in more detail in Section 14.1.

12.4 Catalysis of Condensation Reactions by Zinc-containing Enzymes

Whereas hydrolases (with or without zinc) serve in the cleavage of peptide or (phosphate)ester bonds, there are also Zn-containing enzymes which catalyze the reverse reaction; that is, the linking of smaller molecules through condensation reactions (12.2). Though these enzymes have not been as well characterized as the proteases, the better known representatives include metalloaldolases, which reversibly catalyze aldol condensation reactions (12.17), 5-aminolevulinic acid dehydratase (ALAD) (12.18) and DNA or RNA polymerases and synthetases (ligases).

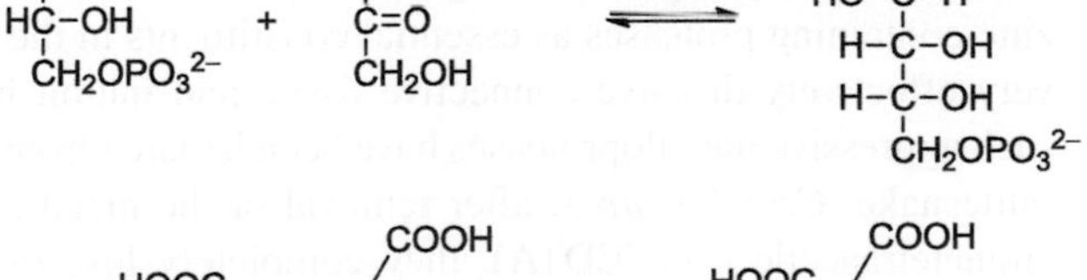

(12.17)

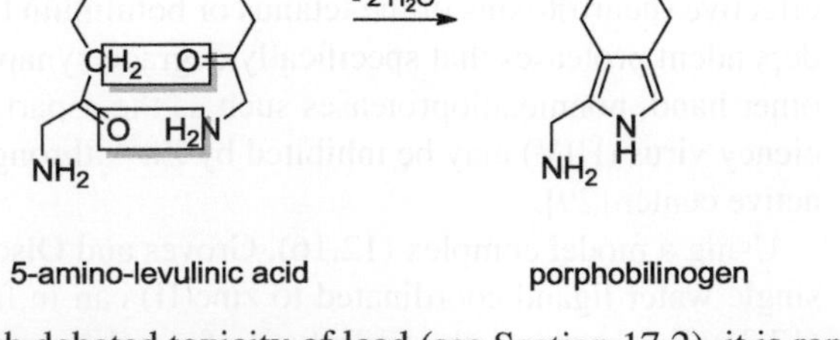

(12.18)

In view of the much debated toxicity of lead (see Section 17.2), it is remarkable that the heavy-metal ion Pb^{2+} inhibits a zinc-requiring enzyme which serves in the formation of an essential precursor for the biosynthesis of tetrapyrroles. This enzyme, ALAD, catalyzes the condensation of two molecules of 5-aminolevulinic acid to give the functionalized pyrrole "porphobilinogen" (12.18) [32] and contains triply Cys^--coordinated Zn^{2+} in each subunit of the octameric protein aggregate.

12.5 Alcohol Dehydrogenase and Related Enzymes

Primary alcohols, particularly ethyl alcohol (ethanol), are metabolized in two steps: zinc-containing ADHs produce aldehyde intermediates, which are then oxidized by aldehyde dehydrogenases (ALDHs; also called alcohol oxidase, see Table 11.1) to give carboxylates (e.g. acetate from ethanol). Enzymes which produce or metabolize alcohols are particularly important for grazing animals and fermenting microorganisms such as yeast; furthermore, the stereospecific enantioselective catalysis (see Figure 12.5) of the reverse process of carbonyl *reduction* by ADH has rendered these enzymes interesting to organic synthetic chemistry.

The ADH enzyme that has been most thoroughly studied is dimeric liver alcohol dehydrogenase (LADH) isolated from horses. It is a metalloenzyme which features two very different zinc ions per subunit and has a molecular mass of about 2 × 40 kDa. The catalytically active metal center is anchored in the protein by one histidine residue and two negatively charged cysteinate groups; in the resting state it also exhibits a labile water ligand at the fourth coordination site [33]. Coordination with two thiolate ligands results in a *neutral* overall charge at the metal site (cf. the charged Zn^{II} centers in CA or CPA) and in restriction to a coordination number of 4 even in the transition state due to the steric demand and the mutual repulsion of the large sulfur donor atoms. The second zinc ion in each subunit is coordinated by four cysteinate ligands and does not directly participate in the enzymatic catalysis; like the similarly coordinated zinc centers of aspartate transcarbamoylase, or of hormone receptor proteins (Section 12.6), it probably has a structural function. Certain bacterial ADHs may contain presumably six-coordinate high-spin iron(II) with three or four histidine and two or three oxygen ligands [34] instead of Zn^{II} with its typical distorted tetrahedral configuration. Since zinc itself is not redox-active, the ADH enzyme requires a dehydrogenase coenzyme, the NAD^+/NADH system (3.12). For a primary alcohol, the

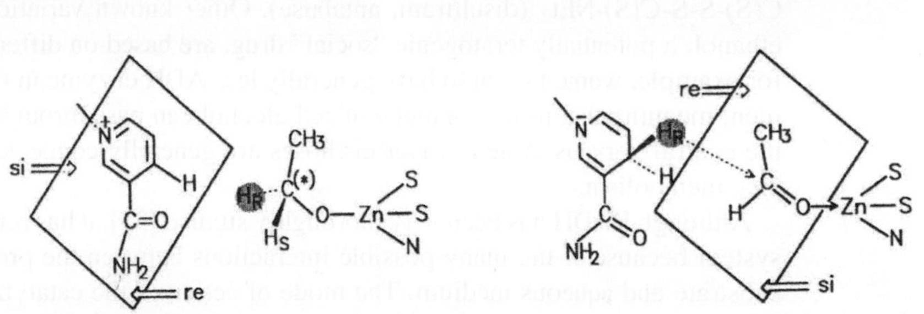

Figure 12.5
Stereospecificity of ADH-catalyzed oxidation/reduction, using the ethanolate/acetaldehyde substrate pair as an example. Due to the fixed orientation and restricted mobility of substrate and coenzyme in the metalloenzyme, only the hydrogen atom H_R, which, for example, could be labeled ^2H (D), can be transferred as "hydride" (formally: $H^- = H^+ + 2e^-$). The potential chirality (*) at the alcohol is thus transferred to the C(4) center of the (dihydro)pyridine ring (enantioselectivity). The labels "R/S" and "re/si" with regard to chiral centers and planes correspond to organic-chemical convention (according to [35]).

chemical equation of the enzyme-catalyzed (reversible) reaction thus has to be written as in (12.19).

$$R{-}CH_2{-}OH + NAD^+ \xrightleftharpoons{ADH} R{-}CHO + NADH + H^+ \qquad (12.19)$$

As suggested by Figure 12.5, the function of the metal is to bind the oxygen atom of the substrate, thus effecting a spatial fixation and possibly an additional electronic activation for this strictly stereospecific reaction.

Widely varying reaction rates have been observed with different substrates, which include mainly short-chain primary and secondary aliphatic alcohols (aldehydes or ketones as products). From these results one can deduce structures for the arrangement of the substrate in the transition state, involving for example a spatial differentiation between large and small substituents at the O-bound carbon atom (12.20).

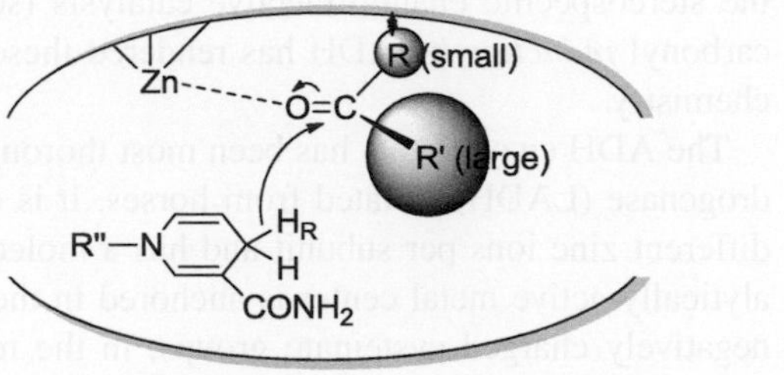

(12.20)

For the most important natural substrates, such as simple primary alcohols, there exists a specificity that is dependent on the type of ADH enzyme. While methanol is only slowly oxidized to formaldehyde (→ neurotoxicity of methanol), the higher primary alcohols n-propanol and n-butanol are oxidized quite rapidly to the respective aldehydes; however, further oxidation of these higher aldehydes is relatively sluggish, which leads to unpleasant physiological effects such as the "hangover" syndrome. Slow follow-up oxidation of the acetaldehyde by aldehyde oxidases (see Table 11.1) may be genetically caused and occurs quite frequently among people of East Asian origin [36]. High aldehyde intermediate concentrations causing very unpleasant symptoms may also result during alcohol abuse therapy following administration of SH-modifying drugs such as tetraethylthiuram disulfide, Et_2N-C(S)-S-S-C(S)-NEt_2 (disulfiram, antabuse). Other known variations in the tolerance for ethanol, a potentially teratogenic "social" drug, are based on different ADH availabilities; for example, women seem to have generally less ADH enzyme in their gastric system than men, meaning that more nonmetabolized alcohol can pass through their bloodstream into the central nervous system. Liver cirrhoses are generally connected with disorders in the zinc metabolism.

Although LADH has been very thoroughly studied [33], it has remained a rather intricate system because of the many possible interactions between the protein, metal, coenzyme, substrate and aqueous medium. The mode of action of the catalytic zinc center in LADH can be described as follows. After coenzyme binding, the dimeric enzyme changes its conformation in terms of a rotation from an open to a closed form. The hydrophilic nucleotide section and the actual redox-active nonpolar 1,4-dihydropyridine ring of NADH are then surrounded by different regions of the enzyme subunits. The active site is located about 2 nm within the protein, at the bottom of some rather hydrophobic channels, which allows for access of substrate. Coenzyme binding and conformational change lead to the displacement of water molecules from the active center; this and the lack of free charges at the metal site favor this catalysis in view of the (formal) transfer of a hydride.

While zinc is not necessary for coenzyme binding and the conformational change, it obviously plays an important role in the binding, orientation and activation of the substrate. Binding occurs via the substrate oxygen atom after displacement of the water ligand, and a metal alkoxide complex, $[(His)(Cys)_2Zn(OR)]^-$, is formed as intermediate. The highly specific orientation allows two electrons and one proton to be transferred directly and enantiospecifically from the coenzyme to the substrate (Figure 12.5). As experiments with model complexes between zinc and macrocyclic polyamine ligands have shown [37], the activating function of the metal consists in the electrostatic stabilization of negative partial charges in the transition state by the Lewis-acidic metal center. Both the transfer of hydroxide, OH^-, by CA and CPA and the transfer of electrons or a hydride, H^-, require such a stabilization.

Zinc-containing glyoxalase I, which participates in the reductive degradation of potentially toxic α-dicarbonyl compounds (12.21; [38]), contains a different organic redox coenzyme, glutathione (GSH; see Section 16.8), instead of NADH.

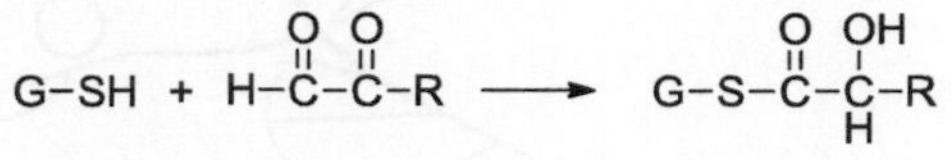

(12.21)

G–SH: glutathione, γ-Glu-Cys-Gly

12.6 The "Zinc Finger" and Other Gene-regulatory Zinc Proteins

The empirically well-established importance of zinc for the growth of organisms, and in particular its high concentration in the reproductive organs, indicates that this metal participates not only in the catalysis of essential metabolic (anabolic, catabolic) reactions but also in the reliable transfer of genetic information (transcription) and, therefore, in its replication. However, only since the early 1980s have special proteins become known which recognize DNA base sequences and thus serve in the selective activation and regulatory control of genetic transcription; the prototypical forms contain several protein domains of about 30 amino acids long which exhibit invariable zinc-coordinating residues [4]. Because of their generally assumed peptide conformation (Figure 12.6), these modular units have been called "zinc fingers" (*zif*) [39].

The zinc content of these factors which regulate gene transcription was discovered by the fact that their stability and function strongly decrease when metal complexing agents like EDTA are added to the buffer solution. The first transcription factor (TF), IIIA, formulated as a zinc finger protein was isolated from the ovaries of immature South African clawed frogs (*Xenopus laevis*) and was found to contain approximately nine domains with the amino acid sequence depicted in (12.22); these domains are typically arranged in strings, each being conformationally stabilized through one Zn^{2+} center. Since then, such zinc finger motifs, with variable "chain" lengths and DNA binding modes, have been detected in many other organisms. About 3% of the human DNA encodes for various zinc finger proteins, and the therapeutic uses of "engineered" zinc fingers for the control of gene expression is being investigated [40].

$$\text{Cys–}X_{2,4}\text{–Cys–}X_3\text{–Phe–}X_5\text{–Leu–}X_2\text{–His–}X_{3,4}\text{–His} \qquad (12.22)$$

Each one of these domains (which can also be synthesized as single units (12.23)) shows a high affinity for the binding of Zn^{2+} and to a smaller extent of Co^{2+}, Ni^{2+} and Fe^{2+}

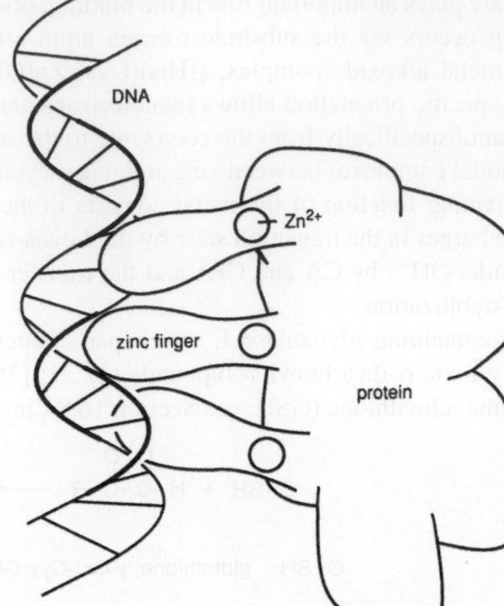

Figure 12.6
Schematic representation of the interaction between DNA and a zinc finger protein.

[41]; in the classical variant of zinc fingers, the coordination of two deprotonated cysteine and two neutral histidine side chains to the metal results in a typical compact folding (structural function of Zn^{2+}). The thus formed protrusions, the zinc "fingers", feature a characteristic β-sheet structure on the Cys^- side and an α-helical arrangement on the His side. According to structure analyses [42,43], the zinc finger proteins wrap closely around the double-stranded DNA, with the actual *zif* regions making multiple specific protein–base pair contacts and then allowing for mutual recognition (Figure 12.6).

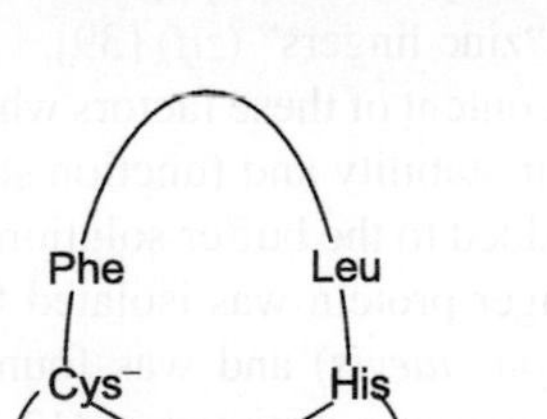

(12.23)

zinc finger modular unit from TF IIIA

The advantages of a modular design using zinc finger domains lie in their variability and stability. Specifically folded protein loops can also be held together by, for example, disulfide bridges, which, however, can be easily cleaved via reduction, unlike the bonds between Zn^{2+} and His or Cys^{-}.

The transcription-activating protein GAL4 from yeast also requires Zn^{2+}; in contrast to the classical zinc finger, with its neutral mononuclear $Zn^{II}(His)_2(Cys^{-})_2$ entity, the GAL4 factor exhibits two neighboring metal centers, which are coordinated to a total of six cysteinate residues (12.24) [44].

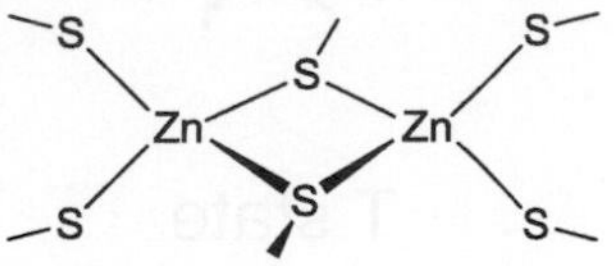

(12.24)

zinc coordination by 6 Cys^{-} in the yeast transcription factor GAL4

Another, structurally related group of zinc-dependent transcription and recognition factors is formed by those hormone receptor proteins which are able to activate genes only after the binding of thyroid, glucocorticoid or other steroid hormones [3,39]. In these cases, one zinc atom per domain has been found to be surrounded by four cysteinate centers. Zinc binding through cysteinate residues has also been found in proteins with repair [45] or developmentally regulatory functions, as well as in nucleic acid-binding proteins of the HIV retrovirus [46].

While the usefulness of coordinatively inert metal ions in maintaining specific protein structures is quite obvious, the high affinity of many gene-regulatory proteins [47] for zinc is probably caused by the typical tolerance of zinc(II) for severe distortions of the tetrahedral coordination; there is no ligand field stabilization energy (compare (12.4)) for the Zn^{2+} ion, with its fully occupied d shell.

12.7 Insulin, hGH, Metallothionein and DNA Repair Systems as Zinc-containing Proteins

A structural function – even if it has a different objective than the examples mentioned previously in this chapter – can be attributed to Zn^{2+} in its complexes with peptide hormones, where it serves to connect two or more single hormone units in oligomeric "storage forms". For instance, the dimerization of human growth hormone (hGH) is induced by coordination (2 His, Glu) with two zinc centers [3]. A similar metal-induced oligomerization has been observed for the pancreatic hormone insulin, which already consists of two relatively short peptide chains. Different modifications of the allosteric insulin hexamers with two or four zinc ions (besides Ca^{2+} ions) have been established as storage forms for insulin in the pancreas and have been crystallographically characterized [48]. The six-coordinate metal in the T-state of the 2-Zn form is connected to three histidine ligands of *different* insulin dimers (Figure 12.7); the three remaining coordination sites are occupied with water, which should allow a facile and reversible removal of Zn^{2+} (e.g. through chelating agents) in order to mobilize individual insulin dimers. Another, less activated form of the allosteric insulin hexamer is the R-state, which features a more closed arrangement and a lower coordination number at the metal.

A possible transport and storage system for zinc exists in the form of the small, cysteine-rich metallothionein proteins, which will be discussed in more detail in Section 17.3. Their preference for soft-metal dications with the favored coordination number of 4 results

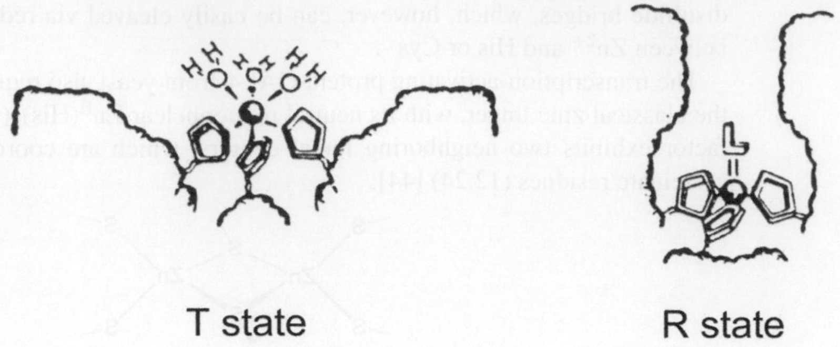

Figure 12.7
Arrangement of the imidazole rings of histidine ligands from three different insulin peptide chains around one structurally coordinating Zn^{2+} center in T- and R-states of a 2-Zn/insulin hexamer (according to [48]).

in particularly efficient binding of Cd^{2+}, the heavier homologue of Zn^{2+}; a heavy-metal detoxification function has therefore been discussed as another possible major role for these fairly ubiquitous proteins. The large concentration of partially noncoordinated cysteine residues suggests still a third function for metallothioneins, namely the efficient trapping of damaging oxidizing radicals such as $OH^{\bullet}$.

More than a structural role is played by fourfold cysteinate-coordinated zinc(II) in the *Ada* DNA repair protein of *E. coli*. This protein recognizes mutagenic methylated forms such as phosphate triesters an 6-O-methylated guanine bases in DNA and removes the methyl group by transfer to a special, previously zinc-bound cysteine residue (12.25), which in turn triggers the activation of methylation resistance genes [45]. Both the activation of the active cysteine and the conformational change ("switching") following methyl transfer can directly involve the metal center.

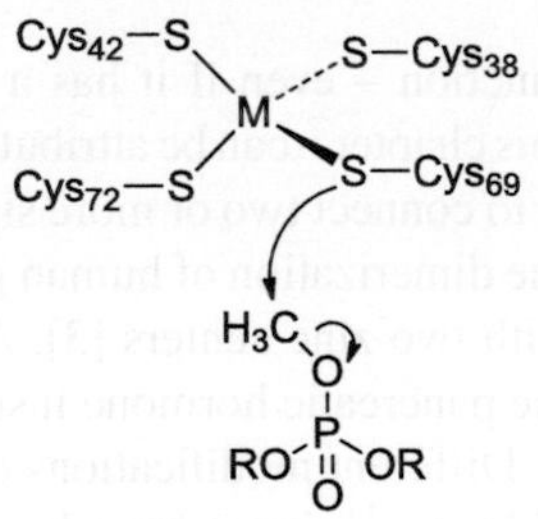

(12.25)

References

1. B. L. Vallee, D. S. Auld, *Acc. Chem. Res.* **1993**, *26*, 543–551: *Zinc: biological functions and coordination motifs*.
2. A. I. Anzellotti, N. P. Farrell, *Chem. Soc. Rev.* **2008**, *37*, 1629–1651: *Zinc metalloproteins as medicinal targets*.
3. B. C. Cunningham, M. G. Mulkerrin, J. A. Wells, *Science* **1991**, *253*, 545–548: *Dimerization of human growth hormone by zinc*.
4. D. Jantz, B. T. Amann, G. J. Gatto, Jr., J. M. Berg, *Chem. Rev.* **2004**, *104*, 789–799: *The design of functional DNA-binding proteins based on zinc finger domains*.

5. F. Botrè, G. Gros, B. T. Storey (eds.), *Carbonic Anhydrase*, VCH, Weinheim, **1991**.
6. Y.-J. Zheng, K. M. Merz, Jr, *J. Am. Chem. Soc.* **1992**, *114*, 10 498–10 507: *Mechanism of the human carbonic anhydrase II catalyzed hydration of carbon dioxide*.
7. J. Y. Liang, W. N. Lipscomb, *Proc. Natl. Acad. Sci. USA* **1990**, *87*, 3675–3679: *Binding of substrate CO_2 to the active site of human carbonic anhydrase II: a molecular dynamics study*.
8. A. Liljas, K. K. Kannan, P. C. Bergsten, I. Waasa, K. Fridborg, B. Strandberg, U. Carlbom, L. Järup, S. Lövgren, M. Petef, *Nature (London)* **1972**, *235*, 131–131: *Crystal structure of human carbonic anhydrase c*.
9. D. Silverman, R. McKenna, *Acc. Chem. Res.* **2007**, *40*, 669–675: *Solvent-mediated proton transfer in catalysis by carbonic anhydrase*.
10. A. Vedani, D. W. Huhta, S. P. Jacober, *J. Am. Chem. Soc.* **1989**, *111*, 4075–4081: *Metal coordination, H-bond network formation, and protein-solvent interactions in native and complexed human carbonic anhydrase I: a molecular mechanics study*.
11. A. Looney, R. Han, K. McNeill, G. Parkin, *J. Am. Chem. Soc.* **1993**, *115*, 4690–4697: *Tris(pyrazolyl)hydroboratozinc hydroxide complexes as functional models for carbonic anhydrase: on the nature of the bicarbonate intermediate*.
12. D. W. Christianson, W. N. Lipscomb, *Acc. Chem. Res.* **1989**, *22*, 62–69: *Carboxypeptidase A*.
13. D. C. Rees, J. B. Howard, P. Chakrabarti, T. Yeates, B. T. Hsu, K. D. Hardman, W. N. Lipscomb, *Crystal structures of metallosubstituted carboxypeptidase A*, in I. Bertini, C. Luchinat, W. Maret, M. Zeppezauer (eds.), *Zinc Enzymes*, Birkhäuser, Boston, **1986**.
14. E. W. Petrillo, M. A. Ondetti, *Med. Res. Rev.* **1982**, 2, 1–41: *Angiotensin-converting enzyme inhibitors: medicinal chemistry and biological actions*.
15. W. Lipscomb, *Chem. Soc. Rev.* **1972**, *1*, 319–336: *Three-dimensional structures and chemical mechanisms of enzymes*.
16. M. W. Makinen, *Assignment of the structural basis of catalytic action of carboxypeptidase A*, in I. Bertini, C. Luchinat, W. Maret, M. Zeppezauer (eds.), *Zinc Enzymes*, Birkhäuser, Boston, **1986**.
17. B. M. Britt, W. L. Peticolas, *J. Am. Chem. Soc.* **1992**, *114*, 5295–5303: *Raman spectral evidence for an anhydride intermediate in the catalysis of ester hydrolysis by carboxypeptidase A*.
18. A. Taylor, *FASEB J.* **1993**, *7*, 290–298: *Aminopeptidases: structure and function*.
19. S. K. Burley, P. R. David, A. Taylor, W. N. Lipscomb, *Proc. Natl. Acad. Sci. USA* **1990**, *87*, 6878–6882: *Molecular structure of leucine aminopeptidase at 2.7-Å resolution*.
20. M. W. Crowder, J. Spencer, A. J. Vila, *Acc. Chem. Res.* **2006**, *39*, 721–738: *Metallo-β-lactamases: novel weaponry for antibiotic resistance in bacteria*.
21. D. T. King, L. J. Worrall, R. Gruninger, N. C. J. Strynadka, *J. Am. Chem. Soc.* **2012**, *134*, 11362–11365: *New Delhi metallo-β-lactamase: structural insight into β-lactam recognition and inhibition*.
22. W. Bode, F. X. Gomis-Rüth, R. Huber, R. Zwilling, W. Stöcker, *Nature (London)* **1992**, *358*, 164–167: *Structure of astacin and implications for activation of astacins and zinc-ligation of collagenases*.
23. H. D. Foda, S. Zucker, *DDT* **2001**, *6*, 478–482: *Matrix metalloproteinases in cancer invasion, metastasis and angiogenesis*.
24. K. Miyazaki, M. Hasegawa, K. Funahashi, M. Umeda, *Nature (London)* **1993**, *362*, 839–841: *A metalloproteinase inhibitor domain in Alzheimer amyloid protein precursor*.
25. B. W. Matthews, *Acc. Chem. Res.* **1988**, *21*, 333–340: *Structural basis of the action of thermolysin and related zinc peptidases*.
26. J. B. Bjarnason, A.T. Tu, *Biochemistry* **1978**, *17*, 3395–3404: *Hemorrhagic toxins from western diamondback rattlesnake (Crotalus atrox) venom: isolation and characterization of five hemorrhagic toxins*.
27. G. Schiavo, F. Benfenati, B. Poulain, O. Rossetto, P. Polverino de Laureto, B. R. DasGupta, C. Montecucco, *Nature (London)* **1992**, *359*, 832–835: *Tetanus and botulinum-B neurotoxins block neurotransmitter release by proteolytic cleavage of synaptobrevin*.
28. B. Willis, L. M. Eubanks, T. J. Dickerson, K. D. Janda, *Angew. Chem. Int. Ed. Engl.* **2008**, *47*, 8360–8379: *The strange case of the botulinum neurotoxin: using chemistry and biology to modulate the most deadly poison*.
29. Z.-Y. Zhang, I. M. Reardon, J. O. Hui, K. L. O'Connell, R. A. Poorman, A. G. Tomasselli, R. L. Heinrickson, *Biochemistry* **1991**, *30*, 8717–8721: *Zinc inhibition of renin and the protease from human immunodeficiency virus type I*.

30. J. T. Groves, J. R. Olson, *Inorg. Chem.* **1985**, *24*, 2715–2717: *Models of zinc-containing proteases. Rapid amide hydrolysis by an unusually acidic Zn^{2+}-OH_2 complex.*
31. J. B. Vincent, M. W. Crowder, B. A. Averill, *Trends Biol. Sci.* **1992**, *17*, 105–110: *Hydrolysis of phosphate monoesters: a biological problem with multiple chemical solutions.*
32. B.-X. Tian, E. Erdtman, L. A. Eriksson, *Phys. Chem. B* **2012**, *116*, 12 105–12 112: *Catalytic mechanism of porphobilinogen synthase: the chemical step revisited by QM/MM calculations.*
33. M. Zeppezauer, *The metal environment of alcohol dehydrogenase: Aspects of chemical speciation and catalytic efficiency in a biological catalyst*, in I. Bertini, C. Luchinat, W. Maret, M. Zeppezauer (eds.), Zinc Enzymes, Birkhäuser, Boston, **1986**.
34. P. Tse, R. K. Scopes, A. G. Wedd, *J. Am. Chem. Soc.* **1989**, *111*, 8703–8706: *Iron-activated alcohol dehydrogenase from Zymomonas mobilis: isolation of apoenzyme and metal dissociation constants.*
35. E. S. Cedergren-Zeppezauer, *Coenzyme binding to three conformational states of horse liver alcohol dehydrogenase*, in I. Bertini, C. Luchinat, W. Maret, M. Zeppezauer (eds.), *Zinc Enzymes*, Birkhäuser, Boston, **1986**.
36. H. W. Goedde, D. P. Agarwal, *Enzyme* **1987**, *37*, 29–44: *Polymorphism of aldehyde dehydrogenase and alcohol sensitivity.*
37. E. Kimura, M. Shionoya, A. Hoshino, T. Ikeda, Y. Yamada, *J. Am. Chem. Soc.* **1992**, *114*, 10 134–10 137: *A model for catalytically active zinc(II) ion in liver alcohol dehydrogenase: a novel "hydride transfer" reaction catalyzed by zinc(II)-macrocyclic polyamine complexes.*
38. B. Mannervik, S. Sellin, L. E. G. Eriksson, *Glyoxalase I – an enzyme containing hexacoordinate Zn^{2+} in its active site*, in I. Bertini, C. Luchinat, W. Maret, M. Zeppezauer (eds.), *Zinc Enzymes*, Birkhäuser, Boston, **1986**.
39. D. Rhodes, A. Klug, *Sci. Am.* **1993**, *268*(2), 56–63: *Zinc fingers.*
40. A. Klug, *FEBS Lett.* **2005**, *579*, 892–894: *Towards therapeutic applications of engineered zinc finger proteins.*
41. B. A. Krizek, J. M. Berg, *Inorg. Chem.* **1992**, *31*, 2984–2986: *Complexes of zinc finger peptides with Ni^{2+} and Fe^{2+}.*
42. N. P. Pavletich, C. O. Pabo, *Science* **1993**, *261*, 1701–1707: *Crystal structure of a five-finger GLI-DNA complex: new perspectives on zinc fingers.*
43. R. S. Brown, *Curr. Opin. Struct. Biol.* **2005**, *15*, 94–98: *Zinc finger proteins: getting a grip on RNA.*
44. P. J. Kraulis, A. R. C. Raine, P. L. Gadhavi, E. D. Laue, *Nature (London)* **1992**, *356*, 448–450: *Structure of the DNA-binding domain of Zinc GAL4.*
45. L. C. Myers, M. P. Terranova, A. E. Ferentz, G. Wagner, G. L. Verdine, *Science* **1993**, *261*, 1164–1167: *Repair of DNA methylphosphotriesters through a metalloactivated cysteine nucleophile.*
46. A. J. van Wijnen, K. L. Wright, J. B. Lian, J. L. Stein, G. S. Stein, *J. Biol. Chem.* **1989**, *264*, 15 034–15 042: *Human H4 histone gene transcription requires the proliferation-specific nuclear factor HiNF-D.*
47. T. V. O'Halloran, *Science* **1993**, *261*, 715–725: *Transition metals in control of gene expression.*
48. M. L. Brader, M. F. Dunn, *Trends Biochem. Sci.* **1991**, *16*, 341–345: *Insulin hexamers: new conformations and applications.*

13 Unequally Distributed Electrolytes: Function and Transport of Alkali and Alkaline Earth Metal Cations

13.1 Characterization and Biological Roles of K^+, Na^+, Ca^{2+} and Mg^{2+}

All metals discussed so far can be categorized as "trace metals" if the criterion used is a daily requirement of less than 25 mg for an adult human being (see Table 2.3). The ions of magnesium and calcium (divalent) and of sodium and potassium (monovalent) certainly do not belong in this category, as can also be inferred from their share in the elemental composition of the human organism (Table 2.1). Because of their high abundance in the earth's crust and in sea water, these cations are predestined for noncatalytic functions. In combination with anions such as chloride, they are often referred to as "electrolytes", "mass elements" or "macronutrients". For the four metal ions K^+, Na^+, Ca^{2+} and Mg^{2+}, three main biological functions can be defined. The least significant of these is the catalytic role (Chapter 14). More important are the following, which require large amounts of material:

1. The construction of supporting and confining **structures**. In addition to silicates, the calcium-containing biominerals play a particularly important role as components of endo- and exoskeletons, teeth and (egg) shells (see Chapter 15). Less obvious is the participation of alkali and alkaline earth metal cations in the stabilization of cell membranes and enzymes or polynucleotide (DNA^{n-}, RNA^{n-}) conformations via electrostatic interactions and osmotic effects; many biomolecules are thus denatured in deionized water.
2. The binding of alkali and alkaline earth metal cations to ligands is generally weak and thus often requires elaborate molecular constructions. On the other hand, the high ionic mobility resulting from the weak binding can be used for **information transfer** through the free diffusion of these electrically charged particles, which can occur extremely rapidly along a deliberately created concentration gradient.

Information transfer relies on electrical membrane potentials, which are created according to the Nernst law (see (8.5)) where the concentrations of an individual metal ion differ across a biological membrane. While a direct charge separation leads to high electrochemical potentials, which could be used for chemical synthesis (redox enzymes; see e.g. Chapter 4), the concentration differences of individually different ions (mainly Na^+, K^+, Ca^{2+}, Mg^{2+} and Cl^-) generate smaller potentials suitable for signal generation (but not chemical reactivity). The required specificity of the ions is given mainly through the

Bioinorganic Chemistry: Inorganic Elements in the Chemistry of Life – An Introduction and Guide, Second Edition.
Written and Translated by Wolfgang Kaim, Brigitte Schwederski and Axel Klein.

Table 13.1 Characteristics of the ions K^+, Na^+, Ca^{2+} and Mg^{2+}.

Property	K^+	Na^+	Ca^{2+}	Mg^{2+}
ionic radius r (pm)[a]	138	102	100	72
r/q (pm/elemental charge)[a]	138	102	50	36
relative value r/q ($Mg^{2+} = 1$)	3.83	2.83	1.39	1.00
surface area $O = 4\pi r^2$ (pm^2)	239 300	130 700	125 700	65 100
relative value O/q ($Mg^{2+} = 1$)	$7.35 \sim 2^3$	$4.02 \sim 2^2$	$1.93 \sim 2^1$	$1.00 = 2^0$
preferred coordination number	6–8	6	6-8	6
preferred donor atoms	O	O	O	O, N, $O{=}PO_3^{3-}$
preferred ligand type	multidentate chelate ligands, particularly macrocycles		bidentate, anionic ligands, e.g. (bridging) polycarboxylates	
distribution (mmol/kg)[b]				
human erythrocytes (intracellular)	92	11	0.1	2.5
human blood plasma (extracellular)	5	152	2.5	1.5
squid nerve (inside)	300	10	0.0 005	7
squid nerve (outside)	22	440	10	55

[a]Ionic radii and radius/charge ratios for octahedral coordination (from [1]).
[b]From [2].

radius/charge and surface-area/charge ratios that characterize simple atomic (i.e. spherical) ions; the four cations discussed in this chapter differ significantly with respect to just these aspects, as shown in Table 13.1. These cations are also significantly different from protons, which are ubiquitous due to the autodissociation of water; proton gradients play a very important role in biological energy transfer (chemiosmotic effect; see Section 14.1 on ATP synthesis).

Also very obviously, the maximal possible rate of a chemical reaction, the "diffusion control" limit, is desirable for many processes requiring information transfer: information reception in the sensory organs, information transfer and processing in the nervous system and reaction or motion control by muscles.

A main requirement for the utilization of ionic diffusion as one of the most rapid "chemical" processes is a concentration gradient, which has to be maintained in a continuously energy-consuming "entatic state" (see insertion Section 2.3.1). This situation can be illustrated by a pump/storage model, as shown in Figure 13.1. Here the ions are actively "pumped" through the biological membrane against the concentration gradient until

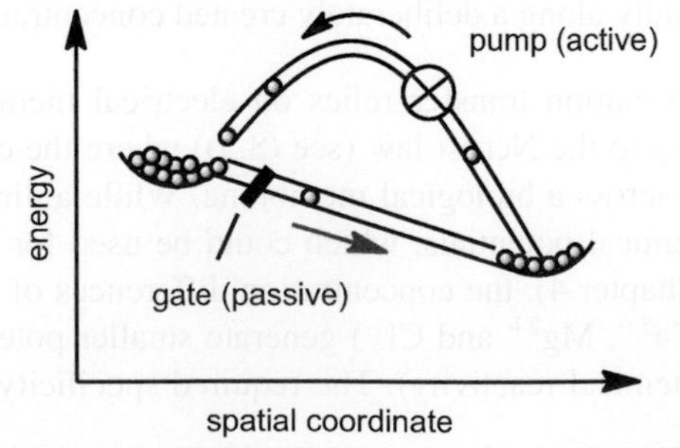

Figure 13.1
Pump, gate and storage model for the maintenance of a local non-equilibrium ion concentration.

a certain stationary non-equilibrium state is reached; the diffusion-controlled concentration equilibration can then occur passively via ion channels with differently (chemically, electrically) regulated gate functions.

In addition, ions can also be transported through membranes via passive diffusion with the help of special complex ligands (see Section 13.2). Before taking a detailed look at the transport mechanisms, we shall compare some characteristic features of the four main cations according to their various functions.

From the radius/charge (*r/q*) and, in particular, surface-area/charge (*O/q*) ratios, it is evident that the four cations can have distinct functions in cooperation with size- and charge-specific ligands. The larger cations in each group of the periodic system tend to prefer even higher coordination numbers than 6 with lower coordination symmetry; they are thus well suited for structural, conformation-specific functions in enzymes. Numerous enzymes are thus activated by the coordination of K^+ [3] or Ca^{2+} (see Section 14.2). Except for the most polarizing Mg^{2+} ion, with its tendency to form strong bonds with N-ligands (e.g. in chlorophyll; see Chapter 4) and with phosphates (see Section 14.1), the other "hard" cations from Table 13.1 almost exclusively prefer ligands with oxygen donor centers. Negatively charged ligands which contain, for example, η^2-carboxylate groups are usually sufficient to tightly bind alkaline earth metal ions inside a protein, due to the strong electrostatic interactions of M^{2+} ions. Alkali metal monocations, on the other hand, can only be retained by multidentate chelate ligands, preferably macrocycles or quasimacrocycles (compare Figure 2.8).

The markedly different concentrations of electrolyte ions in intra- and extracellular compartments are of essential biochemical significance, for the transfer of information for example (Table 13.1). While such unequal spatial distributions have been examined in great detail for the four cations K^+, Na^+, Ca^{2+} and Mg^{2+}, they are also found in anionic electrolytes (Figure 13.2) and in many trace elements: copper, for instance, is mainly found in extracellular regions, and zinc mainly in intracellular regions [4]. For further discussion, it should be kept in mind that K^+ and Mg^{2+}, as well as $HPO_4{}^{2-}$, are more abundant within cells, while Na^+, Ca^{2+} and Cl^- are dominant in the extracellular space. By far the largest transmembrane gradient is that of Ca^{2+}, its intracellular concentration being typically lower by a factor of 10^4 than that outside of cells.

The equilibrium potential of a cell is made up from the potentials of all ions involved. Due to the selectivity of membranes, the main actor is K^+, and the resting cell potential usually lies very close to the potential of K^+. Although concentrations and cell potentials vary between different cell types, a general guideline can be given (Table 13.2).

The resting potential can be influenced by synapses, starting from the normal –90 mV and moving to either the negative regime ("hyperpolarization") or the positive regime ("depolarization") (Figure 13.3).

In case of marginal depolarization of the cell, the K^+ pump and channels are responsible for regulation ("local response"). When the depolarization has reached a certain extent (threshold value), a large number of potential-controlled Na^+ channels are opened and an action potential is produced. The action potential leads to an approximately 1 ms excitation of the nerve cell (Figure 13.3). After opening the Na^+ channels, the permeability of the membrane for sodium increases for a short time and the membrane potential approaches the resting potential of Na^+ (+60 mV), reaching a final value of approximately +30 mV in the so-called *overshoot*. Subsequently, the Na^+ channels are closed and the K^+ channels re-regulate the initial membrane potential (repolarization). This takes time, during which the resting potential of about –70 mV is slightly exceeded (–80 mV post-hyperpolarization).

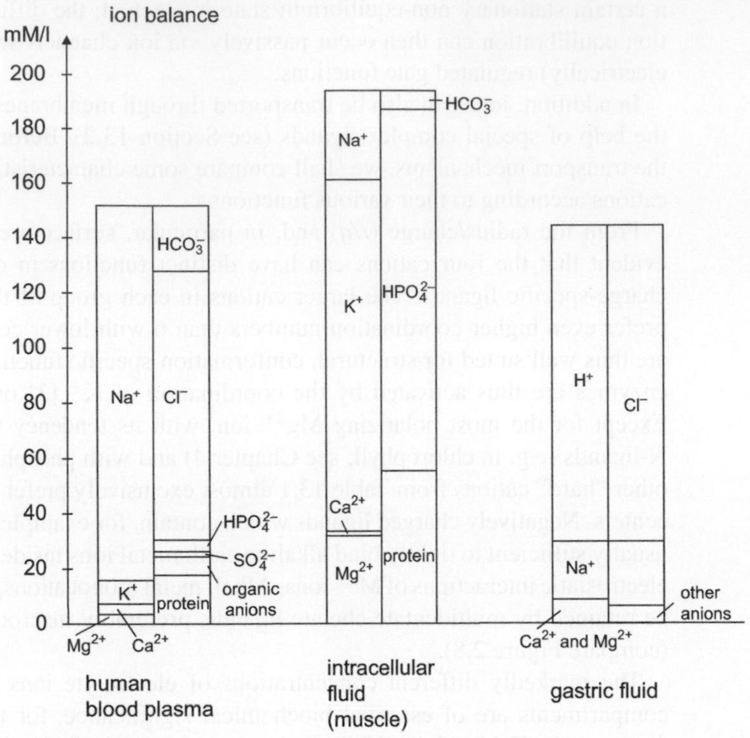

Figure 13.2
Distribution of cationic and anionic electrolytes in three typical fluids of the human body (according to [5]).

The cell potential represents a superposition of the potentials arising from the individual activities of the K^+ channels (and pumps) and of the Na^+ channels. This can be shown by measuring the cell potential while blocking the K^+ channels with tetraethylammonium (= pure Na^+ potential) or in the presence of tetrodotoxin (blocking the Na^+ channels = pure K^+ potential) (see Section 13.3). For an individual cell, the overall shape of the action potential is always the same, since it represents an all-or-nothing response; variable responses are not of any use. On the other hand, the shapes of action potentials can differ greatly from cell to cell; the heart, for instance, shows a total duration of about 300 ms, including a long-lasting plateau phase.

The importance of the four cations under discussion is evident from the fact that disorders in the metabolism of these "electrolytes" can severely affect health (cf. the flow equilibrium;

Table 13.2 Concentration differences and resulting electrochemical potentials.

	Na^+	K^+	Ca^{2+}	Cl^-
extracellular concentration (mmol/l)	150	5	1	120
intracellular concentration (mmol/l)	15	150	10^{-5}	6
ratio	10 : 1	1 : 30	10^5 : 1	20 : 1
equilibrium potential (mV)	+60	−90	+150	−80

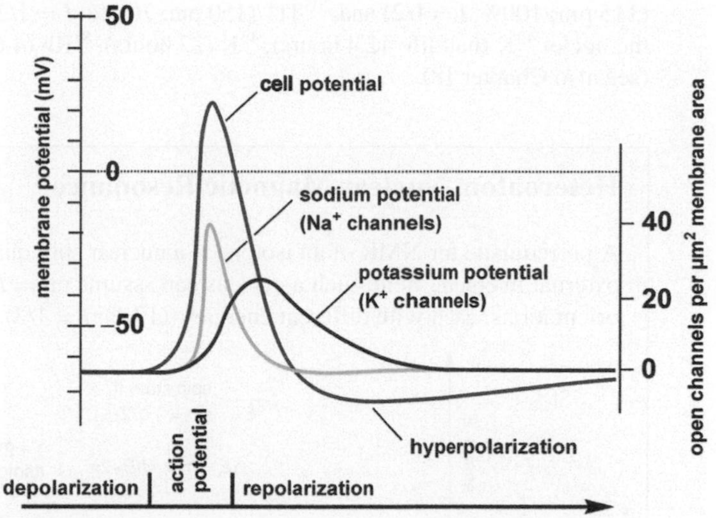

Figure 13.3
Schematic representation of the membrane potential during depolarization (negative potential regime), action potential (towards positive potential), hyperpolarization and repolarization (negative potential regime).

Figure 2.1). An excessive intake of Na^+ in combination with high Cl^- concentrations (e.g. heavily salted food) is a major factor contributing to the development of high blood pressure, for example. On the other hand, it seems to be difficult for the aging organism to prevent excretion of the very labile K^+ ion, due to deteriorating membrane permeability. Magnesium and calcium deficiencies have increasingly become topics of discussion among the health-conscious public, and advertised variety of supplements (involving isotonic beverages) are advertised as dealing with this problem. A deficiency of Mg^{2+} ions can be responsible for diminished mental and physical ability, due to its importance in the ATP and general phosphate metabolism (see Section 14.1). Severe Ca^{2+} deficiency (e.g. through insufficient absorption and utilization due to hormonal disorders) can lead to a variety of symptoms, including skeletal diseases (see Sections 14.2 and 15.1).

Detailed knowledge of biochemical reactions involving the cations from Table 13.1 has been obtained only quite recently, since their analytical detection is rather difficult. Because of their closed electron shells (electronic configuration of the noble gases), the alkali and alkaline earth metal cations are colorless and diamagnetic; furthermore, they are soluble and very mobile due to the labile bonds formed with normal ligands. For this reason, some spectroscopically better suited "ersatz" ions with characteristics as similar as possible to the real ones have been used for physical studies.

Nuclear magnetic resonance (NMR) spectroscopy of ^{23}Na (100% natural abundance), with its nuclear spin of $I = 3/2$, has become a widely used method for obtaining information on the environment of this ion in biological probes [6]. Unfortunately, the poor time resolution of this "slow" spectroscopy (timescale of the order of seconds) gives only statistically averaged information. For the K^+ ion, the NMR method is less applicable due to the very small magnetic moment of ^{39}K (93.1% natural abundance, $I = 3/2$); suitable substitute ions with a comparable radius/charge ratio are the NMR-active $^{107,109}Ag^+$

(115 pm; 100%, $I = 1/2$) and $^{205}Tl^+$ (150 pm; 70.5%, $I = 1/2$), and as radioactive isotopes the nuclei ^{42}K (half life 12.4 hours), ^{43}K (22 hours), ^{81}Rb (4.6 hours) and ^{86}Rb (18.7 days) (see also Chapter 18).

Heteroatom Nuclear Magnetic Resonance

A prerequisite for NMR of an isotope is a nuclear spin quantum number $I \neq 0$. In an external magnetic field, such a nucleus can assume $m_I = I, I-1, \ldots, (-I+1), -I$ orientations, each with different energies ((13.1): $I = 1/2$).

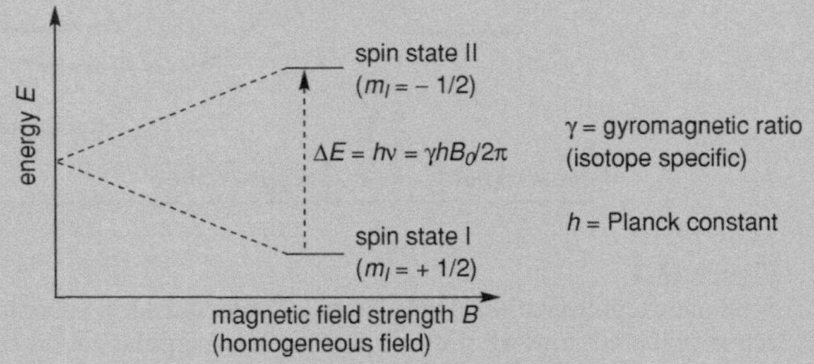

(13.1)

In NMR spectroscopy, transitions are induced between these nuclear spin states which show slightly different occupations according to the Boltzmann distribution; at magnetic field strengths of a few Tesla, the resonances are found in the region of the radio frequencies. The resonance energy depends not only on the nature of the nucleus and its characteristic value γ but also on its electronic and thus chemical environment (→ "chemical shift"). Due to the often very high spectral resolution, even small chemical shift effects of a few ppm or less can be detected and interpreted. Interactions between nuclear spins and between nuclear and electron spin (in paramagnetic species) can provide further details of the electronic and geometric structure.

In heteroatom NMR (i.e. NMR spectroscopy of nuclei other than ^1H [6]), some isotopes can be studied only with great difficulties due to small values of γ or because of a large quadrupole moment, when $I > 1/2$. Another problem is the often low natural abundance of spectroscopically interesting nuclei. In view of the generally low sensitivity of this method, relatively high concentrations or isotopically enriched materials have then to be used in order to obtain spectra of sufficient quality.

From a biological point of view, ions such as $^{23}Na^+$, $^{39}K^+$, $^{43}Ca^{2+}$ and $^{25}Mg^{2+}$ are particularly interesting. $^{23}Na^+$ and $^{39}K^+$ NMR spectroscopy can be used to determine the concentrations of these ions in the intra- or extracellular regions. For instance, the addition of specific "paramagnetic shift reagents" in the extracellular region shifts the signal of these ions relative to the unchanged signal of the ions within the cell [6].

Magnesium ions in natural abundance contain only 10% of the isotope $^{25}Mg^{2+}$ with $I = 5/2$ and a small nuclear magnetic moment. However, Mg^{2+} can often be substituted by the paramagnetic ($S = 5/2$), easily electron paramagnetic resonance (EPR)-detectable Mn^{2+} which features a half-filled 3d shell (see also insertions Sections 3.2.2 and 10.1).

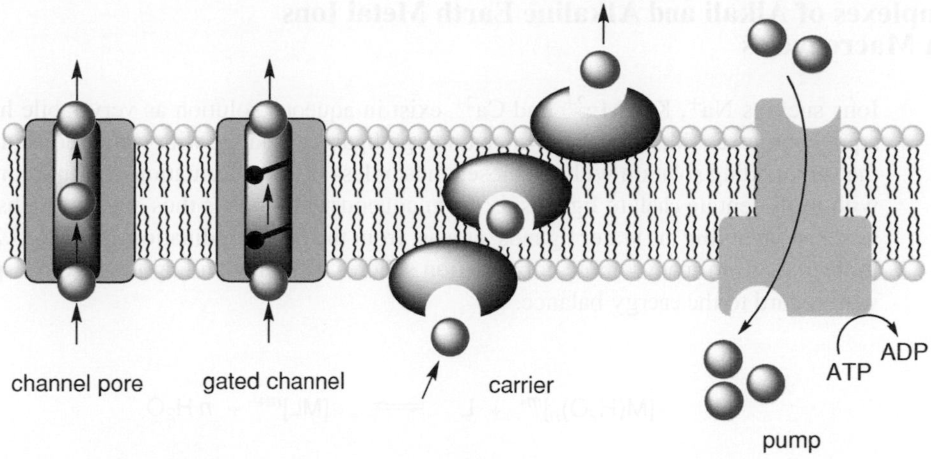

Figure 13.4
Mechanistic alternatives for ion transport across the phospholipid double-layer membrane of a cell (charge compensation not considered).

For the study of calcium ions, there are several alternatives: the NMR-active isotope ^{43}Ca with a nuclear spin of $I = 7/2$ and only 0.13% natural abundance is suitable for biochemical studies only after isotopic enrichment (see Section 14.2); the europium(II) ion Eu^{2+} with a half-filled 4f shell as a substitute, with a slightly larger ionic radius (117 pm) than that of Ca^{2+}, can be studied via EPR or Mössbauer spectroscopy. For Ca^{2+}, it is particularly important to monitor rapid concentration changes on a micromolar scale; Ca^{2+}-specific chelating agents with short response times are used for this purpose as color or fluorescence indicators (see (14.11)(Figure 13.4)).

We shall now describe the four essential solutions to the transport problem for alkaline earth and, in particular, alkali metal cations through cell membranes, as depicted in Figure 13.4 [7].

Plasma membranes consist of a phospholipid double layer and serve to separate an enclosed space (cell) within an aqueous surrounding. Phospholipids (e.g. lecithin) are characterized by a polar (hydrophilic) and a nonpolar (hydrophobic) terminus. The membrane is stabilized by cholesterol molecules and the specific function of a membrane is defined by specific proteins integrated in the membrane double layer (membrane proteins) [5]. Their location varies, depending on their function, between sitting on the outer or the inner part of the membrane and crossing the entire membrane with contacts outside and inside of the cell (or cell compartment). Examples are receptors (usually outside), **channels** or **pores** (crossing) (Section 13.3) and **pumps** (crossing) (Section 13.4) designed for specific ions or molecules, or for interconnections between cells. Additionally, specific **carrier** molecules (ligands) can transport metal ions across the membrane (Figure 13.4). Due to their charge, the metal ions alone cannot penetrate through the highly lipophilic membrane (e.g. in the form of hydrophilic "aqua" complexes $[M(H_2O)_n]^{m+}$). When coordinated to suitable carrier molecules (Section 13.2), the resulting complexes can become highly lipophilic due to outward-facing, nonpolar $-CH_3$, $-CH_2-CH_2-$ and similar organic groups.

13.2 Complexes of Alkali and Alkaline Earth Metal Ions with Macrocycles

Ions such as Na^{+}, K^{+}, Mg^{2+} and Ca^{2+} exist in aqueous solution as very labile hydrated aqua complexes $[M(H_2O)_n]^{m+}$ which undergo ligand exchange with water molecules from the surrounding solution within nanoseconds or less [8]. The formation of stable complexes with multidentate chelate ligands, L, from aqueous solutions is thus always a substitution reaction involving the water ligands, at least from the first coordination sphere (13.2); H_2O molecules from more remote coordination spheres also play an important role, especially with regard to the energy balance.

$$[M(H_2O)_n]^{m+} + L \rightleftharpoons [ML]^{m+} + n\,H_2O \qquad (13.2)$$

The generally observed kinetic and thermodynamic stabilization through the formation of chelate complexes can be attributed to several factors. First of all, an increase in the number of free particles typically occurs during reaction (13.2), resulting in increased entropy; the strongly charge-dependent long-range hydration involving the outer coordination spheres must also be considered here. Furthermore, an intelligent architecture of the chelate complex can lead to a number of conformationally and electrostatically favorable interactions between donor atoms and the metal cation in a fixed chelate ring structure and thus to a contribution from the side of enthalpy change. Finally, a "statistical" (i.e. kinetic) stabilization results because of the low probability of "simultaneous" cleavage of all metal-donor bonds which would be necessary for the dissociation of a chelate complex. As long as there is only a partial breaking of bonds, the "virtual" concentration of donor atoms around the metal center is very high due to the spatial proximity of uncoordinated, but not really free, donor centers [9–12]. Therefore, both the probability and the equilibrium effects favor recombination and thus nondissociation of the chelate complex.

For the dipositively charged alkaline earth metal ions, there is a number of long-established multidentate open-chain chelate ligands such as ethylenediaminetetraacetate (EDTA) available (see (2.1)). On the other hand, efficient synthetic complex ligands for the alkali metal monocations have been available since about 1970, with the design of multidentate macrocyclic ligands (13.3), (13.4) such as crown ethers, cryptands and related components for "supramolecular chemistry" and "molecular recognition" (Nobel Prize in Chemistry 1987 for C. J. Pedersen, J.-M. Lehn and D. J. Cram) [3,7,9–13].

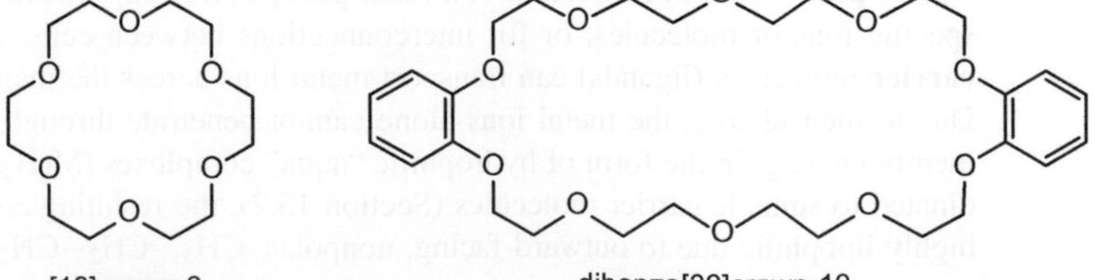

(13.3)

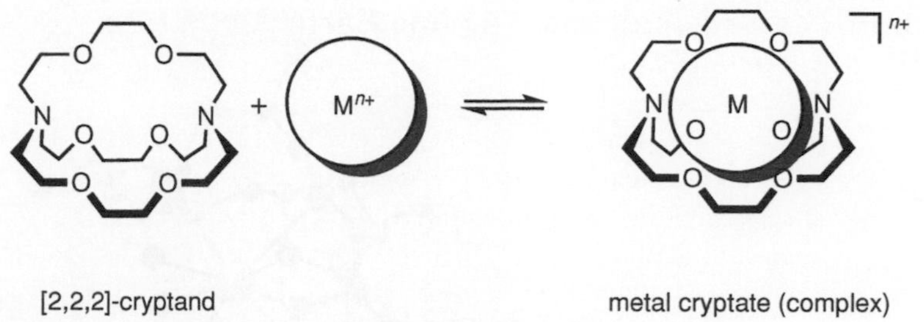

(13.4)

In synthetic systems such as (13.3) and (13.4), and in the numerous naturally occurring "ionophore" complexes that can be isolated from fungi or lichen [3,14], metal ion coordination occurs via several strategically placed heteroatom donor centers. These include ether, alcohol and carbonyl oxygen centers in carboxylate, carboxylic ester and carboxamide groups; furthermore, the sulfur analogs of these oxygen donor groups and nitrogen components such as amine (NR_3), pyridine and imine (RN=C) are also suitable. A particular feature of macrocyclic complexing ligands is that the ring size can be tailored to fit the metal's ionic radius (size selectivity) [10–13]. In contrast to the essentially planar, tetradentate and size-selective tetrapyrrole macrocycles (Table 2.6), the efficient ligands for alkali metal monocations feature a three-dimensional encapsulation of the ion in the complex (see Figure 13.5).

For this reason, the polycyclic cryptands (13.4), with their largely preformed cavity, are superior to the monocyclic crown ethers, *cyclo*-$(OCH_2CH_2)_n$, with regard to complex stability and ion size selectivity [10–13]. Once a suitable molecular architecture has been designed, the number of individually weak coordinative interactions can lead to a collectively strong and inert coordinative binding of the metal ion by the macrocycle. The sometimes considerable conformational change illustrates the structuring effect of even weakly coordinating alkali metal cations on suitable polyfunctional substrates, which can for example wrap around such an ion (metal ion as template; Figure 13.5).

The biological and physiological [3,10,11,13,14], as well as organic-synthetic significance of such complexes is related to the fact that the polar heteroatom donor centers of the

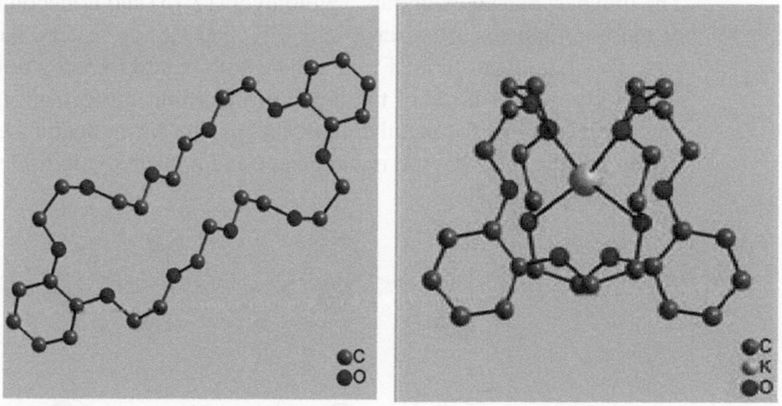

Figure 13.5
Molecular structures of dibenzo[30]crown-10 (left) and its K^+ complex (right), from single crystal x-ray diffraction. Reprinted with permission from [13] © 1985, American Chemical Society.

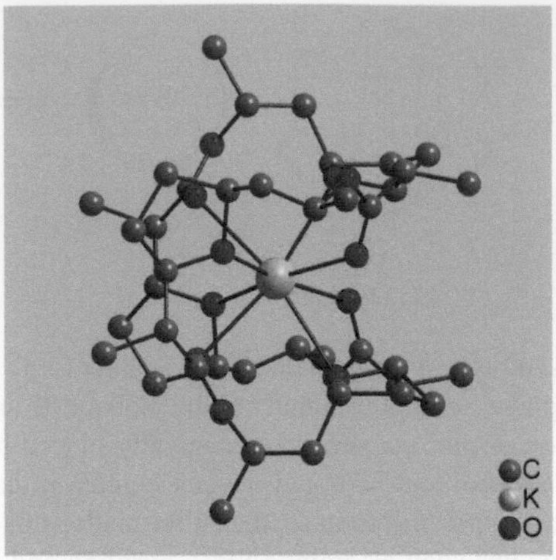

Figure 13.6
Molecular structure of the K^+/nonactin complex in the crystal. Reprinted with permission from [15] © 1967, Elsevier.

macrocycle face towards the inside and to the metal cation, while the outside features alkyl or aryl groups and is thus rather lipophilic. Using such ligands, it is therefore possible to at least partly dissolve ionic compounds such as $KMnO_4$ in nonpolar organic solvents; on the other hand, this situation allows for a mediated transport of metal cations through biological membranes (Figure 13.4), with their hydrophobic, approximately 5–6 nm-spanning phospholipid double layer. The complexation by macrocycles is thus one of the of possible ways of effecting passive transport of hydrophilic metal cations through membranes.

Natural ionophores like those shown in (2.15), (13.5) and (13.6) have meanwhile been isolated in large numbers as potential sensor components and as pharmacologically active natural products from fungi, lichen and marine organisms [3,13–15]. Many of these molecules act as antibiotics, because they can perturb the stationary ionic non-equilibrium and thus the membrane function of bacteria without affecting the more complex ion-transport mechanisms of their higher (host) organisms (defense function of antibiotics). The two best known examples, valinomycin (2.15) and nonactin (13.5), illustrate that many of these ionophores are macrocyclic oligopeptides or –esters, with a number of chiral centers. This latter feature is of importance with regard to recognition and a possible receptor selectivity; furthermore, of the numerous possible conformations, there is frequently only one that is optimal for metal complexation. The K^+/nonactin system, with its high coordination number of 8 for the metal center and a macrocycle folded around this ion, is quite typical (Figure 13.6).

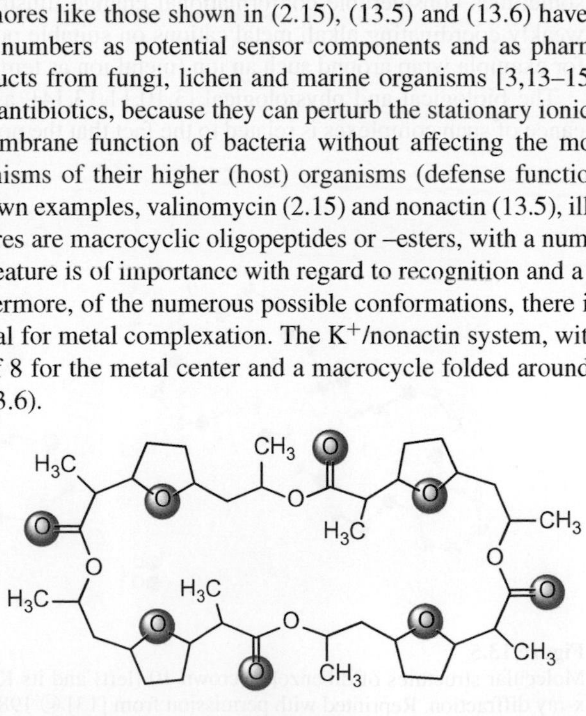

(13.5)

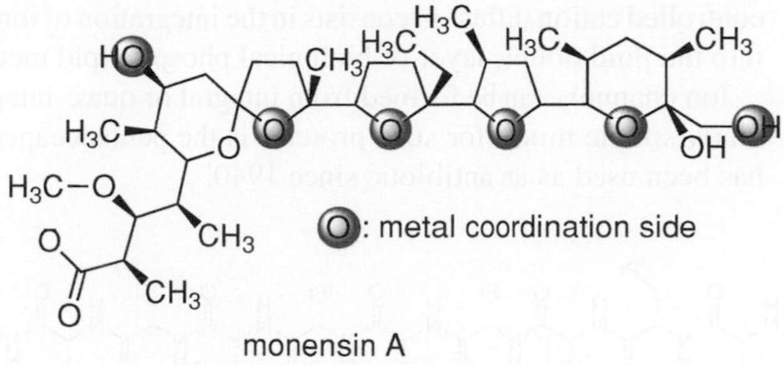

(13.6)

In the case of the dodecadepsipeptide valinomycin (2.15), astonishingly high selectivities of $>10^3$ are observed for the discrimination between K^+ and Na^+; however, this differentiation may strongly depend on the solvent (remember (13.2)). Some important representatives, such as the acyclic, Na^+-specific polyether monensin A (13.6) [3,16], show a cyclization which gives a quasimacrocycle only following metal coordination; hydrogen bonds are then formed between both ends of the open-chain ligand (Figure 13.7).

The discovery of many more natural and physiologically active ionophores and the development of synthetic analogs are being pursued because of the potential pharmacological activity of such compounds. Both the selectivity towards the inside (i.e. regarding the charge and size of the metal ion) and towards the outside (with regard to a membrane-bound receptor) have to be taken into account [3,7].

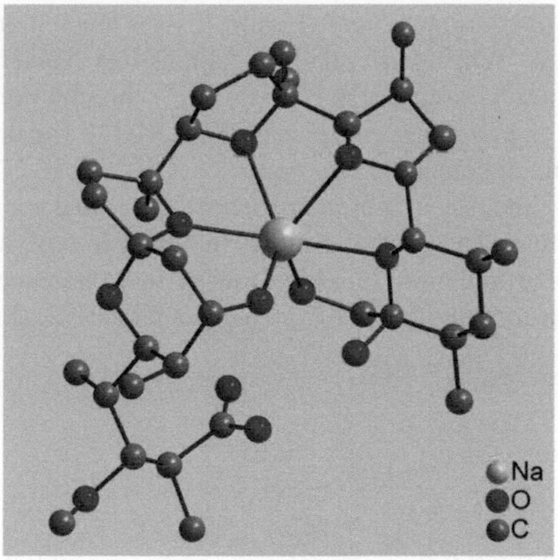

Figure 13.7
Molecular structure of the Na^+/monensin A complex in the crystal. Reprinted with permission from [16] © 1978, International Union of Crystallography.

13.3 Ion Channels

The passive cation transport along a concentration gradient via a carrier mechanism (Figure 13.4) proceeds relatively slowly, due to the necessary steps of complexation, migration and decomplexation. A more efficient, albeit biosynthetically more complex, realization of controlled cation diffusion consists in the integration of ion channels of various complexities into the fluid double layer of biological phospholipid membranes (Figure 13.4).

Ion channels can be formed from integral or quasi-integral membrane proteins; a particularly simple model for such proteins is the pentadecapeptide gramicidin A (13.7), which has been used as an antibiotic since 1940.

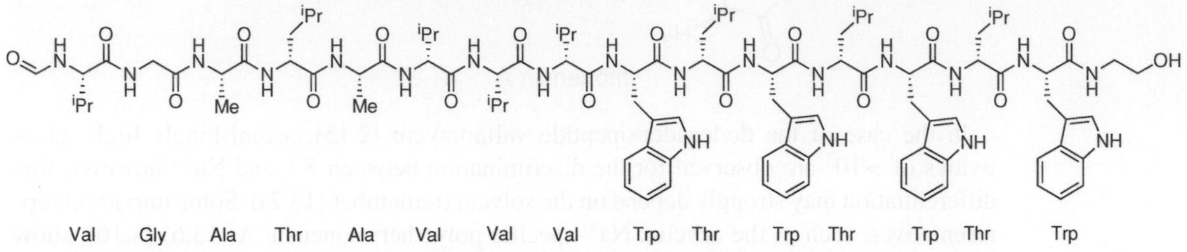

(13.7)

Due to an antiparallel helical aggregation of two molecules, gramicidin A from *Bacillus brevis* forms a channel structure approximately 3 nm long, with an inner diameter of 385–547 pm [17]. Since the thickness of biological phospholipid double-layer membranes is about 5–6 nm, two gramicidin dimer channels have to be arranged in sequence in the fluid membrane in order to permit a very rapid transmembrane cation flow. Ionic transport via such membrane "pores" (Figure 13.4) is much more efficient than the ionophore-mediated process. While some alkali metal ions (Na^+) and other monopositive ions of suitable size can pass consecutively ("single file") through the channel in rudimentary solvated form (e.g. as monoaqua complexes $[M{-}OH_2]^+$), the dipositively charged Ca^{2+} ion blocks the gramicidin channel.

As integral membrane proteins, the actual ion channels have a much more complex structure. The general construction principle of several groups of ion channels seems to be an arrangement in which four or more transmembrane protein sections form a bundle which confines the pore from the membrane side (Figures 13.4 and 13.8). Not unexpectedly,

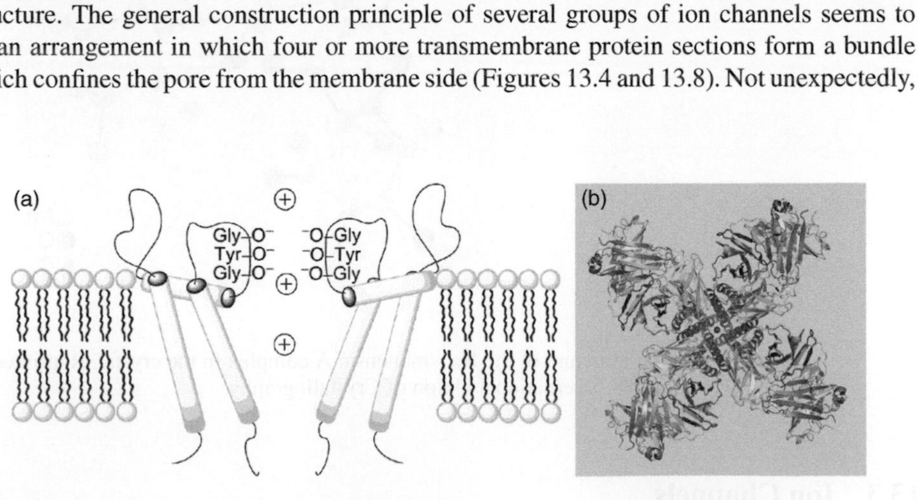

Figure 13.8
(a) Schematic representation of a voltage-dependent K^+ ion channel. (b) Protein structure (top view) of a channel consisting of four homologous transmembrane proteins. The top view shows a potassium ion (violet) in the channel (PDB code 1K4C).

those parts of the helices which function as an immediate lining of the pore contain polar amino acid residues such as serine [18–20]. Of special importance is the entrance region of the channel, where negatively charged amino acid residues promote cation diffusion and thus can contribute to the selectivity and to the gate mechanisms which control the ion flux.

The major breakthrough in characterizing these transmembrane proteins using single crystal X-ray diffraction was achieved by MacKinnon and others, who crystallized them in the presence of bipolar agents (detergents) which mimicked the cell membrane [20,21]. Without such adequately modeled surroundings, the proteins cannot adjust their tertiary structure and thus their function. In 2003, the Nobel Prize in Chemistry was awarded to R. MacKinnon for structural and mechanistic studies of ion channels [21,22].

Due to the essential physiological importance of ion channels, the external control of the ion-selective gates through development of suitable inhibitors ("blockers") or stimulating agents has become one of the most active fields of pharmaceutical and medical research (cardiology, oncology, neurology) [23–25], including the development of synthetic ion channels [26,27]. The 1991 Nobel Prize in Medicine was thus awarded to E. Neher and B. Sakmann for the development of the "patch-clamp" method [28], which allows the preparation of membrane segments containing a single ion channel; studies into the electrical conductivity, opening time and reaction of ion channels towards external stimuli have since been greatly facilitated.

In channel proteins, the gates (Figure 13.4) are normally closed in order to guarantee maintenance of the concentration gradient. The opening of these gates can be influenced by exogeneous or endogeneous low-molecular-weight compounds ("ligands", e.g. neurotransmitters such as glutamate, nucleotides or toxins such as nicotine), by released Ca^{2+} (see Section 14.2), by other proteins or by a change in the electrical potential difference (voltage) across the membrane. Voltage-controlled channels are thus biological switching elements which serve in the transformation of electrical into chemical signals. The development of receptor-specific organic compounds (compare (14.10)) for the blocking of ion channels is one of the main targets of "molecular modeling"; on the other hand, the unspecific blocking of, for example, K^+ channels by $[N(C_2H_5)_4]^+$, Cs^+ or Ba^{2+} can be easily understood on the basis of size and charge effects. The blocking of K^+ channels in the taste receptors by H^+ is probably responsible for the sensing of "sour" tastes [29,30]. An as yet unresolved problem is the preference of K^+ channels for H^+ over Na^+ [30]. Since the size cannot account for this phenomenon, the ability of K^+ (and H^+) to form $K^+ \cdots \pi$(phenyl) interactions with phenyl-containing amino acids such as phenylalanine or tyrosine is being discussed [31,32].

The blocking of channels in the "disk"-containing rod cells of the retina that are permeable for Na^+ in their "resting" state is an essential step in the transformation of a light stimulus into (electrical) nerve impulses [5]. The very sensitive rod cells for black and white vision contain a membrane pigment, "rhodopsin", formed from the polyene retinal and the protein opsin. The polyene, which is connected to the opsin via a protonated azomethine function, $-C{=}NH^+-$, undergoes a single-photon-induced isomerization of a double bond (*Z*-11,12 → *E*-11,12), and via several consecutive steps this charge shift results in a degradation of cyclic guanosine monophosphate (cGMP). Only in the presence of cGMP, however, is the continuous, energy-consuming flow of Na^+ through the inner membrane of the rod cells maintained ("dark current": entatic state). The cGMP degradation leads to a blocking of the Na^+ channels and thus to a marked ionic hyperpolarization (→ amplification), which finally creates an electric signal from this sensory cell.

13.4 Ion Pumps

Passive ion diffusion along transmembrane concentration gradients is possible only because the concentration differences in the stationary "resting states" are maintained through active, energy-consuming ion-transport mechanisms against the tendency for rediffusion and thus against increasing entropy. The complex protein systems required for this task, the ion pumps, have to operate against controlled (Section 13.3) and uncontrolled charge-compensation processes ("leakage"); due to their high energy consumption in the form of hydrolysable ATP equivalents, they belong to the group of ATPase enzymes (Section 14.2). A large part of the basic, continuous energy requirement of the nonactive cell (flow equilibrium; see Figure 2.1) is used to maintain the non-equilibrium situation with the help of ion pumps. Even when resting, the daily turnover of continuously recycled ATP in an adult human being corresponds to about half of the body mass. The ion pumps are large, complex systems and thus very sensitive; their activity, for example, is strongly temperature dependent. There are usually several kinds of ion pumps for one individual ion; two alternatives can be realized, due to the always necessary charge compensation: the "symport" process, wherein cations and anions are simultaneously transported in the same direction, and the "antiport" process, in which ions of the same charge sign are exchanged by movement in opposite directions. When balancing the charges, the protons formed in the hydrolysis of ATP must also be taken into account (see (13.8)), which allows for the existence of some "uniport" processes involving only one inorganic ion apart from H^+. Incidentally, the comparable problem of an $H^+ + e^-$ transport across membranes has already been mentioned in the context of electron transfer in photosynthesis and respiration (Sections 4.1 and 10.4).

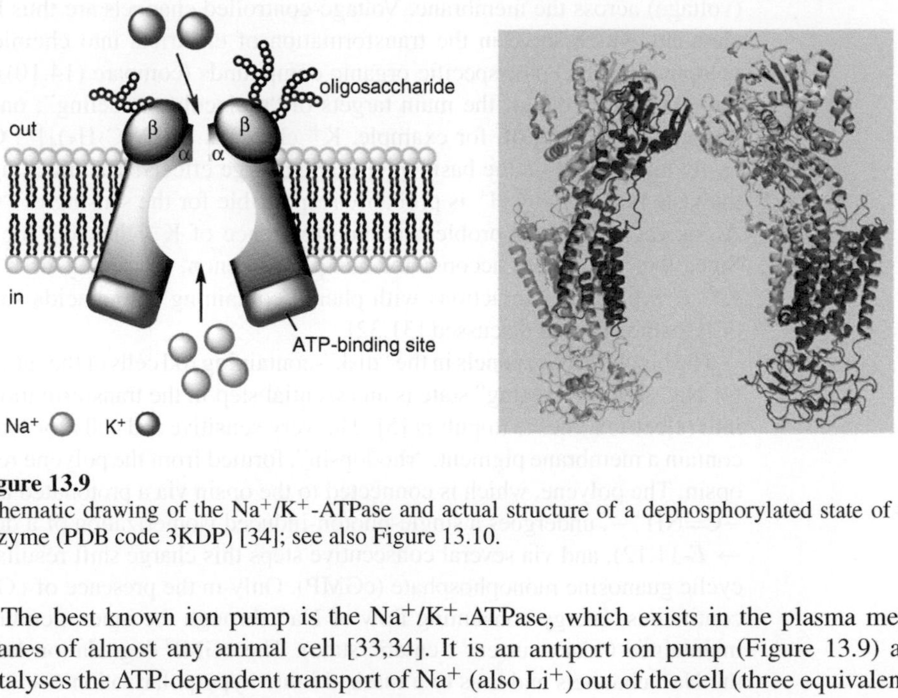

Figure 13.9
Schematic drawing of the Na^+/K^+-ATPase and actual structure of a dephosphorylated state of the enzyme (PDB code 3KDP) [34]; see also Figure 13.10.

The best known ion pump is the Na^+/K^+-ATPase, which exists in the plasma membranes of almost any animal cell [33,34]. It is an antiport ion pump (Figure 13.9) and catalyses the ATP-dependent transport of Na^+ (also Li^+) out of the cell (three equivalents) while K^+ is transported into it (two equivalents) (13.8). Mg^{2+} catalyzes the hydrolysis of ATP.

$$
\begin{array}{c}
3\,Na^+(ic) + 2\,K^+(ec) + ATP^{4-} + H_2O \\
\big\downarrow\ Mg^{2+} \\
3\,Na^+(ec) + 2\,K^+(ic) + ADP^{3-} + HPO_4^{2-} + H^+
\end{array}
\qquad
\begin{array}{l}
(ic) = \text{intracellular} \\
(ec) = \text{extracellular}
\end{array}
\qquad (13.8)
$$

A number of small molecules are able to inhibit the Na^+/K^+-ATPase enzyme. Extracellular binding steroids such as ouabain and digitoxigenin from *Digitalis* (13.9) inhibit the enzyme function, which leads to an increase in the Na^+ concentration in heart muscle cells. As a consequence, the Na^+/Ca^{2+} exchange through a corresponding antiport system is slowed down [35], leading to increased contractility (cardiotonic activity) due to a rising intracellular Ca^{2+} concentration.

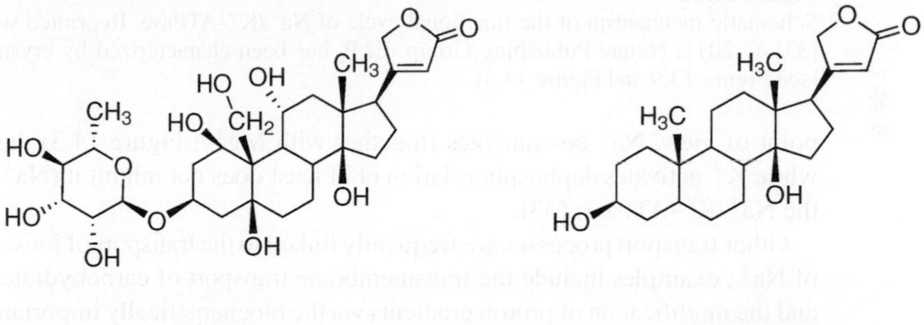

ouabain digitoxigenin (13.9)

Trace amounts of intracellular vanadate (VO_4^{3-}) also strongly inhibit the Na^+/K^+-ATPase; due to its larger size and the resulting preference for a coordination number of 5, the group 5 transition metal vanadium blocks the necessary ATP hydrolysis via stabilization of the enzymatic transition state with five-coordinated pentavalent phosphorus/vanadium (see (14.7) and Figure 14.3).

The integral membrane protein Na^+/K^+-ATPase is very similarly structured in all eucaryotic organisms; it consists of two subunit peptide pairs (heterodimer) with an overall molecular mass of $2 \times 112(\alpha) + 2 \times 35(\beta) \approx 294$ kDa. The cation binding sites are localized at the larger α protein [33,34]. The structural resemblance of the Na^+/K^+-ATPase α subunit to a Ca^{2+}-ATPase [35,36] is surprisingly high, even in the cation-binding pocket, raising the fundamental issue of how the specific cation selectivity is determined [33–35]. The function of this protein has not yet been defined in full detail, but it is believed that the protein can assume at least two markedly different conformations, *E1* and *E2*, in which the binding of the metal ions must be very different (Figure 13.10). Because of the alternatives (Na^+/K^+-filled (there is either K^+- or Na^+-filling), ATP- or ADP-bound, conformations *E1* or *E2*) there must be at least $2^3 = 8$ essentially different states of this protein system. Functional requirements are the possibility for the transport of the ions between intracellular and extracellular regions (while *E1* opens to the extracellular space, *E2* opens to the intracellular space) and the (intracellular) energetic coupling with ATP hydrolysis; the dimeric structure of the protein is characteristic and points to a "flip-flop" mechanism. From an alternative

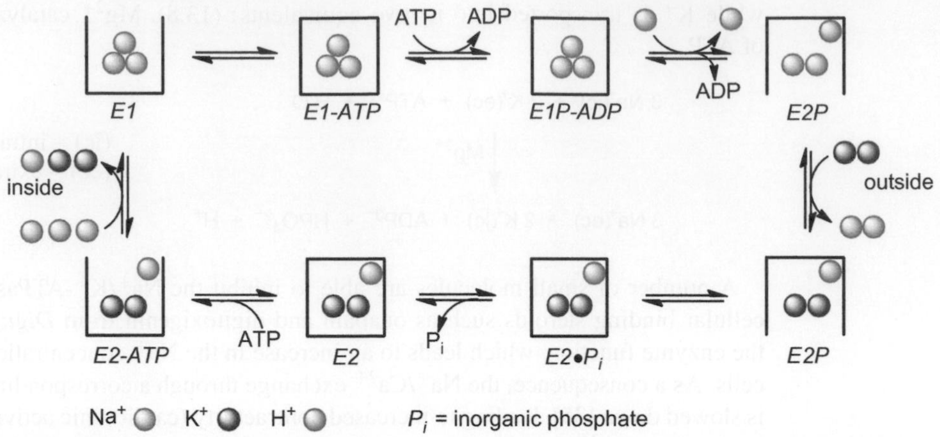

Figure 13.10
Schematic mechanism of the functional cycle of Na^+/K^+-ATPase. Reprinted with permission from [33] © 2011, Nature Publishing Group. E2·P_i has been characterized by crystal structure analysis (see Figure 13.9 and Figure 14.3).

point of view, Na^+ co-catalyzes (together with Mg^{2+}; Figure 14.3) the phosphorylation, while K^+ activates dephosphorylation or at least does not inhibit it (Na^+ pump function of the Na^+/K^+-ATPase) [33].

Other transport processes are frequently linked to the transport of ions, particularly to that of Na^+; examples include the transmembrane transport of carbohydrates and amino acids and the modification of proton gradients via the bioenergetically important Na^+/H^+ antiport system [37]. The glucose/Na^+ symport protein (Na^+/glucose cotransporter SGLUT1) uses the Na^+ gradient to transport glucose into the cell (coupled transport), against the glucose gradient [38]. On the other hand, hormones such as steroids, small peptides and the thyroid hormones (16.4) can effectively stimulate the function of Na^+-dependent ATPases. This capacity confers a high pharmaceutical potency to these low-molecular-weight compounds due to the great importance of the Na^+ transport for the biological energy balance [39], for electrical membrane potentials and for processes mediated by Ca^{2+}.

Another extremely efficient cation-pump system is the H^+/K^+-ATPase, which, through an antiport process coupled to a K^+/Cl^- symport, is responsible for the extraordinary, more than 10^6-fold enrichment of H^+ in the stomach (pH $\approx$ 1).

There are also anion-specific ion pumps (see Section 16.5) [40], transporters [41] and passive channels [42] for various anions, and a passive (no ATPase) antiport system HCO_3^-/Cl^- important for respiration (Figure 13.11) has been identified in erythrocytes [43]. The main purpose of this pump is the disposal of CO_2. At the same time, the underlying process supplies HCO_3^- (which has several uses, such as the formation of solid structures; Chapter 15) and regulates the pH.

The relatively common hereditary disease cystic fibrosis is known to result from a genetically caused misregulation of chloride channels.

Efforts are being made to understand and reconstitute the extraordinarily effective Ca^{2+}-specific pumps, which are concentrated in the sarcoplasmic reticulum of muscle cells (Section 14.2) [35,36,44]. In order to generate the huge concentration gradient between the inside (approximately 10^{-7} M) and the outside of cells (about 10^{-3} M), these particular ion pumps have to be present in high concentrations and function very efficiently.

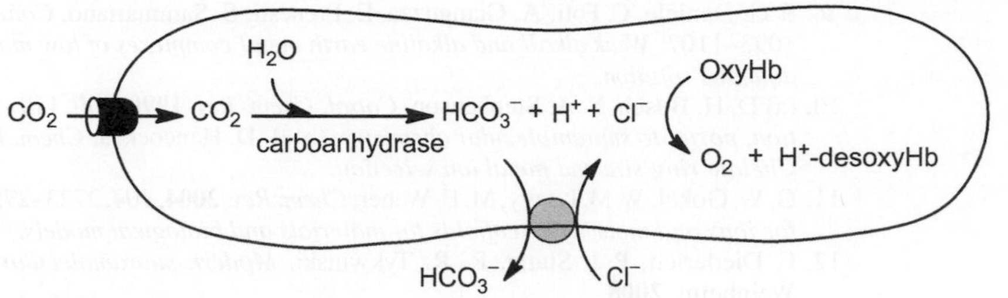

Figure 13.11
The HCO_3^-/Cl^- antiport system in erythrocytes. Hb: hemoglobin.

The investigation of ion pumps/transporters is not only motivated by the fundamental wish to understand their function; some recent findings also point to very interesting medical applications of such knowledge. For instance, chemotherapy can fail due to the development of tumor cell resistance to multiple drugs, a phenomenon known as multidrug resistance (MDR). It is assumed that essentially all cancer-related deaths are a result of chemotherapy failure (see Chapter 19). MDR can have many causes, but one important mechanism of drug resistance is the expression of active drug efflux pumps in the membranes of cancer cells [45]. The investigation of such efflux pumps is thus essential in cancer research.

Further Reading

R. Sutton, B. Rockett, P. G. Swindells, *Chemistry for the Life Science*, 2nd Edition, CRC Press, Boca Raton, **2009**.

References

1. P. Atkins, L. Jones, *Chemical Principles – The Quest for Insight*, 5th Edition, McGraw-Hill College, New York, **2010**.
2. M. N. Hughes, *The Inorganic Chemistry of Biological Processes*, 2nd Edition, John Wiley & Sons, Chichester, **1981**.
3. M. J. Page, E. Di Cera, *Physiol. Rev.* **2006**, *86*, 1049–1092: *Role of* Na^+ *and* K^+ *in enzyme function*.
4. J. J. R. Fraústo da Silva, R. J. P. Williams, *The Biological Chemistry of the Elements – The Inorganic Chemistry of Life*, 2nd Edition, Oxford University Press, Oxford, **2001**.
5. D. L. Nelson, M. M. Cox, *Lehninger Principles of Biochemistry*, 5th Edition, W H Freeman & Co., New York, **2009**.
6. L. Ronconi, P. J. Sadler, *Coord. Chem. Rev.* **2008**, *252*, 2239–2277: *Applications of heteronuclear NMR spectroscopy in biological and medicinal inorganic chemistry*.
7. G. W. Gokel, M. M. Daschbach, *Coord. Chem. Rev.* **2008**, *252*, 886–902: *Coordination and transport of alkali metal cations through phospholipid bilayer membranes by hydraphile channels*.
8. (a) D. J. Miller, J. M. Lisy, *J. Am. Chem. Soc.* **2008**, *130*, 15393–15404: *Entropic effects on hydrated alkali-metal cations: infrared spectroscopy and ab initio calculations of* $M^+(H_2O)_{x=2-5}$ *cluster ions for M = Li, Na, K, and Cs*; (b) M. F. Bush, J. T. O'Brien, J. S. Prell, C.-C. Wu, R. J. Saykally, E. R. Williams, *J. Am. Chem. Soc.* **2009**, *131*, 13270–13277: *Hydration of alkaline earth metal dications: effects of metal ion size determined using infrared action spectroscopy*.

9. P. G. Daniele, C. Foti, A. Gianguzza, E. Prenesti, S. Sammartano, *Coord. Chem. Rev.* **2008**, *252*, 1093–1107: *Weak alkali and alkaline earth metal complexes of low molecular weight ligands in aqueous solution*.
10. (a) D. H. Busch, N. A. Stephenson, *Coord. Chem. Rev.* **1990**, *100*, 119–154: *Molecular organization, portal to supramolecular chemistry*; (b) R. D. Hancock, *J. Chem. Educ.* **1992**, *69*, 615–621: *Chelate ring size and metal ion selection*.
11. G. W. Gokel, W. M. Leevy, M. E. Weber, *Chem. Rev.* **2004**, *104*, 2723–2750: *Crown ethers: sensors for ions and molecular scaffolds for materials and biological models*.
12. F. Diederich, P. J. Stang, R. R. Tykwinski, *Modern supramolecular chemistry*, Wiley-VCH, Weinheim, **2008**.
13. B. Dietrich, *J. Chem. Educ.* **1985**, *62*, 954–964: *Coordination chemistry of alkali and alkaline-earth cations with macrocyclic ligands*.
14. J. P. Michael, G. Pattenden, *Angew. Chem. Int. Ed. Engl.* **1993**, *32*, 1–23: *Marine metabolites and the complexation of metal ions: facts and hypotheses*.
15. B. T. Kilbourn, J. D. Dunitz, L. A. Pioda, W. Simon, *J. Mol. Biol.* **1967**, *30*, 559–563. *Structure of the K^+ complex with nonactin, a macrotetrolide antibiotic possessing highly specific K^+ transport properties*.
16. D. L. Ward, K.-T. Wei, J. G. Hoogerheide, A. I. Popov, *Acta Cryst. B* **1978**, *34*, 110–115: *The crystal and molecular structure of the sodium bromide complex of monensin, $C_{36}H_{62}O_{11}Na^+Br^-$*.
17. (a) B. M. Burkhart, N. Li, D. A. Langs, W. A. Pangborn, W. L. Duax, *Proc. Natl. Acad. Sci USA* **1998**, *95*, 12950–12955: *The conducting form of gramicidin A is a right-handed double-stranded double helix*; (b) D. A. Langs, *Science* **1988**, *241*, 188–191: *Three dimensional structures at 0.86 Å of the uncomplexed form of the transmembrane ion channel peptide gramicidin A*; (c) B. A. Wallace, K. Ravikumar, *Science* **1988**, *241*, 182–187: *The gramicidin pore: crystal structure of a cesium complex*.
18. F. Hucho, C. Weise, *Angew. Chem. Int. Ed. Engl.* **2001**, *40*, 3100–3116: *Ligand-gated ion channels*.
19. D. A. Dougherty. H. A. Lester, *Angew. Chem. Int. Ed. Engl.* **1998**, *37*, 2329–2331: *The crystal structure of a potassium channel – a new era in the chemistry of biological signaling*.
20. S. B. Long, X. Tao, E. B. Campbell, R. MacKinnon, *Nature (London)* **2007**, *450*, 376–383: *Atomic structure of a voltage-dependent K^+ channel in a lipid membrane-like environment*.
21. R. MacKinnon, *Angew. Chem. Int. Ed. Engl.* **2004**, *43*, 4264–4277: *Potassium channels and the atomic basis of selective ion conduction* (Nobel lecture).
22. R. Dutzler, E. B. Campbell, R. MacKinnon, *Science* **2003**, *300*, 108–112: *Gating the selectivity filter in ClC chloride channels*.
23. (a) N. P. Franks, *Nat. Rev. Neurosci.* **2008**, *9*, 370–386: *General anaesthesia: from molecular targets to neuronal pathways of sleep and arousal*; (b) N. P. Franks, W. R. Lieb, *Nature (London)* **1994**, *367*, 607–614: *Molecular and cellular mechanisms of general anaesthesia*.
24. H. Wulff, B. S. Zhorov, *Chem. Rev.* **2008**, *108*, 1744–1773: *K^+ channel modulators for the treatment of neurological disorders and autoimmune diseases*.
25. H. C. Hartzell, *Science* **2008**, *322*, 534–535: *CaCl-ing channels get the last laugh*.
26. G. W. Gokel, I. A. Carasel, *Chem. Soc. Rev.* **2007**, *36*, 378–389: *Biologically active, synthetic ion transporters*.
27. T. M. Fyles, *Chem. Soc. Rev.* **2007**, *36*, 335–347: *Synthetic ion channels in bilayer membranes*.
28. (a) E. Neher, *Angew. Chem. Int. Ed. Engl.* **1992**, *31*, 824–829: *Ion channels for communication between and within cells*; (b) B. Sakmann, *Angew. Chem. Int. Ed. Engl.* **1992**, *31*, 830–841: *Elemental ion flow and synaptic transition* (Nobel lectures).
29. J. Chandrasekhar, M. A. Hoon, N. J. P. Ryba, C. S. Zuker, *Nature (London)* **2006**, *444*, 288–294: *The receptors and cells for mammalian taste*.
30. J. A. DeSimone, V. Lyall, *Am. J. Physiol. Gastrointest. Liver Physiol.* **2006**, *291*, G1005–G1010: *Taste receptors in the gastrointestinal tract III. Salty and sour taste: sensing of sodium and protons by the tongue*.
31. (a) R. Wu, T. B. McMahon, *J. Am. Chem. Soc.* **2008**, *130*, 12554–12555: *Investigation of cation-π interactions in biological systems*; (b) C. Miller, *Science* **1993**, *261*, 1692–1693: *Potassium selectivity in proteins: oxygen cage or π in the face*.
32. J. Hu, L. J. Barbour, G. W. Gokel, *Proc. Natl. Acad. Sci. USA* **2002**, *99*, 5121–5126: *Probing alkali metal-π interactions with the side chain residue of tryptophan*.

33. J. P. Morth, B. P. Pedersen, M. J. Buch-Pedersen, J. P. Andersen, B. Vilsen, M. G. Palmgren, P. Nissen, *Nat. Rev. Mol. Cell Biol.* **2011**, *12*, 60–70: *A structural overview of the plasma membrane* Na^+,K^+*-ATPase and* H^+*-ATPAse ion pumps.*
34. J. P. Morth, B. P. Pedersen, M. S. Toustrup-Jensen, T. L.-M. Sørensen, J. Petersen, J. P. Andersen, B. Vilsen, P. Nissen, *Nature (London)* **2007**, *450*, 1043–1050: *Crystal structure of the sodium–potassium pump.*
35. J. Lytton, *Biochem. J.* **2007**, *406*, 365–382: Na^+/Ca^{2+} *exchangers: three mammalian gene families control* Ca^{2+} *transport.*
36. C. R. D. Lancaster, *Nature (London)* **2004**, *432*, 286–287: *Ion pump in the movies.*
37. I. T. Arkin, H. Xu, M. Ø. Jensen, E. Arbely, E. R. Bennett, K. J.,Bowers, E. Chow, R. O. Dror, M. P. Eastwood, R. Flitman-Tene, B. A. Gregersen, J. L. Klepeis, I. Kolossváry, Y. Shan, D. E. Shaw, *Science* **2007**, *317*, 799–803: *Mechanism of* Na^+/H^+ *antiporting.*
38. E. M. Wright, B. A. Hirayama, D. F. Loo, *J. Intern. Med.* **2007**, *261*, 32–43: *Active sugar transport in health and disease.*
39. A. Y. Mulkidjanian, P. A. Dibrov, M. Y. Galperin, *Biochim. Biophys. Acta* **2008**, *1777*, 985–992: *The past and present sodium energetics: may the sodium-motive force be with you.*
40. G. A. Gerencser, J. Zhang, *Biochim. Biophys. Acta* **2003**, *1618*, 133–139: *Existence and nature of the chloride pump.*
41. E. Ohana, D. Yang, N. Shcheynikov, S. Muallem, *J. Physiol.* **2009**, *587.10*, 2179–2185: *Diverse transport modes by the solute carrier 26 family of anion transporters.*
42. A. S. Verkman, L. J. V. Galietta, *Nat. Rev. Drug Des.* **2009**, *8*, 153–171: *Chloride channels as drug targets.*
43. W. F. Boron, *Biochim. Biophys. Acta* **2010**, *1804*, 410–421: *Evaluating the role of carbonic anhydrase in the transport of* HCO_3^-*-related species.*
44. M. Brini, E. Carafoli, *Physiol. Rev.* **2009**, *89*, 1341–1378: *Calcium pumps in health and disease.*
45. P. D. W. Eckford, F. J. Sharom, *Chem. Rev.* **2009**, *109*, 2989–3011: *ABC efflux pump-based resistance to chemotherapy drugs.*

14 Catalysis and Regulation of Bioenergetic Processes by the Alkaline Earth Metal Ions Mg^{2+} and Ca^{2+}

14.1 Magnesium: Catalysis of Phosphate Transfer by Divalent Ions

Among the bioessential metal cations without redox functions, Mg^{2+} is distinguished by its small ionic radius (see Table 13.1). Due to its rather low radius/charge ratio and the resulting Lewis acidity, this ion prefers multiply negatively charged ligands, especially polyphosphates [1–3]. In contrast to the related Zn^{2+} ion, with its sometimes similar catalytic activity, Mg^{2+} is definitely a "hard" electrophile that does not form inert bonds to simple N and S donor ligands such as histidine or deprotonated cysteine. Furthermore, Mg^{2+} strongly prefers the coordination number 6 with close to octahedral configuration, while other ions with comparable biological functions prefer either lower (Zn^{2+}) or higher (Ca^{2+}) numbers. However, the example of Mg^{2+}-dependent enolase (see (14.9)) shows that deviations from this rule are possible due to the "entatic strain" imposed in an enzyme.

The magnesium ion as a constituent of the chlorophylls has already been introduced in Section 4.2. Further important roles are its carbamate-stabilizing function in the photosynthetic CO_2 fixation by the most abundant enzyme on earth, ribulose-1,5-bisphosphate carboxylase ("rubisco") [4], and its Ca^{2+}-analogous function in exo- and endoskeletons (see Chapter 15) and in the stabilization of cell membranes. While a long-term magnesium deficiency thus inhibits growth, a temporary lack of Mg^{2+} causes a relative overabundance of Ca^{2+} within cells due to the antagonistic relation between Mg^{2+} and Ca^{2+}. Higher intracellular Ca^{2+} concentrations induce increased muscle excitability (cramps); other effects of Mg^{2+} deficiency are reduced mental and physical performance due to insufficient energy production via phosphate transfer and to the inhibition of protein metabolism. Therefore, a hormonal control of Mg^{2+} transport exists, for example, in heart muscle cells, and serious Mg^{2+} deficiency requires a "magnesium therapy" [5].

With regard to enzymatic activity, Mg^{2+} is an essential factor for the biochemical transfer of phosphates and for many nonoxidative cleavage reactions of nucleic acids through nucleases (cleavage) and polymerase proteins (bond formation) [6] or ribozymes [6–8]. Blocking of corresponding viral (RNA) enzymes has emerged as an elegant issue in drug design [9]. Mono-, di- and triphosphate groups are not only among the nucleotide components

Bioinorganic Chemistry: Inorganic Elements in the Chemistry of Life – An Introduction and Guide, Second Edition.
Written and Translated by Wolfgang Kaim, Brigitte Schwederski and Axel Klein.

of RNA and DNA but also essential constituents of intermediate energy carrier molecules (ATP/ADP) in organisms which can be converted by "simple" hydrolysis, through "PO_3^{n-}" transfer between a substrate and water (14.1) [10].

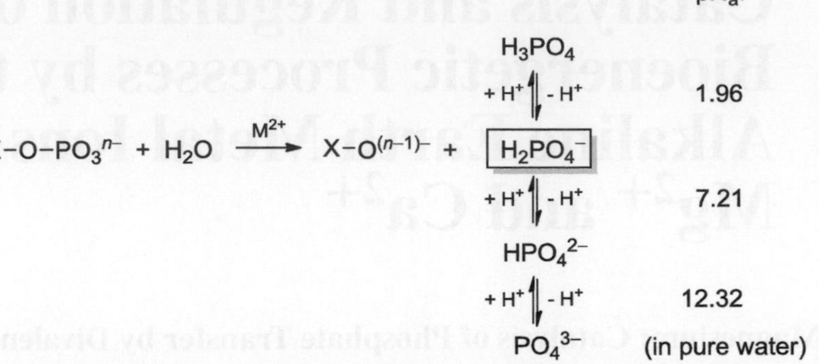

(14.1)

In addition to the well-known ATP^{4-} [1] (14.2), creatine phosphate (14.3) must also be mentioned in this context, because it is important with regard to short-term *anaerobic* hydrolysis and can be detected through *in vivo* ^{31}P-nuclear magnetic resonance (NMR) spectroscopy of muscle tissue.

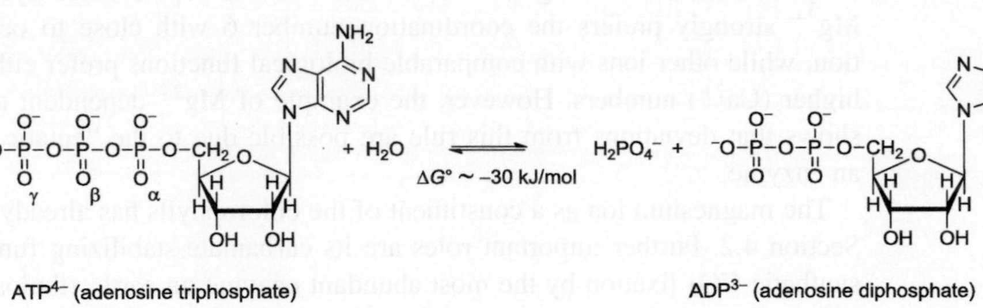

(14.2)

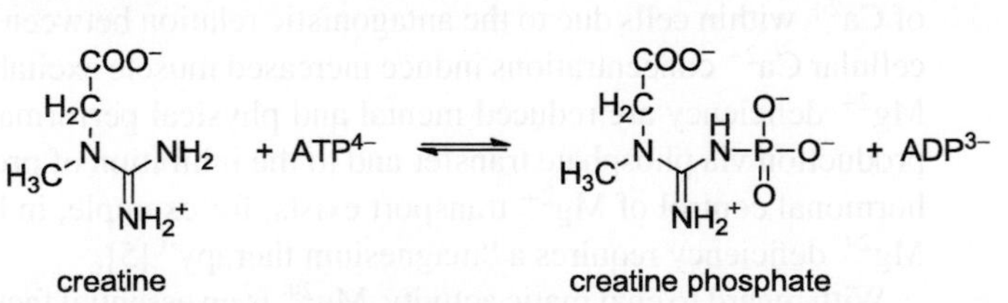

(14.3)

On average, a normally active human adult synthesizes and uses an amount of ATP corresponding to their own body weight each day. Overall, the components of equation (14.2) participate in more chemical reactions than any other compound on the surface of the earth.

[1] The charges of ATP^{4-} and ADP^{3-} are not always given; instead, ATP and ADP are denoted. This should be kept in mind when setting up reaction equations.

All biological phosphate transfer reactions, such as phosphorylations by kinases and dephosphorylations by phosphatases [11], require the presence of catalyzing dipositively charged metal ions. In addition to Mg^{2+} (ionic radius 72 pm for a coordination number 6), Zn^{2+} (74 pm) in alkaline phosphatase (Section 12.3), high-spin Fe^{2+} (78 pm) in purple acid phosphatases (Section 7.6.3) and the comparatively large ions high-spin Mn^{2+} (83 pm) [12] and Ca^{2+} (100 pm) can perform this task *in vivo*. In principle, Cd^{2+} (95 pm) and Pb^{2+} (119 pm) would also be suitable; however, due to their rather soft character, they tend to form strong bonds with sulfur ligands (see Sections 17.2 and 17.3).

Several aspects have to be considered when looking at catalysis by metal ions. First, the function of dipositive metal catalysts in phosphate transfer, including hydrolysis, can be attributed to an effective compensation of the high negative charge, which is a consequence of the ionization of mono- and polyphosphates at physiological pH. The charge compensation by M^{2+} ions concerns both sides of the reaction and therefore contributes to a reduction in activation energy [2]. Trivalent metal ions, M^{3+}, can compensate negative charges even better, but they no longer catalyze the reaction efficiently due to the unproductive stabilization of reaction intermediates (see Section 17.6). Furthermore, the metallic electrophiles M^{2+} activate weak Lewis bases such as water and thus create nucleophiles (M^{2+})–OH via "Umpolung" (see (12.3)) under physiological conditions. It is further obvious that a strongly polarizing dication can coordinate polyphosphates in a chelating manner by binding to the oxygen centers of several phosphate moieties; the result is a spatial fixation, including an activating ring strain (see (14.5)). Finally, metal ions can generally lower the transition state of an associative reaction through intermediate coordination of both reactants (14.4) (see also (14.7)).

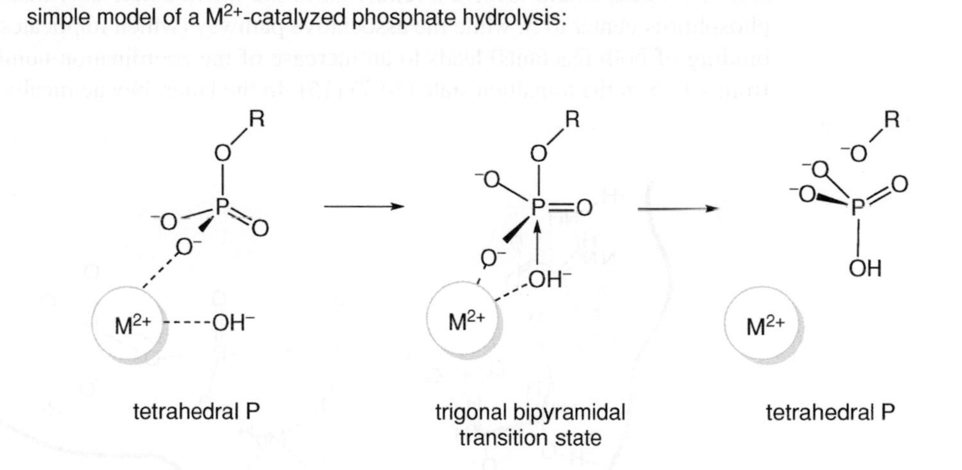

(14.4)

The following conclusions regarding the general reaction mechanism for the hydrolysis of ATP and other nucleoside triphosphates have been drawn from numerous model studies [12–14].

A (partially hydrated) metal dication can typically coordinate to one oxygen center of each of the α, β and γ phosphate groups and – in a free nucleotide – to the imine nitrogen center of the purine heterocycle (macrochelate structures (14.5)).

proposed hydrolysis-productive $(ATP^{4-})(M^{2+})$ structures [10]:

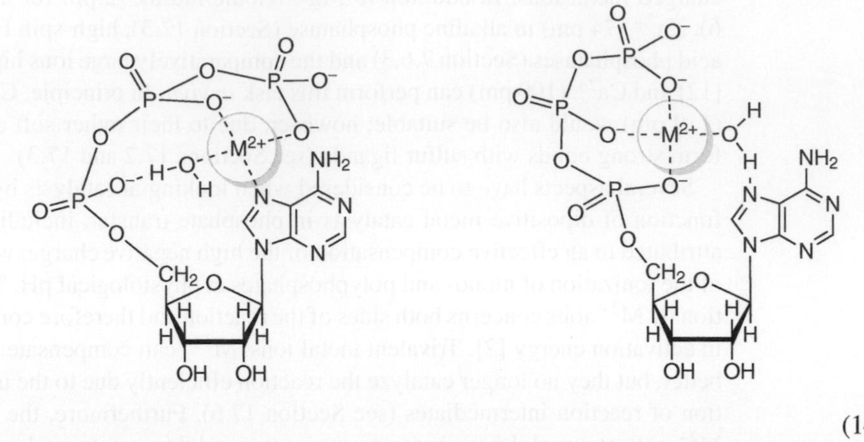

(14.5)

In the enzyme, this coordinative variability may be reduced (Figure 14.1); an additional activation is conceivable after dimerization of such complexes, as in (14.5), which involves stacking of the heterocyclic bases [13]. The reactive species may even require two metal ions (Figure 14.1) [14], one of which attacks at the more basic γ phosphate group and provides a bound hydroxide ion; that is, a deprotonated water ligand (14.4).

Since the overall phosphate transfer reaction (14.6) is a nucleophilic substitution, it can mechanistically proceed as a dissociative (S_N1) or an associative (S_N2) process. A dissociative process would involve a reduction of the intermediate coordination number at the phosphorus center to 3, while the associative pathway (which implicates the simultaneous binding of both reactants) leads to an increase of the coordination number at phosphorus from 4 to 5 in the transition state (14.7) [15]. In the latter, biochemically relevant case, the

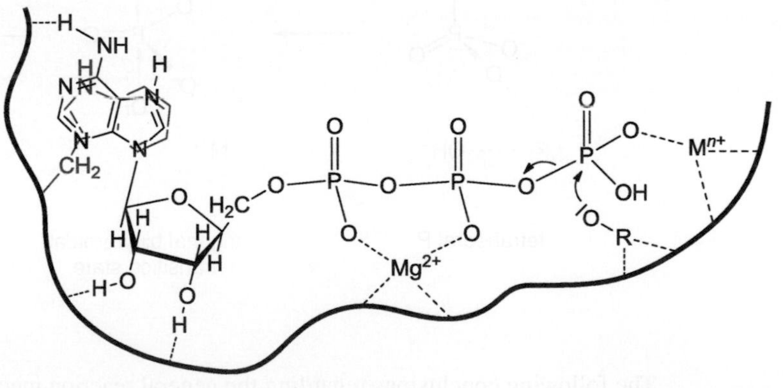

Figure 14.1
Hypothetical arrangement of a reactive $Mg(ATP)^{2-}$ complex in the enzyme. Partially enzyme-bound metal ions activate the triphosphate chain for a nucleophilic attack, in this case by an alkoxide or ester at the terminal phosphate. Binding of the adenine heterocycle may involve π–π interactions with a tryptophan. Eventual dissociation of $Mg(ADP)^{-}$ can be visualized as the consequence of a stronger bond between Mg^{2+} and the (then terminal) β phosphate and thus of weakened new Mg^{2+}/enzyme binding (according to [13]).

reaction is susceptible to stereochemical control. A compelling inorganic-chemical indication for a coordination number 5 in the transition state is the inhibition of ATPases by trace amounts of vanadate(V); the larger transition metal vanadium from group 5 in the periodic table has a higher tolerance of coordination number 5 than the smaller phosphorus atom in phosphates and can thus stabilize five-coordinate intermediate or transition states up to the point of inhibiting the catalytic cycle [16]. The aggregated oligovanadates present in equilibrium at pH 7 (see Figure 11.2) are also known inhibitors of phosphate-transferring enzymes.

$$X{-}PO_3^{2-} + Y \longrightarrow Y{-}PO_3^{2-} + X \qquad (14.6)$$

X, Y: carboxyl functions, phosphates, guanidines, alcohols, water

mechanistic alternatives for the substitution at a tetrahedral center:

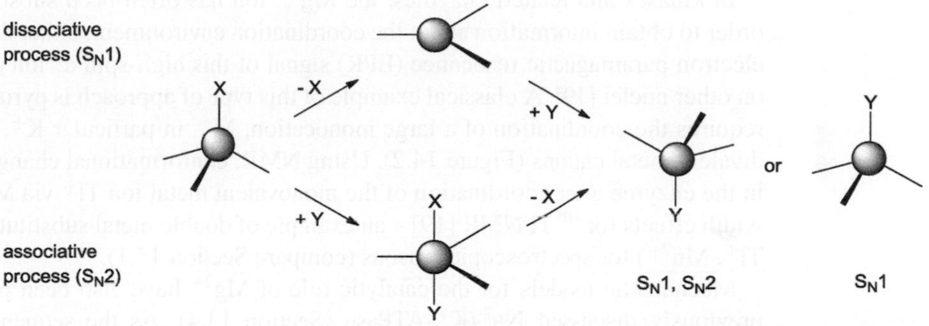

(14.7)

Within metabolic cycles, the "kinase" enzymes catalyze the transfer (14.8) of phosphoryl groups from ATP to other substrates, X, such as carbohydrates (e.g. glucose), carboxylates (e.g. pyruvate, $CH_3{-}C(O){-}COO^-$; see (14.9)) and guanidines (e.g. creatine (14.3)). For the elucidation of the extensive regulatory functions of protein kinases and phosphatases, the 1992 Nobel Prize in Medicine was awarded to E. G. Krebs and E. H. Fischer [17]. Available crystal structure determinations of kinases show ATP binding to Mg^{2+} via one oxygen atom from each of the three phosphate groups and completion of the six-coordination through water and amino acid side chains [18]. During actual catalysis, the Mg^{2+} ion may migrate between αβ and βγ phosphate groups.

$$ATP^{4-} + X{-}H \xrightarrow{\text{kinase}} ADP^{3-} + X{-}PO_3^{2-} + H^+ \qquad (14.8)$$

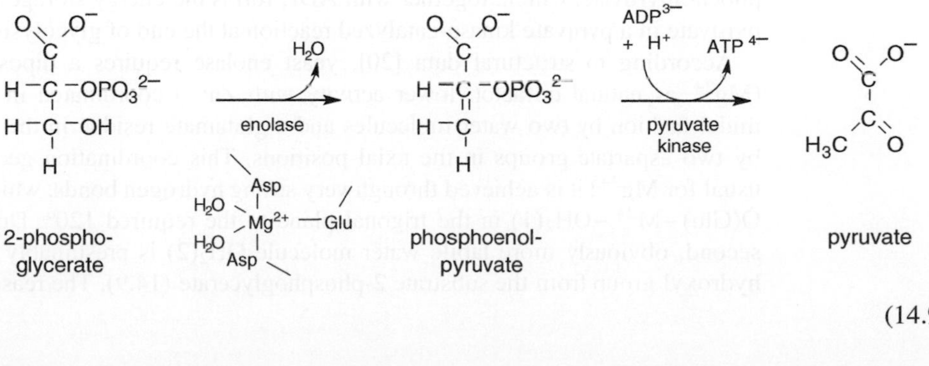

(14.9)

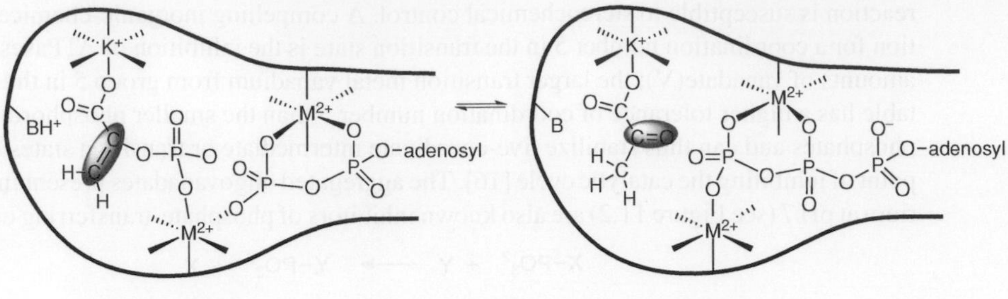

Figure 14.2
Reaction mechanism for metal-dependent pyruvate kinase.

In kinases and related enzymes, the Mg^{2+} ion has often been substituted by Mn^{2+} in order to obtain information about the coordination environment, either directly through the electron paramagnetic resonance (EPR) signal of this high-spin d^5 ion or via its influence on other nuclei [19]. A classical example of this type of approach is pyruvate kinase, which requires the coordination of a large monocation, M^+, in particular K^+, in addition to two divalent metal cations (Figure 14.2). Using NMR, conformational changes can be detected in the enzyme after coordination of the monovalent metal ion Tl^+ via Mn^{2+}-induced line-width effects for ^{205}Tl NMR [19] – an example of double metal substitution (K^+, $Mg^{2+} \rightarrow Tl^+$, Mn^{2+}) for spectroscopic reasons (compare Section 13.1).

Mechanistic models for the catalytic role of Mg^{2+} have also been put forward for the previously discussed Na^+/K^+-ATPase (Section 13.4). As the sequence in Figure 14.3 illustrates, the role of the Mg^{2+} ion presumably consists in chelate-type coordination to the triphosphate oxygen centers (see Figure 14.1), with a resulting activation of the terminal phosphate for esterification by an amino acid side chain of the protein such as Glu^-. In the transition state, the "$P(O)_5$" system is assumed to be five-coordinate in a trigonal-bipyramidal arrangement; its isomerization through a "pseudorotation" may be linked to the Na^+-translocating conformational change in the protein, which in turn gives rise to the hydrolysis to ADP (see Figure 13.10). By pseudorotation back to the original configuration at the phosphorus center, the monophosphate magnesium complex still anchored in the protein can then trigger the reverse conformational change and thus a translocation of K^+. In this process, the monophosphate is dissociated through hydrolysis of the phosphate ester bond and the starting point is reached again.

Magnesium can also occur as an essential component of non-phosphate-transferring enzymes, such as carbohydrate isomerases, DNA-activating topoisomerases and enolases. In an elimination reaction (dehydration), the latter catalyze the synthesis of reactive phosphoenolpyruvate, which, together with ADP, forms the energy storage molecule ATP and pyruvate in a pyruvate kinase-catalyzed reaction at the end of glycolysis (14.9).

According to structural data [20], yeast enolase requires a dipositive metal center (Mg^{2+} as natural cofactor, lower activity with Zn^{2+}) coordinated in a trigonal bipyramidal fashion by two water molecules and a glutamate residue in the trigonal plane and by two aspartate groups in the axial positions. This coordination geometry is very unusual for Mg^{2+}; it is achieved through very strong hydrogen bonds, which widen the angle $O(Glu){-}M^{2+}{-}OH_2(1)$ in the trigonal plane to the required 120°. During catalysis, the second, obviously more labile water molecule $OH_2(2)$ is presumably substituted by the hydroxyl group from the substrate 2-phosphoglycerate (14.9). The reason for this unusual

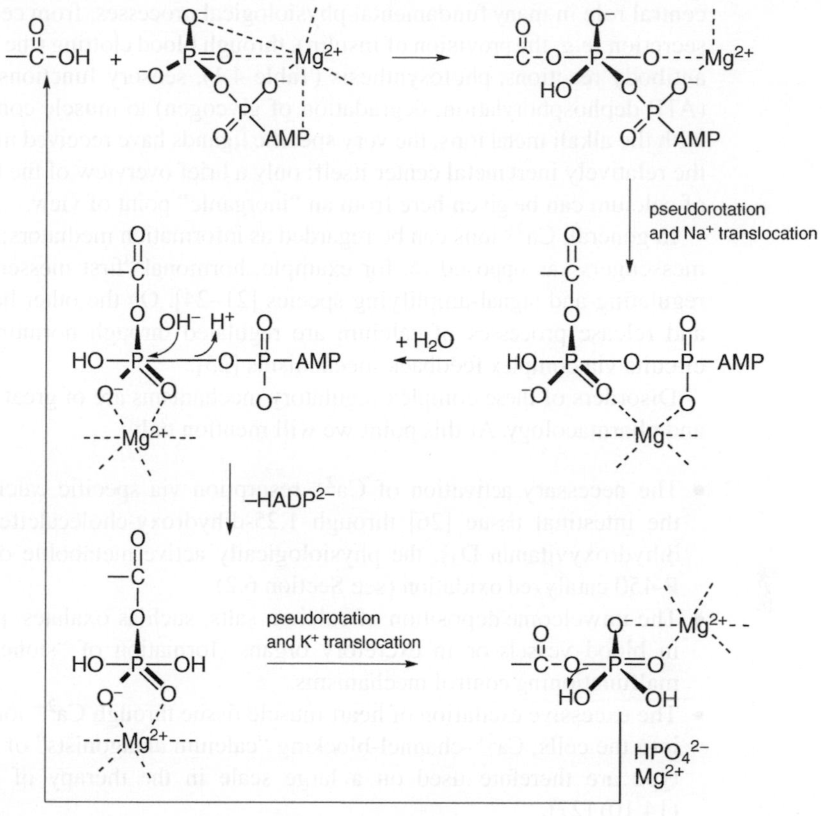

Figure 14.3
Mechanistical hypothesis regarding the role of Mg^{2+} in Na^{+}/K^{+}-ATPase. AMP: adenosine monophosphate.

and certainly high-energy coordination arrangement is an acceleration of the substitution (→ entatic state). With Ca^{2+} instead of Mg^{2+}, there is stronger binding to the apoenzyme and an increase of the coordination number to 6 (size effect); however, the enzyme then becomes inactive [20]. An unusual metal coordination for a dehydrating catalytic center has already been observed in the case of aconitase (Section 7.4).

14.2 The Ubiquitous Regulatory Role of Ca^{2+}

Besides iron, and probably even surpassing it, calcium (in ionic form as Ca^{2+}) is the most important and most versatile "bioinorganic" element. Its wide distribution in bound form in the earth's crust and in dissolved form in sea water has certainly facilitated its numerous uses in biology (→ bioavailability). There are various inorganic calcium compounds, with often strongly pH-dependent solubility (see Table 15.1). Their importance for biological solid-state materials, such as for exo- and endoskeletons, is presented separately in Chapter 15.

Compared to the large amounts of Ca^{2+} stored in the skeleton (about 1.2 kg in an adult human, turnover up to 0.7 g/day), the approximately 10 g of calcium that is not a constituent of solid-state material seems rather unpretentious. However, Ca^{2+} ions play a central role in many fundamental physiological processes, from cell division via hormonal secretion (e.g. the provision of insulin), through blood clotting (the "coagulation cascade"), antibody reactions, photosynthesis (Table 4.1), sensory functions and energy generation (ATP dephosphorylation, degradation of glycogen) to muscle contraction. With Ca^{2+}, as with the alkali metal ions, the very specific ligands have received much more attention than the relatively inert metal center itself; only a brief overview of the biochemical importance of calcium can be given here from an "inorganic" point of view.

In general, Ca^{2+} ions can be regarded as information mediators; that is, as "second/third messengers" as opposed to, for example, hormonal "first messengers", or as triggering, regulating and signal-amplifying species [21–24]. On the other hand, the uptake, storage and release processes of calcium are regulated through hormonally influenced control circuits via complex feedback mechanisms [25].

Disorders of these complex regulatory mechanisms are of great importance in medicine and pharmacology. At this point we will mention only:

- The necessary activation of Ca^{2+} resorption via specific calcium-binding proteins in the intestinal tissue [26] through 1,25-dihydroxy-cholecalciferol (Calcitriol or 1,25-dihydroxyvitamin D_3), the physiologically active metabolite of vitamin D formed by P-450 catalyzed oxidation (see Section 6.2).
- The unwelcome deposition of calcium salts, such as oxalates, phosphates and steroids, in blood vessels or in excretory organs (formation of "stones"; Chapter 15) due to malfunctioning control mechanisms.
- The excessive excitation of heart muscle tissue through Ca^{2+} ions permeating too easily into the cells; Ca^{2+}-channel-blocking "calcium antagonists" of the 1,4-dihydropyridine type are therefore used on a large scale in the therapy of cardiovascular diseases (14.10) [27].

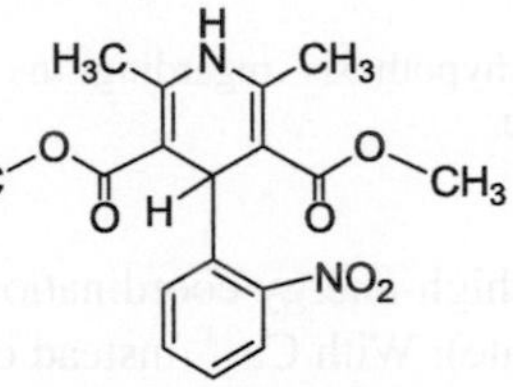

(14.10)

Several neuronal diseases are also believed to be caused by an endogenously disturbed calcium metabolism or through exogenous toxic substances.

The control of the Ca^{2+} concentration in body fluids is crucial as this ion can show extremely high concentration differences of more than four orders of magnitude across cellular and other membranes (Table 13.1). Within the cell (in the cytosol), the concentration of "free" calcium is normally very low (about 10^{-7} M), while the extracellular value is approximately 10^{-3} M. Ca^{2+} concentration gradients do not exist only between the inside and outside of cells but also between compartments of more complex cells, such as the mitochondria and the cell nucleus [28]. The pH-dependent anion concentrations of phosphates and carbonates have to be considered here, as the solubility product may otherwise be exceeded, resulting in undesired precipitation (see Table 15.1). Only the extremely low intracellular Ca^{2+} concentration maintained by the continuously operating Ca^{2+} pumps allows for the diverse control functions, and in particular the amplification of protein activity.

The quantitative assay of calcium ions, particularly during rapidly proceeding Ca^{2+} exchange processes, has been enormously facilitated by the development of Ca^{2+}-specific ligands, which show a rapidly (within milliseconds) decaying coordination-dependent fluorescence. These compounds permit microscopic studies with high temporal and spatial resolution in concentrations of 10^{-1} to 10^{-5} M Ca^{2+}. Synthetic reagents such as "Quin 2AM" (14.11) [29] and the protein aequorin from bioluminescent organisms (jellyfish) are in use today. In a related approach to biological cellular imaging, the 2008 Nobel Prize in Chemistry was awarded to M. Chalfie, O. Shimomura and R. Y. Tsien for the development of the green fluorescent protein (GFP) [30].

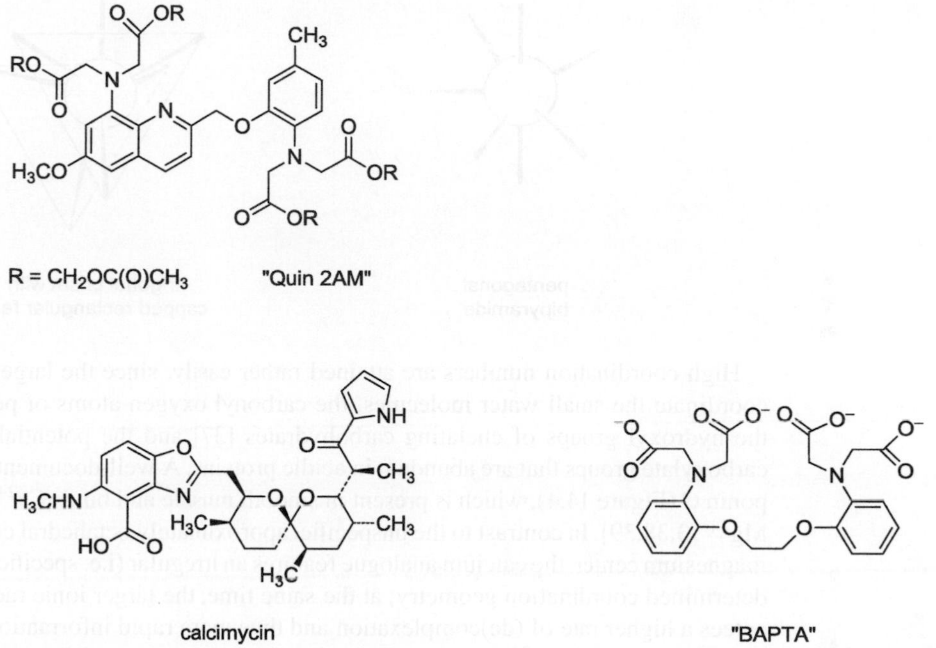

(14.11)

Of course, there are also nonluminescent Ca^{2+}-specific ionophores, such as calcimycin and similar substances from *Streptomyces* strains [31] and the synthetic 1,2-bis(o-aminophenoxy)ethane-N,N,N',N'-tetraacetate ("BAPTA") (14.11). Other means of quantitative calcium determination, such as precipitation with oxalate and detection by ion-sensitive microelectrodes, have become less popular; ^{43}Ca NMR spectroscopy is dependent on the availability of isotope-enriched material, due to the low natural abundance of 0.13% [32].

In order to maintain the large concentration difference, various Ca^{2+} pumps are required; for example, these are the main constituents of the sarcoplasmic reticulum of muscle cells. Among the well-known systems are the Ca^{2+}-dependent ATPase and the already mentioned Na^{+}/Ca^{2+} antiport system (Section 13.4).

Bioavailability and the obviously possible control notwithstanding – why is Ca^{2+} so well suited for effecting transfer, conversion and amplification of information? Ca^{2+} is a divalent ion without redox function which typically exhibits high coordination numbers and often irregular coordination geometry in its complexes, due to its ionic radius of about 100–120 pm [26,31–33]. The Cd^{2+} (95 pm) and Pb^{2+} (119 pm) (Sections 17.2 and 17.3) ions are similar to Ca^{2+} but are biologically harmful, due to their strong coordination with thiolates (Cys^{-});

Mn^{2+} (83 pm) and the heavier homologue Sr^{2+} (118 pm) are less toxic as calcium "substitutes", but the possible biological importance of Sr^{2+} is perhaps obscured by the much more abundant Ca^{2+} (see Table 2.1). Coordination numbers of 7 or 8 are quite common for Ca^{2+} in proteins, in contrast to the strong preference by Mg^{2+} for octahedral configuration. Typically realized geometries for the Ca^{2+} ion, with a coordination number of 7, are the pentagonal bipyramid [33,34], as in α-lactalbumin of milk, a trigonal prismatic arrangement with a capped rectangular face (14.12) [33] and a distorted octahedron with an additional coordination site through (η^2-)carboxylate chelate coordination at one corner [3,26,35].

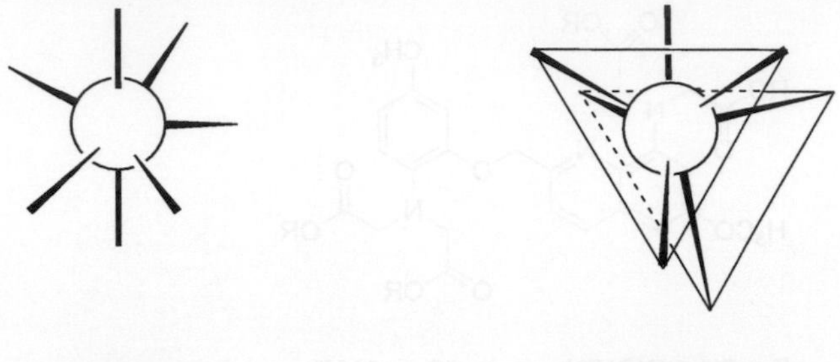

(14.12)

High coordination numbers are attained rather easily, since the large Ca^{2+} ion likes to coordinate the small water molecules, the carbonyl oxygen atoms of peptide bonds [36], the hydroxyl groups of chelating carbohydrates [37] and the potentially chelating (2.7) carboxylate groups that are abundant in acidic proteins. A well-documented example is troponin C (Figure 14.4), which is present in smooth muscle and binds Ca^{2+} but also tolerates Mg^{2+} [3,38,39]. In contrast to the unspecific, approximately octahedral configuration of the magnesium center, the calcium analogue features an irregular (i.e. specific) and thus protein-determined coordination geometry; at the same time, the larger ionic radius of Ca^{2+} guarantees a higher rate of (de)complexation and thus more rapid information transfer [3,38].

Several types of Ca^{2+}-containing protein are now rather well understood with regard to their functions. Calcium-releasing regions close to membranes contain very acidic Ca^{2+} storage proteins, the calsequestrins (approximately 40 kDa), each of which can bind up

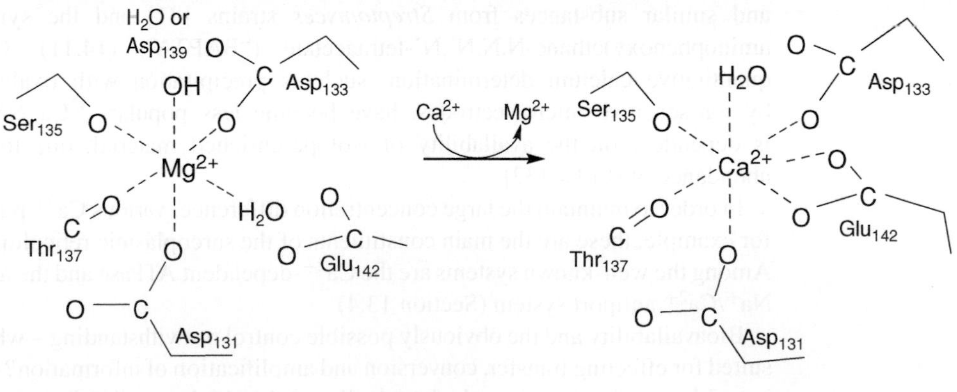

Figure 14.4
Metal coordination in Mg^{2+}- (left) and Ca^{2+}- (right) containing troponin C. Reprinted with permission from [39] © 2008, Elsevier.

to 50 calcium ions [21,40]. The large amounts of Ca^{2+} ions required for information transmission and amplification and for the triggering of muscle contractions are released from such proteins. The activation of stored calcium proceeds via mechanisms that are not yet fully understood; nucleotides whose formation can be influenced by Ca^{2+} (→ feedback) presumably serve as anionic "second messengers". Membrane depolarization through an electric nerve impulse, as well as local hormone–receptor interactions, can also cause the release of Ca^{2+}.

In addition to a protein-structure stabilizing function, as found for example in thermolysin (see Figure 2.4) and proteinase K [2,21,41], Ca^{2+} ions can also show hydrolysis-catalyzing activity. One of the best documented examples is the phosphodiester-cleaving nuclease from staphylococcal strains [2,21,42], which features an "Mg^{2+}-like" coordination sphere of the catalytic metal center (2 η^1-Asp, 1 Thr, 2 H_2O, 1 substrate-O).

Amino acid sequences and structures have been determined for another group of ubiquitous, very stable and, from an evolutionary point of view, very old Ca^{2+}-specific proteins, the "calmodulins" [21,25,28,43,44]. These are rather small proteins (approximately 17 kDa) with a number of calcium-binding acidic regions with carboxylate-containing side chains: glutamate and aspartate. The function of such Ca^{2+} receptor proteins, which in turn serve to activate many "calcium-dependent" enzymes [21], is to co-operatively bind two to four Ca^{2+} ions and thus change the conformation so that the recognition [21,44] and activation of an enzyme can take place through specific calmodulin–protein interactions (Figure 14.5).

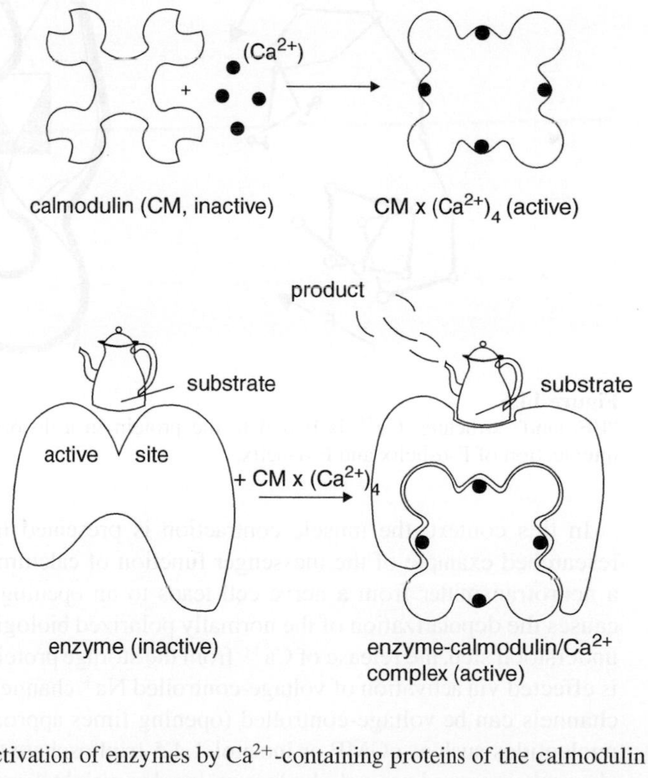

Figure 14.5
Model for the activation of enzymes by Ca^{2+}-containing proteins of the calmodulin type (according to [25]).

Among the enzymes which are activated by calmodulin/Ca^{2+} complexes are:

- adenylate and guanylate cyclases for the formation of cAMP and cGMP;
- NO synthase (see Section 6.5);
- Ca^{2+} ATPase (see Chapter 13);
- NAD kinase for the synthesis of NADP (3.12);
- phosphorylase kinase, which contributes to the degradation of the energy storage molecule glycogen.

Parvalbumins, which are present in smooth muscle and presumably assist in muscle relaxation, the troponins of skeletal (striped) muscle (Figures 14.4 and 14.7) and the S100 proteins [21,32,45], which are found in the nervous system, all belong to the extended calmodulin family, with its typical "EF-hand" protein structure (Figure 14.6) [21,25]. About 200 proteins are known, in which often several neighboring Ca^{2+}-selective "EF-hand" binding sites exist.

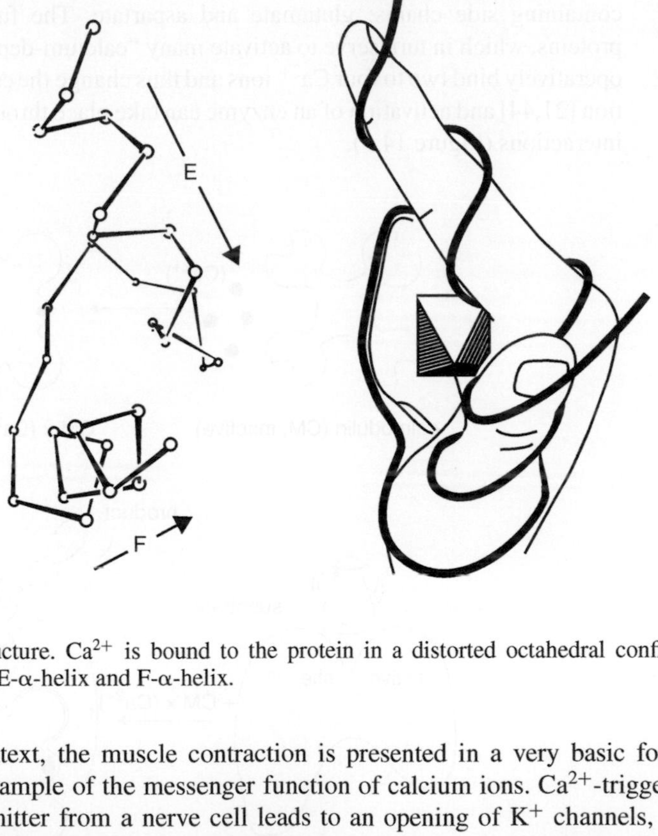

Figure 14.6
"EF-hand" structure. Ca^{2+} is bound to the protein in a distorted octahedral configuration at the intersection of E-α-helix and F-α-helix.

In this context, the muscle contraction is presented in a very basic form as a well-researched example of the messenger function of calcium ions. Ca^{2+}-triggered release of a neurotransmitter from a nerve cell leads to an opening of K^+ channels, which in turn causes the depolarization of the normally polarized biological membrane. In an as yet little understood step, the release of Ca^{2+} from the storage proteins in the sarcoplasmic reticulum is effected via activation of voltage-controlled Na^+ channels. Incidentally, calcium-specific channels can be voltage-controlled (opening times approximately 1 ms) or controlled by nucleotides such as cGMP or inositol-1,4,5-triphosphate (IP_3) [21–24]. Blocking of these channels is caused not only by low-molecular-weight "antagonist" molecules (see (14.10)) but also by other metal cations such as Co^{2+} and the series of lanthanoid ions La^{3+}–Lu^{3+} (ionic radii decrease from 103 to 86 pm) or Y^{3+} (101 pm), which are quite similar in size to Ca^{2+} (100 pm) [46].

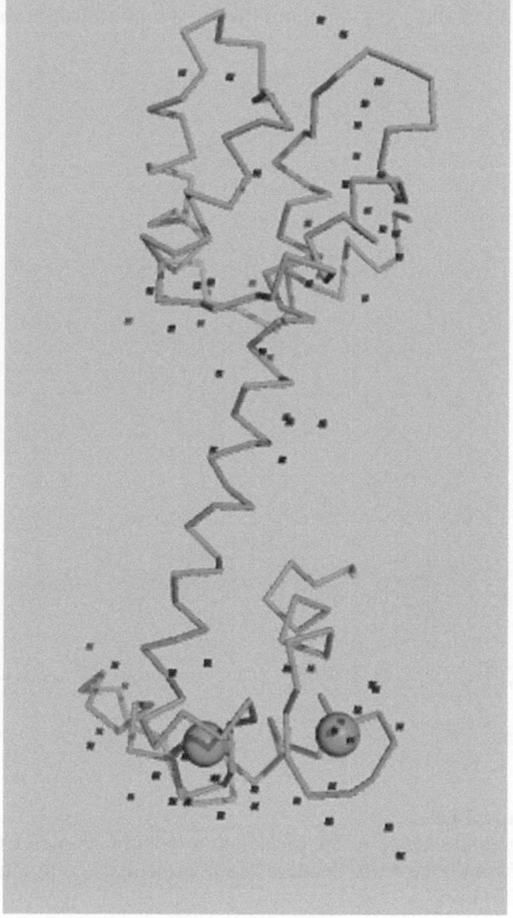

Figure 14.7
Structure of crystallized troponin C from chicken skeletal muscle: α-C backbone representation. Water molecules are depicted as red circles; both seven-coordinate Ca^{2+} ions in the lower half of the protein appear as green circles (PDB code 4TNC) [35a].

The large increase in the concentration of Ca^{2+} ions after their release from the sarcoplasmic reticulum leads to the binding of these ions by troponin C [21,35,38a,45–48]. This protein, with a molecular mass of approximately 18 kDa and "EF-hand" binding sites (Figure 14.7), is similar to calmodulin. It can adopt various conformations with different Ca^{2+} affinities (Figure 14.8). Through deblocking of the initial entatic-state situation, Ca^{2+}-activated troponin C causes an interaction of the thin, also M^{2+}-binding actin fiber with the thick myosin fiber in the muscle cell, combined with a spatial displacement. The simultaneous displacement of the fibers triggered by Ca^{2+} binding creates a macroscopic contraction of the muscle fiber, followed by the release of bound ADP and phosphate. In simple terms, chemical energy in the form of ATP is thus transformed into mechanical energy through an electric impulse (membrane depolarization), using Ca^{2+} for rapid amplification and transformation into a specific chemical signal. Only after Mg^{2+}-dependent binding and hydrolysis of ATP is Ca^{2+} released and pumped back to the storage proteins. This energy-consuming process leads to a separation of myosin and actin fibers; that is, to a lifting of the rigor state and thus to a restoration of the entatically blocked starting situation.

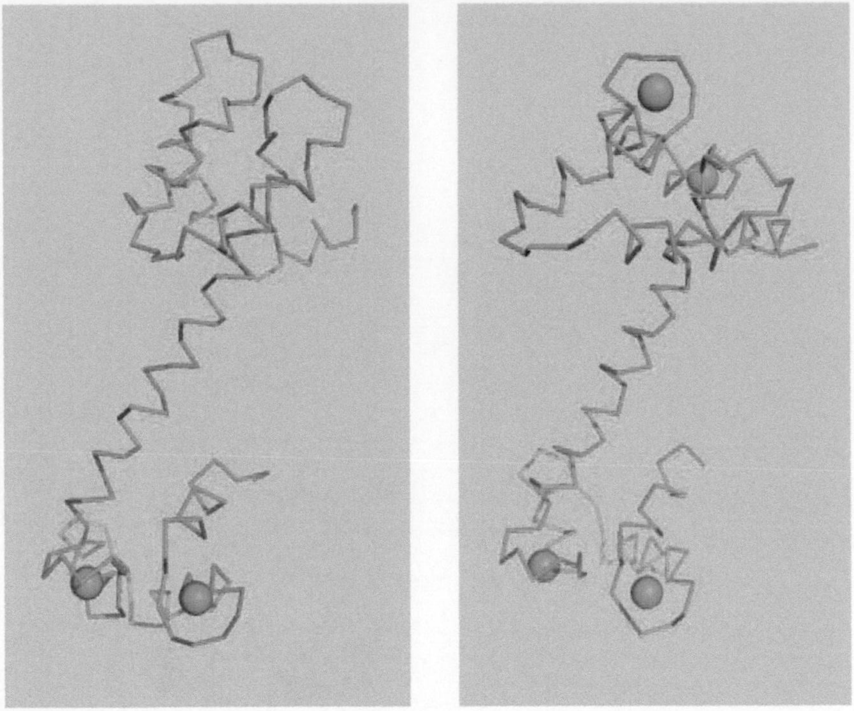

Figure 14.8
Structural change of the dicalcium/troponin C complex (left) after additional Ca^{2+} binding (right). α-C backbone representation of the protein; calcium ions as green spheres (PDB codes 4TNC and 1TCF) [47].

For longer-lasting muscle contraction, a continuous generation of ATP for the myosin ATPase and rapid pumping of Ca^{2+} by the membrane pumps are necessary. A constant cyclic Ca^{2+} flow has been established as prerequisite for continuous contraction in the smooth-muscle system [21,24]. The activation of phosphate-transferring kinases (14.8) through Ca^{2+} plays an important role in combination with muscle contraction and other Ca^{2+}-controlled processes. As triggers of reaction "cascades", Ca^{2+} ions and their calmodulin complexes are important for the generation of ATP through the degradation of glycogen, for blood coagulation and for other secretory processes. In some cases, such as the visual process, where calcium ions were once regarded as actual activators, most recent studies have instead indicated the activation of specific nucleotides through dephosphorylation (→ anionic second/third/fourth . . . messengers). However, in almost all cases there are complex feedback mechanisms between calcium ions and nucleotides which are very sensitive to the ratios of reaction rates.

Further Reading

Y. Nishizawa, H. Morii, J. Durlach (eds.), *New Perspectives in Magnesium Research*, Springer, London, **2007**.

M. Gielen, E. R. T. Tiekink (eds.), *Metallotherapeutic Drugs and Metal-Based Diagnostic Agents – The Use of Metals in Medicine*, John Wiley & Sons, Chichester, **2005**.

J. J. R. Fraústo da Silva, R. J. P. Williams, *The Biological Chemistry of the Elements – The Inorganic Chemistry of Life*, 2nd Edition, Oxford University Press, Oxford, **2001**.

References

1. F. I. Wolf, A. Cittadini, *Mol. Asp. Med.* **2003**, *24*, 3–9: *Chemistry and biochemistry of magnesium.*
2. (a) M. E. Maguire, J. A. Cowan, *BioMetals* **2002**, *15*, 203–210: *Magnesium chemistry and biochemistry*; (b) J. A. Cowan, *BioMetals* **2002**, *15*, 225–235: *Structural and catalytic chemistry of magnesium-dependent enzymes.*
3. (a) T. Dudev, C. Lim, *Ann. Rev. Biophys.* **2008**, *37*, 97–116: *Metal binding affinity and selectivity in metalloproteins: insights from computational studies*; (b) T. Dudev, C. Lim, *Chem. Rev.* **2003**, *103*, 773–787: *Principles governing Mg, Ca, and Zn binding and selectivity in proteins.*
4. I. Andersson, A. Backlund, *Plant Physio. Biochem.* **2008**, *46*, 275–291: *Structure and function of rubisco.*
5. H. Schmidbaur, H. G. Classen, J. Helbig, *Angew. Chem. Int. Ed. Engl.* **1990**, *29*, 1090–1103: *Aspartic and glutamic acid as ligands to alkali and alkaline-earth metals: structural chemistry as related to magnesium therapy.*
6. W. Yang, J. Y. Lee, M. Nowotny, *Mol. Cell* **2006**, *22*, 5–13: *Making and breaking nucleic acids: two-*Mg^{2+}*-ion catalysis and substrate specificity.*
7. J. C. Cochrane, S. A. Strobel, *Acc. Chem. Res.* **2008**, *41*, 1027–1035: *Catalytic strategies of self-cleaving ribozymes.*
8. R. K. O. Sigel, A. M. Pyle, *Chem. Rev.* **2007**, *107*, 97–113: *Alternative roles for metal ions in enzyme catalysis and the implications for ribozyme chemistry.*
9. D. Rogolino, M. Carcelli, M. Sechi, N. Neamati, *Coord. Chem. Rev.* **2012**, *256*, 3063–3086: *Viral enzymes containing magnesium: metal binding as a successful strategy in drug design.*
10. F. H. Westheimer, *Science* **1987**, *235*, 1173–1178: *Why nature chose phosphates.*
11. L. N. Johnson, R. J. Lewis, *Chem. Rev.* **2001**, *101*, 2209–2242: *Structural basis for control by phosphorylation.*
12. (a) N. Mitic, S. J. Smith, A. Neves, L. W. Guddat, L. R. Gahan, G. Schenk, *Chem. Rev.* **2006**, *106*, 3338–3363: *The catalytic mechanisms of binuclear metallohydrolases*; (b) D. E. Wilcox, *Chem. Rev.* **1996**, *96*, 2435–2458: *Binuclear metallohydrolases.*
13. H. Sigel, R. Griesser, *Chem. Soc. Rev.* **2005**, *34*, 875–900: *Nucleoside 5'-triphosphates: self-association, acid–base, and metal ion binding properties in solution.*
14. C. Liu, M. Wang, T. Zhang, H. Sun, *Coord. Chem. Rev.* **2004**, *248*, 147–168: *DNA hydrolysis promoted by di- and multi-nuclear metal complexes.*
15. (a) W. A. Beard, S. H. Wilson, *Chem. Rev.* **2006**, *106*, 361–382: *Structure and mechanism of DNA polymerase* β; (b) H. Pelletier, M. R. Sawaya, A. Kumar, S. H. Wilson, J. Kraut, *Science* **1994**, *264*, 1891–1903: *Structures of ternary complexes of rat DNA polymerase* β*, a DNA template-primer, and ddCTP.*
16. S. Bhattacharyya, A. S. Tracey, *J. Inorg. Biochem.* **2001**, *85*, 9–13: *Vanadium(V) complexes in enzyme systems: aqueous chemistry, inhibition and molecular modeling in inhibitor design.*
17. E G. Krebs, E. H. Fischer, *Angew. Chem. Int. Ed. Engl.* **1993**, *32*, 1122–1129, 1130–1137: *Protein phosphorylation and cellular regulation I + II* (Nobel lectures).
18. K. Niefind, J. Raaf, O.-G. Issinger, *Cell. Mol. Life Sci.* **2009**, *66*, 1800–1816: *Protein kinase CK2: from structures to insights.*
19. J. L. Buchbinder, J. Baraniak, P. A. Frey, G. H. Reed, *Biochem.* **1993**, *32*, 14 111–14 116: *Stereochemistry of metal ion coordination to the terminal thiophosphoryl group of adenosine 5'-O-(3-thiotriphosphate) at the active site of pyruvate kinase.*

20. L. Lebioda, B. Stec, *J. Am. Chem. Soc.* **1989**, *111*, 8511–8513: *XRD of holoenolase trigonal-bipyramidal geometry of the cation binding site.*
21. J. Burgess, E. Raven, *Adv. Inorg. Chem.* **2009**, *61*, 251–366: *Calcium in biological systems.*
22. J. Liu, Z.-J. Xie, *Biochim. Biophys. Acta* **2010**, *1802*, 1237–1245: *The sodium pump and cardiotonic steroids-induced signal transduction protein kinases and calcium-signaling microdomain in regulation of transporter trafficking.*
23. G. C. R. Ellis-Davies, *Chem. Rev.* **2008**, *108*, 1603–1613: *Neurobiology with caged calcium.*
24. M. Brini, E. Carafoli, *Physiol. Rev.* **2009**, *89*, 1341–1378: *Calcium pumps in health and disease.*
25. S. Klumpp, J. E. Schultz, *Pharm. Unserer Zeit* **1985**, *14*, 19–26: *Calcium und Calmodulin.*
26. D. M. E. Szebenyi, K. Moffat, *J. Biol. Chem.* **1986**, *261*, 8761–8777: *The refined structure of vitamin D-dependent calcium-binding protein from bovine intestine.*
27. S. Dai, D. D. Hall, J. W. Hell, *Physiol. Rev.* **2009**, *89*, 411-452: *Supramolecular assemblies and localized regulation of voltage-gated ion channels.*
28. O. Bachs, N. Agell, E. Carafoli, *Biochim. Biophys. Acta* **1992**, *1113*, 259–270: *Calcium and calmodulin function in the cell nucleus.*
29. B. N. G. Giepmans, S. R. Adams, M. H. Ellisman, R. Y. Tsien, *Science* **2006**, *312*, 217–224: *The fluorescent toolbox for assessing protein location and function.*
30. R. Y. Tsien, *Angew. Chem. Int. Ed. Engl.* **2009**, *48*, 5612–5626: *Constructing and exploiting the fluorescent protein paintbox* (Nobel lecture).
31. A. M. Albrecht-Gary, S. Blanc-Parasote, D. W. Boyd, G. Dauphin, G. Jeminet, J. Juillard, M. Prudhomme, C. Tissier, *J. Am. Chem. Soc.* **1989**, *111*, 8598–8609: *X-14885A: an ionophore closely related to calcymicin (A-23187). NMR, thermodynamic and kinetic studies of cation selectivity.*
32. Y. Ogoma, H. Kobayashi, T. Fujii, Y. Kondo, A. Hachimori, T. Shimizu, M. Hatano, *Int. J. Biol. Macromol.* **1992**, *14*, 279–286: *Binding study of metal ions to S100 protein:* ^{43}Ca, ^{25}Mg, ^{67}Zn *and* ^{39}K *n.m.r.*
33. A. L. Swain, E. L. Amma, *Inorg. Chim. Acta* **1989**, *163*, 5–7: *The coordination polyhedron of* Ca^{2+}, Cd^{2+} *in parvalbumin.*
34. N. K. Vyas, M. N. Vyas, F. A. Quiocho, *Nature (London)* **1987**, *327*, 635–638: *A novel Ca binding site in the galactose-binding protein of bacterial transport and chemotaxis.*
35. (a) K. A. Satyshur, S. T. Rao, D. Pyzalska, W. Drendel, M. Graeser, M. Sundaralingam, *J. Biol. Chem.* **1988**, *263*, 1628–1647: *Refined structure of chicken skeletal muscle troponin C in the two-calcium state at 2-Å resolution*; (b) K. A. Satyshur, D. Pyzalska, M. Greaser, S. T. Rao, M. Sundaralingam, *Acta Cryst. D* **1994**, *50*, 40–49: *Structure of chicken skeletal muscle troponin C at 1.78 Å resolution.*
36. P. Chakrabarti, *Biochemistry* **1990**, *29*, 651–658: *Systematics in the interaction of metal ions with the main-chain carbonyl group in protein structures.*
37. W. I. Weis, K. Drickamer, W. A. Hendrickson, *Nature (London)* **1992**, *360*, 127–134: *Structure of a C-type mannose-binding protein complexed with an oligosaccharide.*
38. (a) Z. Grabarek, *Biochim. Biophys. Acta* **2011**, *1813*, 913–921: *Insights into modulation of calcium signaling by magnesium in calmodulin, troponin C and related EF-hand proteins*; (b) M. S. Cates, M. B. Berry, E. L. Ho, Q. Li, J. D. Potter, G. N. Phillips, Jr., *Structure* **1999**, *7*, 1269–1278: *Metal-ion affinity and specificity in EF-hand proteins: coordination geometry and domain plasticity in parvalbumin.*
39. M. Nara, M. Tanokura, *Biochem. Biophys. Res. Commun.* **2008**, *369*, 225–239: *Infrared spectroscopic study of the metal-coordination structures of calcium-binding proteins.*
40. L. Royer, E. Ríos, *J. Physiol.* **2009**, *587.13*, 3101–3111: *Deconstructing calsequestrin. Complex buffering in the calcium store of skeletal muscle.*
41. (a) C.-A. Schoenenberger, H. G. Mannherz, B. M. Jockusch, *Eur. J. Cell Biol.* **2011**, *90*, 797–804: *Actin: from structural plasticity to functional diversity*; (b) P. J. McLaughlin, J. T. Gooch, H. G. Mannherz, A. G. Weeds, *Nature (London)* **1993**, *364*, 685–692: *Structure of gelsolin segment 1-actin complex and the mechanism of filament severing.*
42. F. A. Cotton, E. E. Hazen, M. J. Legg, *Proc. Natl. Acad. Sci. USA* **1979**, *76*, 2551–2555: *Staphylococcal nuclease: proposed mechanism of action based on structure of enzyme-thymidine 3',5'-bisphosphate-calcium ion complex at 1.5-Å resolution.*
43. S. L. Gaffin, *J. Therm. Biol.* **1999**, *24*, 251–264: *Simplified calcium transport and storage pathways.*

44. Z. Grabarek, *J. Mol. Biol.* **2005**, *346*, 1351–1366: *Structure of a trapped intermediate of calmodulin: calcium regulation of EF-hand proteins from a new perspective.*
45. M. C. Schaub, C. W. Heizmann, *Biochem. Biophys. Res. Commun.* **2008**, *369*, 247–264: *Calcium, troponin, calmodulin, S100 proteins: from myocardial basics to new therapeutic strategies.*
46. S. P. Fricker, *Chem. Soc. Rev.* **2006**, *35*, 524–533: *The therapeutic application of lanthanides.*
47. K. Fujimori, M. Sorenson, O. Herzberg, J. Moult, F. C. Reinach, *Nature (London)* **1990**, *345*, 182–184: *Probing the calcium-induced conformational transition of troponin C with site-directed mutants.*
48. M. V. Vinogradova, D. B. Stone, G. G. Malanina, C. Karatzaferi, R. Cooke, R. A. Mendelson, R. J. Fletterick, *Proc. Natl. Acad. Sci. USA* **2005**, *102*, 5038–5043: *Ca^{2+}-regulated structural changes in troponin.*

15 Biomineralization: The Controlled Assembly of "Advanced Materials" in Biology

15.1 Overview

The chemically and morphologically diverse inorganic biominerals belie the impression of life as being monopolized by organic chemistry much more obviously than even the metalloproteins or ionic electrolytes. Even our knowledge about earlier forms of life is largely based on biominerals (i.e. fossils), some of which have accumulated to an enormous, "geological" extent: many mountain ranges, islands and coral reefs consist of biogenic material such as limestone. This immense bioinorganic production across hundreds of millions of years has significantly changed the conditions for life itself; for instance, CO_2 has been bound in the form of carbonates, thus diminishing the early greenhouse effect at the earth's surface. In addition to the well-known calcium-containing shells, teeth and skeletons, a variety of other materials and objects can be classified as biominerals, including the aragonite pearls produced by mollusks, the hulls and spicules of diatoms (Figure 15.1), radiolaria and certain plants, the Ca-, Ba- and Fe-containing crystallites of gravity and magnetic field sensors and some of the pathological "stones" (calculi) formed in the kidney and urinary tract. The iron storage protein ferritin introduced in Section 8.4.2 can also be regarded as a biomineral based on its structure and its content of inorganic material.

The relatively new and highly interdisciplinary research field of biomineralization [1–6] encompasses such diverse areas as geology, classical (descriptive) biology and the modern "biomimetic" material sciences, including nanoparticle research [4–7] (see Section 15.4); chemically, it is concerned with the molecular control mechanisms that operate in biological systems to achieve the formation of well-defined inorganic solid-state materials. The production of morphologically complex minerals according to a genetically determined blueprint mainly results from the necessity for robust support and defense structures. In principle, there is no natural preference for inorganic or organic support materials in endo- or exoskeletons; for instance, the relatively rapidly assembled chitin (polysaccharide) structures of invertebrates and the skeletons of sharks consist largely of organic-chemical material. In fact, most biomineral constructions feature an organic/inorganic composite texture; the bones of vertebrates thus consist of the calcium "mineral" hydroxyapatite and an organic matrix. The advantage of the inorganic component is its hardness and pressure-resistance, which allows for the existence of larger land-living creatures; the organic matrix consisting of collagen fibers, glycoproteins and mucopolysaccharides guarantees elasticity

Bioinorganic Chemistry: Inorganic Elements in the Chemistry of Life – An Introduction and Guide, Second Edition.
Written and Translated by Wolfgang Kaim, Brigitte Schwederski and Axel Klein.

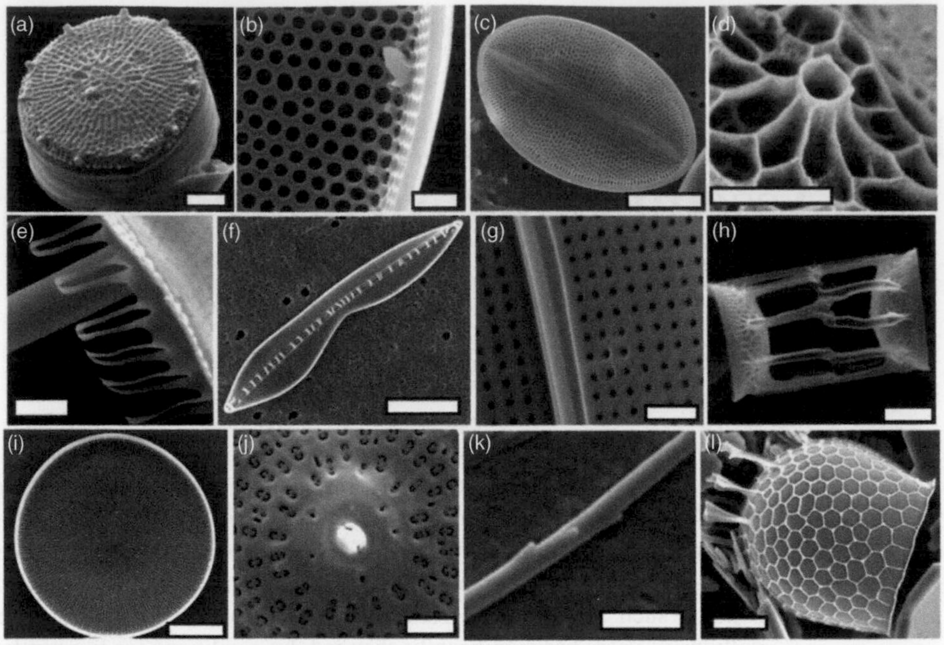

Figure 15.1
Diversity of diatom silica structures. (a) Acid-cleaned material from *Thalassiosira pseudonana*, bar = 1 μm. (b) Close up of *Coscinodiscus wailesii*, bar = 5 μm. (c) *Cocconeis* sp., bar = 10 μm. (d) Rimoportula from *Thalassiosira weissflogii*, bar = 500 μm. (e) Corona structure of *Ditylum brightwellii*, bar = 2 μm. (f) *Bacilaria paxillifer*, bar = 10 μm. (g) Close up of pores in *Gyrosigma balticum*, bar = 2 μm. (h) *Skeletonema costatum*, bar = 2 μm. (i) Valve of *C. wailesii*, bar = 50 μm. (j) Close up of pores in *D. brightwellii*, bar = 2 μm. (k) Seta of *Chaetoceros gracilis*, bar = 1 μm. (l) *Stephanopyxis turris*, bar = 10 μm. Reprinted with permission from [1] © 2008, American Chemical Society.

as well as tensile, bending and breaking strength. Because of such advantages, much of the modern material sciences is focused on the development of composite materials, particularly fiber-enhanced materials, and microstructural physical methods of analysis such as high-resolution electron microscopy are now being used in areas of both natural and synthetic composite systems.

The most important biominerals, some in different polymorphic forms, are summarized in Table 15.1 according to their occurrence and main functions. Examples of the morphological complexity of biominerals are depicted in Figure 15.1.

An essential requirement for a biomineral is its low solubility under normal physiological conditions. Table 15.1 lists several solubility product constants K_{sp} (15.1), which, for better comparison, are reduced to the common unit M^{-2} ($= (mol/l)^{-2}$).

$$\text{cation} + \text{anion} \rightleftharpoons \text{solid} + \text{soluble cation/anion aggregates}$$

$$K_{sp} = [\text{cation}] \times [\text{anion}];\ pK_{sp} = -\log K_{sp} \quad (15.1)$$

$[\,]$: molar concentration

Equation (15.1) is valid only for heterogeneous equilibria with unhindered exchange between solid and solution phases.

Table 15.1 The most important biominerals.

Chemical composition	Mineral form (phase)	Solubility exponent $pK_{sp}*^{a}$ at pH 7	Occurrence and function (examples)
calcium carbonate			
$CaCO_3{}^{b}$	calcite	8.42	exoskeletons (e.g. egg
	aragonite	8.22	shells, corals, mollusk
	vaterite	7.6	shells), spicules, gravity
	amorphous	7.4	sensor
calcium phosphates			
$Ca_{10}(OH)_2(PO_4)_6$	hydroxyapatite	$\approx 13.0^{a}$	endoskeletons (vertebrate bones and teeth)
calcium oxalate	fluoroapatite	$\approx 14.0^{a}$	
$CaC_2O_4(\times n\,H_2O)$	whewellite	8.6	calcium storage and
$n = 1{,}2$	weddelite		passive defense of plants, calculi of excretory tracts
metal sulfates			
$CaSO_4 \times 2\,H_2O$	gypsum	4.2	gravity sensors
$SrSO_4$	celestite	6.5	exoskeletons (*Acantharia*)
$BaSO_4$	baryte	10.0	gravity sensors
amorphous silica			
$SiO_2 \times n\,H_2O = SiO_n(OH)_{4-2n}$	amorphous	solubility < 100 mg/l	valves of diatoms and radiolarians, defense functions in plants
iron oxides			
Fe_3O_4	magnetite		magnetic sensors, teeth of chitons
α,γ-Fe(O)OH	goethite, lepidocrocite		teeth of chitons
"$5\,Fe_2O_3 \times 9\,H_2O$"	ferrihydrite (see Figure 8.7)		teeth of chitons, iron storage (see Section 8.4.2)

[a]Solubility products $K_{sp}*$ in pure water have been reduced to the unit M^{-2} for sake of comparison.
[b]Often occurring with variable small amounts of $MgCO_3$ ($pK_{sp}* = 5.2$).

The biominerals can occur as pure or mixed phases, in amorphous or (micro)crystalline form, or as composites with polymeric organic "matrix" materials such as proteins, lipids and polysaccharides. They can be formed intracellularly, at the cell surface (epicellularly) or in the extracellular space. Table 15.1 only partly reflects the possible variety of inorganic components; for example, the more exotic fluoride and sulfide biominerals have not been included. Overall, the following biominerals are predominant:

- the calcium carbonate $CaCO_3$ phases aragonite and calcite, which often contain some Mg^{2+};
- the calcium phosphates, most frequently as hydroxyapatite $Ca_5(PO_4)_3(OH)$;
- amorphous silica SiO_2;
- the iron oxides/hydroxides ferrihydrite ($Fe_2O_3 \times 0.5H_2O$) and magnetite Fe_3O_4.

All other biominerals occur either as minor components or in very few species.

In addition to their already mentioned **mechanical support function**, the biominerals also assume a **storage function**, due to the large amounts usually present. On the other hand, the formation of biominerals can have a **deposition** character (placer gold [2,8]) or even a **detoxification function** (e.g. CdS; see Section 17.3), particularly in microorganisms [2]. This implies the existence of active regulatory and transport systems (see Chapter 8 and Section 14.2) which control (de)mineralization and regeneration. Magnetite, Fe_3O_4, for instance, is a widely distributed mineral in the earth's crust that is formed geologically at high temperatures and pressures. In contrast, "magnetotactic" bacteria are able to synthesize this mineral in biologically useful form under physiological conditions (see Section 15.3.4). In another example, some marine unicellular organisms utilize celestite, $SrSO_4$, for their exoskeletons [4,9]. Sea water is undersaturated with respect to this mineral, so that only active accumulation mechanisms within the organism can guarantee the existence of these solids. After the death of the organism, these celestite structures dissolve rapidly.

Further important functions of the biominerals consist in:

- their use as essential constituents of mechanically robust **instruments** and **weapons** (e.g. of teeth, for the killing and processing of food);
- the formation of **sensor components** (e.g. of specifically heavy crystallites in gravity-sensitive organs or of magnetic microcrystallites in magnetotactic bacteria);
- the passive **mechanical protection** of animals (e.g. the shells of mollusks) and plants (e.g. silica-containing spikes) against predators or climatic effects.

The last example clearly shows that both chemical composition (→ hardness) and morphology contribute to the full functionality of a biomineral.

Due to the usually large amounts of material required, the chemical composition of a biomineral is determined mainly by the availability of the components, which is also true for "geological" mineral formation. The difference, however, is that the material properties in the biological realm are determined by the chemical composition and by the carefully controlled ("enforced") morphology. The organic components in particular can have a "matrix" or "template" function with regard to vectorial (i.e. directly phase-dependent) crystal growth via specific catalysis and control of nucleation. Biological calcite, for example, usually does not form rhomboid crystals, as is the case for the unrestrained system, but is synthesized in functionally much more useful forms. In contrast to the "geological" minerals, biominerals also have to be formed and redissolved ("demineralized") in a much shorter, biologically acceptable time period; therefore, relatively small crystalline domains or single crystals with large surface areas such as spicules are often formed, due to their better accessibility by the solvent. Nevertheless, the half-life of the calcium exchange in an adult human being amounts to several years, due to the relatively slow turnover in the solid-state skeleton. Pathological effects due to deviations from the normal metabolic rates for biominerals are quite common; these include calcium-containing deposits in blood vessels and the formation of calculi in the excretory organs (CaC_2O_4, apatite, $MgNH_4PO_4$), but also insufficient skeletal mineralization in children (rickets) and unwanted demineralization processes such as dental caries or bone resorption (osteoporosis) in the aging organism.

Different degrees of biomineralization can be distinguished, depending on the type and complexity of the already mentioned control mechanisms [3–5]. The most primitive type is biologically induced mineralization, which occurs mainly in bacteria and algae. In these cases, the biominerals are formed by spontaneous crystallization, via supersaturation through the action of ion pumps (see Section 13.4); polycrystalline aggregates with random orientations are then formed in the extracellular space. Gases formed in biological processes (e.g. by bacteria) frequently react with metal ions from the external medium to form such biomineral deposits. As an example, biomineralization results from the decreasing CO_2 content in water as caused by photosynthetically active algae (15.2).

$$Ca^{2+} + 2\,HCO_3^- \rightleftharpoons CaCO_3\,(s) + \boxed{CO_2\,(g)} + H_2O \qquad (15.2)$$

(CO_2 (g) → photosynthetic assimilation)

Equilibrium (15.2) is shifted to the right by photosynthetic CO_2 assimilation and $CaCO_3$ precipitates. As in the process of global iron(II) oxidation through the action of biogenic O_2, such photosynthetic activity has caused geologically and climatically important chemical changes on the earth's surface. Carbonate biominerals may thus be viewed as a long-term CO_2 sink (see also Section 15.3.2). The reversible process (15.2) is commonly known as "water hardness" equilibrium.

Of greater interest with regard to material sciences are the biologically better controlled processes. The resulting "advanced" bioinorganic solid materials are mostly formed as defined composites from inorganic and organic matter. The organic phase may consist of fibrous proteins, lipids or polysaccharides, its properties being relevant for the resulting morphology and the structural integrity of the composite material. Four types of biocomposite can be distinguished, according to the degree of participation of the organic phase:

- Type I (example: iron oxide-containing teeth of chiton mollusks) consists of randomly arranged crystallites, the structures of which are determined by the physicochemical properties of the mineralization zone. The organic matrix confers only mechanical stability.
- Type II (e.g. avian egg shells) shows matrix-supported crystal formation at predetermined sites but little control over the actual crystal growth.
- Type III (silica deposits in plants or the valves of diatoms) features an amorphous mineral phase, the organic matrix directing nucleation and vectorial growth of the inorganic phase.
- Type IV (bones, teeth, shells of mollusks) involves a high degree of control through the organization of the matrix with regard to nucleation as well as oriented (e.g. epitaxial) crystal growth.

15.2 Nucleation and Crystal Growth

Nucleation and crystal growth are processes that occur in supersaturated media and have to be carefully controlled in a directed mineralization process. These requirements can be met in an organism through highly regulated active transport mechanisms and through specific modulation of the surface reactivity. The transport mechanisms may include transmembrane ion flux (see Chapter 13), ion (de)complexation, enzymatically catalyzed gas exchange (CO_2, O_2 or H_2S), local changes in redox potential (e.g. $Fe^{II/III}$) or pH and variations in the ionic strength of the medium. All these factors can create and sustain a supersaturated solution in a biological compartment. Nucleation, on the other hand, is linked to the kinetics of (heterogeneous) surface reactions; processes such as cluster formation, anisotropic crystal growth and phase transformation are also determined by the properties of the surface. In the biological realm, there are a number of surface structures which specifically prevent undesired nucleation; for instance, fish in polar waters can protect themselves from ice formation in their bodily fluids at temperatures slightly below 0°C in this way [10].

The growth of a crystal or of an amorphous solid from the nucleus formed in an activation step (crystal nucleation energy) can proceed directly from the surrounding solution or through the continuous supplementation of the necessary ions or molecules (Figure 15.2a). On the other hand, the diffusion of particles can be drastically altered through significant changes in the viscosity of the medium; for example, through gel formation of a biological matrix – a mechanism which is presumably responsible for the deposition of amorphous silica in plants [11].

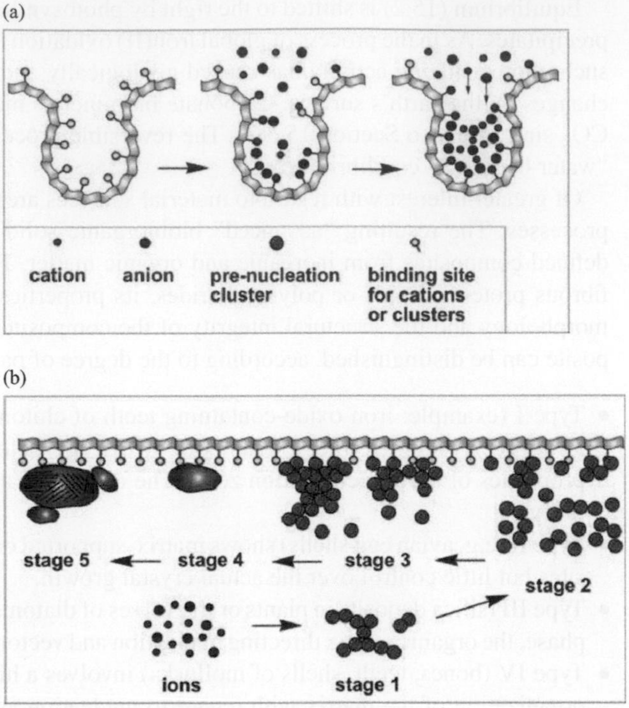

Figure 15.2
(a) Classical model representation of the nucleation and growth limitation of a microcrystallite (according to [12]). (b) "Nonclassical" involvement of prenucleation clusters (according to [3]). In stage 1, aggregates of prenucleation clusters are in equilibrium with ions in solution. The clusters approach a surface with chemical functionality. In stage 2, prenucleation clusters aggregate near the surface, with loose aggregates still in solution. In stage 3, further aggregation causes densification near the surface. In stage 4, nucleation of amorphous spherical particles occurs at the surface only. In stage 5, crystallization occurs in the region of the amorphous particles directed by the surface.

The controlled growth of biominerals can also proceed stepwise, with small activation energies for the individual steps (compare Figure 2.5), through aggregation and phase transitions of solid-state precursors (also called prenucleation clusters; Figure 15.2b) [3–5]. In most instances, these processes are chemically induced; for example, the redox potential seems to play an important role in the transformation of ferrihydrite to magnetite in magnetotactic bacteria or in the teeth of chiton mollusks (Section 15.3.4). A comparable function can be attributed to the pH in many condensation processes.

Two limiting cases can be visualized for the formation of actual microstructures: (1) purely epitactical crystal growth on (organic) matrices and (2) the linking of preformed inorganic crystallites by organic "mortar" material [5]. The importance of the matrix thus lies in the control of nucleation, in the orientation and limitation of crystal growth and in the immobilization of the crystallites.

Dimensions

- *atomic, molecular:* atom sizes vary from 37 pm (H) to 140 pm (U) (or 0.04 to 0.14 nm). Bond lengths vary from 96 pm (0.96 Å) for H−H to >300 pm in metals (or

0.1 to >0.3 nm). Ca–OPO_3 bond lengths in apatite are about 250 pm. The sizes of molecules vary enormously. The diameter of the smallest molecule, H–H, is about 0.1 nm, typical organic dye molecules range at about 2 nm. The diameter of DNA is about 2 nm, its length depending on whether the molecule is stretched ($\approx$ 2 m), or coiled ($\approx$ 1.4 μm). Large rigid organic molecules, so-called dendrimers, can reach diameters of 20–30 Å (2000–3000 pm or 2–3 nm), while larger and nonrigid molecules are folded and should be defined by their hydrodynamic radius (depending on the interaction with the environment, e.g. the solvent).

- mesoscopic: a few nm up to 10 nm.
- nanoscopic: a few nm up to 10 nm (occasionally up to 100 nm).
- macroscopic: "visible" dimensions, beyond 400 nm.
- cellular: the size of a cell nucleus is about 5000 nm = 5 μm (up to 16 μm); sizes of cells are very variable, usually about 1–30 μm. Ova can be much larger (e.g. humans: 110–140 μm; ostrich: 7 cm).

One of the most important and most fascinating aspects of biominerals is the fact that their shapes do not have to coincide at all with the regular crystallographic forms of the inorganic material. Far more significant for the final morphology are the spatial limitations (see Figure 15.2) induced by biopolymers, membranes and vesicles during their formation, even if the completed structures are ultimately assembled outside such confinements. Of course, modifications of crystal growth may also be induced nonbiologically; in fact, rather simple chemicals in the growth medium can influence the shape of the mineral formed. For example, the spindle shape of calcite crystals in gravity sensors can be reproduced by crystallization in the presence of 5 mM malonic acid, and monomolecular layers of stearic acid induce the crystallization of $CaCO_3$ as disk-shaped vaterite crystals rather than as rhombic calcite [13].

15.3 Examples of Biominerals

15.3.1 Calcium Phosphate in the Bones of Vertebrates and the Global P Cycle

The dry support structure of vertebrate long bones consists of elastic fibrous proteins (approximately 30%, mainly collagen) and the inorganic components which are imbedded in "cementing" glycoproteins: calcium phosphate, microcrystallized mainly as hydroxyapatite (approximately 55%), and small amounts of calcium carbonate, silica, magnesium carbonate, other metal ions and citrate (remaining 15%). Collagens are fibrous proteins with a molecular mass of about 300 kDa. Three polypeptide chains are wound together in the fibrils as a superhelix (dimensions approximately 2 × 300 nm) [4,14]. During the development of the bones, the crystallization of the solid calcium phosphate is thought to occur at collagen fibers covered with O-phosphoserine and O-phosphothreonine groups, which provides a template (Figure 15.3). The phosphoproteins are arranged at the collagen fibers in such a way that Ca^{2+} can be bound in regular intervals corresponding to the inorganic crystal structure, presenting a condition for the crystallinity of the inorganic phase [14–16]. The main building blocks of the inorganic component are small crystallites (approximately 2 × 30 × 50 nm) of hydroxyapatite [17].

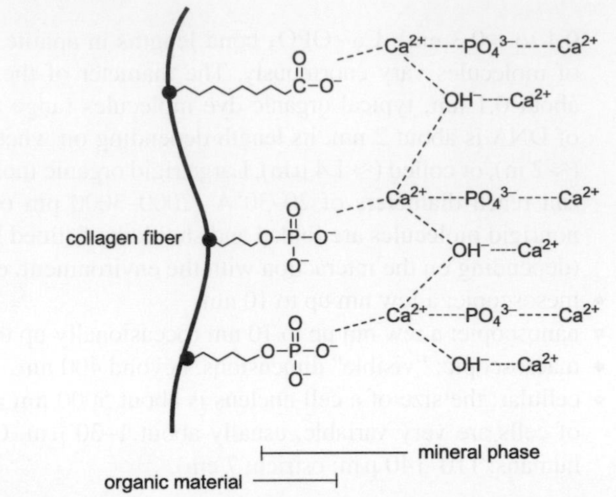

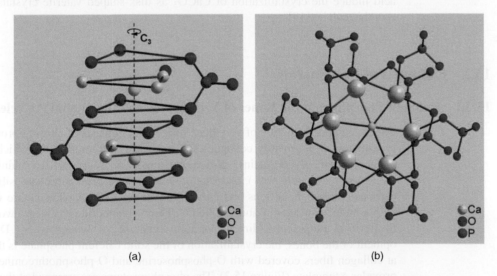

Figure 15.3
Schematic representation of the connection between collagen and hydroxyapatite through carboxylate and phosphate groups (molecular complementarity of the organic–inorganic interface).

Apatite is well known as a nonbiological mineral. It has assumed great importance as a fertilizer, as a basic material for phosphorus chemistry and as an ion exchanger, and it is nature's key player in the global phosphorous cycle (P cycle). The inorganic material of bones, the bone meal, is also used as a fertilizer, while the main organic component can be converted to collagen glue. The complex crystal structure [18] of hydroxyapatite and other apatites is illustrated in Figure 15.4.

The hydroxyl ions are located on threefold axes (hexagonal structure, space group $P6_3/m$). Phosphate oxygen atoms are arranged around these axes in such a way that the packing

Figure 15.4
View of the C_3 symmetric anion channel of hexagonal apatite with half-occupied OH^- positions (blue), from the side (left) and from the top (right). Reprinted with permission from [18a] © 2005, Elsevier.

leaves holes for the Ca^{2+} ions. Each hydroxyl ion is then surrounded by a triangular arrangement of three calcium ions. However, the OH^- groups are randomly distributed 30 pm above or below the plane spanned by the Ca^{2+} ions. This hexagonal structure is particularly susceptible to substitution and defect formation (ion-exchange behavior, H^+ diffusion), which is very advantageous for a high metabolic rate (i.e. the (de)mineralization of the "living" endoskeleton). Hydroxide can be substituted by fluoride or chloride, phosphate by carbonate or sulfate and Ca^{2+} by other divalent ions such as Sr^{2+}. In human tooth enamel, F^- is thus present in amounts between 30 and 3000 ppm, and there is an established but poorly understood dependence of the resistance against dental caries on the fluoride ion content (see Section 16.6). Chloride is present in even larger amounts (0.1–0.5%) than F^-, but there is no established function for the chloride in dental apatite [17,19].

Crystallization of the complex and little soluble hydroxyapatite structures proceeds favorably through the kinetically controlled formation of metastable intermediates within the concept of the Ostwald rule. The *in vitro* transformation of initially precipitated amorphous calcium phosphate to hydroxyapatite (HAP) occurs via octacalcium phosphate (OCP) at higher pH values; at lower values, dicalcium phosphate dihydrate (DCPD) may be an intermediate.

The bones as a supporting scaffold of the vertebrate body can feature different types of integration of organic and inorganic material, which results in a considerable variability of mechanical properties [6,17,20]. The ratio of these components reflects the compromise between hardness (high inorganic content) and elasticity or breaking strength (low inorganic content). The hitherto only partially successful attempts to synthesize suitable (i.e. physiologically tolerated, "biocompatible") and long-term stable bone substitutes for medical purposes ("bioceramics") [21] have substantiated the superiority and complexity of the natural structure. In addition to the microstructural composition, the macroscopic architecture (lightweight construction) determines the mechanical properties of a bone; for instance, the human femur can tolerate loads of up to 1650 kg. This so-called "hierarchical structuring" spans from the nanoscale of the apatite crystals to the macroscale (up to meters) of entire bones (Figure 15.5).

The continuous formation of bone tissue takes place in a peripheral zone consisting of an outer and an inner layer of connective tissue containing osteoblast cells. These osteoblasts are rich in phosphatases and excrete a gelatinous substance, the osteoid; through gradual deposition of inorganic material, the osteoid hardens and the thus walled-in osteoblasts turn into actual bone cells (osteocytes). For the purpose of transformation and to prevent excessive growth of the bone, degradation processes occur simultaneously with the bone formation. Multinucleate giant cells (the osteoclasts) catabolize bones, perhaps using citrate as a chelating agent; the control of osteoclastic activity occurs via the parathyroid hormone (which promotes demineralization) and its antagonist thyreocalcitonin (see Section 14.2). Thus, bone tissue is continuously remodeled by the activity of osteoblasts (formation) and osteoclasts (disintegration), which has two important medical consequences: (1) if bones are not "used" due to long-term bedfast or lack of gravity (space exploration), they will be catabolized by the osteoclasts, resulting in decreased stability; (2) it is possible to design "bioactive" (osteoinductive) materials for medical implants, since they will be used and remodeled by the bone cells. The latter is a very active field in biomineralization research [22].

The Ca^{2+} deposited and thus stored in the skeleton continuously exchanges with dissolved calcium ions. In order for bone growth to occur, a relative excess of Ca^{2+} and corresponding anions such as phosphate and carbonate must be actively created in the bone matrix. This

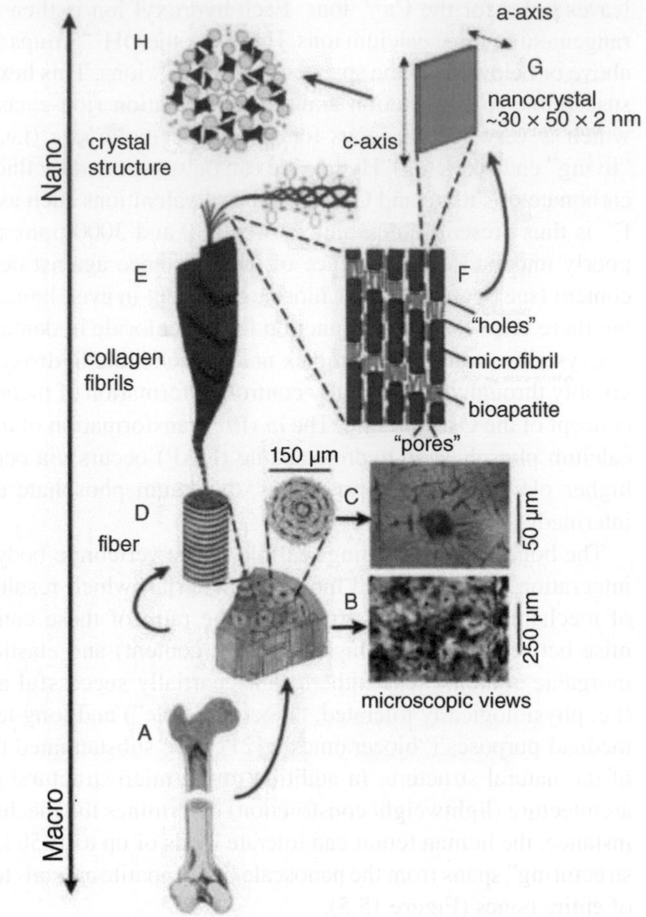

Figure 15.5
Sketches of the hierarchical levels of typical cortical bone (e.g. femur), starting with the entire bone (A), progressing through incrementally finer and finer well-organized structures and culminating with the individual collagen molecule (E) and the unit cell of the mineral (H). Reprinted with permission from [17] © 2008, the Mineralogical Society of America.

is guaranteed through the action of efficient ATP-consuming ion pumps such as the Ca^{2+} ATPases for active calcium transport. Physiologically, carbonate and phosphate exist as hydrogen anions, HCO_3^-, HPO_4^{2-} and $H_2PO_4^-$; when these are incorporated into the bone, protons are released, which are mobile within the bone tissue (hydrogen ion conductivity) and can thus be readily removed from the area of nucleation and mineralization.

Favored by the continuous metabolism of the bone substance and by the substitution-prone structure of hydroxyapatite (Figure 15.4), a chronic heavy-metal poisoning (Sections 17.2 and 17.3) can lead to the substitution of Ca^{2+} by ions such as Cd^{2+} and Pb^{2+} in the bone structure. The results are altered mechanical properties (brittleness) or even very painful bone deformations.

The permanent teeth of higher vertebrates feature tooth enamel as their outer layer. In an adult organism, this enamel no longer contains any living cells; up to 90% can consist of inorganic material, mainly hydroxyapatite [4,22]. It is the enamel which experiences the most extensive changes during tooth development. Initially, it is deposited with a mineral content of only 10–20%, the remaining 80–90% being special matrix proteins and fluids; in later developmental stages, the organic components of enamel are nearly completely substituted by the biomineral. The special feature of tooth enamel as compared to bone material is the significantly larger crystalline domains in the form of long, highly oriented enamel prisms made from hydroxyapatite [4,22]. This "dead" solid-state material is unsurpassed in the biological realm with regards to hardness and durability; however, as is generally (and often painfully) known, regeneration is impossible.

The role of the trace amounts of fluoroapatite, $Ca_{10}F_2(PO_4)_6$, in the prevention of the microbially enhanced decay of tooth enamel (i.e. dental caries) is still a matter of dispute. The mechanisms discussed include surface enamel hardening (anticorrosion effect), enhanced ionic remineralization and the deactivation of acid-forming enzymes through the fluoride deposited on the tooth surface (see also Section 16.6) [23].

The Global P Cycle

Phosphorus is a limiting nutrient for all organisms. Global biological productivity is thus strongly related to the dissolution of phosphate minerals, among which apatite is the most common. The transport of terrestrial phosphate (dissolved or in particles) to the oceans is the major marine source of phosphate (Figure 15.6). In contrast to the natural N (see Section 11.2) and C cycles (see Section 15.3.2), the natural P cycle has

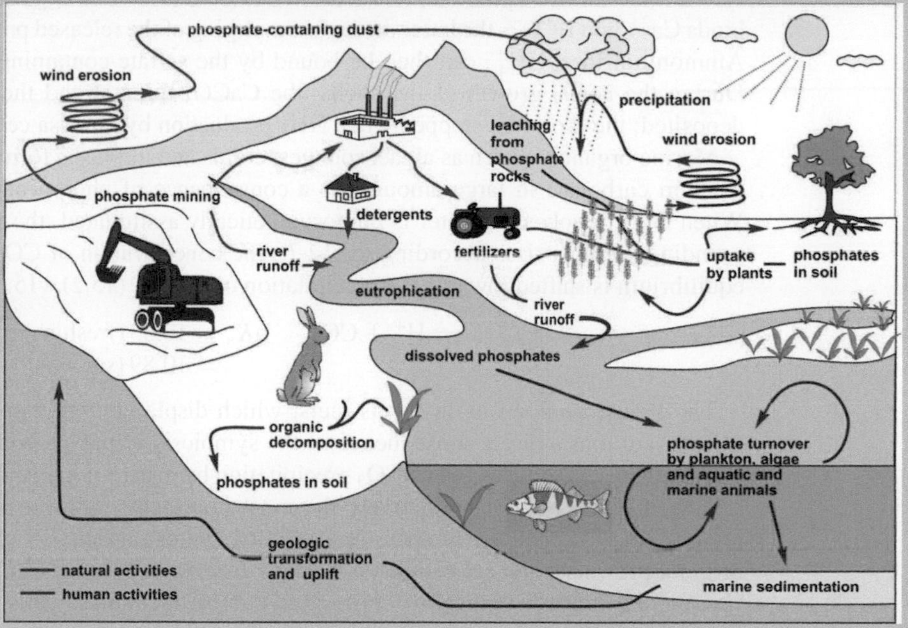

Figure 15.6
Sketch of essential pathways in the natural (arrows in blue) and synthetic (arrows in red) P cycle.

no gaseous atmospheric component (compare N_2, NO_2 and CO_2). Thus, dust particles are the only contributors to the global transport of phosphorous through the atmosphere. Through long-term geological transformations (plate tectonics), sedimented phosphates are retransported to the earth's surface [24–27].

Technically, in the so-called anthrosphere, apatite is mined to produce fertilizers, detergents and phosphoric acid, thus contributing additionally to biological production. While this is highly appreciated in farming, the eutrophication (over-fertilization) of rivers and oceans is a serious ecological problem.

15.3.2 Calcium Carbonate and the Global Inorganic C Cycle

In egg and mollusk shells, the $CaCO_3$ crystals grow in a preformed arrangement of proteins and polysaccharides. The pure organic matrix can be obtained from these shells after dissolution of the $CaCO_3$ through chelating agents such as ethylenediamine tetraacetate (EDTA, Table 2.4) [4]. The matrix is composed of a water-soluble protein and oligosaccharide component and of water-insoluble hydrophobic proteins, also with significant poly-saccharide content. The soluble proteins and sulfate-containing oligosaccharides are strongly acidic and therefore well suited to bind Ca^{2+}. Significant amounts of carboanhydrase are also present (see Section 12.2); this is obviously necessary to produce HCO_3^--supersaturated solutions, allowing for rapid shell growth.

The mineralization process starts at the insoluble matrix in the egg shell membrane, which binds Ca^{2+} and HCO_3^-, the latter through scavenging of the released protons with ammonia. Ammonium ions, NH_4^+, can then be bound by the sulfate-containing soluble matrix [4]. During the actual growth of the shells, the $CaCO_3$ mineral and the soluble protein are deposited; the growth is stopped when NH_3 production by mucosa cells is halted.

Marine organisms such as algae, sponges, corals and mollusks form $MgCO_3$-containing calcium carbonate in large amounts as a consequence of photosynthetic activity (15.2). When CO_2 dissolved in water is photosynthetically assimilated, the pH value of the surrounding medium rises according to (12.7), the concentration of CO_3^{2-} increases and the equilibrium is shifted towards the precipitation of $CaCO_3$ (15.2), (15.3).

$$HCO_3^- \rightleftharpoons H^+ + CO_3^{2-} \quad pK_a = 10.33\,\text{(fresh water)} \\ = 10.89\,\text{(sea water)} \tag{15.3}$$

The limestone deposits in corals reefs, which display a highly species-specific architecture, are thus a direct consequence of the symbiosis of polyps with photosynthetically active algae. CO_2 uptake and $CaCO_3$ precipitation by marine organisms – coccolithophores (plants) usually being quantitatively the most important, followed by foraminifera (animals) – may depend on temperature, salinity, buffer capacity of the medium and the original pH value, and are estimated to range from 0.7 to 1.4 Pg $CaCO_3$–C per year [28]. $CaCO_3$ deposition is particularly favored in compartmentalized cells with small volumes, as diffusion effects are negligible in such cases. Because of the slightly higher solubility of $MgCO_3$, its share is often so small that the typical $CaCO_3$ structures such as calcite and aragonite dominate. However, even very small amounts of "fiber-enhancing" proteins can be sufficient to transform the brittle calcite into, for example, the morphologically and mechanically far more functional spikes of the sea urchin [4,5,13].

The Global C Cycle and the Marine Inorganic C Cycle

While still disputed by a few, it seems clear that anthropogenic activity (burning of fossil fuels) increases the abundance of CO_2 (pCO_2) in the atmosphere far above the levels established for the last 350 000 years (200–300 ppm), with present values at about 400 ppm [28]. In most of the models discussed today, it is assumed that before the appearance of massive anthropogenic emissions (in the last 200 years), the biotic pool (green plants), the pedologic pool (soil) and the oceanic pool provided constant rates of uptake and release of CO_2 into the atmosphere. Furthermore, it is believed that the recent anthropogenic emissions of about 9.1 Petagram (9.1×10^{15} g) carbon per year are essentially balanced by retention in the atmosphere (4.1 Pg/year or 45%, leading to the discussed rise in concentration), additional uptake by the ocean (2.5 Pg/year or 27.5%) and absorption by an unidentified terrestrial sink (2.5 Pg/year or 27.5%) (Figure 15.7) [28].

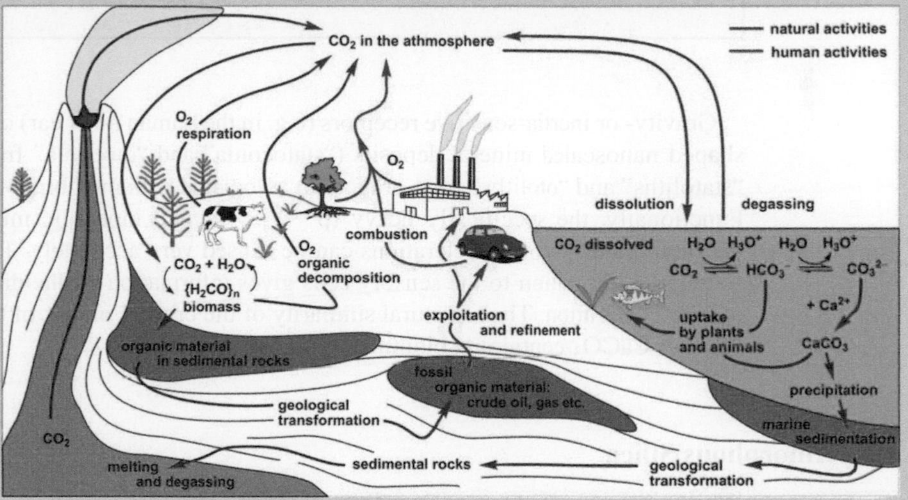

Figure 15.7
Sketch of essential pathways in the natural (arrows in blue) and synthetic (arrows in red) C cycle.

Several consequences may arise from the increased pCO_2 for the marine inorganic C cycle, for which (15.2) describes the biomineralization (deposition) of $CaCO_3$ in shells (from right to left) and the supply of Ca^{2+} and HCO_3^- from the dissolution of sediments. If increasing pCO_2 in the atmosphere goes along with an increasing concentration of CO_2 in sea water, it can be assumed from (15.2) that more $CaCO_3$ will be dissolved and the pH will decrease. Higher H^+ concentration will additionally lead to dissolution of $CaCO_3$ ((15.2) and (15.3)) and formation of HCO_3^-. Thus, in the future carbonate skeletal structures will likely be weaker and more susceptible to erosion and dissolution (especially if they contain increased amounts of Mg^{2+}). With an increase in pCO_2, the future surface oceans might become undersaturated with respect to aragonite [28,29].

In summary, the decreased concentration of carbonate ions may slow the biomineralization of $CaCO_3$ and eventually cause the dissolution of $CaCO_3$ in reefs and coastal sediments. Thus, the huge amounts of CO_2 produced by anthropogenic activities will not by and large be "absorbed" by the oceans (CO_2 sink). However, this pessimistic estimation does not take into account the (possibly beneficial) contributions from organisms in the deep sea. It has been reported that marine fish in shallow water (to 1000 m depth) produce precipitated carbonates (rich in Mg) within their intestines and excrete these at high rates. Thus, marine fish may contribute 3–15% of total oceanic carbonate production and it can be assumed that this production might rise along with increasing pCO_2, thus providing a sink for CO_2 [30].

Even the basic assumptions outlined here are far from being quantitatively backed. The frequently accepted "constant" carbon uptake to the biotic pool through biosynthesis (gross primary production, GPP) of about 120 Pg/year has thus come under discussion. New results show that green plants may take up far more carbon than has been assumed so far, with an estimated GPP of 150–175 Pg/year [31].

Gravity- or inertia-sensitive receptors (e.g. in the human inner ear) often contain spindle-shaped nanoscaled mineral deposits ("statoconia" and "otoconia" from calcite, or larger "statoliths" and "otoliths" from aragonite) associated with membrane-linked sensory cells. Functionally, the specifically heavy ($\rho \approx 2.9$ g/cm^3) inorganic minerals lend mass to the membrane so that accelerations can be sensed very accurately. The movement of the statoconia in relation to the sensory cells gives information on the direction and intensity of the acceleration. The structural similarity of the organic matrix of the statoconia to that of other $CaCO_3$-containing biominerals was demonstrated.

15.3.3 Amorphous Silica

While silicon, in the form of the silicates, comes second relative to all other elements in the earth's crust with regard to quantity, it plays only a marginal role in the biosphere (Table 2.1, Section 16.3). This can be partially attributed to the low solubility of "silicic acid", H_4SiO_4, and its oligomeric condensation product "amorphous silica", $SiO_n(OH)_{4-2n}$ (see Figure 15.8). In water of pH 1–9, this solubility is about 100–140 ppm; in the presence of cations such as calcium, aluminum or iron, the solubility decreases markedly; and in sea water it is only about 5 ppm. In the biosphere, dissolved amorphous silica is resorbed by organisms and then polymerized or connected to other solid structures [1,4,5,20].

As a biomineral, amorphous silica is featured mainly in unicellular organisms [4,5,20], siliceous ("glass") sponges [6,11] and several plants [5,11], where it occurs in the cell membranes of grains, grass and horsetail in the form of "phytolithes" with a passive deterrent function; the brittle tips of the stinging hairs of various nettle plants are also made of amorphous silica.

Accumulated in the deposits of diatomaceous earth, the siliceous remains of diatoms (Figure 15.1) illustrate that this is a rather old group of diverse unicellular organisms, which form two silica-containing box-and-cover structures – the valves – as exoskeletons. After each cell division, two new valves must be formed from amorphous, polymeric

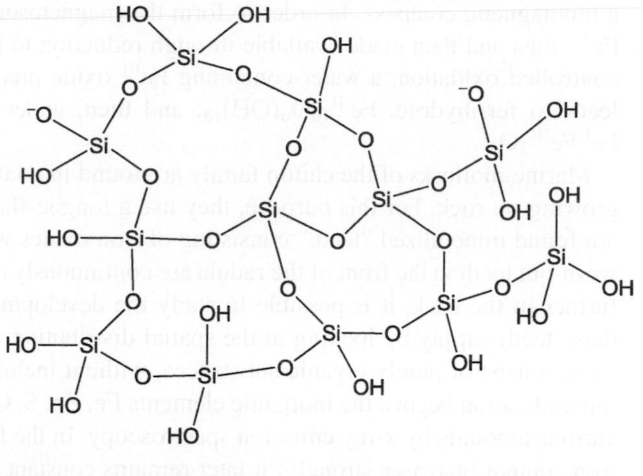

Figure 15.8
Representative linkage of atoms in amorphous silica.

$SiO_n(OH)_{4-2n}$ (Figure 15.8) in order to protect the daughter cells. The morphology of biogenic silica is determined by membrane proteins in which nucleation can take place through condensation (H_2O elimination) between silicic acid or its derivatives and the hydroxyl groups of the organic matrix.

Brittleness and the chemical surface properties of polycondensated silica can be dangerous to human beings, beyond silica's function in nettle plants. The frequent occurrence of oesophageal cancer in certain areas may be connected to dust from silica-containing grains [11,32], in which the fibrous microstructure (passive defense function) of the silica particles can resemble that of the carcinogenic asbestos mineral fibers. The chemical basis of the membrane-dissolving and carcinogenic effect of only slowly degraded mineral or biogenic silicas as fibers or dust particles is still unclear; acid–base effects, irreversible condensation processes and even surface-based oxygen radicals are being discussed [11,32].

15.3.4 Iron Biominerals

The biomineralization of iron oxides is a relatively well-documented field of bioinorganic chemistry (compare Chapter 8), not in the least because of the methods available, in particular Mössbauer spectroscopy. Phosphate- and silica-containing biominerals, on the other hand, have become spectroscopically easily accessible only recently, through solid-state nuclear magnetic resonance (NMR) methods (^{29}Si, ^{31}P). In addition to the iron(III) hydroxide condensation products mentioned in Section 8.4.2, biogenic iron oxide also exists in the form of magnetite (Fe_3O_4). This iron(III,II)-containing mineral has been detected in the fairly ubiquitous magnetotactic bacteria [4,5,33–35], as well as in chitons, mollusks, pigeons, bees, fish and even humans [36]. The magnetotactic bacteria can exhibit an orientation along the earth's magnetic field based on these iron minerals. Under an electron microscope, dark, membrane-coated, iron-containing particles, the magnetosomes, are observed in these bacteria. The particles consist mainly of magnetite, Fe_3O_4, and in some instances of greigite (Fe_3S_4) [5,33–35]; their size (40–120 nm) corresponds to that of the single magnetic domains of Fe_3O_4. The magnetosomes are normally arranged as chains along the direction of bacterial movement, so that this string of particles can function as a biomagnetic compass. In order to form the magnetosomes, iron is taken up as chelated Fe^{3+} ions and then made available through reduction to Fe^{2+} (compare Chapter 8). After controlled oxidation, a water-containing Fe^{III} oxide precipitates (8.20); dehydration first leads to ferrihydrite, $Fe^{III}{}_{10}O_6(OH)_{18}$, and then, under partial reduction, to magnetite, $Fe^{II}Fe^{III}{}_2O_4$.

Marine mollusks of the chiton family are found in tidal zones, where they feed on algae growing on rock. For this purpose, they use a tongue-shaped organ, the radula, on which are found mineralized "teeth" consisting of iron oxides with organic inclusions. Since the worn-out teeth in the front of the radula are continuously replaced by fresh material formed further in the back, it is possible to study the developmental phases of the formation of these teeth simply by looking at the spatial distribution of material on the radula. Young teeth consist of purely organic substances, without inclusion of an inorganic phase. After mineralization begins, the inorganic elements Fe, Zn, S, Ca, Cl, P and K can be detected in various amounts by x-ray emission spectroscopy. In the first phases of mineralization, the iron content increases strongly; it later remains constant at about 10%. Again, metastable ferrihydrite is found as the first unambiguously detected mineral (Figure 8.7). This is then replaced with goethite or lepidocrocite, Fe(O)OH, and finally magnetite. The different minerals occur simultaneously but at spatially different sites of the radula. Adult chiton teeth contain magnetite as the main iron-containing mineral, but calcium and phosphate are also found, presumably for anchoring purposes, in a structure similar to that of apatite [4].

15.3.5 Strontium and Barium Sulfates

The unicellular plankton organism *Acantharia*, which belongs to the radiolarians, features an exoskeleton consisting of exactly 20 $SrSO_4$ single crystals, which can assume very complex shapes. The origin of these forms is connected to the morphology of the cell, since nonbiogenic strontium sulfate (celestite) crystallizes as flat rhombs when left undisturbed. Vesicles in which strontium sulfate crystals grow are formed along radial filaments starting from the center of the cell. Some species show a nearly perfect D_{4h} symmetry of the 20-spicule system [9]. While the general structure is determined by the position of the vesicles, the inherent crystal structure of $SrSO_4$ is responsible for the angles. The high symmetry and simplicity of this special mineral have made the *Acantharia* a preferred object for studies of biomineralization [4,5].

Certain unicellular algae (*Desmidiacea*) contain vesicles with barium sulfate in their crystalline form as baryte ($BaSO_4$, $\rho = 4.5$ g/cm^3). The cells of *Closterium*, for example, are bent in a crescent-like shape, with two baryte crystal containing vesicles at opposing ends of the cell, which suggests a function as inertia sensors. The crystals show the normal rhombic form of baryte; thus, the morphology of the cell does not influence the crystal shape, unlike with the *Acantharia*.

15.4 Biomimetic Materials

Biological materials, as described in this chapter, constitute the major part of the bodies of plants and animals. Although made from a relatively small selection of chemical elements, they provide a huge range of biological functions of great interest to modern materials science. An understanding of the structure and formation of shells, teeth and skeletons, for example, has helped to create medical implants for use in teeth and bone repair, augmentation and regeneration [3–6,21,22].

Another very interesting technological field in which biomimetic materials play an increasingly important role is surface technology. Here, nature provides inspiring examples of superhydrophilic surfaces (wettability), adhesion and light refraction.

Learning from nature and copying its materials is not a new idea. From the myth of Daedalus and Icarus and the sketches of Leonardo da Vinci up to modern lightweight constructions inspired by insect bodies and wings, humans have always tried to copy nature [6,7,37]. However, in recent years, against the background of an understanding of the hierarchical structure of biomaterials, their assembly and their genetic blueprint and of advances in nanoscience (inorganic biological materials are all micro- to nanoscaled) [38], the field of biomimetic materials has become particularly innovative and extremely interesting.

The most advanced and most developed area of biomimetic materials is surely the field of medical implants. In recent years, the materials used to replace or support bones have developed from so-called bio-inert materials (which do not undergo undesired chemical degradation or reaction with surrounding tissue), such as glazed titanium, titanium oxide (TiO_2), aluminium oxide, gold and polytetrafluoroethylene (PTFE) to biocompatible materials (which have a positive impact on surrounding tissue) such as as apatite-covered TiO_2, calcium phosphate and hydroxyl apatite [21,39]. Even more advanced materials (bioactive materials) are beneficial to tissue growth or regeneration (healing) and can be completely incorporated into the concerned tissue. Examples are amorphous ceramics (glasses) of a biorelevant material such as calcium carbonate or phosphate, polyester and synthetic collagens [21].

A number of strong adhesion forces found in nature (e.g. in geckos and mussels) are caused by coordinative interactions between iron (in materials) and catechol-containing biopolymers [40]. This has inspired very interesting applications far beyond the gluing of a gecko's foot to a pane of glass. Such iron–catechol interactions can be used to create very strong or stiff composites [7b,41], catechol polymers which can address cancer cells [42], immobilize DNA [43] or act as hemostatic materials [44], catechol crosslinked self-healing polymers [45] and smart core-shell nanoparticles [46].

Further Reading

A. Sigel, H. Sigel, R. K. O. Sigel (eds.), *Biomineralization: From Nature to Application*, John Wiley & Sons, Chichester, **2008**.

E. Bäuerlein, P. Behrens, M. Epple (eds.), *Handbook of Biomineralization, Vol. 1: Biological Aspects and Structure Formation; Vol. 2: Biomimetic and Bioinspired Chemistry; Vol. 3: Medical and Clinical Aspects*, John Wiley & Sons, Chichester, **2007**.

References

1. M. Hildebrand, *Chem. Rev.* **2008**, *108*, 4855–4874: *Diatoms, biomineralization processes, and genomics.*
2. M. Gadd, *Microbiology* **2010**, *156*, 609–643: *Metals, minerals and microbes: geomicrobiology and bioremediation.*
3. H. Cölfen, *Nat. Mater.* **2010**, *9*, 960–961: *A crystal-clear view.*

4. L. B. Gower, *Chem. Rev.* **2008**, *108*, 4551–4627: *Biomimetic model systems for investigating the amorphous precursor pathway and its role in biomineralization.*
5. F. C. Meldrum, H. Cölfen, *Chem. Rev.* **2008**, *108*, 4332–4432: *Controlling mineral morphologies and structures in biological and synthetic systems.*
6. (a) F. Nudelman, N. A. J. M. Sommerdijk, *Angew. Chem. Int. Ed. Engl.* **2012**, *51*, 6582–6596: *Biomineralization as an inspiration for materials chemistry*; (b) P. Fratzl, *J. R. Soc. Interface* **2007**, *4*, 637–642: *Biomimetic materials research: what can we really learn from nature's structural materials?*
7. (a) Y. Zhao, Z. Xie, H. Gu, C. Zhu, Z. Gu, *Chem. Soc. Rev.* **2012**, *41*, 3297–3317: *Bio-inspired variable structural color materials*; (b) A. R. Studart, *Adv. Mater.* **2012**, *24*, 5024–5044: *Towards high-performance bioinspired composites.*
8. (a) F. Reith, L. Fairbrother, G. Nolze, O. Wilhelmi, P. L. Clode, A. Gregg, J. E. Parsons, S. A. Wakelin, A. Pring, R. Hough, G. Southam, J. Brugger, *Geology* **2010**, *38*, 843–846: *Nanoparticle factories: biofilms hold the key to gold dispersion and nugget formation*; (b) G. Southam, M. F. Lengke, L. Fairbrother, F. Reith, *Elements* **2009**, *5*, 303–307: *The biogeochemistry of gold.*
9. C. C. Perry, J. R. Wilcock, R. J. P. Williams, *Experientia* **1988**, *44*, 638–650: *A physicochemical approach to morphogenesis: the roles of inorganic ions and crystals.*
10. S. N. Patel, S. P. Graether, *Biochem. Cell Biol.* **2010**, *88*, 223–229: *Structures and ice-binding faces of the alanine-rich type I antifreeze proteins.*
11. H. Ehrlich, K. D. Demadis, O. S. Pokrovsky, P. G. Koutsoukos, *Chem. Rev.* **2010**, *110*, 4656–4689: *Modern views on desilicification: biosilica and abiotic silica dissolution in natural and artificial environments.*
12. S. Mann, J. Webb, R. J. P. Williams (eds.), *Biomineralization*, VCH, Weinheim, **1990**.
13. N. A. J. M. Sommerdijk, G. de With, *Chem. Rev.* **2008**, *108*, 4499–4550: *Biomimetic $CaCO_3$ mineralization using designer molecules and interfaces.*
14. A. George, A. Veis, *Chem. Rev.* **2008**, *108*, 4670–4693: *Phosphorylated proteins and control over apatite nucleation, crystal growth, and inhibition.*
15. (a) J. Anwar, D. Zahn, *Angew. Chem. Int. Ed. Engl.* **2011**, *50*, 1996–2013: *Uncovering molecular processes in crystal nucleation and growth by using molecular simulation*; (b) A. Kawska, O. Hochrein, J. Brickmann, R. Kniep, D. Zahn, *Angew. Chem. Int. Ed. Engl.* **2008**, *47*, 4982–4985: *The nucleation mechanism of fluorapatite-collagen composites: Ion association and motif control by collagen proteins.*
16. M. J. Glimcher, *Rev. Min. Geochem.* **2006**, *64*, 223–282: *Bone: nature of the calcium phosphate crystals and cellular, structural, and physical chemical mechanisms in their formation.*
17. J. D. Pasteris, B. Wopenka, E. Valsami-Jones, *Elements* **2008**, *4*, 97–104: *Bone and tooth mineralization: why apatite?*
18. (a) B. Wopenka, J. D. Pasteris, *Mater. Sci. Eng.* **2005**, *C25*, 131–143: *A mineralogical perspective on the apatite in bone*; (b) K. Sudarsanan, R. A. Young, *Acta Cryst.* **1978**, *B34*, 1401–1407: *Structural interactions of F, Cl and OH in apatites.*
19. S. J. Omelon, M. D. Grynpas, *Chem. Rev.* **2008**, *108*, 4694–4715: *Relationships between polyphosphate chemistry, biochemistry and apatite biomineralization.*
20. P. Fratzl, R. Weinkamer, *Prog. Mat. Sci.* **2007**, *52*, 1263–1334: *Nature's hierarchical materials.*
21. M. Vallet-Regí, E. Ruiz-Hernández, *Adv. Mater.* **2011**, *23*, 5177–5218: *Bioceramics: from bone regeneration to cancer nanomedicine.*
22. R. Z. LeGeros, *Chem. Rev.* **2008**, *108*, 4742–4753: *Calcium phosphate-based osteoinductive materials.*
23. A. Wiegand, W. Buchalla, T. Attin, *Dent. Mater.* **2007**, *23*, 343–362: *Review on fluoride-releasing restorative materials – fluoride release and uptake characteristics, antibacterial activity and influence on caries formation.*
24. G. M. Filippelli, *Elements* **2008**, *4*, 89–95: *The global phosphorus cycle: past, present, and future.*
25. Y. Liu, G. Villalba, R. U. Ayres, H. Schroder, *J. Ind. Ecol.* **2008**, *12*, 229–247: *Global phosphorus flows and environmental impacts from a consumption perspective.*
26. A. Paytan, K. McLaughlin, *Chem. Rev.* **2007**, *107*, 563–576: *The oceanic phosphorus cycle.*
27. E. H. Oelkers, E. Valsami-Jones, T. Roncal-Herrero, *Mineral. Mag.* **2008**, *72*, 337–340: *Phosphate mineral reactivity: from global cycles to sustainable development.*
28. (a) F. J. Millero, *Chem. Rev.* **2007**, *107*, 308–341: *The marine inorganic carbon cycle*; (b) R. Lal, *Energy Environ. Sci.*, **2008**, *1*, 86–100: *Sequestration of atmospheric CO_2 in global carbon pools.*

29. J. W. Morse, R. S. Arvidson, A. Lüttge, *Chem. Rev.* **2007**, *107*, 342–381: *Calcium carbonate formation and dissolution.*
30. R. W. Wilson, F. J. Millero, J. R. Taylor, P. J. Walsh, V. Christensen, S. Jennings, M. Grosell, *Science* **2009**, *323*, 359–362: *Contribution of fish to the marine inorganic carbon cycle.*
31. (a) M. Cuntz, *Nature (London)* **2011**, *477*, 547–548: *A dent in carbon's standard*; (b) L. R. Welp, R. F. Keeling, H. A. Meijer, A. F. Bollenbacher, S. C. Piper, K. Yoshimura, R. J. Francey, C. E. Allison, M. Wahlen, *Nature (London)* **2011**, *477*, 579–582: *Interannual variability in the oxygen isotopes of atmospheric CO_2 driven by El Nino.*
32. (a) B. Fubini, I. Fenoglio, *Elements* **2007**, *3*, 407–414: *Toxic potential of mineral dust*; (b) B. Fubini, C. Otero Aréan, *Chem. Soc. Rev.* **1999**, *28*, 373–381: *Chemical aspects of the toxicity of inhaled mineral dusts.*
33. A. Komeili, *FEMS Microbiol. Rev.* **2012**, *36*, 232–255: *Molecular mechanisms of compartmentalization and biomineralization in magnetotactic bacteria.*
34. D. A. Bazylinski, R. B. Frankel, *Nat. Rev. Microbiol.* **2004**, *2*, 217–230: *Magnetosome formation in prokaryotes.*
35. D. Faivre, D. Schüler, *Chem. Rev.* **2008**, *108*, 4875–4898: *Magnetotactic bacteria and magnetosomes.*
36. J. L. Kirschvink, A. Kobayashi-Kirschvink, B. J. Woodford, *Proc. Natl. Acad. Sci. USA* **1992**, *89*, 7683–7687: *Magnetite biomineralization in the human brain.*
37. W. Barthlott, K. Koch, *Beilstein J. Nanotechnol.* **2011**, 2, 135–136: *Biomimetic materials.*
38. S. Hall, *Educ. Chem.* **2011**, *48*, 116–119: *Biomimetics.*
39. C. Ohtsuki, M. Kamitakahara, T. Miyazaki, *J. Roy. Soc. Interface* **2009**, *6*, 349–360: *Bioactive ceramic-based materials with designed reactivity for bone tissue regeneration.*
40. J. J. Wilker, *Angew. Chem. Int. Ed. Engl.* **2010**, *49*, 8076–8078: *The iron-fortified adhesive system of marine mussels.*
41. (a) L. M. Hamming, X. W. Fan, P. B. Messersmith, L. C. Brinson, *Compos. Sci. Technol.* **2008**, *68*, 2042–2048: *Mimicking mussel adhesion to improve interfacial properties in composites*; (b) A. Miserez, T. Schneberk, C. Sun, F. W. Zok, J. H. Waite, *Science* **2008**, *319*, 1816–1819: *The transition from stiff to compliant materials in squid beaks.*
42. J. Su, F. Chen, V. L. Cryns, P. B. Messersmith, *J. Am. Chem. Soc.* **2011**, *133*, 11850–11853: *Catechol polymers for pH-responsive, targeted drug delivery to cancer cells.*
43. H. O. Ham, Z. Liu, K. H. A. Lau, H. Lee, P. B. Messersmith, *Angew. Chem. Int. Ed. Engl.* **2011**, *50*, 732–736: *Facile DNA immobilization on surfaces through a catecholamine polymer.*
44. J. H. Ryu, Y. Lee, W. H. Kong, T. G. Kim, T. G. Park, H. Lee, *Biomacromolecules* **2011**, *12*, 2653–2659: *Catechol-functionalized chitosan/pluronic hydrogels for tissue adhesives and hemostatic materials.*
45. N. Holten-Andersen, M. J. Harrington, H. Birkedal, B. P. Lee, P. B. Messersmith, K. Y. C. Lee, J. H. Waite, *Proc. Nat. Acad. Sci. USA* **2011**, *108*, 2651–2655: *pH-induced metal-ligand cross-links inspired by mussel yield self-healing polymer networks with near-covalent elastic moduli.*
46. K. C. L. Black, Z. Liu, P. B. Messersmith, *Chem. Mater.* **2011**, *23*, 1130–1135: *Catechol redox induced formation of metal core-polymer shell nanoparticles.*

16 Biological Functions of the Nonmetallic Inorganic Elements

16.1 Overview

As members of the group of nonmetals in the periodic table (Figure 1.4), the elements carbon, hydrogen, nitrogen, oxygen, sulfur, phosphorus and chlorine [1] traditionally belong to "conventional" biochemistry. Among the other less metallic elements, the noble gases and the rare, nonradioactive elements germanium, antimony, bismuth and tellurium have not been found to possess any natural–biological significance. For the remaining nonmetallic elements – boron, silicon, arsenic, selenium, fluorine, bromine and iodine – details concerning their biological role are only partially known. However, it should be noted that, historically, the elements iodine and phosphorus were first obtained from biogenic material.

16.2 Boron

Only little is known about the biochemical function of boron, although a boron-containing natural product, the antibiotic "boromycin", was isolated in 1967, and borates seem to be essential for the normal growth and infection resistance of (domestic) plants, particularly of tobacco, beet and cabbage species [2]. Its role in the crosslinking of carbohydrates in cell membranes has been shown [3,4], confirming the well-known tendency of boron to form five-membered ring chelates with deprotonated *cis*-hydroxy functions (16.1). Due to its relatively high concentration in sea water, boron has been found accumulated in some algae and sponges.

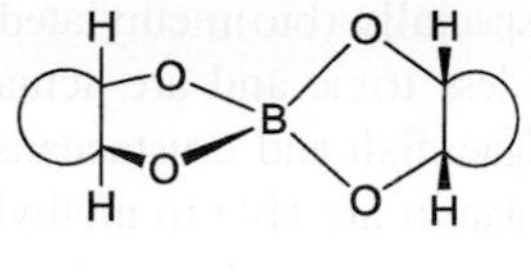

(16.1)

16.3 Silicon

Amorphous silica, $SiO_2 \times n\,H_2O$, was introduced in Section 15.3.3 as a solid constituent of marine organisms, land-living plants and animal skeletons. The pathogenic effect of silica-containing dust and fibers was also mentioned. As is the case with many other trace elements, severe deficiency symptoms of silicon become particularly evident during

Bioinorganic Chemistry: Inorganic Elements in the Chemistry of Life – An Introduction and Guide, Second Edition.
Written and Translated by Wolfgang Kaim, Brigitte Schwederski and Axel Klein.

growth periods [2,5]. Detailed molecular information is not readily available on the binding of silicates (e.g. during bone formation), mainly due to analytical problems; a regulatory function with regard to crosslinking and an association with proteins and polysaccharides (compare Figure 15.3) can be envisaged [6]. There are antagonistic relationships between silicon and boron, and the interaction between silicates and highly charged cations (Ca^{2+}, Fe^{3+}, Al^{3+}) is considered essential [7].

16.4 Arsenic and Trivalent Phosphorus

Although a complete lack of arsenic can clearly lead to disorders in the reproduction and growth of land-living animals, particularly with an impaired metabolism of the limiting amino acid methionine [2], arsenic in the soluble form, $As(OH)_3$, of arsenic(III) oxide, As_2O_3, is a potent poison and a carcinogenic substance for human beings. There is a close relationship between arsenates(V) and phosphates as a result of their relative positions in the periodic system; however, due to the "thiophilicity" (i.e. the affinity of As^{III} in particular for negatively charged S^{-II} ligands), the coordinatively unsaturated arsenic compounds are able to affect enzymes by blocking the redox-active sulfhydryl groups, which are then no longer able to form conformation-determining disulfide bridges. In case of acute arsenic poisoning, as manifested by gastrointestinal cramps, a therapy with thiolate-containing chelate agents such as dimercaprol (2.1) is indicated.

The potential toxicity of arsenic, as documented by real and fictitious poisonings [8], is characterized by a broad variance, including the development of resistance e.g. against arsenic-containing drinking water. Methylation and oxidoreductase enzymes (see Section 11.1) are involved in these defense mechanisms. On major chronic arsenic poisoning with concentrations above 0.010 mg/l, typical skin lesions and later tumor development (skin, lungs, liver, bladder) are observed. In addition to poisonings in regions of Argentina, Chile and Taiwan, the construction of deepwater wells in an attempt to improve the water quality in the delta of the Ganges River (Bangladesh, West Bengal) has resulted in widespread arsenic intoxication, "humanity's biggest mass poisoning" [9]. A complex geochemical redox mechanism involving $Fe^{II/III}$, $As^{III/V}$, $S^{-I/-II}$ and O_2/H_2O is believed to be responsible for mobilizing the poison from the arsenic-containing pyrite ($Fe^{II}S_2$), an underground mineral.

In organic, especially (bio)methylated, form (compare Sections 3.3.2 and 17.3), arsenic compounds are less toxic and are actually quite common, particularly in marine organisms such as algae, fish and crustaceans (including lobster) [10]. S-adenosyl methionine and methylcobalamin are able to methylate trivalent arsenic rapidly to give the following substances:

- $O{=}As(CH_3)_2R$, R = OH, CH_3, CH_2CH_2OH, 5'-deoxyribosyl and derivatives;
- $(CH_3)_3As^+{-}R$, R = 5'-deoxyribosyl and derivatives;
- $(CH_3)_3As^+{-}CH_2COO^-$ ("arsenobetaine").

The low concentration of phosphate in sea water, caused by precipitation with multivalent cations, makes efficient uptake and accumulation mechanisms necessary; however, these may not be able to distinguish between phosphates and the related, only marginally less abundant arsenates (or vanadates; compare Section 11.4). Based on their redox potentials and reactivities, reduction and biomethylation are possible for arsenic(+V,+III) compounds but not for phosphates(+V). This differentiation may be the reason for the existence of peralkylated arsenous oxides, arsonium salts and arsenobetaines as stable, excretable natural products. An alternative strategy for detoxification (see Section 17.1) consists in energy-requiring transport out of the cell through anion-specific ATPases within the membrane. Such an "oxy anion pump" has been identified in arsenic-resistant microorganisms; this effects the excretion of arsenite, antimonite and arsenate, the latter being formed by a molybdenum-containing oxidase (see Table 11.1) [11].

It has been reported that phosphate, which, like arsenate, is carried specifically as mono-hydrogen dianion, HEO_4^{2-}, by a transport protein (E = P, As) [12], can be converted to mutagenic phosphane (PH_3) according to (16.2), under the special reducing conditions of methane formation from CO_2/HCO_3^- (16.3) (see Figure 1.2 and Section 9.5).

$$HPO_4^{2-} + 10\,H^+ + 8\,e^- \rightleftharpoons PH_3 + 4\,H_2O \quad (16.2)$$

$$HCO_3^- + 9\,H^+ + 8\,e^- \rightleftharpoons CH_4 + 3\,H_2O \quad (16.3)$$

Implications of this result include the potential self-ignition of marsh gas (via traces of pyrophoric P_2H_4), the role of volatile biogenic PH_3 in the global phosphorus cycle [13] (see also Figure 15.6) and the possible connection between a phosphate-rich diet and the incidence of intestinal cancer [14].

16.5 Bromine

Bromides have been used as sedatives in the treatment of nervous disorders since the 19th century. A modification of the transmembrane ionic non-equilibrium state is possible (see Chapter 13) when introducing this heavier homologue of chloride, but molecular details of the effect of Br on ion channels or pumps are not yet known [1]. Due to the relative abundance of soluble bromide (and also iodide) in sea water, algae and other marine organisms often contain rather large amounts of organic bromine and iodine compounds [15], which are synthesized with the help of heme- or vanadium-containing haloperoxidases (Section 11.4).

16.6 Fluorine

The cariostatic effect of trace amounts of fluorides has been established for a long time, but a definitive mechanism of action of this trace component cannot yet be determined unambiguously despite numerous studies. A more effective remineralization, a "hardening" of the tooth enamel surface (e.g. through formation of a particularly compact, acid-resistant crystalline layer under participation of fluoroapatite (see Section 15.3.1)) and the inhibition of caries-promoting enzymes by fluoride, which dissolves from solid-state depots, are among the most commonly accepted mechanisms [1,16]. In this context, the usefulness of fluoridation of drinking water (1 ppm) remains widely controversial [17]. In contrast to the practice in many European countries, this caries-preventive measure is common in parts of the USA, the UK, Canada and Australia, despite concerns over possible toxicity (fluorosis). The ability of fluorides to efficiently bind to many (heavy) metal ions in unphysiological high oxidation states and thus render them bioavailable is an additional argument against fluoridation of drinking water, which has led to its substitution in some places by fluorosilicate, SiF_6^{2-} [18].

Fluoride is used in trace amounts in toothpastes and other medical preparations for the prevention of dental caries. Fluoride-containing compounds are preferred, such as monofluorophosphate PO_3F^{2-}, SnF_2, AlF_3 and organoammonium fluorides, which promise efficient binding by the hydroxyapatite surface of teeth. For children, fluoride-containing tablets and gels ensure an effective incorporation into the growing tooth enamel. However, fluoride exhibits a rather small bio-optimal concentration range (Figure 2.2) of only about one order of magnitude, and the switch from beneficial activity to toxic effect is therefore a matter of concern. For instance, organisms which are rich in fluoride, such as the Antarctic krill (a planktonic crustacean), are not suitable for human consumption. Some plants, bacteria and caterpillars accumulate toxic fluoroorganic substances [19,20] such as $FH_2C\text{-}COO^-$ and use it as a deterrent.

Due to the importance of Ca^{2+} for so many biochemical processes (Section 14.2), the marked insolubility of calcium fluoride, CaF_2 ($pK_{sp} = 10$), renders acute (→ tissue necrosis) and chronic fluoride poisoning very serious. Fluorosis manifests itself through discoloration of the teeth, deformation of the skeleton, renal failure and muscle weakness; calcium gluconate is indicated as an antagonist in cases of acute poisoning. In spite of the only moderately acidic character of hydrofluoric acid ($pK_a \approx 3.2$), the action of acidic fluoride solutions on tissues, particularly on skin, leads to poorly healing wounds, which have to be treated immediately. Here again the reason is the deactivation of Ca^{2+}, which is essential for several steps of the coagulation mechanism. Numerous metalloenzymes are inhibited by binding of F^- to the metal center.

16.7 Iodine

The heaviest (and rarest) stable halogen, which was first isolated from the ashes of marine algae by B. Courtois around 1811, was already recognized as an essential element for higher organisms by the middle of the 19th century [1]. This observation was facilitated by the pronounced accumulation of this very characteristic element in the thyroid gland, where it is present in the form of *polyiodinated* small organic compounds: the thyroid hormones thyroxine (tetraiodothyronine) and the even more active triiodothyronine (16.4).

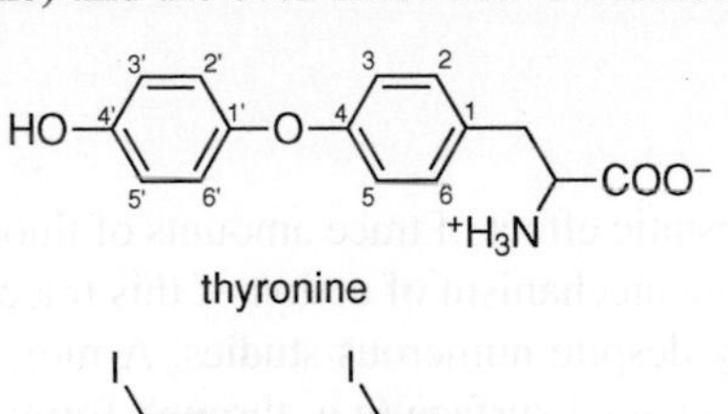

(16.4)

The extreme twofold (physiological *and* intramolecular) enrichment of iodine is quite remarkable, especially considering the central role of thyroid hormones in the control of energy metabolism and associated processes, from the biosynthesis of ATPases to the molting of birds. Familiar physiological disorders occur as a result of reduced thyroid activity (ranging from relatively innocuous conditions such as "feeling cold" or tiredness to the severe deficits of congenital hypothyroidism in newborn infants) or hyperactivity ("feeling hot", restlessness, nervousness). Low thyroid activity due to iodine deficiency may be compensated for by excessive growth of the organ (goiter, struma) with an increased tendency for tumor formation, a disorder that is not uncommon in regions far removed from the sea; it can be counteracted by supplementing iodide preparations. In some countries, the drinking water is therefore iodinated or the common table or cooking salt is made available only in iodinated form (I^-, IO_3^-). Due to the extreme localization of iodine in the human body, tumors of the thyroid can be successfully diagnosed and treated using the radioactive isotopes ^{131}I and ^{123}I (see Section 18.1.4).

What is the particular function of this rare element in the hormones? In its carbon-bound form, iodine is not redox-active at physiological potentials. Furthermore, metal ions are not significantly influenced by the compounds (16.4). Any explanation of the specific role of iodine has to take into account that bound iodide, as the heaviest stable halide, is an unusually large, spherical substituent; the ionic radius of I^- (about 220 ppm) is unmatched by other monovalent elemental ions. In fact, when iodide in the 3,5,3′ positions of thyronine is substituted by approximately spherical methyl groups, or, especially in the 3′ position, by the even larger isopropyl group, $CH(CH_3)_2$, a hormonal activity similar to that of the polyiodinated species is found. Therefore, a receptor structure [21] like that shown in Figure 16.1 is assumed for the thyroid hormones, in which normally *one* large spherical "lock" cavity is preformed for the substituent in the 3′ position of the hormone "key" [22].

A correspondence according to Figure 16.1 may have developed when organisms were still able to utilize the relatively abundant iodide from sea water. The iodination of thyronine,

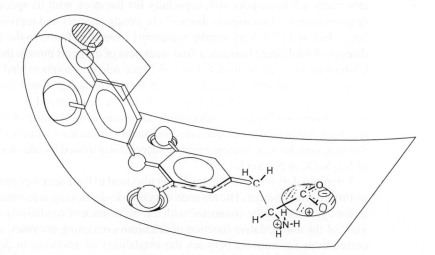

Figure 16.1
Assumed fit of 3,5,3′-triiodothyronine (T_3) hormone into the matching receptor. Reprinted with permission from [22] © 1984, Plenum Press/Springer-Verlag US.

which is derived from the amino acid tyrosine, is an electrophilic substitution at the phenolic (i.e. electron-rich) aromatic nucleus, which, in contrast to alkylations, is obviously still possible under physiological conditions. Heme-containing thyreoperoxidases are mainly responsible for this reaction (see Section 6.3), and the 3,5-diiodotyrosine intermediates are oxidatively coupled. The deiodination of T_4 to give the primarily active T_3 proceeds via a selenium-containing deiodinase [23,24].

16.8 Selenium

As the heavier homologue of sulfur, selenium has recently become the most discussed essential nonmetallic element in biology and medicine [25–28]. Popular scientific and advertising publications make it appear a miracle cure against cancer and aging. Although the chemistry of selenium qualitatively resembles that of sulfur, the selenium analogues of thiols, the selenols, feature a lower redox potential and can thus be more easily oxidized. On the other hand, the tetravalent state of selenous acid is thermodynamically stable, in contrast to the metastable sulfites (see the stability diagrams in Figure 16.2). Finally, the reactivity of selenium compounds is generally higher than that of corresponding sulfur analogues, due to longer bonds from Se to coordinated atoms.

Like fluorine, the trace element selenium features a rather small therapeutic window (compare Figure 2.2); deficiency symptoms (daily dose < 50 mg) and poison effects (>500 mg) lie close together. The high toxicity of selenium compounds has been known for a long time, characteristic symptoms being the loss of hair and the excretion of evil-smelling dimethylselenium, $(CH_3)_2Se$, through the breath or skin. Dimethylselenium is formed via biomethylation and has a garlic-like odor, which is noticeable even in the smallest of concentrations. Disorders of the central nervous system and characteristic degenerations of keratinous (i.e. disulfide-bridge-containing) tissue such as hairs and hooves have been observed for grazing animals which feed on particularly selenium-rich soil in Central Asia and in parts of the western USA, while selenium deficiency symptoms can occur on extremely selenium-poor soil, especially for livestock with its specialized diet; muscular degeneration ("white muscle disease") in young animals and reproductive disorders have been observed [25]. Very similar symptoms have been found in the form of the "Keshan" disease of adolescent humans, a fatal weakness of the heart muscle that occurs in regions of China with very selenium-deficient soil. Since selenium tends to bind soft heavy-metal ions to an even greater extent than does sulfur, there is an antagonism between the selenium and the heavy-metal (e.g. copper) content of soil. A heavy-metal detoxification function has not yet been established for selenoproteins, presumably because of the very small amounts involved; other sulfur-rich proteins, the metallothioneins (Section 17.3), are available for this function. Keshan disease might be successfully treated by administering trace quantities of Na_2SeO_3 or Na_2SeO_4 [25].

Selenium deficiency in mammals can also lead to liver necroses and an increased susceptibility for liver cancer. The absence of peroxide-destroying selenium enzymes (see below) in the eye lens may be connected with the occurrence of oxidatively induced glaucoma. In view of the antioxidative function of selenium-containing enzymes, some epidemiological correlations for humans between the availability of selenium in drinking water and the abundance of breast or colon cancers have been noted. The antimutagenic effect of selenium compounds has also been established in bacterial tests such as the Ames test, even

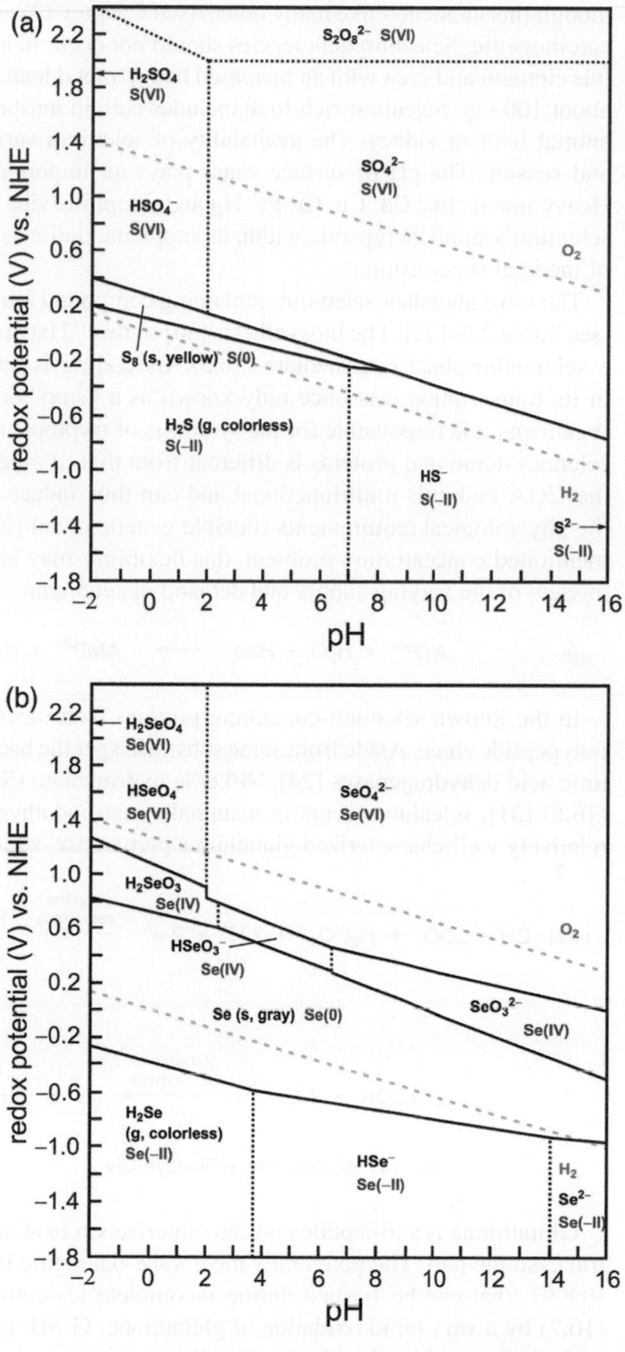

Figure 16.2
Stability (Pourbaix) diagrams for (a) sulfur and (b) selenium (according to [29]).

though this element – like many others (see Chapter 19) – has previously been described as carcinogenic. Selenium deficiencies should not occur in a normal diet, despite the rarity of this element and even with an increased heavy-metal load, as the yearly requirement is only about 100 mg. Selenium-rich food includes certain mushrooms, garlic, asparagus, fish and animal liver or kidney. The availability of selenium varies considerably with geography and season. The pH of surface water plays an important role, as does the soil content. Heavy metals like Cd, Cu, Cr, Pb, Hg and thiophilic zinc are selenium antagonists. Due to selenium's small therapeutic width, its supplementation is not recommended in the absence of medical supervision.

The most abundant selenium-containing compound found in organisms is selenocysteine (see Table 2.5) [28]. The biosynthesis [30] of this "21st proteinogenic amino acid" involves a selenophosphate intermediate (16.5). Its coding is unusual, as the codon UGA used in its transcription was once only known as a "stop" or "nonsense" command. In some organisms it is responsible for the synthesis of tryptophan. Furthermore, the integration of selenocysteine into proteins is different from that of other amino acids. It is assumed that the UGA codon is multifunctional and can thus induce different responses according to the physiological requirements (flexible genetic code) [24]. With regard to the previously mentioned concentration problem, this flexibility may be part of a regulatory mechanism in view of the varying supply and demand of selenium.

$$ATP^{4-} + H_2O + HSe^- \longrightarrow AMP^{2-} + H_2PO_4^- + HPO_2Se^{2-} \qquad (16.5)$$

In the known selenium-containing proteins, selenocysteine appears only once in each polypeptide chain. Aside from some substrate-specific bacterial formate, xanthine and nicotinic acid dehydrogenases [24], Ni/Fe/Se hydrogenase (Section 9.3) and glycine reductase (16.6) [31], selenium occurs in mammals in an iodothyronine deiodinase [23] and in the relatively well-characterized glutathione peroxidase, which catalyzes reaction (16.7).

$$^{+}H_3N{-}CH_2{-}COO^- + H_2PO_4^- + 2\,H^+ + 2\,e^- \xrightarrow{\text{glycine reductase}} {}^{+}NH_4 + CH_3{-}COOPO_3^{2-} + H_2O \qquad (16.6)$$

$$ROOH + 2\,G{-}SH \xrightarrow{\text{glutathione peroxidase}} G{-}S{-}S{-}G + H_2O + ROH \qquad (16.7)$$

G−SH: glutathione, γ-Glu-Cys-Gly

Glutathione is a tripeptide and can dimerize via oxidative disulfide formation at the central cysteine part. The potentially membrane-damaging lipid hydroperoxide intermediates, ROOH, that can be formed during incomplete O_2 conversion are consumed in reaction (16.7) by a very rapid oxidation of glutathione, G-SH, to the disulfide G-S-S-G. Together with the tripeptide glutathione, this Se-containing peroxidase enzyme thus functions as an antioxidant, like other peroxidases, superoxide dismutases (10.5), the vitamins C and E (16.9) and other heteroatom-rich compounds. A list of some major antioxidants [32], arranged according to their specific functions, is given in Table 16.1 (compare (4.6), (5.2) and (16.8)).

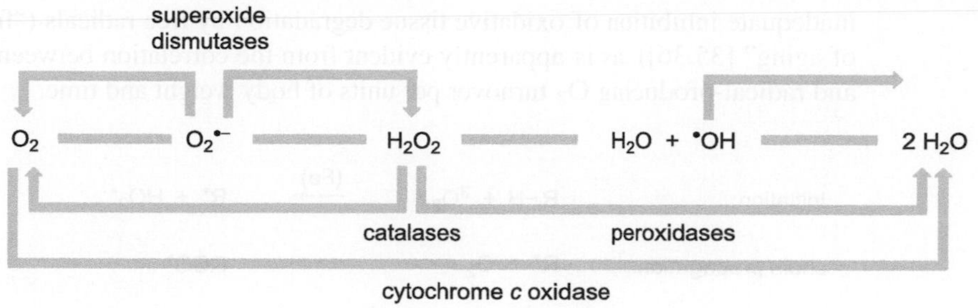

(16.8)

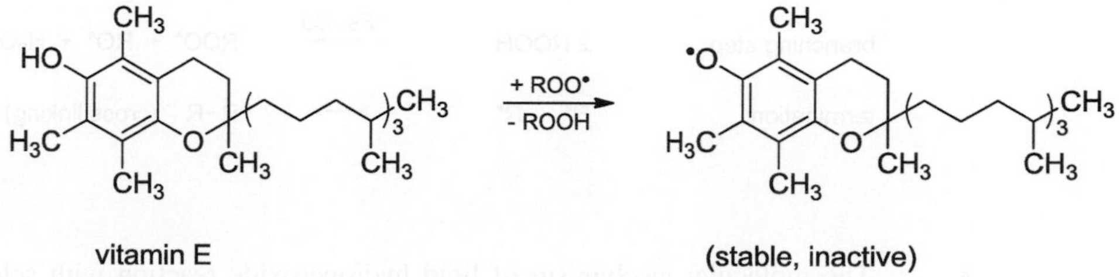

(16.9)

The glutathione peroxidases are tetrameric proteins with subunits of 21 kDa each. These selenoproteins were first isolated from erythrocytes, which are cells with high "oxidative stress". Despite various precautions (Table 16.1), there is a considerable amount of uncontrolled (e.g. iron(II)-catalyzed) oxidation of the long alkyl chains of fatty acids through $^3O_2$ and its transformation products (4.6), (5.2). According to the simplified scheme (16.10), a particularly effective autoxidation of lipid molecules can proceed via a radical chain reaction, with branching made possible by peroxides and transition metals [33,34]. Lipid autoxidation significantly impairs membrane stability and function (e.g. through crosslinking). Since the functioning of compartmentalizing membranes is absolutely essential for all organisms, the corresponding disorders have severe consequences. The inhibition of autoxidative chain reactions has thus become an important requirement for the existence of organisms, particularly in an oxygen-containing atmosphere. According to recent hypotheses, the process of aging and some forms of uncontrolled tumor growth are caused by

Table 16.1 Biological antioxidants.

Antioxidatively active compounds	Section or formula	Targets
peroxidases, catalases (Fe, Mn, V, Se + glutathione)	6.3, 11.4, 16.8	ROOH, HOOH
superoxide dismutases (Ni, Cu/Zn, Fe, Mn)	9.6, 10.5	$O_2^{\bullet-}$, $HO_2^{\bullet}$
vitamin C (ascorbate)	(3.12)	•OH
ceruloplasmin (in plasma)	10	
vitamin E (α-tocopherol)	(16.9)	ROO•
β-carotene (in membranes)		
transferrin	8.4.1	•OH
S-rich compounds, e.g. metallothionein	17.3	•OH
uric acid	(11.11)	•OH
gold-containing compounds (therapeutic)	19.4.1	$^1O_2$ (hypothetical)

inadequate inhibition of oxidative tissue degradation by free radicals ("free radical theory of aging" [35,36]), as is apparently evident from the correlation between average lifespan and radical-producing O_2 turnover per units of body weight and time.

initiation:	$R{-}H + {}^3O_2$	$\xrightarrow{(Fe)}$	$R^\bullet + HO_2^\bullet$
chain propagation:	$R^\bullet + O_2$	$\longrightarrow$	$ROO^\bullet$
	$ROO^\bullet + R{-}H$	$\longrightarrow$	$R^\bullet + ROOH$ (alkyl hydroperoxide)
branching step:	$2\ ROOH$	$\xrightarrow{Fe,\ Cu}$	$ROO^\bullet + RO^\bullet + H_2O$
termination:	$R^\bullet + R^\bullet$	$\longrightarrow$	$R{-}R$ (cross linking)

(16.10)

The molecular mechanism of lipid hydroperoxide reaction with selenium-containing glutathione peroxidase is assumed to involve the ionized (i.e. the selenolate) form, R-Se^-, of selenocysteine at pH 7. The degradation of ROOH proceeds very rapidly (nearly diffusion-controlled), thereby justifying the biological use of the otherwise problematic trace element selenium. A primary oxygen abstraction may lead to the state of selenenic acid, RSeOH. Scheme (16.11) shows a catalysis mechanism which does not include further oxidation to tetravalent selenium, even though that state can be produced by the action of excess peroxide on glutathione peroxidase *in vitro*.

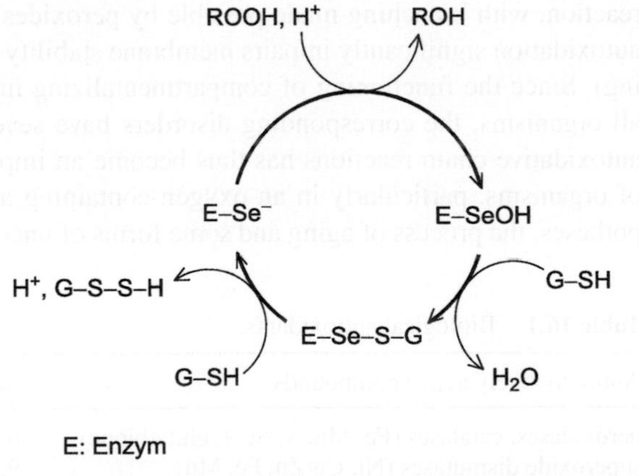

(16.11)

A diselenide formation can be excluded because of the distances in the structurally sufficiently characterized tetrameric protein [37]. As an intermediate of mechanism (16.11), a selenide/sulfide bridge between enzyme and glutathione substrate must be considered; model compounds with Se–S bonds could be synthesized. Glutathione and the protein are fixed relative to each other by charged amino acid residues, resulting in a high specificity of the enzyme for this particular thiol; a specificity with respect to the peroxide substrate does not exist and would not be biologically useful.

References

1. K. L. Kirk: *Biochemistry of the Elemental Halogens and Inorganic Halides*, Plenum Press, New York, **1991**.
2. F. H. Nielsen, *FASEB J.* **1991**, *5*, 2661–2667: *Nutritional requirements for boron, silicon, vanadium, nickel, and arsenic: current knowledge and speculation*.
3. M. A. O'Neill, S. Eberhard, P. Albersheim, A. G. Darvill, *Science* **2001**, *294*, 846–849: *Requirement of borate cross-linking of cell wall rhamnogalacturonan II for arabidopsis growth*.
4. H. Höfte, *Science* **2001**, *294*, 795–797: *A baroque residue in red wine*.
5. C. Exley, *J. Inorg. Biochem.* **1998**, *69*, 139–144: *Silicon in life: a bioinorganic solution to bioorganic essentiality*.
6. R. Tacke, *Angew. Chem. Int. Ed. Engl.* **1999**, *38*, 3015–3018: *Milestones in the biochemistry of silicon: from basic research to biotechnological applications*.
7. C. C. Perry, T. Keeling-Tucker, *J. Inorg. Biochem.* **1998**, *69*, 181–191: *Aspects of the bioinorganic chemistry of silicon in conjunction with the biometals calcium, iron and aluminium*.
8. (a) H. Sun (ed.), *Biological Chemistry of Arsenic, Antimony and Bismuth*, John Wiley & Sons, New York, **2010**; (b) W. R. Cullen, *Is Arsenic an Aphrodisiac?*, RSC Publishing, Cambridge, 2008 .
9. Y. Bhattacharjee, *Science* **2007**, *315*, 1659–1661: *A sluggish response to humanity's biggest mass poisoning*.
10. K. A. Francesconi, J. S. Edmonds, *Adv. Inorg. Chem.* **1997**, *44*, 147–189: *Arsenic and marine organisms*.
11. B. P. Rosen, C.-M. Hsu, C. E. Karkaria, P. Kaur, J. B. Owolabi, L. S. Tisa, *Biochim. Biophys. Acta* **1990**, *1018*, 203–205: *A plasmid-encoded anion-translocation ATPase*.
12. M. Elias, A. Wellner, K. Goldin-Azulay, E. Chabriere, J. A. Vorholt, T. J. Erb, D. S. Tawfik, *Nature (London)* **2012**, *491*, 134–137: *The molecular basis of phosphate discrimination in arsenate-rich environments*.
13. G. M. Filippelli, *Elements* **2008**, *4*, 88–95: *The global phosphorus cycle: past, present, and future*.
14. G. Gassmann, D. Glindemann, *Angew. Chem. Int. Ed. Engl.* **1993**, *105*, 761–763: *Phosphine (PH_3) in the biosphere*.
15. G. W. Gribble, *Chem. Soc. Rev.* **1999**, *28*, 335–346: The diversity of naturally occurring organobromine compounds.
16. C. Dawes, J. M. ten Cate (eds.), *J. Dent. Res.* (special issue) **1990**, *69*, 505–831: *International symposium on fluorides: mechanism of action and recommendations for use*.
17. D. Fagin, *Sci. Am.* **2008**, *January*, 58–65: *Second thoughts about fluoride*.
18. E. T. Urbansky, *Chem. Rev.* **2002**, *102*, 2837–2854: *Fate of fluorosilicate drinking water additives*.
19. D. O'Hagan, C. Schaffrath, S. L. Cobb, J. T. G. Hamilton, C. D. Murphy, *Nature (London)* **2002**, *416*, 279–279: *Biosynthesis of an organofluorine molecule*.
20. J.-P. Bégué, D. Bonnet-Delpon, *Bioorganic and Medicinal Chemistry of Fluorine*, John Wiley & Sons, Hoboken, **2008**.
21. R. L. Wagner, J. W. Apriletti, M. E. McGrath, B. L. West, J. D. Baxter, R. J. Fletterick, *Nature (London)* **1995**, *378*, 690–697: *A structural role for hormone in the thyroid hormone receptor*.
22. N. M. Alexander, *Iodine*, in E. Frieden (ed.), *Biochemistry of the Essential Ultratrace Elements*, Plenum Press, New York, **1984**.
23. M. J. Berry, L. Banu, P. R. Larsen, *Nature (London)* **1991**, *349*, 438–440: *Type I iodothyronine deiodinase is a selenocysteine-containing enzyme*.
24. S. C. Low, M. J. Berry, *Trends Biol. Sci.* **1996**, *21*, 203–208: *Knowing when not to stop: selenocysteine incorporation in eucaryotes*.
25. A. Wendel (ed.): *Selenium in Biology and Medicine*, Springer-Verlag, Berlin, **1989**.
26. G. N. Schrauzer (ed.), *Selenium*, John Wiley & Sons, Chichester, **1990**.
27. G. Mugesh, W.-W. Du Mont, H. Sies, *Chem. Rev.* **2001**, *101*, 2125–180: *Chemistry of biologically important synthetic organoselenium compounds*.
28. T. C. Stadtman, *J. Biol. Chem.* **1991**, *266*, 16257–16260: *Biosynthesis and function of selenocysteine-containing enzymes*.
29. B. Douglas, D. H. McDaniel, J. J. Alexander, *Concepts and Models of Inorganic Chemistry*, 2nd Edition, John Wiley & Sons, New York, **1983**.

30. X.-M. Xu, B. A. Carlson, H. Mix, Y. Zhang, K. Saira, R. S. Glass, M. J. Berry, V. N. Gladyshev, D. L. Hatfield, *PloS Biol.* **2007**, *4*, 96–105: *Biosynthesis of selenocysteine on its tRNA in eukaryotes.*
31. R. A. Arkowitz, R. H. Abeles, *J. Am. Chem. Soc.* **1990**, *112*, 870–872: *Isolation and characterization of a covalent selenocysteine intermediate in the glycine reductase system.*
32. E. D. Harris, *FASEB J.* **1992**, *6*, 2675–2683: *Regulation of antioxidant enzymes.*
33. H. Sies, *Eur. J. Biochem.* **1993**, *215*, 213–219: *Strategies of antioxidant defense.*
34. J. M. C. Gutteridge, B. Halliwell, *Trends Biochem. Sci.* **1990**, *15*, 129–135: *The measurement and mechanism of lipid peroxidation in biological systems.*
35. A. Bachi, I. Dalle-Donne, A. Scaloni, *Chem. Rev.* **2013**, *113*(1), 596–698: *Redox proteomics: chemical principles, methodological approaches and biological/biomedical promises.*
36. T. Finkel, N. J. Holbrook, *Nature (London)* **2000**, *408*, 239–247: *Oxidants, oxidative stress and the biology of ageing.*
37. B. Ren, W. Huang, B. Åkeson, R. Ladenstein, *J. Mol. Biol.* **1997**, *268*, 869–885: *The crystal structure of seleno-glutathione peroxidase from human plasma at 2.9 Å resolution.*

17 The Bioinorganic Chemistry of the Quintessentially Toxic Metals

17.1 Overview

The previous chapters have demonstrated that many "inorganic" elements (in the form of their chemical compounds) are essential for life. However, even such essential substances will be poisonous if the dosage is high enough (the "Paracelsus principle"; compare Figure 2.2). With regard to toxicity [1–9], two other, nonbioessential groups of inorganic elements can be distinguished: those which have not (yet) been recognized as relevant for life due to low abundance or bioavailability (e.g. insolubility at pH 7) and those for which exclusively negative effects have been found to date (Figure 17.1). Among the latter group are the "soft", thiophilic heavy metals mercury, thallium, cadmium and lead.

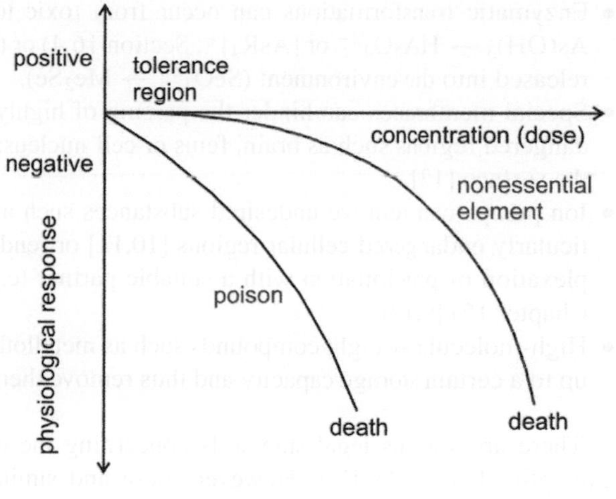

Figure 17.1
Representative dose–effect diagrams for nonessential elements (compare Figure 2.2).

Many potentially toxic heavy metals, such as Sb, Sn, Bi, Zr, the lanthanoides, Th and Ag, are quite insoluble in their normal oxidized chemical forms under physiological conditions, which include an oxidizing atmosphere, pH 7 and a rather high chloride concentration. This is also true for the light metals titanium and aluminum; however, in areas where the soil has locally suffered marked acidification by anthropogenic emissions of sulfur dioxide

Bioinorganic Chemistry: Inorganic Elements in the Chemistry of Life – An Introduction and Guide, Second Edition.
Written and Translated by Wolfgang Kaim, Brigitte Schwederski and Axel Klein.

SO_2 or nitrogen dioxide NO_2 ("acid rain"), the solubility and thus bioavailablity of Al^{3+} is increased. Since Al^{3+} is a strongly polarizing "hard" cation with a high charge/radius ratio, it is able to tightly coordinate and thus deactivate proteins and nucleic acids (Section 17.6). Enzymes can also be blocked by the coordination of other harmful metals, to active sites or to essential sulfhydryl groups for example; furthermore, there is the possibility of substitution of "natural" metal centers in metalloenzymes by "foreign" metals with similar but not identical chemical characteristics: Zn $\leftrightarrow$ Cd; Ca $\leftrightarrow$ Pb, Cd or ^{90}Sr; K $\leftrightarrow$ Tl or $^{134/137}$Cs; Mg $\leftrightarrow$ Be or Al; Fe $\leftrightarrow$ ^{239}Pu (see Chapter 18).

Even before the additional introduction of toxic substances into the environment by human beings, organisms had to cope with such "poisonous" elements as cadmium and mercury, either in the form of a continuous stress situation or during suddenly occurring catastrophic events such as volcanic eruptions. Thus, for *each* element there is a global cycle, which is usually also influenced by the biosphere; the ecological system as a whole exists in an energetic and material flow equilibrium – just as individuals do (Figure 2.1). For some elements, such as lead, mercury, arsenic and chromium, the type of compound (e.g. the ligation or oxidation state) plays a crucial role for the toxicity. For instance, in many but not all cases (arsenic; see Section 16.4) the (bio)alkylated cations R_nM^+ are more toxic than the metals proper or their hydrated cations, M^{m+}. Organisms have developed various, generally energy-consuming detoxification strategies in order to remove unwanted inorganic substances [7–14]:

- Enzymatic transformations can occur from toxic to less toxic states ($Hg^{2+} \rightarrow Hg^0$; $As(OH)_3 \rightarrow HAsO_4{}^{2-}$ or $[AsR_4]^+$; Section 16.4) or to volatile compounds which can be released into the environment ($SeO_3{}^{2-} \rightarrow Me_2Se$).
- Special membranes can hinder the passing of highly charged ions into particularly endangered regions such as brain, fetus or cell nucleus; the toxic species may be bound at the surface [13].
- Ion pumps can remove undesired substances such as $AsO_4{}^{3-}$ (Section 16.4) from particularly endangered cellular regions [10,11] or render them less harmful through complexation or precipitation with a suitable partner (e.g. $Cd^{2+} + S^{2-} \rightarrow CdS\downarrow$; see also Chapter 15) [9,12].
- High-molecular-weight compounds such as metallothionein proteins can bind toxic ions up to a certain storage capacity and thus remove them from circulation [14].

There are various legal standards concerning the toxicity of heavy-metal species, as exemplified in Table 17.1. However, these and similar standards have to be judged as values derived in the course of a *political assessment*; they can sometimes be exceeded even in natural soil and often do not take into consideration the chemical form of the element (speciation). According to Figure 2.3, the reaction of a population to varying concentrations of a substance is not uniform but shows a typical S-shape for the idealized case of only quantitative differences; statistical relations of this kind form the basis for the LD_{50} values (lethal dose for 50% of the population) used in toxicological evaluations. Within a population of more complex organisms, and certainly in comparisons between different species, there may also be qualitative differences in the reaction towards "toxic elements".

Therapeutic detoxification, for example through complexation of toxic metal ions, is particularly indicated in acute cases of poisoning. There are different chelating agents (2.1) for each of the metal ions, according to their characteristics. The preferences of Zn^{2+}, Cd^{2+}

Table 17.1 Typical legal standards for the metal content of drinking water.

Metal (ionic form)	EPA MCL values for drinking water[a]
zinc (Zn^{2+})	5000 ppb
copper (Cu^{2+})	1300 ppb
aluminum (Al^{3+})	200 ppb
chromium (Cr^{3+})	100 ppb
tungsten ($WO_4{}^{2-}$)	50 ppb
nickel (Ni^{2+})	20 ppb
lead (Pb^{2+})	15 ppb
beryllium (Be^{2+})	4 ppb
arsenic (As^{3+})	10 ppb
cadmium (Cd^{2+})	5 ppb
mercury (Hg^{2+})	2 ppb
thallium (Tl^{+})	2 ppb

[a]US Environmental Protection Agency (EPA) allowable maximum contaminant level (MCL) (Drinking Water Contaminants, 2012, http://water.epa.gov/drink/contaminants/index.cfm#Inorganic; last accessed 8 March 2013); ppb = parts per billion, 10^{-9}.

and especially Cu^{2+} for N,S-ligands, of arsenic and mercury for exclusive S-coordination and of Pb^{2+} and Cd^{2+} for mainly S-containing polychelating ligands (compare (17.1)) are quite typical. As the complexes have to be stable in the physiological pH range and should be excretable with urine, most practical chelating ligands for detoxification contain additional hydrophilic groups. However, many of these ligands show only limited selectivity and thus undesired side effects; the application of chelate drugs can thus be considered only as an emergency measure [15].

In the following, the bioinorganic basis of toxic effects of the four "soft" metals lead, cadmium, thallium and mercury and of the "hard" metals aluminum, beryllium, chromium and its heavier homologue tungsten will be discussed in detail; radioactive isotopes will be described in Chapter 18. A brief account will also be given of the toxicity of nanomaterials, which is important in view of their strongly increasing applications in everyday life.

17.2 Lead

Historically, lead is the "oldest" recognized toxic metal, and it is also the one which has been most extensively spread into the environment by humans [13–19]. In contrast to mercury and cadmium, it is not particularly rare in the earth's crust. Its relatively easy mining and processing, its apparent resistance to corrosion and the not easily recognizable toxicity made it a highly valuable metal in ancient civilizations. A logarithmic plot of the world lead production (Figure 17.2) illustrates the correspondence between the mining of the classical precious metals silver and gold and the inevitable coproduction of lead; in the process of "cupellation", heated raw metal is treated with air to remove the "less" noble lead as liquid and rather volatile lead(II) oxide, PbO (m.p. 888 °C) [16].

After exhaustion of the known lead reserves in the final centuries of the Roman Empire, the "discovery" of the "New World", the Americas, with its huge deposits of noble metals, caused a significant increase in world lead production, which was furthered by an increasing demand for the element in printing and weapons technology (lead ammunition). Additional demand arose following the Industrial Revolution, when lead began to be

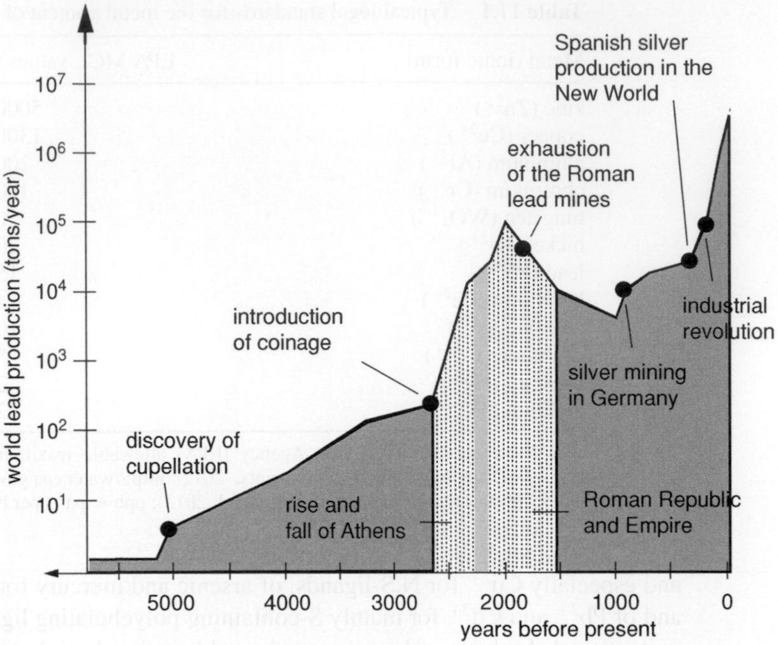

Figure 17.2
Logarithmic representation of the world lead production since the beginning of civilization (according to [16]).

used in accumulators ("batteries"), bearing alloys, solder, optical glasses, pigments ("white lead", $PbCO_3 \times Pb(OH)_2$; "red lead", $Pb_3O_4 = Pb_2[PbO_4]$), radiation protection material and fuel additives. Only since the 1980s has lead production leveled off, but it still ranks sixth behind iron, aluminum, copper, manganese and zinc with regard to worldwide metal production.

Core samples taken from the inland ice of Greenland show that global air pollution by lead compounds has increased more than 100-fold during the last 3000 years. From an estimated concentration of 0.4 ng Pb/m^3 in prehistoric times, the lead content has risen to as much as 500–10 000 ng/m^3 in population centers today [16–18]. In terms of global cycles of the elements, lead has thus experienced the largest anthropogenic increase; its global elemental flux today has risen by a factor of about 20 over the "natural" (i.e. prehistoric) state.

Organometallic species such as the fuel additive tetraethyl lead, $Pb(C_2H_5)_4$ [20], and some inorganic lead compounds are distinguished by a rather high volatility and a tendency for rapid global distribution through the atmosphere; the same holds for cadmium and mercury compounds. However, the major part of lead pollution occurs in the form of oxides, which are bound to very small particles and may thus be removed (e.g. from food) by careful washing.

In aqueous solution, the behavior of lead is quite peculiar: with a redox potential of $E(Pb^{2+}/Pb) = -0.13$ V versus normal hydrogen electrode (NHE), lead is not a noble metal; however, at pH 7 the potential $2\ H^+/H_2$ is -0.42 V, so lead is not attacked by oxygen-free water. Although oxygen from air would be able to oxidize lead even in neutral solution, there are insoluble basic lead carbonates and lead sulfates which form a protective layer at the metal surface; hydrogen carbonate and sulfate stem from natural waters, while carbon dioxide comes from air. In the days of the Roman Empire, a popular method of wine "improvement" was to take low-grade wine containing fruit acids and evaporate it in order to obtain a sweetening grape syrup, *sapa*., A considerable amount of the lead lining of the vessels used was apparently dissolved in the resulting acidic solution (up to 30 mg/l). Corresponding effects can be observed even for Pb^{2+}-containing crystal glass. According to contemporary studies, this method of sweetening may have led to chronic lead poisoning in excess of the present-day limit of 500 mg/day.

Historically, lead poisoning ("saturnism") is connected not only with ancient Rome but also with the intense mining and smelting activity in medieval Central Europe. As a counter measure, the consumption of butter was practiced at that time; more recent attempts at detoxification include the combination of 2,3-dimercapto-1-propanol (BAL) with Ca(EDTA) (Table 2.4) and the development of specific, thiohydroxamate-containing ligands (chelate therapeutics (17.1)) [15,21]. The list of examples of lead poisoning through painters' pigments (e.g. F. Goya, 1746–1828), in the printing business and in highly polluted mining areas (e.g. of Eastern Europe) is long. In the 1840s, a large expedition to find the Northwest Passage was doomed not in the least due to slow lead poisoning of the crew resulting from faulty soldering of the newly developed food cans [22]. In addition to the established neurotoxicity [13,19,23], potential carcinogenic effects [5,6,13,15] led to the classification of lead compounds as "probably carcinogenic to humans" (Group 2A) by the International Agency for Research on Cancer (IARC) and as Group 2 (considered to be carcinogenic to humans based on long-term animal studies) by the German MAK commission.

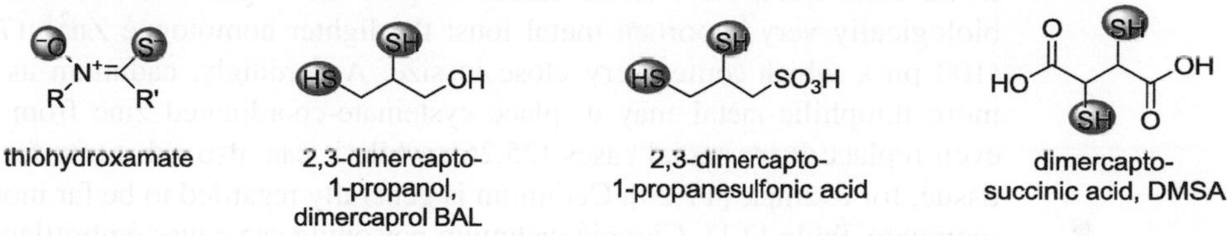

(17.1)

As with many other elements, the physiological retention time of lead and its compounds depends strongly on the location in the body. In blood and soft tissues (liver, kidney), retention times of about 1 month are observed; the lead compounds are excreted with urine, sweat or as components of (sulfide-containing) hair and nails [15,17]. The strong bonding between heavy metals and sulfide-rich keratin in hair and nails (which are very persistent) allows for good forensic proof of heavy-metal poisoning; in some cases, the slow growth of these keratinous tissues may even allow to trace the history of the poisoning. The major part of incorporated lead is stored in the bone tissue however, due to the similar (in)solubility properties of Pb^{2+} and Ca^{2+} compounds [15,17,23]; the residence time there may reach 30 years or more, with possible effects on the development of degenerative processes such as osteoporosis.

The triethyl lead cation, $(C_2H_5)_3Pb^+$, is formed from the earlier fuel additive tetraethyl lead by the dissociation of a carbanion. Like the related organomercury cations, RHg^+ (Section 17.5), and triorganotin cations, R_3Sn^+, these organometallic compounds may cause severe disorders of the central and peripheral nervous system (cramps, paralysis, loss of coordination). The toxicity of organometallic cations results from the permeability of membranes, including the very discriminating blood–brain barrier [13,23,24], for such lipophilic species. Poisoning with inorganic lead compounds on the other hand primarily causes hematological and gastrointestinal symptoms such as colics. Even relatively low concentrations of lead inhibit the zinc-dependent 5-aminolevulinic acid dehydratase (ALAD; Chapter 12 (12.18)) [15,19,23], which catalyzes an essential step in porphyrin and thus in heme biosynthesis, namely the pyrrole ring formation to porphobilinogen (Section 12.4). The occurrence of unreacted 5-aminolevulinic acid (ALA) in the urine is thus a very characteristic indication of lead poisoning [15]. Since the incorporation of iron into the porphyrin ligand with the help of -SH group-containing heme synthetase (ferrochelatase) [15,19] is also inhibited by Pb^{2+}, there are also typical porphyrin precursors, the "protoporphyrins", which can be detected in the urine after chronic lead poisoning. Accordingly, a form of anemia results before long-term neurotoxic symptoms – especially a mental retardation in children – become evident [6,13,15,17,19,20,23]. Further toxic effects of lead poisoning include reproductive disorders such as sterility and miscarriages. The replacement of organic lead compounds as anti-knock additives in fuel has meanwhile led to a significant decrease in lead pollution, especially in North America and Europe [16–18,20]. Lead exhibits a variable isotopic composition, depending on the source (lead as end product of *two* radioactive decay series; see (18.1)), and thus allows for detailed analyses of origin and distribution [16–18].

17.3 Cadmium

In its ionic form, Cd^{2+} (ionic radius 95 pm) shows great chemical similarity with two biologically very important metal ions: the lighter homologue Zn^{2+} (74 pm) and Ca^{2+} (100 pm), which comes very close in size. Accordingly, cadmium as the "softer" and more thiophilic metal may displace cysteinate-coordinated zinc from its enzymes and even replace it in special cases [25,26], while it can also substitute for calcium in bone tissue, for example [17,26]. Cadmium is generally regarded to be far more toxic than lead (compare Table 17.1). Chronic cadmium poisoning can cause embrittlement of bones and extremely painful deformations of the skeleton; such symptoms have been observed on a large scale as "itai-itai disease" in Japan following the use of cadmium-containing waters to irrigate rice fields during the 1950s. Older women with a disposition for osteoporosis were particularly affected. Although skeletal damage in Cd poisoning is primarily caused indirectly via impairment of renal functions, the biomineral part of the skeletons of itai-itai patients eventually contained up to 1% cadmium, a calcium-deficient diet having aggravated the toxic demineralization effect. Once cadmium is stored in the skeleton, its biological retention time is on the order of decades [17,26].

Cadmium is today used as a component of the Ni/Cd "batteries", in color pigments (CdS or CdSe), in stabilizers for plastics and for metal surface treatment; furthermore, cadmium is a side product of zinc smelting [26]. Cadmium can be incorporated through food, with the liver and kidney of slaughter animals and wild mushrooms being particularly rich in this element. For as yet unknown reasons, cadmium absorption is especially effective via tobacco smoke, the Cd content in the blood of smokers being much higher than that of nonsmokers [17,27]. High concentrations of cadmium compounds have proven to be carcinogenic in animal experiments and the IARC has recently classified cadmium as a human carcinogen (Group 1) [5,6,12,15,26,27].

Cadmium is also concentrated in the liver and kidney in the human body, where, as in many other organisms, small (6 kDa) and unusually cysteine-rich ($\leq$30%) proteins

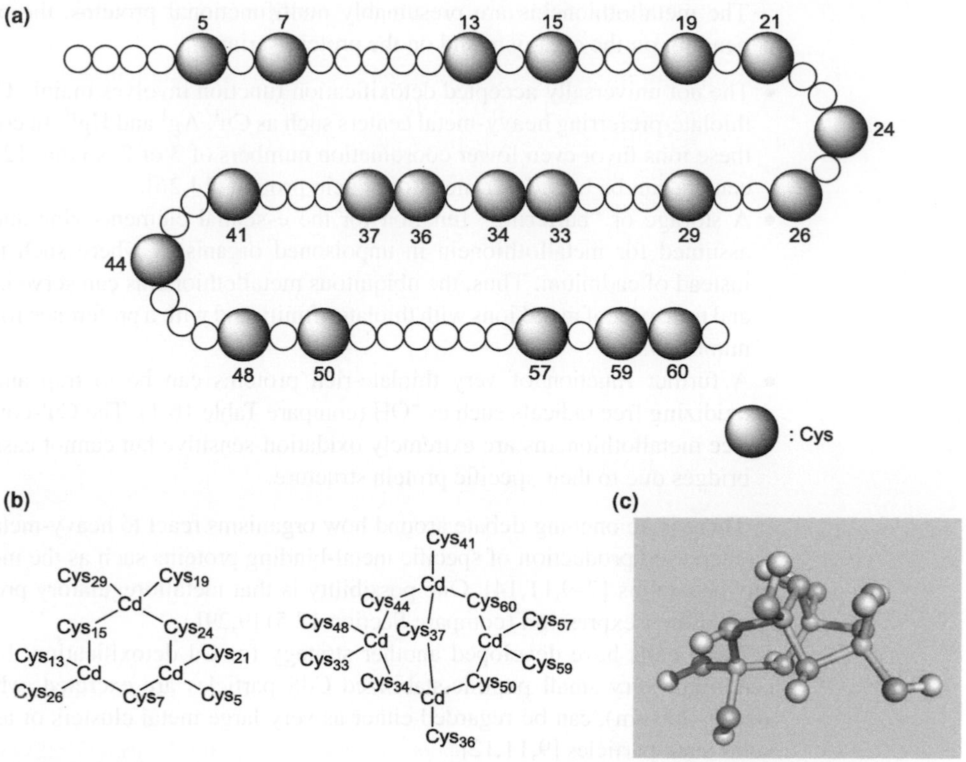

Figure 17.3
(a) Typical cysteine arrangement in the amino acid sequence of mammalian metallothionein. (b) Binding of twelve terminal and eight bridging cysteinate residues to a total of seven Cd centers. (c) Resulting active site (Cd_4S_{11} cluster) from rat metallothionein (PDB code 4MT2) [28].

preferentially bind the soft heavy-metal ion Cd^{2+} in addition to Zn^{2+} and Cu^{+}. These "metallothionein" proteins feature a highly conserved sequence homology, particularly with regard to the cysteine residues (Figure 17.3). It can thus be assumed that they were optimized early in their evolution and that they possess an essential function [13,14,26,28,29] (see also Section 12.7). In addition to the liver and kidneys, the small intestines, the pancreas and the testicles of mammals contain larger amounts of these proteins. Structural studies of the mammalian form have been carried out using x-ray diffraction, extended x-ray absorption fine structure (EXAFS), ultraviolet-visible (UV-vis), circular dichroism (CD) and ^{113}Cd-nuclear magnetic resonance (NMR) spectroscopy (^{113}Cd, $I = 1/2$, 12.3% natural abundance) [25,27–29]. It has been found that a total of seven metal centers, each of them four-coordinate, can be bound in two clusters by nine and eleven cysteinate residues, respectively (Figure 17.3).

Peptide synthesis of the cysteine-rich domains of metallothioneins has demonstrated that small synthetic oligopeptides can bind metals like Cd^{2+} and can thus indicate approaches towards creating modified metallothioneins. Crustaceans and microorganisms such as yeast contain metallothioneins with slightly different cluster compositions and metal selectivities. In some plants and microorganisms there are cysteine-rich peptides (the phytochelatins H–[γ-Glu–Cys]$_n$–Gly–OH (n = 2–11)) which also participate in heavy-metal (Cu, Zn or Cd) homeostasis without forming metallothionein-type clusters [7–9,11].

The metallothioneins are presumably multifunctional proteins, the principal function depending on the organism and on the protein variant:

- The not universally accepted detoxification function involves mainly Cd^{II} but also other thiolate-preferring heavy-metal centers such as Cu^{I}, Ag^{I} and Hg^{II}. In comparison to Cd^{II}, these ions favor even lower coordination numbers of 3 or 2, so that 12 or even 18 metal centers can be bound per metallothionein protein [14,26].
- A storage or "buffering" function for the essential elements zinc and copper may be assumed for metallothionein in unpoisoned organisms where such metals are bound instead of cadmium. Thus, the ubiquitous metallothioneins can serve in the homeostasis and transport of metal ions with thiolate affinity and with a preference for the coordination number 4.
- A further function of very thiolate-rich proteins can be to trap and thus deactivate oxidizing free radicals such as $^{\bullet}OH$ (compare Table 16.1). The Cu^{I}-containing or metal-free metallothioneins are extremely oxidation-sensitive but cannot easily form disulfide bridges due to their specific protein structure.

There is an ongoing debate around how organisms react to heavy-metal poisoning with an increased production of specific metal-binding proteins such as the metallothioneins or phytochelatins [7–9,11,14]. One possibility is that metalloregulatory proteins trigger and control their expression (compare Section 17.5) [9,29].

Yeast cells have developed another strategy for Cd detoxification. In the presence of cadmium, very small peptide-stabilized CdS particles are excreted, which, due to their size ($\sim$200 nm), can be regarded either as very large metal clusters or as semiconducting solid-state particles [9,11,12].

In contrast to many other heavy metals and toxic elements, cadmium *does not* easily pass into the central nervous system or the fetus, because, in its ionized form and under physiological conditions, it cannot be bioalkylated to form stable, membrane-penetrating organometallic compounds such as R_2Cd or RCd^+ [30,31]. The reactivity of such species in aqueous media is related to the electropositive character of Cd, as reflected by the rather low oxidation potential ($E = -0.40$ V) of the element; in contrast, compounds of the elements Hg (+0.85 V), Se (+0.20 V), Te (−0.02 V), Pb (−0.13 V), Sn (−0.15 V), As (−0.22 V) and possibly Tl (−0.34 V) can be bioalkylated (potentials are given for the oxidation to the lowest stable oxidation state of the respective elements at pH 7). As discussed in Chapter 3, the cobalamin-catalyzed biomethylation of the less electropositive (noble) elements proceeds via a carbanion mechanism, and that of the more oxidizable (less noble) elements presumably via radicals.

17.4 Thallium

Like cadmium, thallium is a thiophilic heavy metal; it forms a stable, monovalent cation, Tl^+, under physiological conditions. As a "substitute" for the similar K^+ ion, Tl^+ can penetrate membranes via potassium ion channels and pumps to reach sensitive areas and cause disorders. There are three reasons for this particular behavior: Tl^{3+} (the other stable oxidation state) is easily reduced to Tl^+ ($E_0 = +1.28$), the ionic radii of Tl^+ (159 pm) and K^+ (151 pm) (each for coordination number 8) are quite similar and Tl^+ shows a high affinity for inorganic S^{-II} ligands, like Ag^+, but with a slightly better solubility of

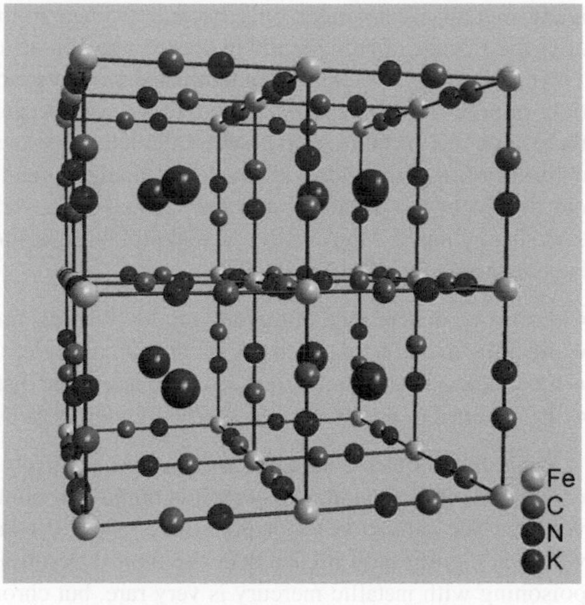

Figure 17.4
Representation of the cubic lattice of Prussian blue. The positions highlighted in purple can be occupied by monocations up to a size of 182 pm; the initially present K^+ ions (151 pm) can hence be replaced by Tl^+ (159 pm) or Cs^+(174 pm).

the chloride. Long-term symptoms of severe thallium poisoning are paralysis and impaired sensory perception (→ neurotoxicity); the first, typical signs, however, are gastroenteritis and the loss of hair.

According to the similarity of K^+ and Tl^+, the latter can reach nearly all regions of an organism, where it can then be trapped and slowly released as a "depot poison". Many K^+-associated enzymes, such as amino acid and enzyme synthetases, the Na^+/K^+-ATPase (Section 13.4) and most of the pyruvate metabolism, which supplies energy to the nerve cells (Section 14.1), are inhibited by thallium [8,32,33]. Suitable countermeasures following thallium poisoning include dialysis and high supplementation with K^+ in combination with administration of large quantities of mixed-valent iron cyanide complexes such as Prussian blue, simplified as $KFe^{III}[Fe^{II}(CN)_6]$ (Figure 17.4), in colloidal form; a chelate therapy is explicitly contraindicated [32,33]. Due to their "open" structure, the Prussian blue particles are able to function as nontoxic cation exchangers; that is, they do not release significant amounts of cyanide but can bind large monocations such as Tl^+ and Cs^+ instead of K^+ and thus remove them from the organism [34].

17.5 Mercury

Like lead, mercury is an ancient environmental contaminant [17]; Pliny the Elder (23–79 CE) described the high mortality rate of workers in contemporary mercury mines. While the metal has been used as a cure in elemental or oxidic form for the treatment of syphilis, in disinfecting preparations and as a component of fungicides, the extreme toxicity of many mercury compounds only became evident following incidents of mass poisoning after the release of Hg-containing wastes into Minamata Bay in Japan between 1948 and 1960 and after grains seeds contaminated with organomercury fungicides were used for flour production in Iraq in 1972. Both these cases became famous worldwide. Accordingly, the use of this metal (e.g. in chloralkali electrolysis cells for the industrial production of chlorine and sodium hydroxide, in dental amalgams and in batteries) is largely being phased out. In recent last decades, mercury has become a very thoroughly studied example of a toxic heavy metal. Many of the typical problems arising from heavy-metal pollution have been examined in detail, such as:

- the strong dependence of toxicity on the chemical form of the element;
- the influence of human activity on the global cycle;
- the kinetics of metabolic transformations and the distribution in the human body;
- the genetics of the latent "resistance" of microorganisms against heavy-metal poisoning.

Regarding the elemental form, mercury is a relatively "noble" heavy metal; that is, it does not corrode in normal atmosphere, it is liquid at room temperature and it is comparatively volatile. The saturation vapor pressure is about 0.1 Pa, corresponding to 18 mg Hg/m^3, which is significantly higher than the typical permissible limit of 0.1 mg/m^3 air. Acute poisoning with metallic mercury is very rare, but chronic inhalation of Hg vapor through the extensive use of this metal in physics and chemistry has been described in detail, often by the affected scientists themselves. The typical neurological symptoms of mercury poisoning gradually appear when working in insufficiently ventilated rooms where spilled metallic mercury had been in contact with air for a long period of time. The symptoms have been described vividly by Alfred Stock (1876–1946), who developed the high-vacuum technique with Hg metal valves [35] for the preparation of very sensitive gases such as the lower boranes. In addition to externally visible signs like a dark lining of the teeth (which also occurs after lead poisoning), diminished blood circulation in the extremities is followed by impaired concentration and coordination, tremors ("mad hatter" syndrome or "hatter's shake", named for the hatters who used to use $Hg(NO_3)_2$ for felt treatment; compare Lewis Carroll's *Alice's Adventures in Wonderland*), memory loss and, at a very high level of exposure, loss of hearing, blindness and death. After recognizing the reasons for his severe illness, Stock installed a ventilation system in his laboratories and, following a period of recuperation, was able to continue some of his scientific work [35]. Despite its "noble" character, metallic mercury dissolves to a small extent in oxygen-containing water and blood. In amalgamated (alloyed) form, particularly as a component of Ag-, Sn-, Zn- and sometimes Cu-containing tooth fillings, mercury is far less volatile and water soluble; the potential long-term health hazards of additional Hg incorporation in amalgam fillings continue to be a subject of controversy [13,17].

In its usual ionic form as Hg^{2+} ion, mercury is immediately toxic, since this species is easily soluble at pH 7 and does not form insoluble compounds with those anions which are abundant in bodily fluids. An especially toxic form is represented by the organometallic cations, RHg^+, and particularly by the methylmercury cation ($H_3C{-}Hg^+$), which can be formed in the organism from Hg^{2+} through biomethylation (see (17.2) and Section 3.3.2) [2,9,13,15,30,31,36,37]. This specific toxicity is connected to the ambivalent lipophilic/hydrophilic character of such water-soluble organometallic cations, which allows them to penetrate the very tightly constructed membrane partitions between the central nervous system or the growing fetus and the rest of the organism. The placenta membrane and the blood–brain barrier [13,24] restrict the access of many substances unless, like the social drugs ethanol and nicotine or the R_nM^+ ions, they are relatively small, feature a hydrophilic *and* lipophilic molecular region and can thus move freely in the aqueous plasma and through the nonpolar membrane barriers. In 1997 the fatal poisoning of the US scientist Karen Wetterhahn shocked the chemical community. Wetterhahn had carried out experiments with dimethyl mercury, Me_2Hg, which was used as a standard probe for ^{199}Hg NMR spectroscopy. It turned out that a few spilled drops of Me_2Hg (a liquid) rapidly penetrated Wetterhahn's protective gloves, causing severe neurotoxic indications many weeks later, leading to coma and finally to death [38].

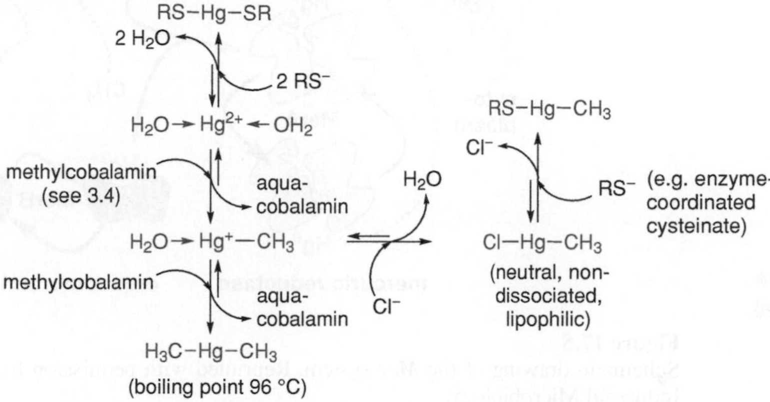

(17.2)

As a noble metal, and like silver and gold, mercury forms relatively covalent bonds with chloride. Following an incorporation of organomercury compounds (e.g. with food), the little-dissociated molecules RHgCl can be formed in the stomach with its high content of hydrochloric acid; due to their lipophilicity, these molecules can then be efficiently resorbed. Mercury compounds are distinguished by a preference for very low coordination numbers of the metal center, the coordination number 2 with a linear arrangement being highly favored. As a consequence, mercury is not very susceptible to simple chelate coordination; even the detoxification function of metallothionein is thus not as efficient for this very thiophilic heavy-metal ion Hg^{2+} as it is for Cd^{2+}, for example, with its strongly preferred four-coordination.

As the equilibria in (17.2) illustrate, the toxicity of mercury compounds is based on the strong affinity of this metal for the deprotonated forms of thiol ligands such as cysteine; therefore, thiols, RSH, with sulfhydryl groups, -SH, are also called mercaptans (*mer*curium *captans*). Even in the absence of chelate effects, the resulting complex formation constants between 10^{16} and 10^{22} are very high; on the other hand, oxygen donor ligands are only weakly bound by Hg^{II}. Mercury compounds affect all protein structures and especially enzymes in which cysteines significantly influence the activity as metal-coordinating, redox-active or conformation-determining groups (via disulfide bridges). It is characteristic, however, that the RHg^+–thiolate bond is kinetically labile despite an extremely favorable equilibrium situation, so that exchange, recomplexation and eventually excretion are possible when better suited detoxification ligands such as dimercaprol or dimercaptosuccinic acid (17.1) are being offered. The kinetic lability of two-coordinate mercury complexes is a consequence of easily attainable transition states for an associative substitution (compare (14.7)) with three- or four-coordinate metal. However, this kinetic lability also implies that toxic mercury compounds are rapidly distributed in the body to those parts which feature the highest affinity towards this heavy metal; in fact, mercury is mainly found in the liver,

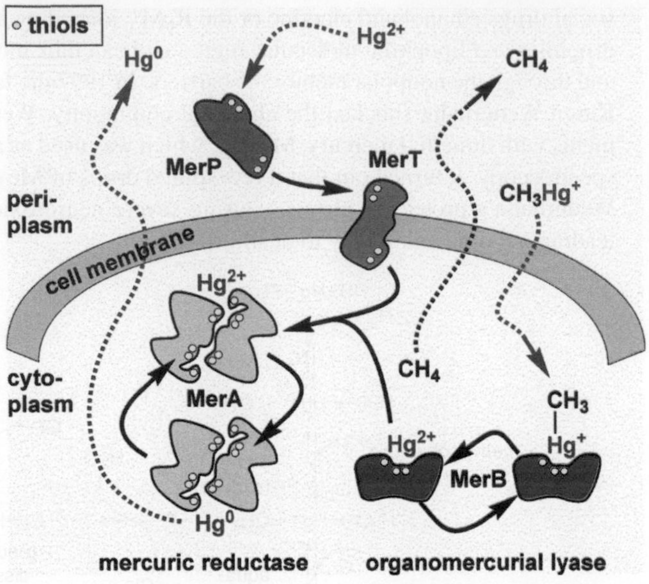

Figure 17.5
Schematic drawing of the *Mer* system. Reprinted with permission from [11] © 2005, Society for Industrial Microbiology.

kidney and nervous system, especially the brain. Furthermore, alkylmercury species can form relatively inert bonds to carbon centers of nucleobases, which may account for the mutagenic effect of organomercurials.

The relatively low immediate toxicity of elemental mercury and its volatility have enabled bacteria to develop a resistance mechanism towards soluble Hg compounds [11,39]. This mechanism has been intensely studied in view of the increasing heavy-metal pollution in the environment; corresponding organisms have thus been isolated from Hg-polluted soil and waters (especially harbor waters). The detoxification system is effective against "heavy-metal stress" and is quite well understood with respect to genetic control; the synthesis of the required proteins (*MerA*, *MerB*, *MerP*, *MerT*) is triggered and controlled by a metal-selective and gene-regulating sensor protein "*MerR*" (Figure 17.5).

MerR is a metalloregulatory protein [11,29,39,40] which is sensitive even to nanomolar concentrations of Hg^{2+} and is remarkably selective for Hg^{2+} over the neighboring ions in the periodic system, such as Cd^{2+} and Au^{+}. *MerR* binds mercury, presumably in a three-coordinate form, using cysteinate ligands [13,39]. It does not only control its own synthesis, but rather, after metal binding, the activation of an RNA polymerase for the synthesis of further proteins. Of these, *MerP* is used in the periplasmic binding of dissolved mercury (Hg^{2+}, RHg^{+}) and *MerT* effects the transmembrane transport into the cell. Two "processing" enzymes are especially interesting: a specific Hg^{II} reductase (*MerA*) [39,40] and organomercury (or organomercurial) lyase (*MerB*) [41].

Organomercury lyase is a relatively small (22 kDa) monomeric protein which accelerates the cleavage of the otherwise kinetically inert Hg–C bonds by a factor of 10^6–10^7 (17.3). It also cleaves some tetraorganotin compounds, albeit with much lower efficiency. It contains four conserved cysteine residues, presumably for mercury binding. A mechanistic hypothesis (17.4) [11,31,41] shows the possible function of several cysteine ligands in a concerted substitution of R^- by Cys^-, which requires an increased metal coordination number in the transition state (from 2 to 3 or 4).

$$RHgX + H^+ + X^- \rightleftharpoons R{-}H + HgX_2 \qquad X = R'S, Hal \tag{17.3}$$

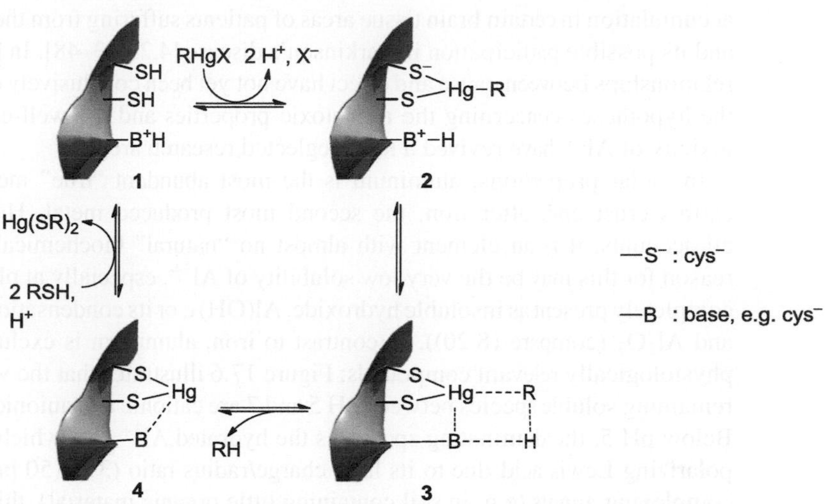

(17.4)

Reduction of coordinated Hg^{II} to the volatile and less toxic elemental mercury is effected by a special Hg^{II} reductase, a dimeric flavin- and NADPH-containing protein with a molecular mass of 2×60 kDa. The equilibrium of the catalyzed reaction (17.5) depends on the entatic stress forced on the normally very stable bis(thiolato)mercury(II) complexes by the enzyme [11,40].

$$Hg(SR)_2 + NADPH + H^+ \rightleftharpoons Hg + NADP^+ + 2\,RSH \tag{17.5}$$

Flavin adenine dinucleotide (FAD; see (3.12)) and four cysteine residues are present in the active center at the interface between two protein subunits, α_1 and α_2. An increase in the metal coordination number and an activating Hg coordination to both subunits are discussed subject of discussion [40,41], preceding the reducing action of the dihydroflavin ($FADH^-$). The mechanism of the final rapid two-electron transfer from $FADH^-$ to specially coordinated Hg^{II} is still unclear.

The global cycle of mercury is largely determined by the volatility of the element and its compounds, particularly of dimethylmercury, Me_2Hg (boiling point 96 °C), which is formed after complete biomethylation; today, the natural and (locally concentrated) anthropogenic sources contribute to a roughly equal extent. The accumulation of organomercury compounds in marine animals, which sometimes amounts to many orders of magnitude, is quite astonishing and involves particularly predatory fish at the end of the food chain; these animals can therefore represent a major source of Hg in the human diet. The biological half-life of $MeHg^+$ in such fish is estimated to be several years. Less well known is the role of bacteria in the transformation of mercury compounds into volatile forms, either through methylation in anoxic environments or through reduction to the metal (17.5), which is the dominating form in the atmosphere (half-life about 1 year).

17.6 Aluminum

Only since about 1975 has aluminum received scientific and public interest as a "harmful" and "toxic" metal. One reason for this concern has been the forest damage partly attributed to the acidification of soil by "acid rain" and the resulting release of the rhizotoxic Al^{3+} [8,9, 42,43]. Another is the still controversial x-ray microanalytical discovery of aluminosilicate accumulation in certain brain tissue areas of patients suffering from the Alzheimer's disease and its possible participation in Parkinson's disease [4,23,43–48]. In both cases, satisfying relationships between cause and effect have not yet been conclusively established; however, the hypotheses concerning the neurotoxic properties and the well-established high plant toxicity of Al^{3+} have revived a long neglected research area.

In molar proportions, aluminum is the most abundant "true" metallic element in the earth's crust and, after iron, the second most produced metal. However, according to all accounts, it is an element with almost no "natural" biochemical function. The main reason for this may be the very low solubility of Al^{3+}, especially at pH 7, when it is almost completely present as insoluble hydroxide, $Al(OH)_3$, or its condensation products, AlO(OH) and Al_2O_3 (compare (8.20)). In contrast to iron, aluminum is exclusively trivalent in its physiologically relevant compounds; Figure 17.6 illustrates that the very small amounts of remaining soluble species between pH 5 and 7 are cationic and anionic hydroxo complexes. Below pH 5, the dominating species is the hydrated Al^{3+} ion, which is a strongly ligand-polarizing Lewis acid due to its high charge/radius ratio (3+)/(50 pm). In the absence of complexing agents (e.g. in soil containing little organic material), this ion can be released through sufficiently acidic precipitation and thus be made bioavailable.

Depending on their organic and mineral composition, soils may be buffered to different extents against added acid: neutral to slightly basic pH values are prevalent in calcareous soil, according to the hydrogen carbonate buffer system, while silicates such as the

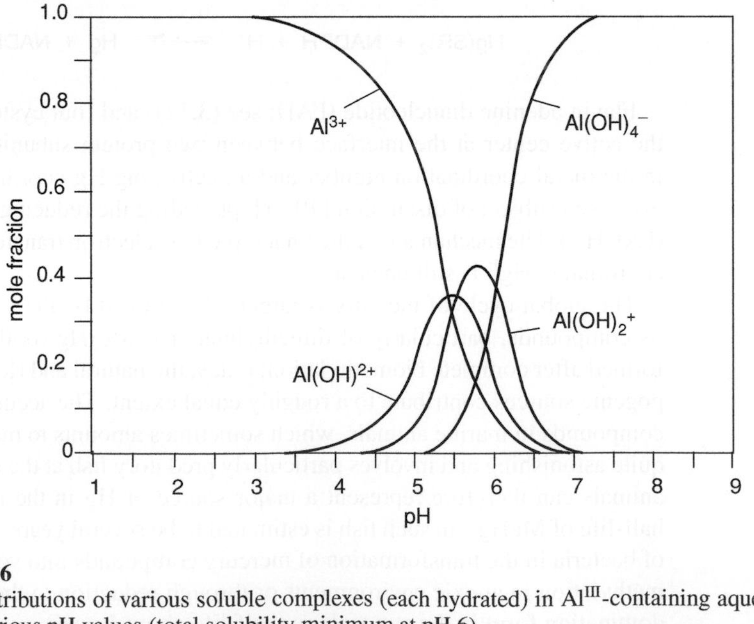

Figure 17.6
Molar contributions of various soluble complexes (each hydrated) in Al^{III}-containing aqueous solutions at various pH values (total solubility minimum at pH 6).

aluminum-containing feldspar buffer at about pH 5.0–6.5. During "weathering" processes, alkali and alkaline earth metal ions are released from the complex silicates and less soluble oxides result. The aluminum-containing clay minerals are only attacked under release of Al^{3+}(aq) in exchange for H^+ at below pH 5; even lower pH values are necessary to dissolve the more compact aluminosilicates and, finally, the aluminum and iron oxides proper.

The hydrated Al^{3+} ion, which can replace Mg^{2+} with great efficiency due to its higher charge and smaller ionic radius, and hydroxyaluminosilicates are widely believed to be soluble tree-damaging agents. The detrimental reaction is not the replacement of the Mg^{2+} ion in chlorophyll, as is often assumed; the far higher Lewis acidity, stronger ligand coordination and much slower ligand substitution rate of Al^{3+} relative to Mg^{2+} cause alterations, in particular inhibitions of Mg^{2+}-regulated biochemical processes. Obviously, the vast number of Mg^{2+}/enzyme-induced phosphate transfer reactions (compare Section 14.1) can be blocked by Al^{3+}. The very high affinity of Al^{3+} towards polyanionic phosphate groups precludes an efficient catalysis; that is, a process with rapid restoration of the substrate-free catalyst system. For instance, the ATP tetraanion binds Al^{3+} more strongly than Mg^{2+}, by a factor of 10^5. Even the exchange of water molecules in the first coordination sphere of the hydrated complex proceeds more than five orders of magnitude more slowly in the aluminum system than in the Mg^{2+} aqua complex. The effective thermodynamic and kinetic inhibition by Al^{2+} of Mg^{2+}-dependent enzymatic processes involving kinases, cyclases, esterases, ATPases or phosphatases is therefore not unexpected.

The reduced growth of tree roots due to increased ratios of aluminum to calcium and aluminum to magnesium is one of the consequences of soil acidification. In addition, both H^+ and atmospheric nitrogen in the form of $NH_4{}^+$ are antagonists with regard to Mg^{2+}, further reducing the bioavailability of this important ion.

With respect to its coordination behavior, Al^{3+} as a small, highly charged metal ion prefers complexation with negatively charged oxygen-containing ligands. In the blood plasma (i.e. at low phosphate concentrations), chelate complexation occurs mainly with the partly or completely deprotonated citrate anion, $^{-}OOC{-}CH_2{-}C(OH)(COO^-){-}CH_2{-}COO^-$ [23,43,44]. With about 5 mg/l plasma fluid, the normal concentration of Al^{3+} in blood is quite low and hard to precisely determine, due to the ubiquity of aluminum. Once the metal ion (e.g. in chelate-ligated form) has passed the first barriers in the gastrointestinal tract, Al^{3+} is stored and transported mainly by transferrin (Section 8.4.1), which is hardly surprising considering the chemical similarity of Al^{3+} and Fe^{3+}. Differences exist mainly with respect to the more rapid release of Al^{3+} versus Fe^{3+} through transferrin [23,43,44]. When an efficient excretion through the kidneys is no longer possible, aluminum may form small, less highly charged complexes, which can apparently cross the membrane barriers that usually hinder a highly charged ion from reaching the nervous system, for example. With regard to a possible role of Al^{3+} in neuropathological symptoms, the strong tendency of this ion for coordination with deprotonated 1,2-dihydroxy aromatic systems such as the catecholamine neurotransmitters epinephrine and DOPA has been established (compare (8.6) and (10.6)) [49]. This is in line with a number of recent reports on animal models of Al intoxication, which show largely altered neurotransmitter metabolism [23]. Therefore, even though there is no clear evidence supporting Al participation in Parkinson's disease, aluminum might interfere with certain key events of striatum neurotransmission and contribute to neurodegeneration.

In connection with Alzheimer's disease, there have been conflicting analytical reports concerning the possible accumulation of aluminum compounds (e.g. as aluminosilicates) in Alzheimer-typical neurofibrillary tangles (NFTs) or β-amyloid (Aβ) protein-containing senile plaques (SPs) in certain regions of the brain (e.g. the hippocampus) [44–48]. However, while manifestation of dementia is always accompanied by the presence of SPs (resulting from an imbalance in Aβ peptide metabolism followed by its aggregation and deposition), SPs are also found in a significant portion of the elderly population not showing any form of dementia. Therefore, it seems that amyloid deposition is only one among several major factors causing the disease [44–47]. Inhibition of amyloid aggregation is thus one of the major issues for the development of drugs aimed at treating Alzheimer's disease, among them rather "nonbiological" metal complexes [47].

Encephalopathy, dementia and even mortality have been established for hemodialysis patients who were subjected to an increased level of aluminum. Some of these patients had received aluminum compounds in the dialysis fluid to prevent hyperphosphataemia; others had accumulated dissolved aluminum, which remained after water treatment to remove phosphates (normal Al^{3+} concentration: 10 mg/l; typical permissible level: 200 mg/l). In view of the large quantities of water required for dialysis and in the absence of functioning kidneys, the high total amounts of Al^{3+} were only insufficiently excreted. Chelate drugs such as deferrioxamine (2.1) (see also Section 8.2), which originally target Fe^{3+} have also been used in cases of Al^{3+} poisoning. Other symptoms of Al^{3+} overload include several forms of anemia (Al^{3+}/Fe^{3+} antagonism) and disorders of the bone metabolism; an accumulation of Al^{3+}, with its very high affinity towards phosphates, has been observed in bone-forming osteoblasts. The tendency of Al^{3+} to form strong bonds with fluoride and the Al/Si antagonism must also be mentioned. There is no reason, however, for an exaggerated anxiety, e.g. with respect to the use of aluminium-containing kitchenware. Even heating solutions containing fruit acids (acidic + chelating agents) in such containers will result in only minute dissolution. Aluminum is not only a major component of many natural soils but is also present in relatively large concentrations in beverages such as tea (tea leaves accumulate Al) and beer, and as a component of processed food. It is normally rapidly excreted via functioning kidneys. Large amounts of aluminum compounds can be found in chewing gum, toothpastes and some strong antacids used to neutralize excess gastric acid.

17.7 Beryllium

There is a diagonal relationship between aluminum and divalent beryllium in the periodic table. Accordingly, the chemistry of both elements is rather similar; however, Be^{2+} is slightly more soluble at pH 6–7 than is Al^{3+}, due to its lower charge and despite its extremely small ionic radius of 27 pm for the preferred coordination number of 4. This very rare element has become environmentally relevant through technical combustion and the use of beryllium-containing compounds and alloys, which exhibit superior mechanical and nuclear properties [5,6,15,50,51]. Exposure to beryllium, typically inhaled in the form of dust, may lead to pulmonary diseases ("beryllosis", "chronic beryllium disease") and lung cancer.

Once incorporated, beryllium is excreted only very slowly from organisms, due to its spatially very concentrated dipositive charge. A long-term deposit of beryllium is the phosphate-containing bone tissue. As a lighter homologue of magnesium, divalent beryllium can reach the cell nucleus, where it may exert proven mutagenic (DNA-altering) as well as carcinogenic (tumor-inducing) effects. Gene transcription and expression are strongly impaired. A beryllium-induced inhibition of RNA or DNA synthesis, which is otherwise catalyzed by Mg^{2+}, ions has also been demonstrated (compare Section 14.1). Be^{2+} can drastically change the activity of phosphatases and kinases [6], the contraproductively strong binding of this ion inhibiting the catalysis of (de)phosphorylation reactions (see Section 14.1). Finally, beryllium interacts with the immune system [50,51], its compounds being allergenic contact toxins. A possible chelate therapy of beryllium poisoning (as for Hg or Cd) is not feasible, since Be-specific (hard) O-containing ligands such as citrate are also strongly binding of other highly abundant (hard) metal cations such as Mg^{2+} and Fe^{3+}.

17.8 Chromium and Tungsten

Under physiological conditions, the persistent oxidation states of chromium are Cr^{III} and Cr^{VI}, the latter as chromate, $CrO_4{}^{2-}$, at pH 7. Like trivalent aluminum and iron, trivalent chromium exists as an insoluble hydroxide at neutral pH and can thus be absorbed by organisms and utilized as an essential element only in specially ligated form (compare Section 11.5). On the other hand, the chromates(VI), which have been recognized as skin irritants ever since the beginning of their industrial use, are classified as potential carcinogenic substances [5,6,15]. The uptake of $CrO_4{}^{2-}$ by organisms seems to be surprising at first because the redox potential for reaction (17.6) is about +0.6 V at pH 7; that is, chromate is only metastable in the presence of reducing organic compounds under physiological conditions. Nevertheless, the required number of three electrons (compare the metastability of permanganate, $MnO_4{}^-$, in water) leaves only a few special biochemical reduction systems as efficient reductants for chromate [15,52]. These include heme- and flavoproteins (NADH-dependent), thiols such as glutathione, GSH (compare 16.7), and ascorbate (3.12). According to electron paramagnetic resonance (EPR) studies, highly reactive Cr^V states with one unpaired 3d electron may be formed in these reactions [53].

$$CrO_4{}^{2-} + 4\,H_2O + 3\,e^- \rightleftharpoons Cr(OH)_3 + 5\,OH^- \qquad (17.6)$$

Due to its structural similarity with the sulfate ion, $SO_4{}^{2-}$, chromate can overcome membrane barriers and reach the cell nucleus unless it is rapidly reduced (Figure 17.7).

Chromate can oxidatively damage genetically important components of the cell nucleus [5,6,15]. The substitutionally more labile and more strongly oxidizing Cr^V or Cr^{IV} intermediates formed during one-electron reduction steps and the simultaneously produced $RS^\bullet$ and $^\bullet OH$ radicals can directly attack the DNA and effect bond cleavage, crosslinking and, as a consequence, faulty gene expression. Furthermore, the resulting substitutionally *inert* Cr^{III} can irreversibly bind to phosphate-containing DNA or free nucleotides and thus also affect genetic functions (compare Section 19.2). Due to the ligand field stabilization for a high-spin d^3 configuration in octahedral symmetry, the Cr^{III} complexes are generally distinguished by an enormous kinetic stability; hydrated Cr^{3+} exchanges its water ligands much more slowly, by a factor of 10^6, than even hydrated Al^{3+}.

Tungsten (the symbol "W" is derived from the German "Wolfram") is the bioelement with the highest atomic number, but it is not a universal bioelement. For several microbiological species, tungsten is essential for growth (e.g. for the hyperthermophilic archaeon *Pyrococcus furiosus* [54]), while for many others it seems to be toxic. Recent reports reveal that the dissolution of metallic tungsten particles can cause adverse environmental effects, including soil acidification and toxic effects to plants, soil microorganisms and invertebrates [55–57]. Interesting properties such as high hardness and density and the highest melting point among all the metals have led to tungsten's broad practical use in industry; its release into environmental systems may occur as the result of natural or anthropogenic

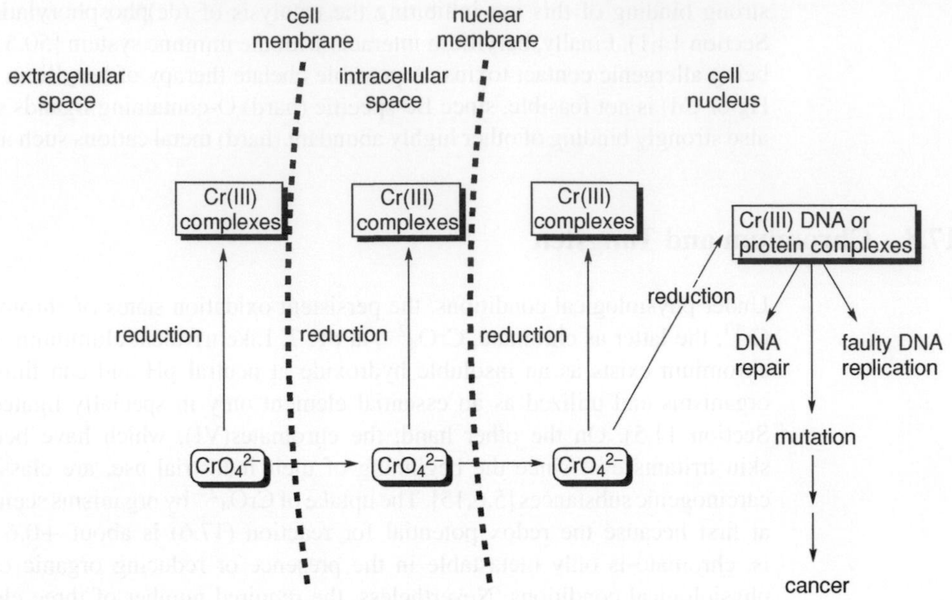

Figure 17.7
Flow diagram for the uptake and reduction of chromate in different cellular compartments.

activities, in particular from mining and smelting of tungsten-containing minerals such as wolframite, scheelite and ferberite. In the USA, occupational exposure to air pollution standards of 5 mg/m^3 for insoluble and 1 mg/m^3 for soluble tungsten compounds exist. While the toxicity to humans of W^{VI} is markedly lower than that of Cr^{VI} (Table 17.1), aquatic organisms seem to be far more susceptible to tungsten intoxication. Low limits for tungsten in drinking water and fishing water reservoirs have therefore been set. Owing to the complexity of tungsten speciation and aquatic chemistry, it is not yet understood which tungsten compounds are responsible for these toxic effects. The stability of the oxidation state, W^{VI}, is markedly higher than that of Cr^{VI}, and the most frequent tungsten compounds in water and soil are tungstate WO_4^{2-} and numerous polymeric polyoxotungstates. Tungstates can polymerize with phosphate, and cellular phosphorylation and dephosphorylation reactions may be disrupted, resulting in negative effects on cellular functions.

17.9 Toxicity of Nanomaterials

The rapidly growing applications of nanomaterials, ranging from structural strength enhancement and energy conservation to antimicrobial properties and self-cleaning surfaces, demand a detailed understanding of their benefits and risks in toxicology. In contrast to classical chemicals (molecules or ions), the mobility of nanoparticles is largely governed by their size (see also Section 15.2). Importantly, agglomeration of the particles during prolonged environmental exposure or uptake into an organism (also during toxicity tests) can change their effective size. The major toxic interactions of nanoparticles are specific catalytic processes inside the cell and enhanced heavy-metal transport into the cytosol through the so-called "Trojan horse mechanism" [58,59]. This latter mechanism can also be used for enhanced crossing of the blood–brain barrier by functionalized nanomaterials (so-called nanocarriers), which are a promising line of study in drug trafficking and neuroimaging [60]. It is proposed that the current safety of nanoparticles in consumer goods can be increased by substituting the currently used persistent oxides with biodegradable materials (see also Chapter 15).

Further Reading

Kirk-Othmer Chemical Technology and the Environment, Vol. 2, John Wiley & Sons, Chichester, **2007**.

M. Lippmann, *Environmental Toxicants*, 3rd Edition, John Wiley & Sons, Chichester, **2009**.

P. Patnaik, *A Comprehensive Guide to the Hazardous Properties of Chemical Substances*, 3rd Edition, John Wiley & Sons, Chichester, **2007**.

G. F. Nordberg, B. A. Fowler, M. Nordberg, L. T. Friberg (eds.), *Handbook on the Toxicology of Metals*, 3rd Edition, Elsevier, Amsterdam, **2007**.

E. Merian, M. Anke, M. Ihnat and M. Stoeppler (eds.) *Elements and their Compounds in the Environment*, 2nd Edition, Wiley-VCH, Weinheim, **2004**.

References

1. J. C. Merrill, J. J. P. Morton, S. D. Soileau, *Metals*, in H. A. Wallace (ed.), *Principles and Methods of Toxicology*, 5th Edition, CRC Press, Boca Raton, **2008**.
2. (a) V. Mudgal, N. Madaan, A. Mudgal, R. B. Singh, S. Mishra, *Open Nutraceuticals J.*, **2010**, *3*, 94–99: *Effect of toxic metals on human health*; (b) S. Mishra, S. P. Dwivedi, R. B. Singh, *Open Nutraceuticals J.*, **2010**, *3*, 188–193: *A review on epigenetic effect of heavy metal carcinogens on human health.*
3. R. A. Howd, A. M. Fan, *Risk Assessment for Chemicals in Drinking Water*, John Wiley & Sons, Hoboken, **2008**.
4. A. Sigel, H. Sigel, R. K. O. Sigel (eds.), *Neurodegenerative Diseases and Metal Ions: Metal Ions in Life Sciences, Vol. 1*, John Wiley & Sons, Chichester, **2006**.
5. S. V. S. Rana, *J. Trace Elem. Med. Biol.* **2008**, *22*, 262–284: *Metals and apoptosis: recent developments.*
6. D. Beyersmann, A. Hartwig, *Arch. Toxicol.* **2008**, *82*, 493–512: *Carcinogenic metal compounds: recent insight into molecular and cellular mechanisms.*
7. S. Clemens, *Biochimie* **2006**, *88*, 1707–1719: *Toxic metal accumulation, responses to exposure and mechanisms of tolerance in plants.*
8. P. Babula, V. Adam, R. Opatrilova, J. Zehnalek, L. Havel, R. Kizek, *Environ. Chem. Lett.* **2008**, *6*, 189–213: *Uncommon heavy metals, metalloids and their plant toxicity: a review.*
9. (a) G. M. Gadd, Y. J. Rhee, K. Stephenson, Z. Wei, *Environ. Microbiol. Rep.* **2012**, *4*, 270–296: *Geomycology: metals, actinides and biominerals*; (b) G. M. Gadd, *Microbiology* **2010**, *156*, 609–643; *Metals, minerals and microbes: geomicrobiology and bioremediation.*
10. J. J. Harrison, H. Ceri, R. J. Turner, *Nat. Rev. Microbiol.* **2007**, *5*, 928–938: *Multimetal resistance and tolerance in microbial biofilms.*
11. S. Silver, L. T. Phung, *J. Ind. Microbiol. Biotechnol.* **2005**, *32*, 587–605. *A bacterial view of the periodic table: genes and proteins for toxic inorganic ions.*
12. C. T. Dameron, R. N. Reese, R. K. Mehra, A. R. Kortan, P. J. Carroll, M. L. Steigerwald, L. E. Brus, D. R. Winge, *Nature (London)*, **1989**, *338*, 596–597: *Biosynthesis of cadmium sulfide quantum semiconductor crystallites.*
13. L. Charlet, Y. Chapron, P. Faller, R. Kirsch, A. T. Stone, P. C. Baveye, *Coord. Chem. Rev.* **2012**, *256*, 2147–2163: *Neurodegenerative diseases and exposure to the environmental metals Mn, Pd and Hg.*

14. (a) C. A. Blindauer, O. I. Leszczyszyn, *Nat. Prod. Rep.* **2010**, *27*, 720–741: *Metallothioneins: unparalleled diversity in structures and functions for metal ion homeostasis and more*; (b) M. Nordberg, G. F. Nordberg, in *Met. Ions Life Sci.* **2009**, *5*, 1–29: *Metallothioneins: historical development and overview.*
15. M. S. Sinicropi, D. Amantea, A. Caruso, C. Saturnino, *Arch. Toxicol.* **2010**, *84*, 501–520: *Chemical and biological properties of toxic metals and use of chelating agents for the pharmacological treatment of metal poisoning.*
16. (a) J. O. Nriagu, *Science* **1998**, *281*, 1622–1623: *Tales told in lead*; (b) J. O. Nriagu, *Science* **1996**, *272*, 223–224: *A history of global metal pollution.*
17. L. Järup, *Brit. Med. Bull.* **2003**, *68*, 167–182: *Hazards of heavy metal contamination.*
18. D. M. Settle, C. C. Patterson, *Science* **1980**, *207*, 1167–1176: *Lead in albacore: guide to lead pollution in Americans.*
19. R. L. Boeckx, *Anal. Chem.* **1986**, *58*, 274A–288A: *Lead poisoning in children.*
20. D. Seyferth, *Organometallics* **2003**, *22*, 2346–2357 (part 1), 5154–5178 (part 2): *The rise and fall of tetraethyllead.*
21. K. Abu-Dari, F. E. Hahn, K. N. Raymond, *J. Am. Chem. Soc.* **1990**, *112*, 1519–1524: *Lead sequestering agents. 1. Synthesis, physical properties and structures of lead thiohydroxamato complexes.*
22. W. Kowal, O. B. Beattie, H. Baadsgaard, P. M. Krahn, *Nature (London)* **1990**, *343*, 319–320: *Did solder kill Franklin's men?*
23. S. V. Verstraeten, L. Aimo, P. I. Oteiza, *Arch. Toxicol.* **2008**, *82*, 789–802: *Aluminium and lead: molecular mechanisms of brain toxicity.*
24. (a) J. Lichota, T. Skjorringe, L. B. Thomsen, T. Moos, *J. Neurochem.* **2010**, *113*, 1–13: *Macromolecular drug transport into the brain using targeted therapy*; (b) A. Boumendjel, J. Boutonnat, J. Robert, *ABC superfamily transporters at the human blood-brain barrier*, in J.-M. Scherrmann (ed.), *ABC Transporters and Multidrug Resistance*, John Wiley & Sons, Hoboken, **2009**.
25. (a) H. Strasdeit, *Angew. Chem. Int. Ed. Engl.* **2001**, *40*, 707–709: *The first cadmium-specific enzyme*; (b) N. M. Price, F. M. M. Morel, *Nature (London)* **1990**, *344*, 658–660: *Cadmium and cobalt substitution for zinc in a marine diatom.*
26. (a) G. F. Nordberg, *Tox. Appl. Pharm.* **2009**, *238*, 192–200: *Historical perspectives on cadmium toxicology*; (b) L. Järup, A. Åkesson, *Tox. Appl. Pharm.* **2009**, *238*, 201–208: *Current status of cadmium as an environmental health problem.*
27. J. Pius, *Tox. Appl. Pharm.* **2009**, *238*, 272–279: *Mechanisms of cadmium carcinogenesis.*
28. W. Braun, M. Vasak, A. H. Robbins, C. D. Stout, G. Wagner, J. H. Kagi, K. Wüthrich, *Proc. Natl. Acad. Sci. USA* **1992**, *89*, 10124–10128: *Comparison of the NMR solution structure and the x-ray crystal structure of rat metallothionein-2.*
29. (a) K. J. Waldron, J. C. Rutherford, D. Ford, N. J. Robinson, *Nature (London)* **2009**, *460*, 823–830: *Metalloproteins and metal sensing*; (b) D. B. Zamble, *Metalloregulatory proteins*, in T. P. Begley (ed.) *Wiley Encyclopedia of Chemical Biology, Vol. 3*, John Wiley & Sons, Hoboken, **2009**.
30. (a) A. V. Hirner, A. W. Rettenmeier, *Methylated metal(loid) species in humans*, in A. Sigel, H. Sigel, R. K. O. Sigel (eds.), *Metal Ions in Life Sciences, Vol. 7*: *Organometallics in Environment and Toxicology*, RSC Publishing, Colchester, **2010**; (b) E. Dopp, L. M. Hartmann, A.-M. Florea, A. W. Rettenmeier, A. V. Hirner, *Crit. Rev. Tox.* **2004**, *34*, 301–333: *Environmental distribution, analysis, and toxicity of organometal(loid) compounds.*
31. J. S. Thayer, *Appl. Organomet. Chem.* **2002**, *16*, 677–691: *Biological methylation of less-studied elements.*
32. B. Blumenthal, K. Sellers, M. Koval, *Thallium and thallium compounds*, in A. Seidel (ed.) *Kirk-Othmer Encyclopedia of Chemical Technology, Vol. 24*, 5th Edition, John Wiley & Sons, Hoboken, **2007**.
33. A. L. J. Peter, T. Viraraghavan, *Environ. Int.* **2005**, *31*, 493–501: *Thallium: a review of public health and environmental concerns.*
34. M. Ware, *J. Chem. Educ.* **2008**, *85*, 612–620: *Prussian Blue: artists' pigment and chemists' sponge.*
35. E. K. Mellon, *J. Chem. Educ.* **1977**, *54*, 211–213: *Alfred E. Stock and the insidious "Quecksilbervergiftung".*
36. R. A. Diaz-Bone, T. Van de Wiele, *Pure Appl. Chem.* **2010**, *82*, 409–427: *Biotransformation of metal(loid)s by intestinal microorganisms.*

37. W. F. Fitzgerald, C. H. Lamborg, C. R. Hammerschmidt, *Chem. Rev.* **2007**, *107*, 641–662: *Marine biogeochemical cycling of mercury.*
38. D. W. Nierenberg, R. E. Nordgren, M. B. Chang, R. W. Siegler, M. B. Blayney, F. Hochberg, T. Y. Toribara, E. Cernichiari, T, Clarkson, *New Engl. J. Med.* **1998**, *338*, 1672–1676: *Delayed cerebellar disease and death after accidental exposure to dimethylmercury.*
39. J. G. Omichinski, *Science* **2007**, *317*, 205–206: *Toward methylmercury bioremediation.*
40. (a) R. Ledwidge, B. Hong, V. Dötsch, S. M. Miller, *Biochemistry* **2010**, *49*, 8988–8998: *NmerA of Tn501 mercuric ion reductase: structural modulation of the pKa values of the metal binding cysteine thiols*; (b) B. Hong, R. Nauss, I. M. Harwood, S. M. Miller, *Biochemistry* **2010**, *49*, 8187–8196: *Direct measurement of mercury(II) removal from organomercurial lyase (MerB) by tryptophan fluorescence: NmerA domain of coevolved γ-proteobacterial mercuric ion reductase (MerA) is more efficient than MerA catalytic core or glutathione*; (c) M. J. Moore, M. D. Distefano, L. D. Zydowsky, R. T. Cummings, C. T. Walsh, *Acc. Chem. Res.* **1990**, *23*, 301–308: *Organomercurial lyase and mercuric ion reductases: nature's mercuric detoxification catalysis.*
41. (a) J. M. Parks, H. Guo, C. Momany, L. Liang, S. M. Miller, A. O. Summers, J. C. Smith, *J. Am. Chem. Soc.* **2009**, *131*, 13278–13285: *Mechanism of Hg-C protonolysis in the organomercurial lyase MerB*; (b) J. G. Melnick, G. Parkin, *Science* **2007**, *317*, 225–227: *Cleaving mercury-alkyl bonds: a functional model for mercury detoxification by MerB.*
42. P. R. Ryan, E. Delhaize, *Funct. Plant Biol.* **2010**, *37*, 275–284: *The convergent evolution of aluminium resistance in plants exploits a convenient currency.*
43. R. B. Martin, *Acc. Chem. Res.* **1994**, *27*, 204–210: *Aluminum: a neurotoxic product of acid rain.*
44. (a) C. Exley, *Coord. Chem. Rev.* **2012**, *256*, 2142–2146: *The coordination chemistry of aluminium in neurogenerative disease*; (b) V. Kumar, K. D. Gill, *Arch. Toxicol.* **2009**, *83*, 965–978: *Aluminium neurotoxicity: neurobehavioural and oxidative aspects.*
45. D. Drago, S. Bolognin, P. Zatta, *Curr. Alzheimer Res.* **2008**, *5*, 500–507: *Role of metal ions in the Aβ oligomerization in Alzheimer's disease and in other neurological disorders.*
46. A. Rauk, *Chem. Soc. Rev.* **2009**, *38*, 2698–2715: *The chemistry of Alzheimer's disease.*
47. (a) D. Valensin, C. Gabbiani, L. Messori, *Coord. Chem. Rev.* **2012**, *256*, 2357–2366: *Metal compounds as inhibitors of β-amyloid aggregation. Perspectives for an innovative metallotherapeutics on Alzheimer's disease*; (b) C. Rodriguez-Rodriguez, M. Telpoukhovskaia, C. Orvig, *Coord. Chem. Rev.* **2012**, *256*, 2308–2332: *The art of building multifunctional metal-binding agents from basic molecular scaffolds for the potential application in neurodegenerative diseases.*
48. P. Zatta, T. Kiss, M. Suwalsky, G. Berthon, *Coord. Chem. Rev.* **2002**, *228*, 271–284: *Aluminium(III) as a promoter of cellular oxidation.*
49. (a) P. Rubini, A. Lakatos, D. Champmartin, T. Kiss, *Coord. Chem. Rev.* **2002**, *228*, 137–152: *Speciation and structural aspects of interactions of Al(III) with small biomolecules*; (b) T. Kiss, I Sovago, R. B. Martin, *J. Am. Chem. Soc.* **1989**, *111*, 3611–3614: *Complexes of 3,4-dihydroxyphenyl derivatives. 9.* Al^{3+} *bonding to catecholamines and tiron.*
50. B. L. Scott, T. M. McCleskey, A. Chaudhary, E. Hong-Geller, S. Gnanakaran, *Chem. Commun.* **2008**, 2837–2847: *The bioinorganic chemistry and associated immunology of chronic beryllium disease.*
51. P. F. Wambach, J. C. Laul, *J. Chem. Health Saf.* **2008**, *15*, 5–12: *Beryllium health effects, exposure limits and regulatory requirements.*
52. M. I. Ramirez-Diaz, C. Diaz-Perez, E. Vargas, H. Riveros-Rosas, J. Campos-Garcia, C. Cervantes, *Biometals* **2008**, *21*, 321–332: *Mechanisms of bacterial resistance to chromium compounds.*
53. (a) A. Levina, L. Zhang, P. A. Lay, *J. Am. Chem. Soc.* **2010**, *132*, 8720–8731: *Formation and reactivity of chromium(V) – thiolato complexes: a model for the intracellular reactions of carcinogenic chromium(VI) with biological thiols*; (b) R. N. Bose, S. Moghaddas, E. Gelerinter, *Inorg. Chem.* **1992**, *31*, 1987–1994: *Long-lived chromium(IV) and chromium(V) metabolites in the chromium(VI) glutathion reaction: NMR, ESR, HPLC, and kinetic characterization.*
54. A.-M. Sevcenco, M. W. H. Pinkse, E. Bol, G. C. Krijger, H. Th. Wolterbeek, P. D. E. M. Verhaert, P.-L. Hagedoorn, W. R. Hagen, *Metallomics*, **2009**, *1*, 395–402: *The tungsten metallome of Pyrococcus furiosus.*
55. N. Strigul, A. Koutsospyros, C. Christodoulatos, *Ecotox. Environ. Safety* **2010**, *73*, 164–171: *Tungsten speciation and toxicity: acute toxicity of mono-and poly-tungstates to fish.*
56. D. R. Johnson, C. Y. Ang, A. J. Bednar, L. S. Inouye, *Tox. Sci.* **2010**, *116*, 523–532: *Tungsten effects on phosphate-dependent biochemical pathways are species and liver cell line dependent.*

57. D. B. Ringelberg, C. M. Reynolds, L. E. Winfield, L. S. Inouye, D. R. Johnson, A. J. Bednar, *J. Environ. Qual.* **2009**, *38*, 103–110: *Tungsten effects on microbial community structure and activity in a soil.*
58. S. Sharifi, S. Behzadi, S. Laurent, M. L. Forrest, P. Stroeve, M. Mahmoudi, *Chem. Soc. Rev.* **2012**, *41*, 2323–2343: *Toxicity of nanomaterials.*
59. J. Lee, S. Mahendra, P. J. J. Alvarez, *ACSNano* **2010**, *4*, 3580–3590: *Nanomaterials in the construction industry: a review of their applications and environmental health and safety considerations.*
60. S. Bhaskar, F. Tian, T. Stoeger, W. Kreyling, J. M. de la Fuente, V. Grazú, P. Borm, G. Estrada, V. Ntziachristos, D. Razansky, *Part. Fibre Tox.* **2010**, *7:3*, 3: *Multifunctional nanocarriers for diagnostics, drug delivery and targeted treatment across blood-brain barrier: perspectives on tracking and neuroimaging.*

18 Biochemical Behavior of Radionuclides and Medical Imaging Using Inorganic Compounds

18.1 Radiation Risks and Medical Benefits from Natural and Synthetic Radionuclides

18.1.1 The Biochemical Impact of Ionizing Radiation from Radioactive Isotopes

Independent of the consequences of human activities and long before humans appeared in biology, most organisms had to coexist not only with "toxic" elements and their compounds but also with naturally occurring radioactive isotopes and the resulting radiation. All forms of high-energy "ionizing" radiation (α, β, γ, x-ray and neutron radiation) can cause the cleavage of chemical bonds, thereby damaging enzymes or genetic material (DNA) either directly or indirectly (e.g. through the hydroxyl radical $^{\bullet}OH$, which is typically formed from H_2O, the main component of all organisms [1]. The inherent radical scavenging and repair mechanisms of organisms that are active in counteracting "oxidative stress" (Section 16.8) are also effective to a certain degree against radiation damage. For instance, exposure of the bacterium *Deinococcus radiodurans* to a dose of 10 kGy of ionizing radiation results in about 100 double strand breaks per chromosome. In this case, the damage is repaired without lethality, mutagenesis or rearrangement, whereas other organisms cannot survive such a dose. The highly efficient DNA repair by *D. radiodurans* is attributed to the expression of a DNA-protecting protein. All forms contain high amounts of iron and can be activated to exhibit ferroxidase activity (Section 8.4), protecting DNA from both hydroxyl radical-induced cleavage and from DNase I-mediated cleavage [2,3]. Copper complexes, including the variably applied superoxide dismutase (compare Section 10.5) [4], and sulfur compounds, such as cysteine and cysteamine (= 2-mercaptoethylamine, $H_2N{-}CH_2{-}CH_2{-}SH$), can contribute to a reduction in biological radiation damage when administered prior to exposure, through either radical scavenging or rapid one-electron reduction of ionized species. Other therapeutic measures taken in the event of imminent incorporation of radioactive substances include the saturation of depots in the body with nonradioactive material (e.g. by taking "iodide tablets") and the

Bioinorganic Chemistry: Inorganic Elements in the Chemistry of Life – An Introduction and Guide, Second Edition.
Written and Translated by Wolfgang Kaim, Brigitte Schwederski and Axel Klein.

undelayed and selective complexation and excretion by use of specific chelate ligands (e.g. for Sr or Pu).

In the early days of the nuclear sciences, radioactive material was handled less cautiously, as evidenced by Marie Curie's death through radiation-induced cancer and similar incidents [5,6]. Today, the large-scale technical production, testing and dismantling of nuclear weapons and the minutely described global effects of the disastrous reactor accidents in Chernobyl (Ukraine) in 1986, Harrisburg (USA) in 1979 [7] and Fukushima Daiichi (Japan) in 2011 have very much heightened the sensitivity of the public towards radiation risks and the need for strict guidelines on the handling of diagnostically and therapeutically useful radionuclides.

Radioactive isotopes are distinguished chemically from stable isotopes of the same element by the generally small isotope effect on reaction rates resulting from the different atomic or molecular mass (expressed in unified atomic units (u) or Dalton (Da)). The extremely low detection limit for many radiating isotopes allows a temporal and spatial observation of both physiologically important (^{31}P, ^{22}Na, ^{35}S) and "unphysiological" elements in organisms (→ diagnostics). On the other hand, even minute amounts of radioactive isotopes can cause severe genetic damage. The extent of such damage and the overall radiobiological effect depend on many factors, such as the volatility of the isotope-containing compound, the possible binding to transporting particles (carriers), the effectiveness of absorption of the organism, the kind of radiation emitted and its energy (α or β particle radiation versus electromagnetic γ radiation), the localization of the isotope-containing metabolite in the organism and the radioactive ("physical") and biological half-lives (i.e. the average time of exposure). The latter is very much determined by the chemical speciation of the corresponding element, the kind of exposure and the distribution in the organism; the variations known from "normal" toxicology (Figure 2.3) within and between populations also apply in this case [1,5,7]. Before discussing the medical uses of radioactive isotopes as tracers in diagnosis or as possible tumor-destroying therapeutics, we shall briefly review the naturally occurring radionuclides and the main "inorganic" isotopes resulting from nuclear fission reactions in their interactions with organisms.

18.1.2 Natural and Synthetic Radioisotopes

A number of radioisotopes (radioactive isotopes, radionuclides) are naturally occurring. These fall into three categories:

1. *Cosmogenic isotopes*, such as ^{3}H (tritium, half-life 12.3 years) and ^{14}C (half-life 5730 years). These are present because they are continually being formed in the atmosphere due to cosmic radiation.
2. *Primordial radionuclides*, which originate mainly from the interiors of stars and, like the elements uranium (U) and thorium (Th), and isotopes such as ^{40}K (0.012% natural abundance, half-life 1.3×10^9 years), are still present because their half-lives are so long that they have not yet completely decayed.
3. *Secondary radionuclides*, such as ^{220}Rn (radon) and ^{210}Po (polonium). These are radiogenic isotopes derived from the decay of primordial radionuclides (18.1). They typically have much shorter half-lives than the primordial radionuclides.

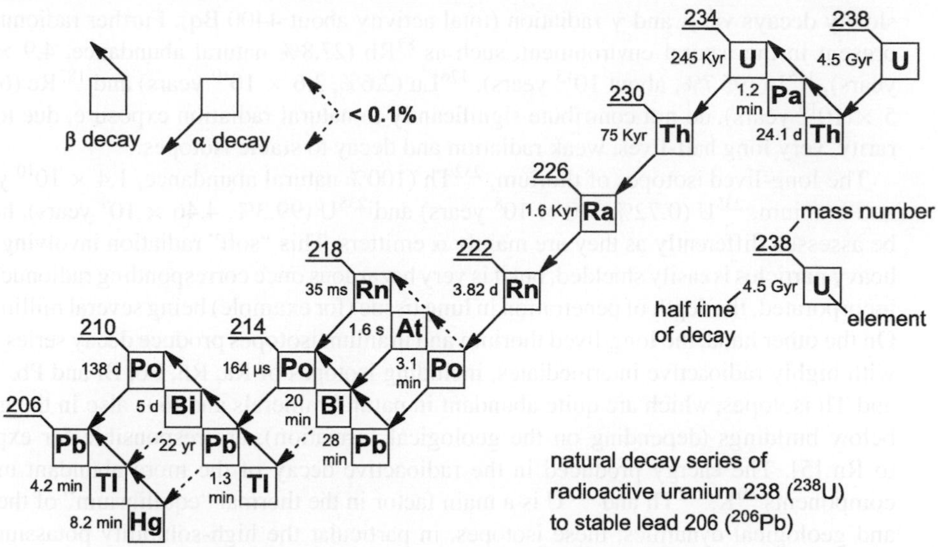

(18.1)

Synthetic radionuclides are produced by nuclear reactors, particle accelerators and radionuclide generators:

- Some radioisotopes produced in *nuclear reactors* exploit the high flux of neutrons. The neutrons transform elements placed within the reactor. A typical product from a nuclear reactor is ^{204}Tl (from natural Tl; one of the occurring reactions is ^{203}Tl(n,γ)^{204}Tl). These processes (also called "transmutations") provide the (theoretical) possibility of transforming highly dangerous radioactive isotopes into less harmful ones (→ radioactive waste management).
- *Particle accelerators*, such as cyclotrons, accelerate particles such as protons, deuterons and even heavier nuclei so that they bombard a target and produce radionuclides. For example, the reaction ^{192}Os(p,n)^{192}Ir produces the isotope ^{192}Ir. The use of protons frequently produces positron-emitting radioisotopes such as ^{18}F.
- *Radionuclide generators* contain a parent isotope, which decays to produce a (daughter) radioisotope. The parent radioisotope is usually produced in a nuclear reactor. A typical example is the ^{99m}Tc generator used in nuclear medicine. The parent produced in the nuclear reactor is ^{99}Mo (reaction: ^{235}U(n,f)^{99}Mo; see Figure 18.2).

Further radionuclides are generated and distributed into the biosphere by nuclear experiments (fission, nuclear breeding, bomb testing and disasters in nuclear power stations). The isotopes generated by fission of ^{235}U usually exhibit masses around 90 and 135 Da, such as ^{85}K, ^{89}Sr, ^{131}I and ^{134}Cs. Nuclear breeding leads to heavy isotopes such as ^{239}Pu.

18.1.3 Bioinorganic Chemistry of Radionuclides

18.1.3.1 Naturally Occurring Radionuclides

In addition to the β-emitting isotopes ^{3}H and ^{14}C, the bioessential inorganic element potassium is naturally present in the form of its radioactive isotope ^{40}K (0.012% natural abundance). Due to its long half-life of 1.3 billion years, there is still enough of this primordial radioisotope in existence that an adult human contains about 15 mg of it, which slowly decays via β and γ radiation (total activity about 4400 Bq). Further radionuclides present in the natural environment, such as ^{87}Rb (27.8% natural abundance, 4.9 × 10^{10} years), ^{115}In (95.7%, about 10^{15} years), ^{176}Lu (2.6%, 3.6 × 10^{10} years) and ^{187}Re (62.6%, 5 × 10^{10} years), do not contribute significantly to natural radiation exposure, due to their rarity, very long half-lives, weak radiation and decay to stable isotopes.

The long-lived isotopes of thorium, ^{232}Th (100% natural abundance, 1.4 × 10^{10} years), and uranium, ^{235}U (0.72%, 7.0 × 10^{8} years) and ^{238}U (99.3%, 4.46 × 10^{9} years), have to be assessed differently as they are mainly α emitters. This "soft" radiation involving fairly heavy particles is easily shielded, but it is very hazardous once corresponding radionuclei are incorporated, the depth of penetration in lung tissue (for example) being several millimetres. On the other hand, the long-lived thorium and uranium isotopes produce decay series (18.1) with highly radioactive intermediates, including isotopes of Ra, Rn, Po, Bi and Pb. The U and Th isotopes, which are quite abundant in natural minerals and thus also in the ground below buildings (depending on the geological formation), are responsible for exposure to Rn [5]. The energy produced in the radioactive decay of the more abundant mineral components ^{40}K, ^{232}Th and ^{238}U is a main factor in the thermal "equilibrium" of the earth and geological dynamics; these isotopes, in particular the high-solubility potassium, are also the main radioactive components in sea water.

18.1.3.2 Naturally Occurring Radionuclides from the Radioactive Decay of ^{232}Th and ^{238}U

The most important longer-lived heavy isotopes from the natural U/Th decay series (18.1) are characterized as follows:

- ^{210}Pb (22.3 years, β): lead; compare Section 17.2.
- ^{210}Po (138 days, α): polonium; heavier homologue of tellurium.
- ^{222}Rn (3.8 days, α): radon. As a noble gas, Rn is not bound to particulate matter. Its distribution in the atmosphere, hydrosphere and body is thus essentially unhindered. The two polonium daughter isotopes of ^{222}Rn, namely ^{218}Po and ^{214}Po (see (18.1)), are very short-lived and thus intense α emitters and so may trigger carcinogenesis of the respiratory organs; after nuclear transformation of the gaseous "intermediate" radon they can remain in the tissues, bound to solid particles (Po is a chalcogen). As in the case of cadmium poisoning (compare Section 17.3), smoking is known to increase the health hazard disproportionally (→ synergism). Radon poisoning [5,8] is not only a problem for mine workers (particularly in the mining of uranium); relatively high concentrations have been measured inside of buildings situated above geologically old, uranium-containing (granite) soil, for example in Central Switzerland, Sweden and the Eastern United States and Canada. A practical corrective measure is to install efficient ventilation below a hermetically sealed basement, but this cannot protect against the traces of radon found in drinking water and natural gas or emanating from building materials. Inevitably, the air ventilation system required in such cases also results in a loss of heating energy. The largest part of radioactive background exposure in the industrial nations comes from "natural" radon [5,8], a fact that has long escaped public attention due to focus on other matters such as the reactor accident in Chernobyl.
- ^{226}Ra (1600 years, α + γ): radium is the heaviest of the alkaline earth metal group. This very radiotoxic element [5,9] thus acts as a calcium analogue and, in case of exposure, is enriched in the bone tissue (Ra^{2+} as "bone-seeker") where it can affect the formation of blood cells and cause leukemia.

18.1.3.3 Plutonium

Although plutonium occurs, to a very small extent, in "natural" uranium ores, it is essentially a "synthetic" element, featuring relatively long-lived isotopes. The use of ^{239}Pu (2.4×10^4 years, α) as a source of radiation, and especially as a nuclear fuel in civil and military areas, has led to a large global production and thus to the potential for the release of significant amounts of this isotope into the environment. Together with ^{226}Ra, ^{222}Rn and ^{90}Sr, ^{239}Pu belongs in the category of the most radiotoxic isotopes [5,10,11]. In addition to the long radioactive half-life of ^{239}Pu (there are other, even longer-lived plutonium isotopes), which still guarantees sufficiently harmful numbers of decay processes, the physiological half-life of about 70 years is so high that self-detoxification through excretion is ineffective. After release via nuclear explosions or a reactor accident, plutonium mainly occurs in oxidic form bound to particulate matter and, following contact with water, as a colloidal hydroxide; its special role in solubilizing highly charged Pu ions is due to coordinating carbonate. Mammals absorb plutonium-contaminated particles through the lungs, and tumor formation can occur at this early stage. Following possible dissolution, plutonium metal can use the iron transport system and thus reach the liver and bone tissue, after which it is very hard to remove.

The special problem of decontamination by chelate therapy is related to the similarity between plutonium in its typical oxidation states, +III/+IV, and the biochemically very important iron(+II/+III) system; in contrast to uranium, plutonium exhibits a rather stable trivalent state. The plutonium redox pair features a higher positive charge than the iron redox system; however, this is compensated for by the larger ionic radius of this actinoid element. A good correspondence therefore results with respect to the charge-to-radius ratios (compare Section 13.1). This correspondence is even more relevant because the redox potentials of both pairs are comparable.

It is thus a challenge for coordination chemistry to develop and provide plutonium-selective chelate ligands which affect the iron metabolism as little as possible. Fairly successful catecholate systems such as "3,4,3-LICAMC" ((2.1) and Table 2.4) and corresponding macrocyclic ligands have been designed to bind large, highly charged metal ions with high coordination numbers [12].

18.1.3.4 Radionuclides from Nuclear Fission

Some longer-lived isotopes resulting from the fission of ^{235}U and ^{239}Pu have become generally known, in connection with early extensive nuclear weapons testing, the reactor accidents in Fukushima and Chernobyl and the discussion about technical reprocessing of nuclear fuel. These are listed here according to increasing nuclear mass. Characteristic for nuclear fission is that these isotopes are concentrated around mass numbers of about 90 and 135 Da.

Fukushima Daiichi, Chernobyl, Hiroshima and Nuclear Weapons Testing

On March 11, 2011 an earthquake and tsunami largely destroyed the cooling facilities of the Fukushima Daiichi Nuclear Power Plant (Fukushima I), resulting in the (partial) meltdown of three of the six reactors, the destruction of reactor buildings by several hydrogen explosions and the overheating (due to insufficient cooling) of all of the reactors and of the storage facilities for used fuel rods. The amount of radioactive material released was immense, although not yet fully quantified. Thus, the Fukushima nuclear disaster represents the second ever event of level 7 on the International Nuclear Event Scale (INES).

The first level 7 event happened on April 26, 1986, at the Chernobyl Nuclear Power Plant in the Ukraine. The total amount of radioactive material released into the environment on that occassion, particularly during the first 10 days, has been estimated to about 12 trillion (10^{18}) international units of radioactivity, termed the "Becquerel". The radioactivity emitted in both cases included over a hundred mostly short-lived radioactive isotopes, but iodine and cesium were of the main relevance from a human health and environmental standpoint. Radioactive material from the Chernobyl plant was detectable at very low levels over practically the entire northern hemisphere. The explosion put 400 times more radioactive material into the earth's atmosphere than the atomic bomb dropped on Hiroshima (August 6, 1945). On the other hand, all of the atomic weapons tests conducted in the 1950s and 1960s taken together are estimated to have put some 100 to 1000 times more radioactive material into the atmosphere than the Chernobyl accident (Source: IAEA: "Ten years after Chernobyl: what do we really know", based on the proceedings of the IAEA/WHO/EC International Conference, Vienna, April 1996; http://iaea.org/Publications/Booklets/Chernoten/index.html, last accessed March 12, 2013).

- ^{85}Kr (half-life 10.7 years, $\beta + \gamma$ radiation): As a chemically inert noble gas, krypton poses a great problem with respect to its retention in reprocessing plants. This nonpolar and thus rather lipophilic substance is distributed in the body such that it is found especially commonly in fatty tissue. As is the case with many other radionuclides, leukemia is a typical consequence of exposure. In contrast to ^{85}Kr, metastable ^{81m}Kr produced in a controlled nuclear process is a very short-lived (13 seconds) pure γ emitter and can thus be used in the scintigraphic depiction of the respiratory organs; its daughter isotope ^{81}Kr (2×10^5 years) also does not emit particles.
- ^{89}Sr (51 days); ^{90}Sr (29 years, only β radiation): Divalent strontium, which may be transported via particles or by leaching, is deposited mainly in the mineralized part of bone tissue, due to its similarity with the lighter homologue calcium. Like Ca^{2+}, Sr^{2+} is very slowly exchanged in the biomineral, the biological half-life being in the order of 1 decade. Therefore, the long-term effect of ^{90}Sr on bone marrow function and blood formation is to cause leukemia, the bone marrow region being generally regarded as one of the most radiation-sensitive parts of the human body. To complete the negative profile of ^{90}Sr, it is only β emitting and, therefore, difficult to detect; furthermore, the similarly β-emitting ^{90}Y (64 hours) is formed as the daughter isotope. The excretion of incorporated ^{90}Sr may be promoted by administering Ca^{2+}–EDTA complexes (compare (2.1)), as long as there has been no binding in the solid material of the bone. Extraction procedures with macrocyclic ligands of the crown ether type (compare (13.3)) have been considered for the separation of ^{90}Sr in reprocessing plants.
- ^{103}Ru (39 days); ^{106}Ru (1 years, $\beta + \gamma$): Ruthenium is the heavier homologue of iron and can partially replace it in biorelevant compounds. According to Table 5.2, radioactive ruthenium is mainly found in the hemoglobin of erythrocytes.

- ^{131}I (8 days, $\beta + \gamma$): The extreme efficiency of the human body regarding iodine accumulation (compare Section 16.7) is disadvantageous here, as a continuously high exposure may eventually cause tumors of the thyroid gland. As iodide (I^-) or iodate (IO_3^-), the element is very soluble in water, allowing an effective access of radioactive iodine via food and drinking water. In case of imminent exposure, however, the resorption may be inhibited by saturating the iodine demand with an excess of nonradioactive iodide. The biological half-life of iodine is about 140 days for humans: much longer than the physical half-life of 8 days (daughter isotope: ^{131}Xe). ^{131}I is an important radioisotope for medical applications, either in ionic form or after covalent binding to organic carrier molecules.
- ^{132}Te (78 hours, $\beta + \gamma$): As the heavier homologue of sulfur and selenium, tellurium has an affinity for heavy metals and is thus found mainly in the liver and kidney.
- ^{133}Xe (5 days, $\beta + \gamma$); similar to ^{85}Kr: Other, exclusively γ-emitting xenon isotopes are used as radiodiagnostics, especially in the imaging of lung ventilation and of the circulatory system (see Section 18.1.4).
- ^{134}Cs (2 years); ^{137}Cs (30 years; $\beta + \gamma$): Cesium is a heavier, larger homologue of the bioessential potassium. Despite the different ionic radii of Cs^+ (174 pm) and K^+ (151 pm) (each for coordination number 8), there is apparently no effective biological "rejection" of this ion, presumably due to the more than million-fold lower abundance of Cs^+ (e.g. in sea water) relative to K^+. In its ionic compounds, the cesium cation is water soluble and therefore very bioavailable through food or drinking water, in a similar fashion to the iodide anion. Like other alkali metal cations, Cs^+ can be bound and thus retained for some time by ionophores (compare Section 13.2) from lichen and mushrooms, such as the anions of (nor)badion A [13]. In mammals, radioactive cesium is found to a large extent in the intracellular space of muscle; however, like the mimicked potassium but unlike iodide, it is widely distributed over the whole body. Astonishingly, the biological half-life of Cs^+ in humans amounts to about 4 months and is thus longer than that of the related potassium ions. As a simple countermeasure to reduce the uptake of $^{137}Cs^+$ of plants, excess K^+ can be supplied to ^{137}Cs-contaminated soil [14,15], while after the Chernobyl accident large-scale cesium separation from minutely contaminated whey was accomplished in Germany using inorganic ion exchangers of the Prussian blue type (compare also the procedure for Tl^+ detoxification; Section 17.4). Since those days, our knowledge of the soil biogeochemistry of cesium has vastly increased [15].
- ^{140}Ba (13 days, $\beta + \gamma$): Barium is the heavier homologue of strontium. Its affinity to bone tissue can thus be anticipated. It should be remembered, however, that Ba^{2+} forms a very insoluble sulfate.
- ^{144}Ce (284 days, $\beta + \gamma$): Cerium can occur in trivalent or tetravalent form. Since the ions of the lanthanoid elements are generally large, it may substitute for iron, due to a similar charge-to-radius ratio.
- ^{147}Pm (2.6 years, $\beta + \gamma$): Also a lanthanoid element. Lanthanoid(III) ions can replace Ca^{2+} in enzymes (see Sections 14.2 and 18.2).

Considering the problem of long-term storage of nuclear waste, the isotopes ^{137}Cs, ^{90}Sr and ^{147}Pm are among the main radioactive decay components of completely exhausted nuclear fuel material up to 200 years after use. The technology by which to transform highly dangerous radioactive isotopes into less harmful ones using neutron sources (partitioning and transmutation) is in principle available, but the high costs have so far prevented its application in radioactive waste management [16]. Instead, radioactive waste containing highly radioactive material is still dumped.

18.1.4 Radiopharmaceuticals

Radiopharmaceuticals are mainly used for diagnostic purposes: to obtain information on the pathological states of organs. In order to accomplish this function without posing an unnecessary radiation risk, the nuclides used should not be α or β particle emitters and their γ energies should preferentially lie between 100 and 250 keV, the region which is best accessible to scintillation counters and thus very sensitive to external detection (radioscintigraphy; single-photon-emission computed tomography, SPECT). In contrast to other imaging techniques, such as magnetic resonance imaging (MRI), ultrasound and computed tomography (CT) (Section 18.2), which are better suited to representing structural details, radioscintigraphy has the advantage of depicting time-resolved physiological processes. The physical half-life of the isotope should therefore be long enough to allow a controlled production, administration and sufficient distribution in the organism (competition between physical and biological half-lives). On the other hand, the radioactive decay time should be short enough (<8 days) that the radiation is sufficiently intense for detection over a short period of time, leveling off rapidly after the actual diagnosis and thus allowing a repetition of the procedure during medical treatment. The administered amount and the additional radiation dose can then be kept as small as possible. Considering the limited number of radionuclides and the wide variation of physical half-lives, there is only a rather small number of really useful isotopes available. A look at the periodic system immediately shows that metallic elements, with their particular coordination chemistry, should provide most of the radioisotopes physically suitable for medicine (see Table 18.1) [17–23]. Furthermore, an uncomplicated handling of radiopharmaceuticals is desirable (simple chemistry), and very short-lived isotopes require fast transport from the reactor source to the clinical institution.

To fulfill its purpose, it is necessary that the radioisotope in its administered form is selectively taken up by single organs or by tumors. It is remarkable that tumor tissue, with its altered and generally increased metabolism, can accumulate certain inorganic compounds (see also Chapter 19); increased membrane permeability, local changes in pH and variations in electrolyte ion or bioligand concentrations are being discussed as possible reasons.

The most common isotopes for radiotherapy and radiodiagnostics are:

1. ^{131}I, mainly in the form of iodide, with exclusive selectivity for the thyroid gland [18,19,22]; see also Section 16.7.
2. ^{67}Ga, which is used in its trivalent form mainly as a slowly hydrolyzing citrate complex in localizing inflammatory processes [24] or cancer cells [22], due to its being transported by transferrins (Section 8.4.1)
3. ^{99m}Tc, in a wide variety of complexes for the imaging of various organs ($\sim$80% of all radiodiagnostic investigations; see Section 18.1.5) [18,19,22,25].

In attempts to improve the selectivity and thus reduce radiation dosage, a few new developments are worth mentioning:

- *Radioimmunoassay, immunoscintigraphy:* These tag radionuclides to monoclonal antibodies and thus combine a very sensitive detection methodology with a specific transport medium. While the nonmetallic radioactive iodine isotopes have to be covalently bound to antibody molecules, labeling with metal isotopes requires antibody-anchored chelate ligands. Since the metal ions involved are typically large (Table 18.1), polydentate chelate ligands such as the potentially octadentate analogue of EDTA (2.1) diethylenetriamine

Table 18.1 Some frequently used radionuclides in diagnostics (γ emitters and positron emission tomography (PET) radioisotopes) and therapy (β emitter).

Nuclide (type of decay)	Physical half-life	Main emission of γ energy (keV)	Typical field of application
^{43}K (β,γ)	22 hours	373	heart diagnostics
^{57}Co (γ)	271 days	122	B_{12} diagnostics
^{67}Cu (β,γ)	62 hours	93, 185	radio(immuno)therapy
^{67}Ga (γ)	78 hours	93, 185, 300	tumor diagnostics
^{81}Rb (β^+,γ)	4.6 hours	190, 446	heart diagnostics
^{90}Y (β,γ)	64 hours	556	radio(immuno)therapy
^{97}Ru (γ)	69 hours	216, 325, 461	tumor, liver diagnostics
^{99m}Tc (γ)	6 hours	140	tumor diagnostics
^{111}In (γ)	67 hours	171, 245	radioimmunology
^{123}I (γ)	13 hours	159	thyroid gland diagnostics
^{129}Cs (γ)	32 hours	372, 411	heart diagnostics
^{131}I (β,γ)	8 days	364	thyroid gland (diagnostics, therapy)
^{153}Sm (β,γ)	46 hours	103	radio(immuno)therapy
^{169}Yb (γ)	32 days	198	diagnostics
^{186}Re (β,γ)	89 hours	137	radio(immuno)therapy
^{188}Re (β,γ)	17 hours	155	radio(immuno)therapy
^{192}Hg (γ)	5 hours	157, 275, 307	diagnostics
^{201}Tl (γ)	73 hours	68–80	heart diagnostics
^{203}Pb (γ)	2 days	279	diagnostics
^{212}Pb (β,γ)	11 hours	239	radio(immuno)therapy
^{212}Bi (α,β,γ)	1 hours	727	radio(immuno)therapy
^{11}C (β^+)	0.34 hours	961	PET
^{13}N (β^+)	0.17 hours	1190	PET
^{18}F (β^+)	1.83 hours	635	PET
^{64}Cu (β^+)	12.7 hours	650	PET
^{124}I (β^+)	77 hours	1530, 2130	PET

pentaacetate (DTPA (18.2); see also Figure 18.4) or functionalized macrocycles are usually employed [17–19,24,25]. Suitable isotopes for such advanced diagnostic use are the γ-emitting and short-lived isotopes ^{67}Ga, ^{97}Ru, ^{99m}Tc and ^{111}In.

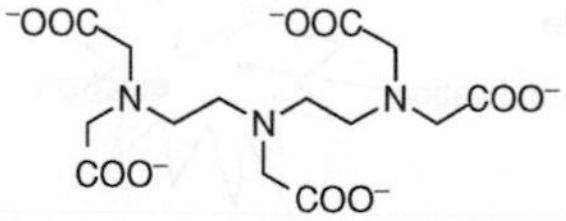

(18.2)

- *Targeted α-therapy (radioimmunotherapy):* Using the same tagging methods, particle (preferably α) emitters such as ^{67}Cu, ^{90}Y, ^{186}Re, ^{188}Re and ^{212}Bi can be selectively incorporated into tumor tissue, allowing very specific tumor cell killing [26].
- *Boron neutron capture therapy (BNCT):* Stable inorganic isotopes such as ^{10}B (19.8% natural abundance), ^{6}Li (7.59%) ^{97}Ru (15.7%) and ^{157}Gd (15.65%) can have an

indirect radiotherapeutic effect. They feature a very high cross-section for the capture of slow, "thermal" neutrons from a neutron source and can thus be selectively converted to α-emitting nuclear-excited isotopes like $^{11}B^*$ in the (tumor) tissue. Boron-modified biomolecules and polyborane clusters (e.g. (18.3)) [27,28] have been tested for this purpose, selective transport to the tumor remaining a major problem [28].

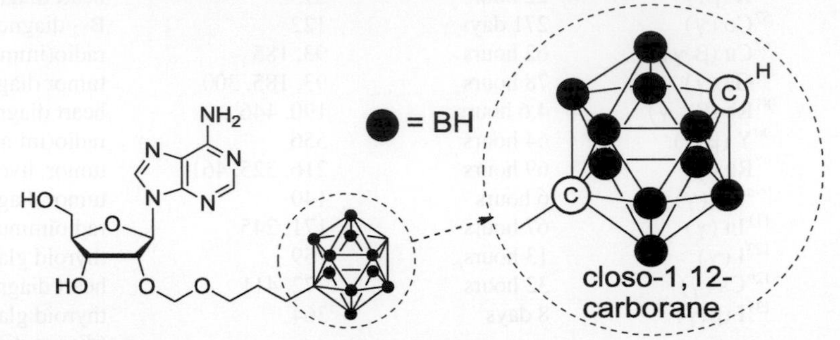

(18.3)

- *Other NCT methods:* Although so far unused in clinical trials, gadolinium neutron capture therapy (GdNCT) has great potential, since ^{157}Gd has an immense cross-section for neutrons (255 000 barns, compared to 3838 for ^{10}B) and its incorporation in tissue using Gd complexes has been widely explored due to the use of Gd in MRI (see Section 18.2). The motivation for using LiNCT (^{7}Li isotope) comes from the application of Li salts in the treatment of bipolar disorder (Section 19.4.2) [28].
- *PET:* β^+-emitting and thus very short-lived isotopes such as ^{82}Rb (76 seconds), ^{62}Cu (9.7 minutes), ^{68}Ga (68 minutes), ^{73}Se (7.1 hours), ^{72}As (26 hours), ^{11}C (20.4 minutes), ^{18}F (110 minutes), ^{15}O (2.9 minutes) and ^{13}N (9.9 minutes) provide exceptional spatial and short-time resolution for this imaging method. Detection is effected via the annihilation process: $\beta^+ + e^- \rightarrow 2\,\gamma$. The two γ quanta with each 511 keV are emitted in opposite directions of exactly 180°, allowing extremely high spatial resolution (Figure 18.1) [19–23].

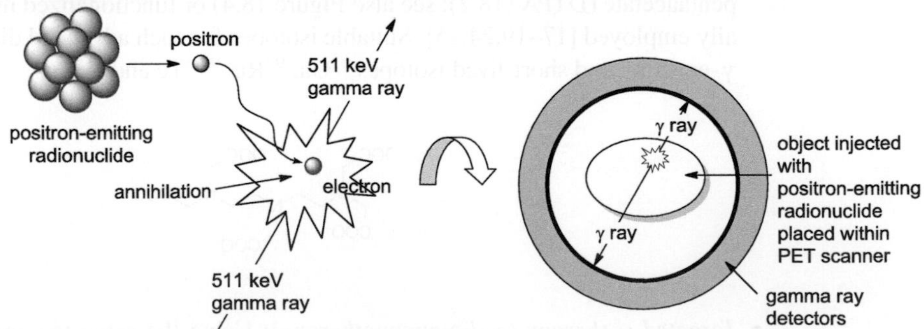

Figure 18.1
Basic physics of positron emission tomography (PET). Reprinted from Biswal, S., Gambhir, S.S., In: Templeton N.S., Lasic D.D., (eds.), *Gene and Cell Therapy: Therapeutic Mechanisms and Strategies, 2nd Ed.* 447–480 © 2003, CRC Press LLC.

18.1.5 Technetium: A "Synthetic Bioinorganic Element"

In quantitative terms, technetium compounds are by far the most important radiopharmaceuticals used today (market share > 80%) [18–23,29]. The longest-lived isotope of this element is ^{98}Tc, with a half-life of about 4 million years. Since technetium does not occur as a product of very slow decay series such as (18.1), the amount produced in nuclear synthetic reactions during the formation of the solar system has totally disappeared. Surprisingly, however, the supply of technetium is higher today than that of the stable heavier homologue rhenium; its price is also lower than that of gold or of the platinum metals. The reason is that the relatively long-lived isotope ^{99}Tc (2.1×10^5 years, only β emitter) makes up about 6% of the products of uranium fission.

For radiomedical purposes, the isotope ^{99}Tc is important not in its slowly β-decaying ground state but in a metastable, nuclear-excited state; that is, as an exclusively γ-emitting ^{99m}Tc with a diagnostically useful half-life of 6 hours. At 140 keV, the γ emission lies in a physiologically and technically very suitable region; also, the daughter isotope ^{99}Tc is a pure β emitter that does not disturb the γ detection. However, the major reason for the popularity of this radioisotope in radiodiagnostics is the availability of an easily operable technetium "reactor" or "generator", which allows the convenient preparation of applicable solutions in a normal clinical environment [23].

The starting isotope is radioactive ^{99}Mo (as molybdate $MoO_4{}^{2-}$; compare Section 11.1), which can be generated from neutron addition to nonradioactive ^{98}Mo (Figure 18.2). The half-life of 66 hours for ^{99}Mo allows a controlled planning of its clinical use; the resulting monoanionic pertechnetate $[^{99m}TcO_4]^-$ is separated from the dianionic molybdate $[^{99}MoO_4]^2$ through elution with physiological saline at an ion-exchange column in nanomolar to micromolar concentrations. The chemical basis of the pertechnetate reactor, which has been used since about 1965, is that technetium differs from its lighter homologue manganese by being relatively stable and only weakly oxidizing in the heptavalent form. In contrast to manganese, the thermodynamically stable oxidation states of technetium in water and in the absence of special ligands include only the very soluble pertechnetate $[TcO_4]^-$, the insoluble hydrous oxide $TcO_2 \times n\ H_2O$ and the metallic element Tc (Figure 18.3). Aqua complexes $[Tc(H_2O)_6]^{n+}$ (with $n = 2$ or 3) are only assumed at very acidic conditions, which underlines the necessity of having suitable ligands to stabilize the low oxidation states (see Table 18.2)

Like radioactive iodide, I^-, monoanionic $[^{99m}TcO_4]^-$ with Tc in the oxidation state +VII can be used in the examination of the thyroid gland. However, for most applications, Tc complexes with ligands such as phosphanes, arsanes, dioximes and isonitriles, with Tc in oxidation states varying from +V to +I, are applied (Table 18.2). The ligands stabilize the oxidation states (reduced Tc exhibits a pronounced tendency for cluster formation – compare the neighboring element molybdenum (11.1)) and define the coordination sphere around the Tc atom, leading to defined species (complexes) in most cases. This is important for their selective application, since these complexes either mimic specific ions, such as K^+ or I^-, or have to meet specific properties, such as certain charge (or none), specific shape or hydrophilic/lipophilic properties. The first-generation radiopharmaceuticals were simple, usually non-substrate-specific ^{99m}Tc-labelled species (complexes, particles). Labeling consisted only in the reduction of $[TcO_4]^-$ in the presence of the ligand. Although side reactions were negligible, impurities (e.g. Sn^{IV} species from the reduction procedure using Sn^{II} compounds), non-substrate-specificity and the fact that the chemical structures of these first radiopharmaceuticals were unknown diminished their performance and initiated

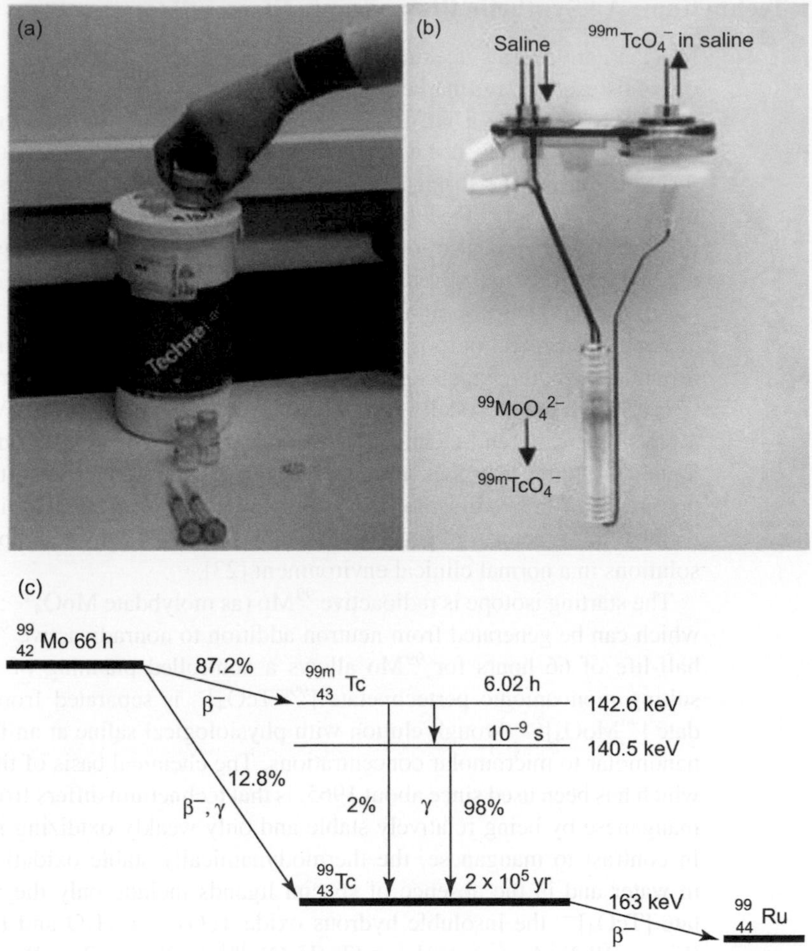

Figure 18.2
(a) Typical ^{99}Mo-^{99m}Tc generator. (b) Its inner components. (c) Decay scheme of ^{99}Mo. Reprinted with permission from [30] © 2011, Royal Society of Chemistry.

further development [19,22,25,29–34]. Remarkably, organometallic complexes containing the $[Cp^{(R)}Tc(I)(CO)_3]$ complex fragment ($Cp^{(R)}$: substituted cyclopentadienide) are meanwhile established (compare also Section 19.5). In addition to complex formation, the specific labeling of biological material such as proteins, carbohydrates and RNA (preparing so-called complex–biomolecule *conjugates*), and incorporation into complete cells such as erythrocytes, have now become common clinical practice [19,22,30,32].

For the imaging of bone tissue, the well-established (although structurally not well characterized) polymeric complexes of technetium with diphosphonate ligands, $^{-2}O_3P{-}CR_2{-}PO_3^{\;2-}$ (R = H, CH_3, OH), are still in use (Table 18.2) [29,34]. Diphosphonates are the analogues of polyphosphates but are more difficult to hydrolyze; they can be excreted rather rapidly, due to their slow enzymatic degradation. It is assumed that the bifunctional phosphonate ligands allow a coordination of the Tc tracer with exposed Ca^{2+} ions of the growing hydroxyapatite crystals.

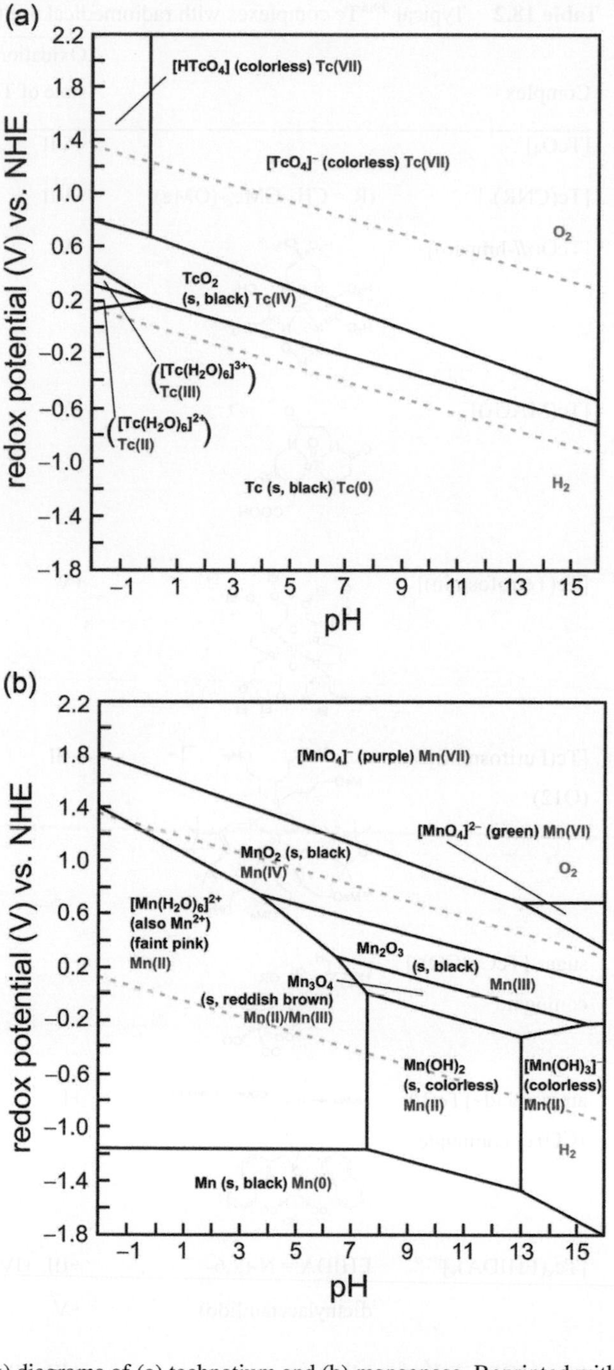

Figure 18.3
Stability (Pourbaix) diagrams of (a) technetium and (b) manganese. Reprinted with permission from Lewis, B.J., *et al.*, J. Nucl. Mater, 340, 69–82 © 2005, Elsevier.

Table 18.2 Typical ^{99m}Tc complexes with radiomedical utility [22,30–34].

Complex		Oxidation state of Tc	Similar to	Target organ/system
$[TcO_4]^-$		+VII	I^-	thyroid gland
$[Tc(CNR)_6]^+$	(R = CH_2-CMe_2-(OMe))	+VII	K^+	heart muscle
[TcO(*dl*-hmpao)]	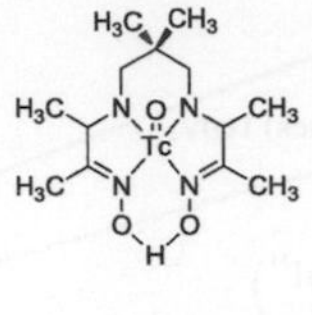	+V	small, lipophilic	brain
$[Tc(MAG_3)]^-$	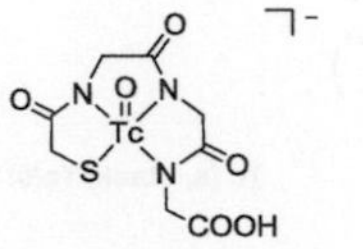	+V	small, highly hydrophilic	kidney
$[Tc(Tetrofosmin)]^+$	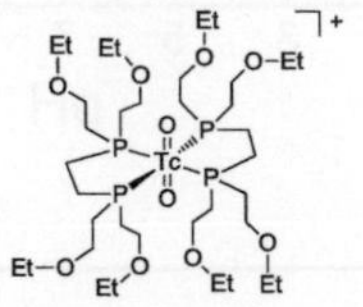	+V	K^+	heart muscle
$[Tc(Furifosmin)]^+$ (Q12)	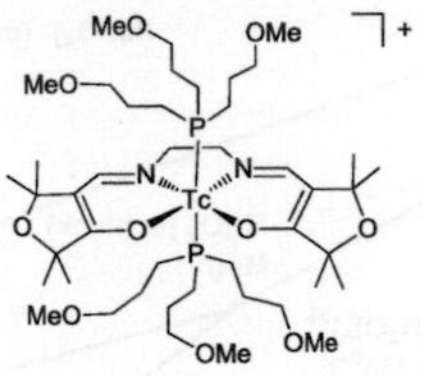	+III	K^+	heart muscle
sugar-$[TcCp(CO)_3]$ conjugate	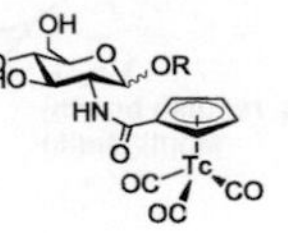	+I	sugar	glucose metabolism
amino acid–[Tc(L) $(CO)_3$] conjugate	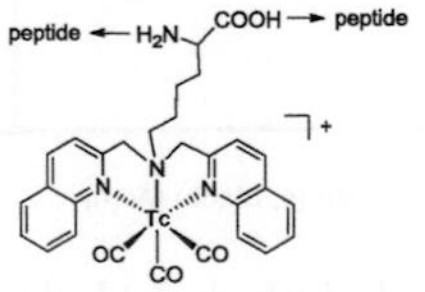	+I	peptide	leukocytes
$[Tc_x(EHIDA)_y]^{n-}$ [a]	EHIDA = N-(2,6-diethylacetanilido)	+III, +IV, +V		liver
$[Tc_x(MDP)_y]^{n-}$ [a,b]	MDP = methylenediphosphonate	+III, +IV, +V		bone

[a]The structures of the complexes and the oxidation state of Tc are not clear. The labeling agents were prepared from the ligands and from $^{99m}TcO_4^-$ in the presence of reducing agents such as $SnCl_2$ and BH_4^- [33,34].
[b]The compound $[Li(H_2O)_3][Tc(OH)(MDP)] \times 1/3H_2O$ has been structurally characterized and shows MDP^{2-} bridging polymeric chains of octahedrally coordinated TcO_6 units with two bridging OH ligands [34].

18.1.6 Radiotracers for the Investigation of the Metallome

In analogy with the established notions of the genome and the proteome, the term "metallome" was originally coined by R. J. P. Williams [35] and refers to the distribution of an element in intracellular compartments, cells or organisms resulting from the element's "paths" (see also insert Chapter 8). These "paths" are the combined and integrated actions on environmentally presented metal ions of cellular sequestering, passive and active transport, storage and release, regulation of expression and insertion into biomacromolecules, all resulting in a particular distribution of metals over compartments in terms of concentration and speciation (coordination chemistry, oxidation state) [36].

The study of a metallome and of the interactions and functional connections of metal ions and their species with genes, proteins, metabolites and other biomolecules within organisms and ecosystems (frequently called "metallomics") [35–39] requires either highly sensitive metal-specific spectroscopic methods directly connected with such metal nuclear properties as specific radioactive decay, emission (Mössbauer or x-ray emission spectroscopy) and nuclear magnetic moments (NMR, electron paramagnetic resonance (EPR)) or highly sensitive mass-spectrometric methods [37–39]. It is clear that radionuclides, as listed in Table 18.1, are well suited to the study of the metallome of the corresponding element or its substitutes (Ru or Ga for Fe, Rb or Cs for K, etc.).

18.2 Medical Imaging Based on Nonradioactive Inorganic Compounds

18.2.1 Magnetic Resonance Imaging

MRI is very probably the most important diagnostic imaging tool in clinical medicine today, offering exquisite anatomical images of soft tissues based on resonance detection of protons (compare ^{1}H NMR) largely in water or lipids. Image contrast is readily manipulated by choosing from a standard set of pulse sequences, which weight signal intensities based on differences in proton densities and longitudinal (T_1) and transversal (T_2) magnetic relaxation times. For many clinical applications, however, it is now common practice to administer an exogenous contrast agent in order to highlight specific tissue regions based on flow or agent biodistribution. MRI contrast agents have so far been largely confined to small, highly paramagnetic metal complexes, typically gadolinium(III) complexes ($S = 7/2$), which alter signal intensity by shortening the relaxation times of the water protons. The first example of such a modern MRI contrast agent was the Gd^{III} complex of DTPA (18.2), approved for clinical use in 1988 [40–43]. Further developments focus on polydentate EDTA derivatives providing a trigonal prismatic coordination of eight N or O ligands and one water molecule in $[(N_xO_y)Gd{-}OH_2]$ ($x + y = 8$) (Figure 18.4) [43–46].

The ability of a contrast agent to change the relaxation rate is represented quantitatively as relaxivity, r_1 or r_2, where the subscript refers to either the longitudinal ($1/T_1$) or the transverse ($1/T_2$) rate. Relaxivity is simply the change in relaxation rate after the introduction of the contrast agent ($\Delta(1/T_1)$), normalized to the concentration of contrast agent or metal ion [M] (18.4).

$$r_1 = \frac{\Delta(1/T_1)}{[\mathrm{M}]} \tag{18.4}$$

The relaxivity can be correlated with a number of physicochemical parameters that characterize the complex structure and dynamics in solution. Those that can be chemically

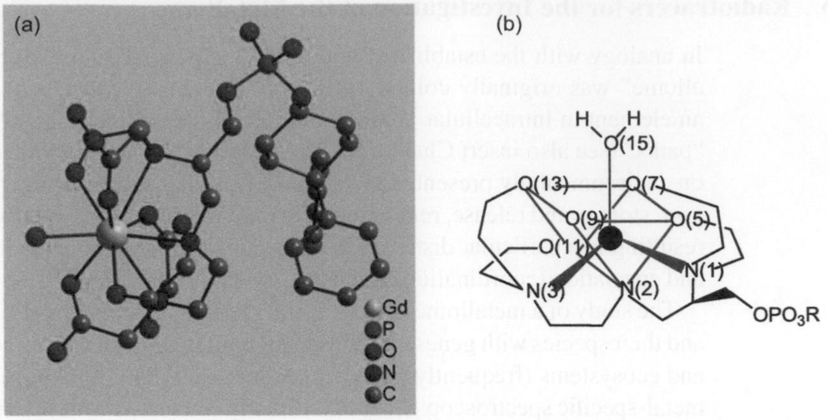

Figure 18.4
(a) Crystal structure of the Gd complex $[Gd(R\text{-}DTPA)(H_2O)]^{2-}$ (DTPA = diethylenetriamine pentaacetate, see (18.2)). (b) Schematic view of the $N_3O_5Gd\text{-}OH_2$ binding motif. The 4,4-diphenylcyclohexylphosphonooxymethyl side group, R, is intended for improved tissue targeting. Reprinted with permission from [43] © 2007, American Chemical Society.

tuned are of primary importance in the ligand design. They are (1) the number of inner-sphere water molecules directly coordinated to the Gd^{III} center, (2) the residence time τ_M of the coordinated water molecule, (3) the rotational correlation time τ_R representing the molecular tumbling time of a complex, (4) the interaction of the complex with water molecules in the second and outer spheres and (5) electronic parameters.

Commercial contrast agents are only effective at high concentrations (0.1 mM). As a result of this, considerable efforts have been made to increase the sensitivity (and selectivity) of the method:

- searches for new Gd compounds with high relaxivity by ligand design [40–46];
- compounds with other lanthanide metals with high relaxivity due to chemical exchange saturation transfer (CEST) spectroscopy [42,45–47];
- complexes with biologically relevant paramagnetic 3d metal ions such as high-spin Fe^{2+} or Cu^{2+} for function-dependant contrast enhancement (e.g. in neurobiology) [32,46,48];
- technical developments of the spectrometers (high-field) and of the detection methods [41,42].

18.2.2 X-ray Contrast Agents

The use of x-ray radiation (discovered by Wilhelm C. Roentgen in 1895) to visualize the internal anatomic structures of a patient without surgery is still one of the most important diagnostic tools in modern medicine, useful in performing the initial diagnosis, the planning of treatment and the post-treatment evaluation. Even with the recent phenomenal growth of MRI and ultrasound procedures, x-ray imaging studies remain the workhorse of modern radiology (75–80% of all diagnostic imaging procedures are currently x-ray related [49]). While the large inherent contrast between electron-dense bones and the surrounding more permeable soft tissues allows high-resolution imaging, the native contrast between the different soft tissues is so small that unenhanced x-ray imaging cannot differentiate between them. In order to better delineate soft-tissue regions such as the cardiovascular system, safe and efficient x-ray contrast agents (also called "radiographic contrast agents", "radiopaque agents" or "roentgenographic agents") have been developed [49–51]. Contrast agents are a class of pharmaceuticals that, when administrated to a patient, enter and pass through anatomic regions of interest to provide transient contrast enhancement, and are preferably completely excreted afterwards without being metabolized. There are two types of x-ray contrast agent currently approved for human use: barium sulfate suspensions, which are used strictly for gastrointestinal (GI) tract imaging, and water-soluble aromatic iodinated contrast agents [50,51]. The earliest use of a barium contrast agent for GI imaging can be traced back to 1910. Today, inexpensive $BaSO_4$ formulations are routinely used in about 5 million x-ray procedures per year in the USA alone.

The development of x-ray imaging using contrast agents proceeds essentially along two lines. The first is the development of new contrast agents, including complexes of specific metals such as Ag, Cs, Ga, Bi, Zr, Sn, W and rare earth metals, all with generally high but varying x-ray densities and varying absorption energies. These compounds comprise nonsoluble metals and metal salts (as $BaSO_4$), as well as soluble but highly stable complexes with macrocyclic ligands, comparable to those discussed for radioimmune and MRI diagnostic purposes [49,51]. In order for the new contrast agents to be competitive with water-soluble iodinated agents, they should possess a number of crucial properties, including high water solubility, stability at physiological pH and a suitable pharmacokinetic profile (organ specificity). They should be effectively excreted, should not exert any immunogenicity, carcinogenicity, mutagenicity or teratogenicity and should allow ready and easy administration [49]. Recently, nanomaterials have also come into the discussion [52].

Due to the increasing availability of monochromatized x-ray radiation from synchrotron facilities, the second route of x-ray development, synchrotron radiation computed tomography (SRCT) using high-Z contrast agents (mainly I and Xe), is evolving rapidly [51].

Further Reading

S. Tabakov, F. Milano, S.-E. Strand, C. Lewis, P. Sprawls, *Encyclopaedia of Medical Physics*, CRC press, Taylor & Francis, Boca Raton, **2012**.

E. Alessio, *Bioinorganic Medical Chemistry*, Wiley-VCH, Weinheim, **2011**.

D. A. Atwood (ed.), *Radionucleides in the environment*, John Wiley & Sons, Chichester, **2010**.

C. Jones, J. Thornback, *Medicinal Applications of Coordination Chemistry*, RSC Publishing, Cambridge, **2007**.

M. J. Welch, C. S. Redvanly (eds.), *Handbook of Radiopharmaceuticals – Radiochemistry and Applications*, John Wiley & Sons, Chichester, **2002**.

References

1. (a) L. R. Dartnell, *Astrobiol.* **2011**, *11*, 551–582: *Ionizing radiation and life*; (b) P. O'Neill, P. Wardman, *Int. J. Radiat. Biol.* **2009**, *85*, 9–25: *Radiation chemistry comes before radiation biology*.
2. M. G. Cuypers, E. P. Mitchell, C. V. Romão, S. M. McSweeney, *J. Mol. Biol.* **2007**, *371*, 787–799: *The crystal structure of the Dps2 from deinococcus radiodurans reveals an unusual pore profile with a non-specific metal binding site*.
3. A. Grove, S. P. Wilkinson, *J. Mol. Biol.* **2005**, *347*, 495–508: *Differential DNA binding and protection by dimeric and dodecameric forms of the ferritin homolog Dps from deinococcus radiodurans*.

4. T. M. Seed, *Health Phys.* **2007**, *89*, 531–545: *Radiation protectants: current status and future prospects.*
5. R. Doll, *Brit. J. Cancer* **1995**, *72*, 1339–1349: *Hazards of ionising radiation: 100 years of observations on man.*
6. (a) R. M. Macklis, *Sci. Am.* **1993**, *269*, 94–99: *The great radium scandal;* (b) *Los Alamos Science*, **1995**, *23*, 224–233: *Radon – the benchmark.*
7. D. V. Gopinath, *Curr. Sci.* **2007**, *93*, 1230–1248: *Radiation effects, nuclear energy and comparative risks.*
8. J. Woodhouse, *Radon*, in R. H. Waring, G. B. Steventon, S. C. Mitchell (eds.), *Molecules of Death*, 2nd Edition, Imperial College Press, London, **2007**.
9. M. J. Atkinson, M. T. Spanner, M. Rosemann, U. Linzner, W. A. Müller, W. Gössner, *Rad. Res.* **2005**, *164*, 230–233: *Intracellular sequestration of ^{223}Ra by the iron-storage protein ferritin.*
10. S. Fukuda, *Curr. Med. Chem.* **2005**, *12*, 2765–2770: *Chelating agents used for plutonium and uranium removal in radiation emergency medicine.*
11. R. L. Kathren, *Rad. Prot. Dos.* **2004**, *109*, 399–407: *A review of contributions of human tissue studies to biokinetics, bioeffects and dosimetry of plutonium in man.*
12. A. E. V. Gorden, J. Xu, K. N. Raymond, *Chem. Rev.* **2003**, *103*, 4207–4282: *Rational design of sequestering agents for plutonium and other actinides.*
13. A. Korovitch, J.-B. Mulon, V. Souchon, I. Leray, B. Valeur, A. Mallinger, B. Nadal, T. Le Gall, C. Lion, N.-T. Ha-Duong, J.-M. El Hage Chahine, *J. Phys. Chem. B* **2010**, *114*, 12655–12665: *Norbadione A: kinetics and thermodynamics of cesium uptake in aqueous and alcoholic media.*
14. E. Marshall, *Science* **1989**, *245*, 123–124: *Fallout from Pacific tests reaches Congress.*
15. (a) H. Vandenhove, L. Sweeck, *Integr. Environ. Assess. Manag.* **2011**, *7*, 374–378: *Soil vulnerability for cesium transfer*; (b) Y.-G. Zhu, E. Smolders, *J. Exp. Bot.* **2000**, *51*, 1635–1645: *Plant uptake of radiocesium: a review of mechanisms, regulation and application.*
16. International Atomic Energy Agency, *Implications of Partitioning and Transmutation in Radioactive Waste Management, IAEA Technical Reports Series no. 435*, International Atomic Energy Agency, Vienna, **2004**.
17. T. A. Kaden, *Dalton Trans.* **2006**, 3617–3623: *Labelling monoclonal antibodies with macrocyclic radiometal complexes. A challenge for coordination chemists.*
18. W. A. Volkert, T. J. Hoffman, *Chem. Rev.* **1999**, *99*, 2269–2292: *Therapeutic radiopharmaceuticals.*
19. P. Blower, *Dalton Trans.* **2006**, 1705–1711: *Towards molecular imaging and treatment of disease with radionuclides: the role of inorganic chemistry.*
20. T. J. McCarthy, S. W. Schwarz, M. J. Welch, *J. Chem. Educ.* **1994**, *71*, 830–836: *Nuclear medicine and positron emission tomography: an overview.*
21. S. M. Ametamey, M. Honer, P. A. Schubiger, *Chem. Rev.* **2008**, *108*, 1501–1516: *Molecular imaging with PET.*
22. G. R. Morais, A. Paulo, I. Santos, *Organometallics* **2012**, *31*, 5693–5714: *Organometallic complexes for SPECT imaging and/or radionuclide therapy.*
23. S. M. Qaim, *Rad. Phys. Chem.* **2004**, *71*, 917–926: *Use of cyclotrons in medicine.*
24. J. Burgess, *Chem. Soc. Rev.* **1996**, *25*, 85–92: *Man and the elements of groups 3 and 13.*
25. T. Storr, K. H. Thompson, C. Orvig, *Chem. Soc. Rev.* **2006**, *35*, 534–544: *Design of targeting ligands in medicinal inorganic chemistry.*
26. W. M. Brechbiel, *Dalton Trans.* **2007**, 4918–4928: *Targeted α-therapy: past, present, future?*
27. (a) N. P. E. Barry, P. J. Sadler, *Chem. Soc. Rev.* **2012**, *41*, 3264–3279: *Dicarba-closo-dodecaborane-containing half-sandwich complexes of ruthenium, osmium and iridium: biological relevance and synthethic strategies*; (b) F. Issa, M. Kassiou, L. M. Rendina, *Chem. Rev.* **2011**, *111*, 5701–5722: *Boron in drug discovery: carboranes as unique pharmacophores in biologically active compounds*; (c) W. Kaim, N. S. Hosmane, *J. Chem. Sci.*, **2010**, *122*, 7–18: *Multidimensional potential of boron-containing molecules in functional materials.*
28. L. M. Rendina, *J. Med. Chem.* **2010**, *53*, 8224–8227: *Can lithium salts herald a new era for neutron capture therapy?*
29. M. J. Clarke, L. Podbielski, *Coord. Chem. Rev.* **1987**, *78*, 253–331: *Medical diagnostic imaging with complexes of ^{99m}Tc.*

30. S. Liu, S. Chakraborty, *Dalton Trans.* **2011**, *40*, 6077–6086: *^{99m}Tc-centred one-pot synthesis for preparation of ^{99m}Tc radiotracers.*
31. R. Alberto, *Eur. J. Inorg. Chem.* **2009**, 21–31: *The chemistry of technetium-water complexes within the manganese triad: challenges and perspectives.*
32. F. Mendes, A. Paulo, I. Santos, *Dalton Trans.* **2011**, *40*, 5377–5393: *Metalloprobes for functional monitoring of tumour multidrug resistance by nuclear imaging.*
33. A. M. Rey, *Curr. Med. Chem.* **2010**, *17*, 3673–3683: *Radiometal complexes in molecular imaging and therapy.*
34. U. Abram, R. Alberto, *J. Braz. Chem. Soc.*, **2006**, *17*, 1486–1500: *Technetium and rhenium – coordination chemistry and nuclear medical applications.*
35. R. J. P. Williams, *Coord. Chem. Rev.* **2001**, *216–217*, 538–595: *Chemical selection of elements by cells.*
36. W. R. Hagen, *Metallomics* **2009**, *1*, 384–391: *Metallomic EPR spectroscopy.*
37. E. A. Permyakov, *Metalloproteomics*, John Wiley & Sons, Chichester, **2009**.
38. S. Mounicou, J. Szpunar, R. Lobinski, *Chem. Soc. Rev.* **2009**, *38*, 1119–1138: *Metallomics: the concept and methodology.*
39. Y.-F. Li, C. Chen, Y. Qu, Y. Gao, B. Li, Y. Zhao, Z. Chai, *Pure Appl. Chem.* **2008**, *80*, 2577–2594: *Metallomics, elementomics, and analytical techniques.*
40. E. J. Werner, A. Datta, C. J. Jocher, K. N. Raymond, *Angew. Chem. Int. Ed. Engl.* **2008**, *47*, 8568–8580: *High-relaxivity MRI contrast agents: where coordination chemistry meets medical imaging.*
41. P. Hermann, J. Kotek, v. Kubíček, I. Lukeš, *Dalton Trans.* **2008**, 3027–3047: *Gadolinium(III) complexes as MRI contrast agents: ligand design and properties of the complexes.*
42. P. Caravan, *Chem. Soc. Rev.* **2006**, *35*, 512–523: *Strategies for increasing the sensitivity of gadolinium based MRI contrast agents.*
43. Z. Tyeklár, S. U. Dunham, K. Midelfort, D. M. Scott, H. Sajiki, K. Ong, R. B. Lauffer, P. Caravan, T. J. McMurry, *Inorg. Chem.* **2007**, *46*, 6621–6631: *Structural, kinetic, and thermodynamic characterization of the interconverting isomers of MS-325, a gadolinium(III)-based magnetic resonance angiography contrast agent.*
44. M. Botta, L. Tei, *Eur. J. Inorg. Chem.* **2012**, *12*, 1945–1960: *Relaxivity enhancement in macromolecular and nanosized Gd^{III}-based MRI contrast agents.*
45. M. Bottrill, L. Kwok, N. J. Long, *Chem. Soc. Rev.* **2006**, *35*, 557–571: *Lanthanides in magnetic resonance imaging.*
46. K. L. Haas, K. J. Franz, *Chem. Rev.* **2009**, *109*, 4921–4960: *Application of metal coordination chemistry to explore and manipulate cell biology.*
47. M. Woods, D. E. Woessner, A. D. Sherry, *Chem. Soc. Rev.* **2006**, *35*, 500–511: *Paramagnetic lanthanide complexes as PARACEST agents for medical imaging.*
48. E. L. Que, D. W. Domaille, C. J. Chang, *Chem. Rev.* **2008**, *108*, 1517–1549: *Metals in neurobiology: probing their chemistry and biology with molecular imaging.*
49. S.-B. Yu, A. D. Watson, *Chem. Rev.* **1999**, *99*, 2353–2377: *Metal-based x-ray contrast media.*
50. A. Rutten, M. Prokop, *Anti-Canc. Agents Med. Chem.* **2007**, *7*, 307–316: *Contrast agents in x-ray computed tomography and its applications in oncology.*
51. H. Chen, M. M. Rogalski, J. N. Anker, *Phys. Chem. Chem. Phys.* **2012**, *14*, 13 469–13 486: *Advances in functional x-ray imaging techniques and contrast agents.*
52. Y. Liu, K. Ai, L. Lu, *Acc. Chem. Res.* **2012**, *45*, 1817–1827: *Nanoparticulate x-ray computed tomography contrast agents: from design validation to in vivo applications.*

19 Chemotherapy Involving Nonessential Elements

19.1 Overview

According to the principles outlined in Section 2.1, it is self-evident that compounds of the essential elements can be therapeutically useful. As with pharmaceuticals which contain only "organic" molecules as active ingredients, it may be that a special chemical compound of such an element is particularly active: either the administered substance itself or a metabolized product (compare Table 6.2). In addition to particle-emitting (i.e. primarily physically effective) radiotherapeutic agents with inorganic isotopes (Chapter 18), there are now an increasing number of chemotherapeutically active compounds of those inorganic elements which are nonessential according to present knowledge. Among these are arsenic compounds [1], such as the salvarsan derivatives $[As(aryl)]_n$, the discovery of which was a landmark in chemotherapy (P. Ehrlich, Nobel Prize in Medicine 1908), some long-known antiseptic preparations containing mercury, silver or boron and certain bismuth complexes which have proven to be specifically effective against infectious forms of gastritis. The clinically used "inorganic" drugs based on nonessential elements [2–8] described in this chapter show a more complex mechanism of action than the early bactericidal preparations. The use of platinum complexes in cancer treatment, of gold compounds in the therapy of rheumatoid arthritis, of lithium salts in the treatment of psychiatric disorders, of bismuth compounds against gastric ulcers and of many more inorganic compounds will be outlined here, along with some of the mechanisms established for these drugs. A plethora of new compounds containing nonessential inorganic elements is still under investigation, with promising potential for therapeutic applications (anticancer, antibiotic, anti-HIV). These too will be briefly discussed, as will the newly evolving field of bioorganometallic chemistry.

19.2 Platinum Complexes in Cancer Therapy

19.2.1 Discovery, Application and Structure–Effect Relationships

The cytostatic effect of *cis*-diamminedichloridoplatinum(II), "cisplatin" (19.1a), a square-planar complex due to the d^8 configuration of the metal (compare Section 3.2.1), was serendipitously discovered by B. Rosenberg in the 1960s [9]. Studying the influence of weak alternating currents on the growth of *E. coli* bacteria, he used ostensibly inert platinum electrodes. The result of his experiments was an inhibition of cell reproduction without simultaneous inhibition of bacterial growth, which eventually led to the formation of long,

Bioinorganic Chemistry: Inorganic Elements in the Chemistry of Life – An Introduction and Guide, Second Edition.
Written and Translated by Wolfgang Kaim, Brigitte Schwederski and Axel Klein.

filamentous cells. In the course of subsequent, more detailed studies, it was found that it was not the electric current itself but trace amounts of *cis*-configured chlorido complexes such as (19.1a), resulting from an oxidation of the platinum electrode, that were responsible for this biological effect [9]. Like gold, oxidized platinum forms very stable complexes with halides and pseudohalides, and the otherwise very noble metals are thus much more easily oxidized in the presence of such ligands. Familiar examples include the dissolution of these metals in concentrated nitric and (chloride-containing) hydrochloric acid, "aqua regia" and the cyanide process in the oxidative leaching of gold ores. In the presence of chloride and ammonium/ammonia as components of typical buffered culture media, sufficient material is apparently oxidatively dissolved from the platinum electrode; the potential for the formation of the primarily resulting complexes $[PtCl_{4,6}]^{2-}$ from Pt is about 0.7 V. The observed filamentous growth of bacteria indicates the potential antitumor activity of the corresponding substances via an inhibition of cell division (cytostatic effect) [10,11]. Of a large number of platinum complexes [3–8,11–19] and other metal compounds [3–8,19–22] tested *in vivo* and in partial clinical screening, cisplatin (19.1a) for a long time shown the best cytostatic results [13].

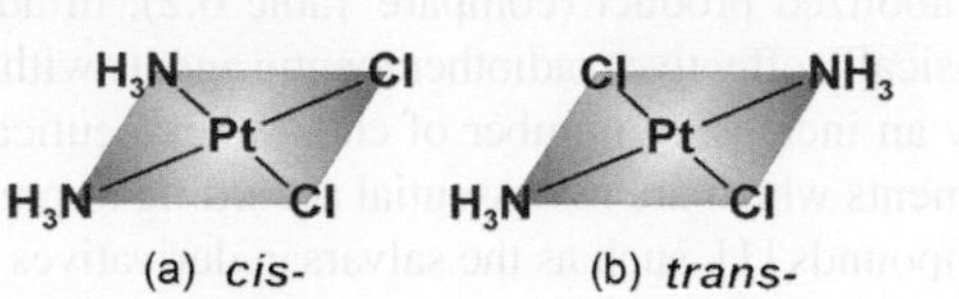

(19.1)

Complex (19.1a), which has been approved as a drug since about 1978, is still being used either alone or in combination with other cytostatic agents such as bleomycin, vinblastin, adriamycin, cyclophosphamide and doxorubicin (19.2) against testicular and ovarian cancers, and increasingly against bladder, cervical and lung tumors and tumors in the head/neck area. Over the years, the prospects for a complete cure of testicular and bladder cancer have vastly improved (>90%; → cyclist Lance Armstrong), mainly due to the use of cisplatin and other platinum-containing drugs. Although a number of new platinum-containing complexes, such as carboplatin, oxaliplatin, spiroplatin, iproplatin, satraplatin and picoplatin (second generation), and more recently Aroplatin and ProLindac (19.3) have entered the market [13,14], cisplatin is still the metal-containing cytostatic drug with the highest turnover worldwide. For a long time now, this compound has topped the list of the most successful patent applications granted to a US university (here Michigan State University).

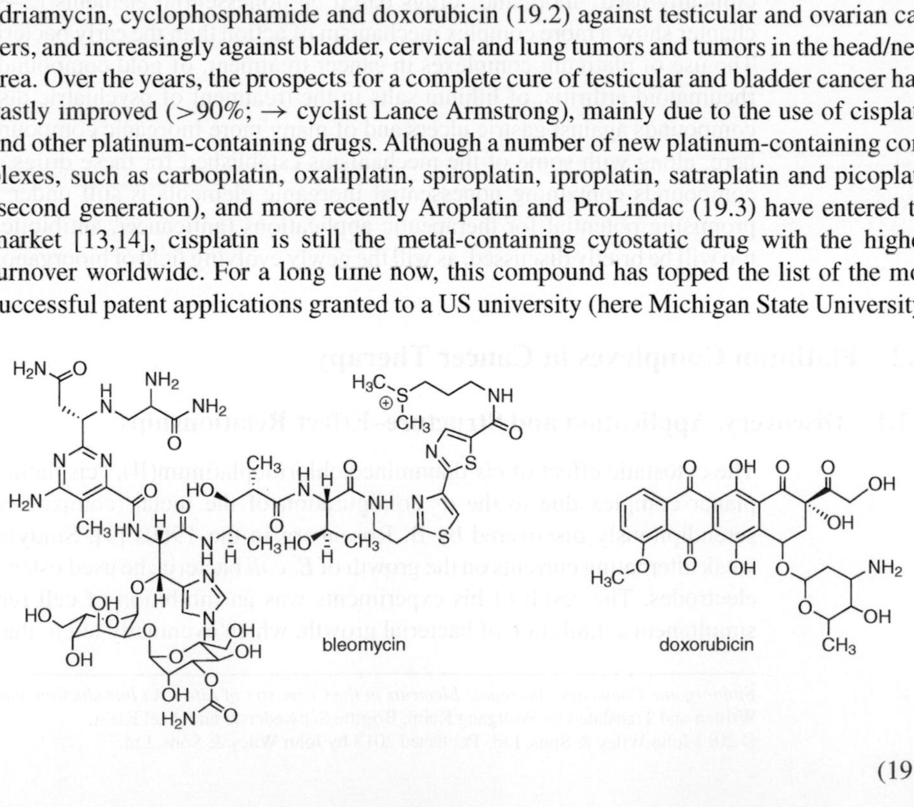

(19.2)

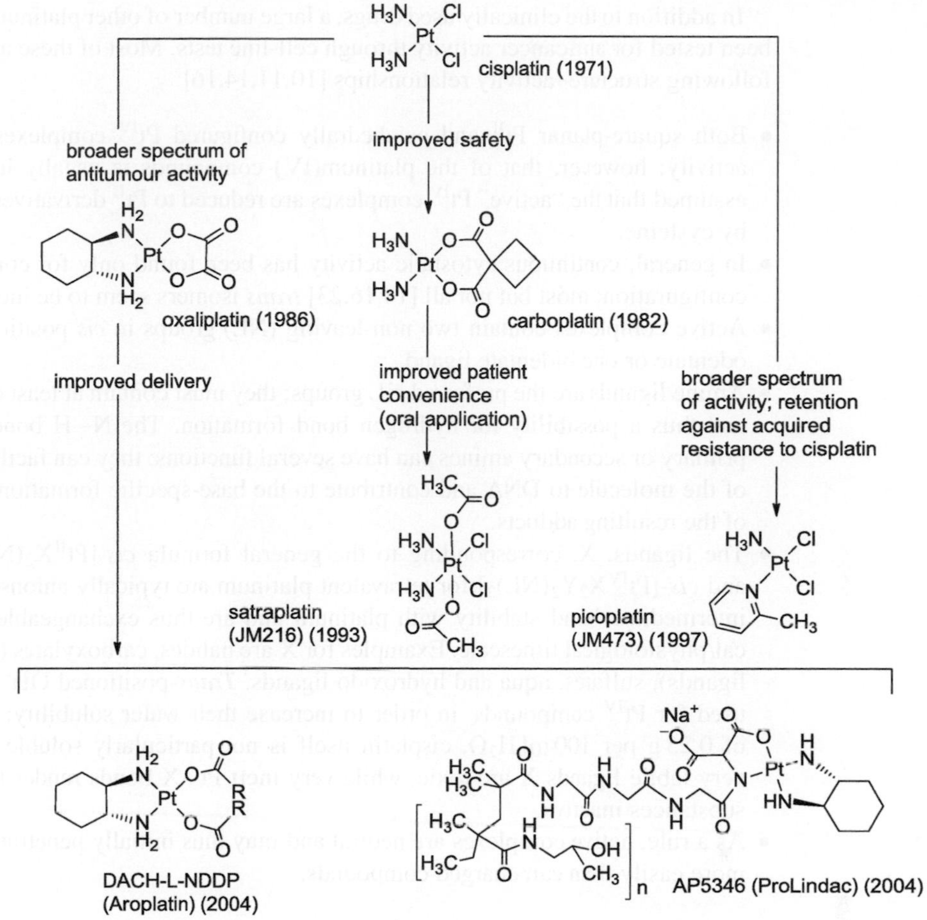

Reprinted with permission from [13] © 2007, Nature Publishing Group. (19.3)

The most common side effects of a cisplatin therapy include kidney and gastrointestinal problems, including nausea, which may be attributed to the inhibition of enzymes through coordination of the heavy-metal platinum to sulfhydryl groups in proteins. Accordingly, a treatment with sulfur compounds, such as sodium diethyldithiocarbamate (19.4) and thiourea, and subsequent diuresis may counteract these symptoms. In contrast to many other cytostatic agents, however, cisplatin causes only minor, reversible damage to the spinal region.

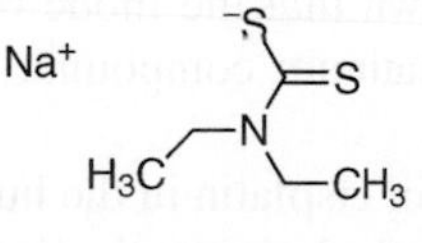

(19.4)

The aim of the development of second- and third-generation analogues of cisplatin was (and still is) to obtain drugs with broader applicability, lower therapeutic dosage, reduced side effects, diminished therapeutic resistance (a number of cisplatin-resistant cell lines have meanwhile developed) and oral administration (cisplatin has to be injected) (19.3) [10,13–15].

In addition to the clinically used drugs, a large number of other platinum complexes have been tested for anticancer activity through cell-line tests. Most of these are in line with the following structure–activity relationships [10,11,14,16]:

- Both square-planar Pt^{II} and octahedrally configured Pt^{IV} complexes show cytostatic activity; however, that of the platinum(IV) compounds is usually lower. It has been assumed that the "active" Pt^{IV} complexes are reduced to Pt^{II} derivatives *in vivo*, possibly by cysteine.
- In general, continuous cytostatic activity has been found only for compounds with *cis* configuration; most but not all [14,16,23] *trans* isomers seem to be ineffective.
- Active complexes contain two non-leaving (NL) groups in *cis* positions and two monodentate or one bidentate ligand.
- Amine ligands are the preferred NL groups; they must contain at least one N–H function and thus a possibility for hydrogen bond formation. The N−H bonds in coordinated primary or secondary amines can have several functions: they can facilitate the approach of the molecule to DNA and contribute to the base-specific formation and stabilization of the resulting adducts.
- The ligands, X, corresponding to the general formula *cis*-$[Pt^{II}X_2(NL)_2]$ for divalent and *cis*-$[Pt^{IV}X_2Y_2(NL)_2]$ for tetravalent platinum are typically anions which exhibit an intermediate bond stability with platinum and are thus exchangeable on a therapeutical/physiological timescale. Examples for X are halides, carboxylates (often as chelating ligands), sulfates, aqua and hydroxido ligands. *Trans*-positioned OH^- groups are often used for Pt^{IV} compounds, in order to increase their water solubility; with a maximum of 0.25 g per 100 ml H_2O, cisplatin itself is not particularly soluble. Complexes with very labile ligands X are toxic, while very inert Pt−X bonds render the corresponding substances inactive.
- As a rule, active complexes are neutral and may thus initially penetrate cell membranes more easily than can charged compounds.

19.2.2 Cisplatin: Mode of Action

Tumor cells are distinguished from normal body cells by the loss of genetic control of their lifespan. Likewise, their feedback mechanisms with regard to the existence of neighboring cells are impaired, which leads to the uncontrolled growth of tumor tissue. In normal cells, these processes are restrained and regulated by proto-oncogenes; cancer may thus result from changes of these genes or their expression. According to this basic concept of carcinogenesis, cisplatin is believed to exert its cytostatic effect primarily through coordination with DNA in the cell nucleus, while reactions in other regions (e.g. with serum proteins) cause undesired side effects. From experiments with second-generation compounds, it is known that the mode of action may actually be quite complex; depending on the type of platinum compound, there is more or less inhibition of DNA, RNA and protein synthesis.

The pathways of cisplatin in the human body are schematically outlined in Figure 19.1. After injection (oral administration is not possible, due to hydrolysis in highly acidic gastric juice), cisplatin can be bound to plasma proteins and then renally excreted (30–70%); the remaining fraction is transported by the blood in unaltered form. After passive transport of neutral cisplatin through the cell membranes of different organs and tumor cells, it is rapidly hydrolyzed (19.5) due to the markedly lower chloride concentration in intracellular

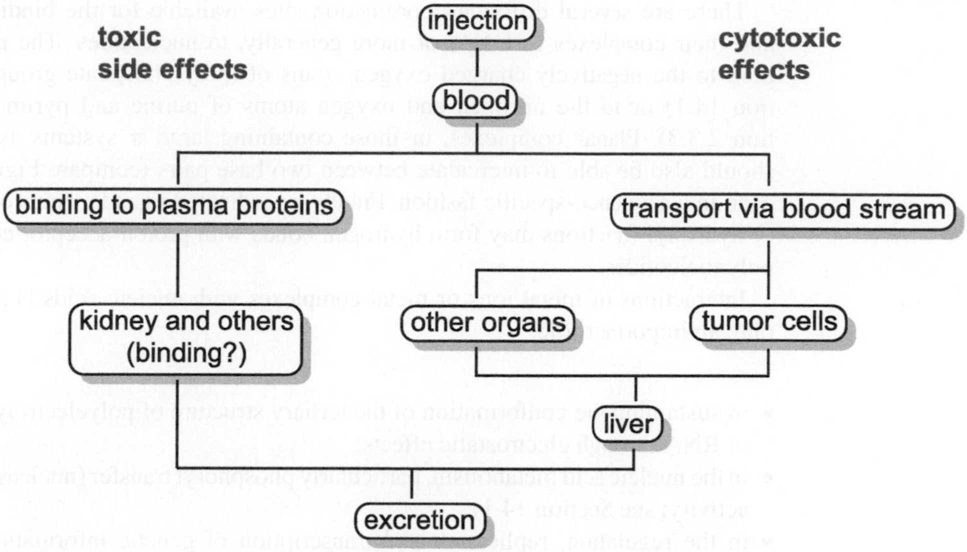

Figure 19.1
Simplified metabolic pathways of cisplatin in the human body.

regions (compare Figure 13.2). Within cells, about 40% of the platinum is present as *cis*-$[Pt(NH_3)_2Cl(H_2O)]^+$. This hydrolysis product (19.5) of cisplatin is kinetically labile as H_2O is a much better leaving group with respect to Pt^{II} than is Cl^-; it is thus assumed that *cis*-$[Pt(NH_3)_2Cl(H_2O)]^+$ is a particularly active form of the cytostatic agent. The positive charge of this substituted complex supports such an assumption, as it will be more likely to approach and coordinate with the negatively charged DNA.

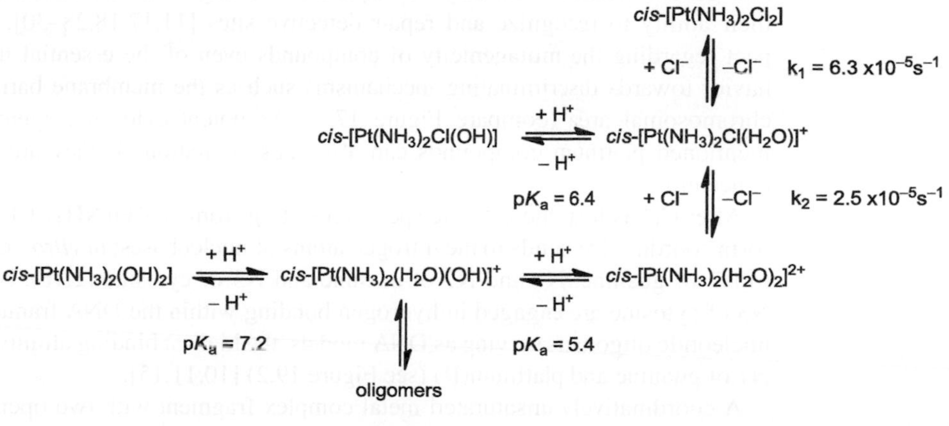

(19.5)

Platinum retention times are different in different organs, decreasing in the order kidney > liver > genitals > spleen > bladder > heart > skin > stomach > brain [11,18]. After interaction with the DNA in the cells of these organs, the degradation products are excreted via the liver and kidney.

There are several different coordination sites available for the binding of metal ions and their complexes to DNA, or more generally, to nucleotides. The metal centers can bind to the negatively charged oxygen atoms of (poly)phosphate groups (compare Section 14.1) or to the nitrogen and oxygen atoms of purine and pyrimidine bases (Section 2.3.3). Planar complexes, or those containing large π systems as ligands [24,25], should also be able to intercalate between two base pairs (compare Figure 2.9), possibly even in a sequence-specific fashion. Finally, coordinated ligands with, for example, amine or hydroxyl functions may form hydrogen bonds with proton-acceptor components of the polynucleotides.

Interactions of metal ions or metal complexes with nucleic acids [11,15,26] generally play an important role:

- in sustaining the conformation of the tertiary structure of polyelectrolytes such as DNA or RNA through electrostatic effects;
- in the nucleic acid metabolism, particularly phosphoryl transfer (nuclease and polymerase activity; see Section 14.1);
- in the regulation, replication and transcription of genetic information (see also Section 12.6);
- in efforts directed at specific DNA cleavage with synthetic probes (restriction enzyme analogues; compare Section 2.3.3);
- in metal-induced mutagenesis [27].

Such mutations can be the result of structural distortions of the DNA caused by unphysiological crosslinking or of the stabilization of an incorrect nucleobase tautomer following metal coordination (see (2.18)) [27]. Metal complex–nucleic acid interactions and their physiological consequences can thus be quite varied; even in the case of the extensively studied platinum compounds, they are far from being fully understood. Replication and genetic transcription may be impaired with regard to their accuracy ("fidelity") or their ability to recognize and repair defective sites [11,17,18,28–30]. A significant aspect regarding the mutagenicity of compounds even of the essential metals is their behavior towards discriminating mechanisms such as the membrane barrier protecting the chromosomal area (compare Figure 17.7). As potent cytostatic agents, the previously mentioned platinum compounds can also cause mutations if they are present in higher concentrations.

After Cl^- is lost, the cationic species resulting from *cis*-$[Pt(NH_3)_2Cl_2]$ can be shown to form coordinative bonds to the nitrogen atoms of nucleobases; *in vitro*, these include bonds to N7 of guanine, N1 and N7 of adenine and N3 of cytosine (2.16). N1 of adenine and N3 of cytosine are engaged in hydrogen bonding within the DNA framework; for various nucleotide oligomers serving as DNA models, the highest binding affinity is found between N7 of guanine and platinum(II) (see Figure 19.2) [10,11,15].

A coordinatively unsaturated metal complex fragment with two open sites in *cis* position may bind in different ways to double-stranded DNA (Figure 19.2). Since complexes with only one labile ligand, such as chlorido(diethylenetriamine)platinum(II) $[Pt(dien)Cl]^+$ (19.6), are therapeutically inactive, the monofunctional platinum species presumably serve only as intermediates. The possible alternatives of coordinative interaction between DNA and $[Pt\,(NL)_2]^{2+}$ include chelate complex formation (e.g. via coordination to the nitrogen and oxygen atoms of one guanine base), intrastrand crosslinking of two nucleobases of a

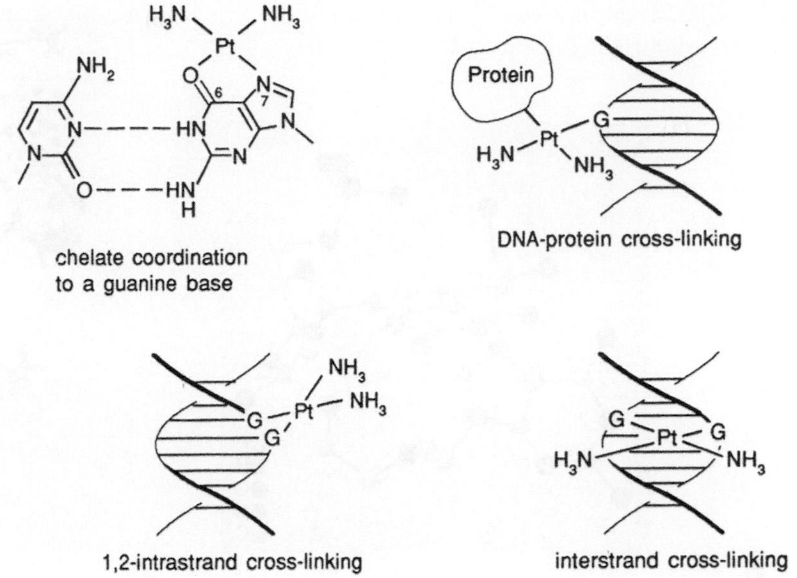

Figure 19.2
Possible kinds of bonding between cis-$[Pt(NH_3)_2]^{2+}$ and guanosine (G) in double-stranded DNA.

single DNA strand, interstrand crosslinking of two different strands of one DNA molecule and the metal-induced attachment of a protein to DNA.

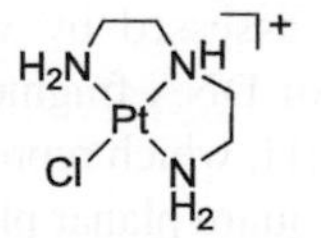

(19.6)

Experiments have shown that the chelate complex formation with O6 and N7 (Figure 19.2) of a free guanine base is possible in principal but is not favored within the DNA double helix. Interstrand crosslinking and protein–DNA interactions also make only minor contributions to the overall platinum/DNA adduct formation in the case of $[Pt(NH_3)_2]^{2+}$, the situation (and the cytostatic spectrum) being different with more recently developed bi- and trinuclear complexes (19.7) [15]. Most of the retained diammine platinum(II) forms bonds with two neighboring N7-coordinated guanosine (G) nucleotides on the same DNA strand (1,2-intrastrand d(GpG) crosslinking; d: deoxy form of ribose; p: phosphate); 1,2-intrastrand d(ApG) crosslinking (each via N7) has also been observed [11–13,15,17].

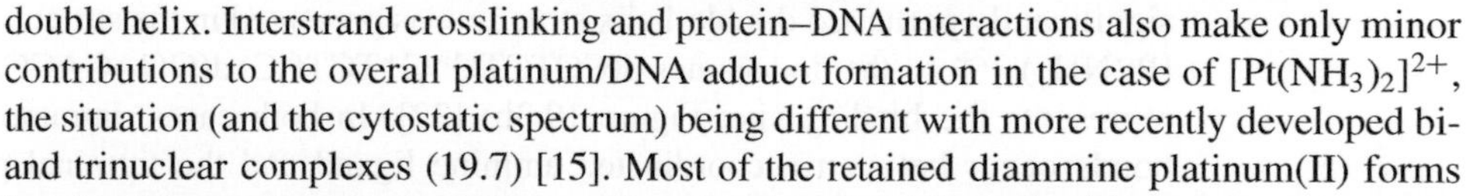

(BBR305) (BBR3464)

(19.7)

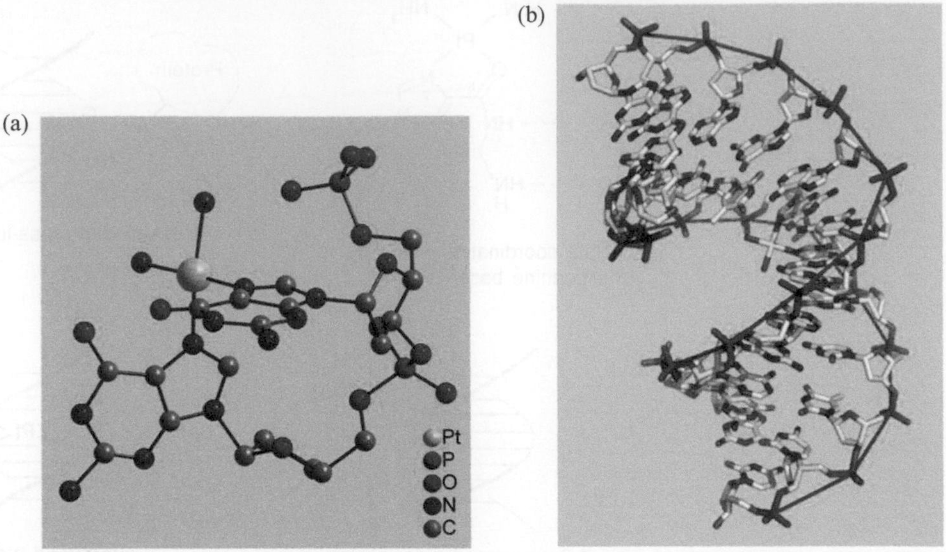

Figure 19.3
Molecular structures of (a) *cis*-[Pt(NH_3)$_2${d(pGpG)}] (one of four crystallographically independent molecules in the unit cell) [31] and (b) d(CCTCTG*G*TCTCC) d(GGAGACCAGAGG) (G*G* represents the binding site of the [Pt(NH_3)$_2$]$^{2+}$ adduct, PDB code 3LPV) [32].

The structural changes in DNA following coordination of [Pt(NL)$_2$]$^{2+}$ fragments have been quantitatively assessed by various physical measurements of single- or double-stranded DNA and of DNA fragments [11–13,15,17,31,32]. In the x-ray structure of *cis*-[Pt(NH_3)$_2${d(pGpG)}], which represents the [Pt(NH_3)$_2$]$^{2+}$-adduct to single-stranded DNA (Figure 19.3a), the square planar platinum center is surrounded by two *cis*-positioned NH_3 ligands and two N7 nitrogen atoms of the two guanine bases [31]. While the nucleobases are situated nearly parallel to each other (stacking) in an intact DNA, they form a dihedral angle of about 80° in this model for platinum-coordinated DNA, and the consequence is a significantly perturbed double-helix structure, as presented in the structure of the adduct of [Pt(NH_3)$_2$]$^{2+}$ to the dodecamer d(CCTCTG*G*TCTCC) d(GGAGACCAGAGG) (G*G* represents the binding site; Figure 19.3b) [32]. In both cases, intramolecular hydrogen bonds occur between a coordinated ammine ligand and the terminal phosphate, which seems to justify the requirement for such ligands in active platinum cytostatica.

Of importance for the cytotoxic behavior are the consequences of the distortions in the Pt-DNA adducts. The adducts can impede cellular processes such as DNA replication and transcription, which require DNA strand separation. Furthermore, the damage is recognized by several cellular proteins (chromatin alternation), some of which are involved in DNA repair, including transcription-coupled repair (TCR). The specific binding of a chromosomal "high-mobility group" (HMG-1) protein to *cis*-[Pt(NH_3)$_2$]$^{2+}$-containing DNA suggests a flawed genetic information transfer, either via altered transcription or through faulty recognition and thus shielding from DNA repair processes [17]. Additionally, in some cases, prolonged G2-phase cell cycle arrest has been observed. If the damage repair fails or has been circumvented by direct signaling (protein–protein interaction),

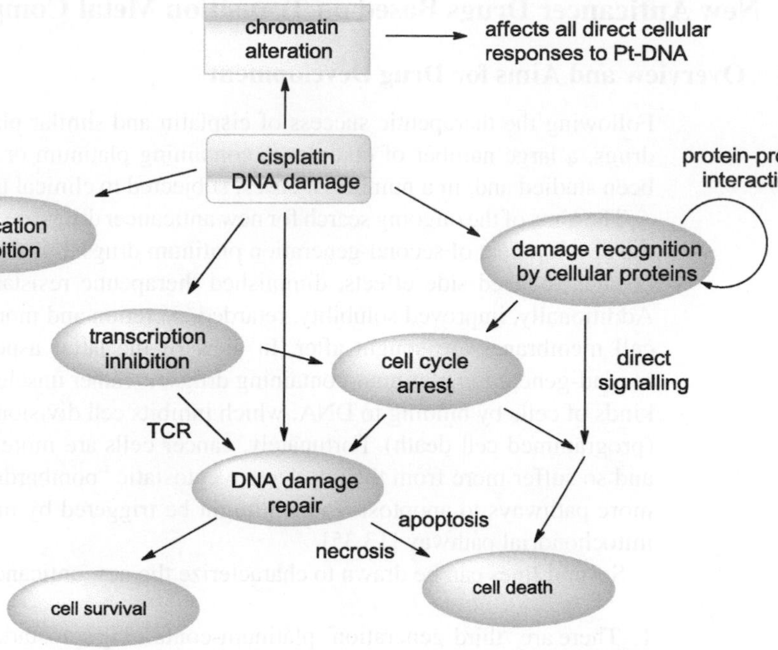

Figure 19.4
Direct cellular responses to platinum adducts. Reprinted with permission from [17] © 2007, American Chemical Society.

the final cellular outcome is generally apoptotic (programmed) cell death (Figure 19.4) [12,13,15,17].

So far the pathways from Pt-DNA binding to apoptosis remain incompletely elucidated; however, it is clear that knowledge of the complex interactions of Pt-DNA adducts with the proteins involved in DNA recognition and repair will help to design new, highly active and tumor-specific metal-containing anticancer drugs. Another important aspect for new drugs is drug resistance [10,30,33,34]. Resistance to chemotherapy represents a major obstacle in the treatment of cancer patients. Many tumors are intrinsically resistant to chemotherapy, whereas others initially respond to treatment but acquire resistance to selected cytotoxic drugs during chemotherapy. Multidrug resistance (MDR) is the phenomenon by which cultured cells *in vitro* and tumor cells *in vivo* show resistance simultaneously to a variety of structurally and functionally dissimilar cytotoxic compounds [33]. Important aspects of drug resistance are increased recognition and repair of DNA adducts [29,30], increased deactivation of drugs (in the case of platinum, by S ligands such as glutathione (GSH)) and increased efflux of drugs by membrane transport proteins such as *Pgp* [30,33,34]. *Pgp* is a 170 kDa plasma-membrane protein which shares high homology with the family of ATP-binding cassette (ABC) membrane transport proteins (Chapter 13); it is thought to hydrolyze ATP in order to affect outward transport of substrates across the cell-surface membrane. There is evidence that membrane-associated domains interact directly with selected cytotoxic agents in order to affect transport. The net effect is that *Pgp* decreases the intracellular concentration of drugs in *Pgp*-expressing MDR cells relative to non-*Pgp*-expressing drug-sensitive cells [30,32,33].

19.3 New Anticancer Drugs Based on Transition Metal Complexes

19.3.1 Overview and Aims for Drug Development

Following the therapeutic success of cisplatin and similar platinum-containing antitumor drugs, a large number of complexes containing platinum or other transition metals have been studied and, in a number of cases, subjected to clinical trials [3–8,10,13–24].

The aims of the ongoing search for new anticancer drugs are the same as those outlined for the development of second-generation platinum drugs: broad applicability, low therapeutic dosage, reduced side effects, diminished therapeutic resistance and oral administration. Additionally, improved solubility, retarded excretion and more selective transport through cell membranes are sought after. In view of the latter aspect, cisplatin, as well as the second-generation platinum-containing drugs, is rather unselective, bringing damage to all kinds of cells by binding to DNA, which inhibits cell division, finally leading to apoptosis (programmed cell death). Fortunately, cancer cells are more dependent on rapid growth, and so suffer more from the unselective cytostatic "bombardment". Importantly, there are more pathways to apoptosis (which might be triggered by metal complexes), such as the mitochondrial pathway [13,35].

Several lines can be drawn to characterize the new anticancer drugs:

1. There are "third-generation" platinum-containing cytostatic agents, in which established Pt complexes are coupled to a functional carrier molecule [10,13–15,36,37]. The carrier should improve the selectivity for tumor tissue by providing required material for the rapidly growing tumor cells. Inherent cytotoxic activity of the ligand (e.g. of the amine doxorubicin (19.2)) may contribute in a synergistic manner.
2. Alternatively, active anticancer drugs can be generated from prodrugs by selective stimulus, as shown for example for the complex $[Pt(N_3)_2(OH)_2(NH_3)_2]$, which is activated by light [7,14,38]. Generally, the use of Pt^{IV} prodrugs seems promising, since they are far more stable than Pt^{II} derivatives, making side reactions less probable and producing longer retention times. The hypoxic surrounding of cancer cells may reduce them to Pt^{II} derivatives, thus activating them [7,36].
3. An increasing number of cytostatic platinum complexes have been found which clearly do not obey the established rules [10]. These so-called "rule breakers" have come into focus because most of them do not show cisplatin-typical resistance; a pathway to apoptosis different from that of cisplatin can thus be assumed [14,16].
4. Complexes of other metals have revealed interesting cytotoxic properties and have been studied. Since bonds to Pt^{II} are quite inert, the NL groups in other, substitutionally more labile complexes have to be coordinated in a kinetically stable fashion: either through polyhapto binding of organic ligands or through chelate ligation. Examples of such complexes are the metallocenes and metallocene dichlorides of Ti^{IV} (19.8) and Mo^{IV} [20], the compound bis(1-phenyl-1,3-butanedionato)diethoxytitanium(IV) (budotitan) (19.9) [20,39] and a number of ruthenium complexes [5,7,8,19,21,24,25,36,37,40], gold complexes [8,22,35] and bis(cyclopentadien)iron(II) (ferrocene) derivatives [20,36,37,41,42,44].

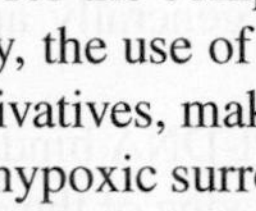

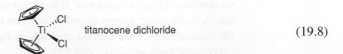

titanocene dichloride (19.8)

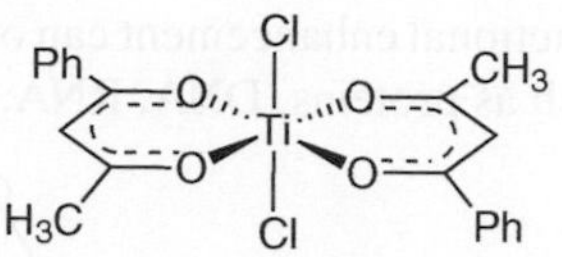

(19.9)

5. Finally, transition metal complexes combining the property of intercalating into DNA with reactive photochemical or electrochemical activity can be used for selective (oxidative) DNA cleavage [15,21,24]. From a photochemical viewpoint, Ru^{II} complexes of intercalating α-diimine ligands are interesting [21,24,25,38], while combinations of established intercalating drugs such as bleomycin (19.2), (19.18) with Fe^{III}, Cu^{II}, Mn^{II} or Co^{III} [4,25,44,45] and Cu^{I} phenanthroline complexes are able to oxidatively damage DNA [15,24,25].

Additionally, and sometimes paralleling the development of new metal-based anticancer drugs, metal complexes active against bacterial diseases (e.g. malaria) or viruses (including HIV) have been developed in recent years.

19.3.2 Nonplatinum Anticancer Drugs

The cytotoxic properties of titanocene dichloride $[Ti(Cp)_2Cl_2]$, Cp = η^5-C_5H_5 (19.8) were established in 1979 [20,42]. Since then, further metallocenes (M = V, Nb, Mo, Fe, Ge or Sn) have been studied and found to be active. $[Ti(Cp)_2Cl_2]$ can be used for a number of human carcinomas, such as intestinal (colon) cancer, lung cancer and breast carcinoma, but, it fails to be active against diverse carcinomas in the head or jugular area. $[Ti(Cp)_2Cl_2]$ was up to phase II of clinical trials but did not enter further stages of development due to serious side effects (liver damage) and a lack of improvement over other treatment regimes. Interestingly, cross-resistance between $[Ti(Cp)_2Cl_2]$ (or other Ti-based drugs) and cisplatin is not observed. From this, a different pathway to apoptosis can be assumed, and recent mechanistic studies have provided a plausible mechanism for delivery of Ti to cancer cells via transferrin-mediated endocytosis. This mechanism requires the presence of labile Cp–Ti bonds that hydrolyze on a timescale to deliver Ti to transferrin (Section 8.4.1) [20,39]. In addition to similarly constructed diorganotin(IV) compounds such as R_2SnX_2, the compound bis(1-phenyl-1,3-butanedionato)diethoxytitanium(IV), budotitan (19.9) [20,39], has turned out to be very promising against colon cancer, and reached phase I clinical trials following an early preclinical evaluation. However, it did not progress any further despite the development of a Cremophor EL-based formulation for it. The reasons were the same as for $[Ti(Cp)_2Cl_2]$: massive side effects and no better results compared to other treatment regimes.

Very promising results have been obtained in the treatment of breast cancer using the so-called ferrocifens (19.10). These are ferrocenyl derivatives of the selective estrogen-receptor modulators tamoxifen and hydroxytamoxifen, which are themselves the leading agents for the treatment of breast cancer. The binding of ferrocenyl markedly enhances the cytotoxic potential [2,36,37,41–43]. Interestingly, ferroquine, the ferrocenyl analogue of chloroquine (19.11), is presently in phase IIb clinical trials for the treatment of malaria [2]. Again, the ferrocene conjugate turns out to be more potent than the original molecule

[36,37,43]. Such functional enhancement can often be encountered for ferrocene conjugates of biomolecules such as proteins, DNA, RNA, carbohydrates and hormones [37,42,43].

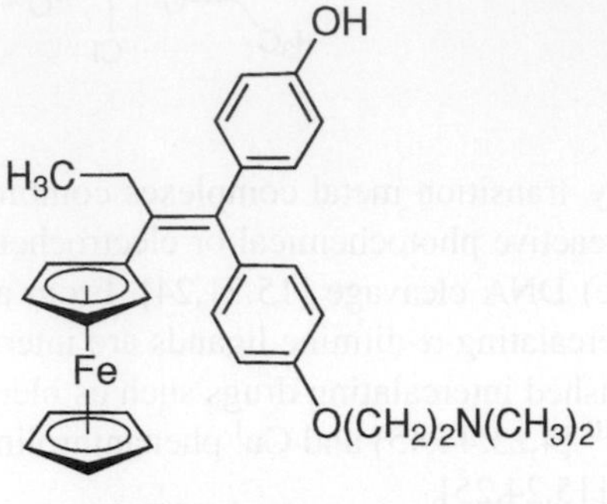

ferrocifen

(19.10)

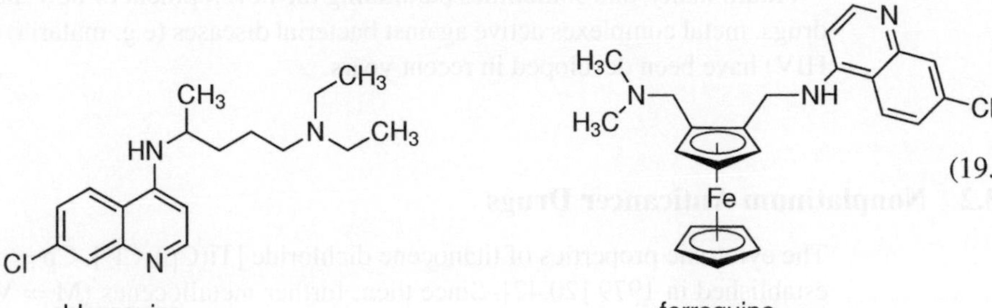

chloroquine ferroquine

(19.11)

A very important contribution to the development of new antitumor drugs comes from a class of substitutionally rather inert complexes with imine, amine, phosphane, carboxylate and porphyrin ligands. The most prominent examples are the ruthenium(III) complexes NAMI-A and KP1019 (19.12).

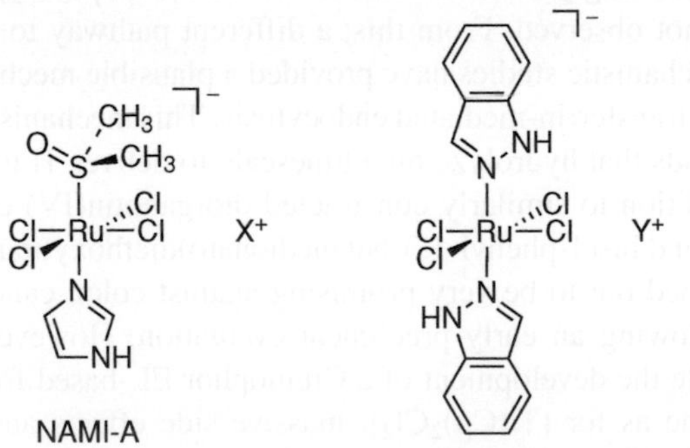

NAMI-A KP1019

(19.12)

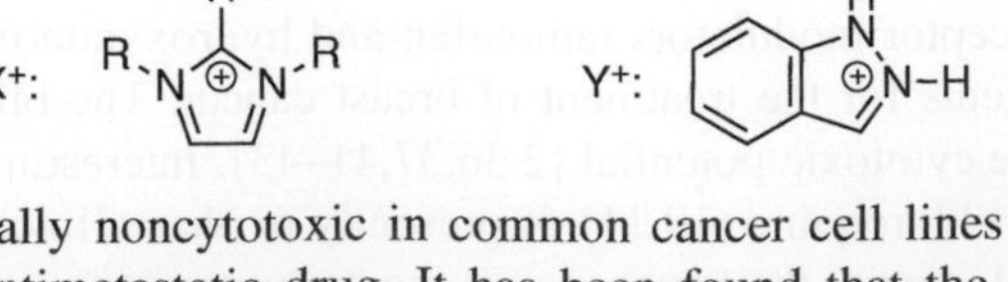

NAMI-A is practically noncytotoxic in common cancer cell lines but has turned out to be a specifically antimetastatic drug. It has been found that the chlorido ligands in

NAMI-A are rapidly hydrolyzed, comparably to those of cisplatin. *Trans*-tetrachloro-bis(1*H*-indazole)ruthenate(III) (KP1019 or FFC14A) is more stable towards aquation and hydrolysis and is more readily taken up by cells than is NAMI-A. It shows a remarkable activity against primary cisplatin-resistant colorectal tumors, but no pronounced antimetastatic activity. Both drugs have completed phase I clinical trials [21,40]. Since both NAMI-A and KP1019 readily react with biological reductants (e.g. ascorbate and glutathione) in protein-free model systems, it has been suggested that the reduction of Ru^{III} prodrugs to Ru^{II} species may be required for their biological activity (as already mentioned for Pt^{IV} complexes). Furthermore, for both ruthenium complexes, a high tendency towards accumulation in tumor cells can be concluded; this is very probably due to the resemblance to Fe (transferrin pathway; see Section 8.4.1).

Another class of cytotoxic ruthenium compounds is the so-called RAPTA complexes (19.13), which are organometallic arene Ru^{II} complexes [7,21,36,42,43]. In line with the assumption that Ru^{II} is the cytotoxic "state", these air-stable compounds exhibit high cytotoxicity against common cell lines.

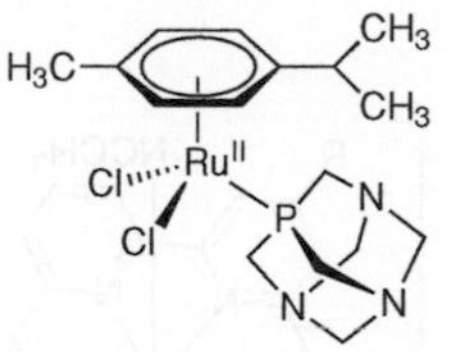

RAPTA-C (19.13)

For all these ruthenium-containing drugs, the binding of the metal to DNA has been found; however, the binding and the associated inhibition of DNA replication are not responsible for the induction of apoptosis. For KP1019, binding to the serum protein transferrin and transport into (and within) the cell via the transferrin pathway have been studied. The selective activation (by reduction) in the tumor cells has also been confirmed. Apoptosis (programmed cell death) is induced at nontoxic administration via the mitochondrial pathway. The drug can thus be applied in cases in which platinum-containing drugs fail. Preliminary *in vitro* and *in vivo* studies have confirmed this [40].

Further complexes of ruthenium, but also rhodium, palladium, iron, gallium, vanadium, molybdenum and gold with various ligands, are under investigation. From these chemically and structurally very variable systems, only a few are in line with established hypotheses drawn from the detailed studies on cisplatin (e.g. Ru^{III}, Fe^{III}, Ga^{III} [19]), while early transition metals in high oxidation states (Ti^{IV}, V^{IV}) [20] can use the transferrin pathway to accumulate in the highly iron-dependant cancer cells. Other Ru^{III} complexes [19], Au^{III} porphyrins [8,22] and Au^{I} complexes with phosphane, thiolate (e.g. auranofin (19.14)) or carbene ligand-containing species such as $[Au(^{i}Pr_2Im)_2]Cl$ (19.15) make use of their redox chemistry (with or without the help of O_2), resulting in oxidative damage to the mitochondria [19,22,35]. A further group of complexes is active in cell-cycle arrest, such as the organometallic Au^{III} complex $[Au(C\hat{}N\hat{}C)(meim\text{-}1)]^+$ (meim-1 = 1-methylimidazole) (19.16) and the pentapyridyl Fe^{II} complex (19.17) [8].

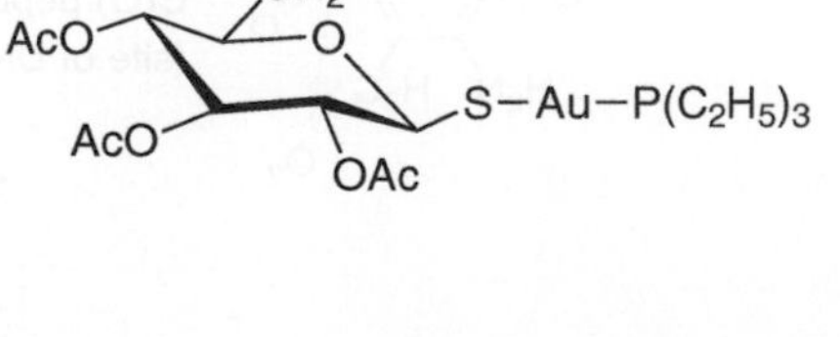

auranofin (19.14)

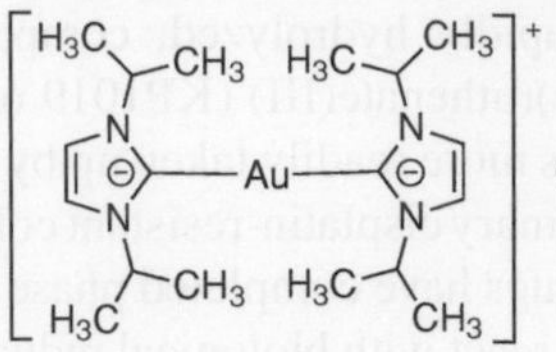

(19.15)

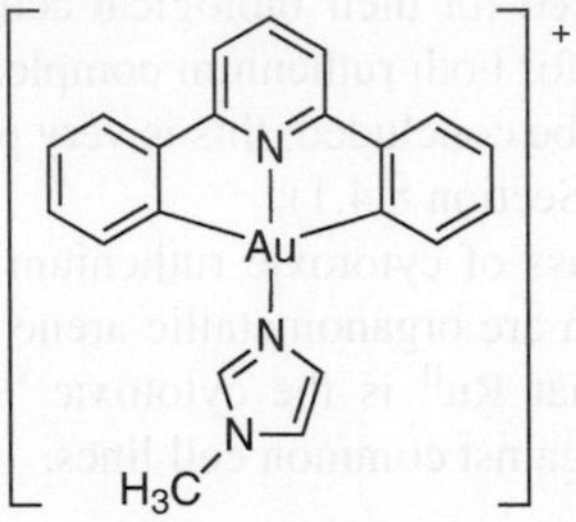

(19.16)

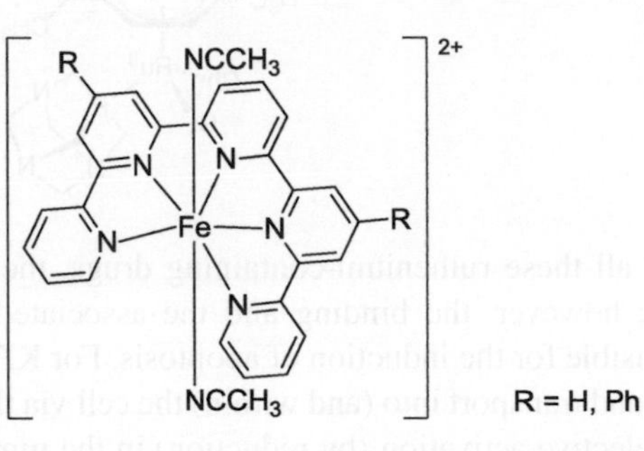

(19.17)

For a large number of other complexes, *in vitro* studies have shown very high cytotoxicity in the lower nanomolar regime (IC_{50} for cisplatin lies in the micromolar range), but the mechanisms remain to be elucidated.

Iron, ruthenium and copper complexes with potentially DNA-intercalating organic ligands are particularly known as complexes with the antibiotic bleomycin (19.2), (19.18) or with ligands such as 1,10-phenanthroline (2.19). The presence of redox-active metal ions is essential for the oxidative or photoinduced DNA cleavage caused by these reagents [24,25,38]; for example, a metal-based activation of radical-forming O_2 involving oxoferryl intermediates (see Sections 6.2–6.4) is assumed for antineoplastic iron(II)/bleomycin complexes (19.18) [44,45].

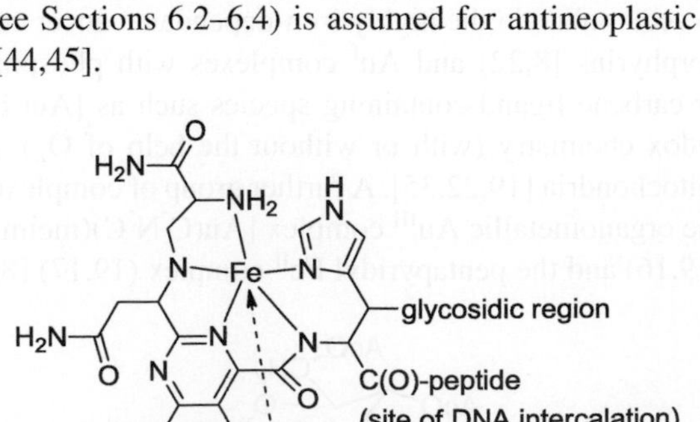

(19.18)

Bis(1,10-phenanthroline)copper(I) is the best known example of a metal-containing chemical nuclease; its peroxide-induced capability for phosphodiester cleavage in DNA or RNA may be applied in sequence-specific fashion by using carrier systems. In similar complexes of ruthenium (2.19), which can be activated by irradiation, the partially selective interaction with parts of the DNA probably occurs via electrostatic effects and shape adaptation ("molecular recognition") rather than direct intercalation (Figure 2.9). Such chemical analogues of restriction enzymes have a potential in various fields requiring DNA analysis and modification [24,25].

19.4 Further Inorganic Compounds in (Noncancer) Chemotherapy

19.4.1 Gold-containing Drugs Used in the Therapy of Rheumatoid Arthritis

Gold was used for therapeutic purposes in ancient civilizations. For instance, a Chinese prescription dating from the 6th century CE describes in detail the dissolution of metallic gold for use in elixirs aimed at achieving immortality. As can be reconstructed today [46], the oxidative dissolution of this noble metal involved potassium nitrate, KNO_3, containing iodate, IO_3^-, as an impurity, which can be reduced to iodide by reductants such as $FeSO_4$ or organic material. In the presence of I^-, the potential for oxidation of gold is lowered by approximately 1 V, and $[AuI_2]^-$ is formed in the process.

Gold was propagated as a universal remedy and its protective function against leprosy has frequently been mentioned. In 1890, Robert Koch found that Au^I cyanide, AuCN, inhibits the growth of bacteria which cause tuberculosis, but a systematic use of this compound was impossible due to its high toxicity. In 1924 Mollgaard applied a thiosulfato complex of monovalent gold in an attempt to cure tuberculosis, and a few years later Au^I thioglucose (19.19) was beginning to be used in the therapy of rheumatic fever. However, it was only much later that similar compounds were subjected to systematic clinical trials and received proper attention [3,22,35,47].

Only the monovalent form of gold has therapeutic importance in antirheumatic agents, while both Au^I and Au^{III} are found in gold-containing anticancer drugs (Section 19.3.2). The aqua complex of Au^I is unstable and disproportionates according to $3\ Au^I \rightleftarrows 2\ Au^0 + Au^{III}$; however, the monovalent state can be stabilized through "soft", polarizable ligands such as CN^-, PR_3 and thiolates, RS^-, under formation of preferentially linear d^{10} metal complexes. Most of the Au^I compounds used as antirheumatic agents (representative examples are given in (19.19)) contain thiolate ligands. Au^{III} will act as a strong oxidant under such conditions.

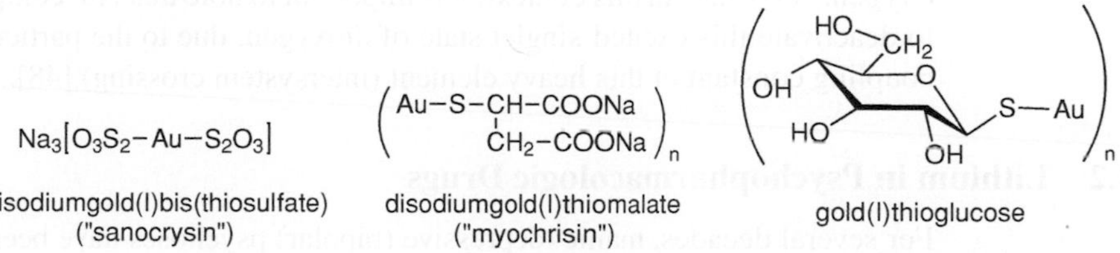

trisodiumgold(I)bis(thiosulfate) ("sanocrysin") disodiumgold(I)thiomalate ("myochrisin") gold(I)thioglucose

(19.19)

The linear arrangement of the sulfur ligands around the two-coordinate metal center occurs in all cases, including solganol and myochrisine; in these latter drugs, sulfur-bridged oligomeric species seem to be present. A number of related compounds have been crystallized and show polymeric $CsNa_2H[Au_2(STm)_2]_n$ (STm = thiomalate), cyclic oligomeric $[Au(SC_6H_2{}^iPr_3)]_6$ or monomeric structures $(NH_4)_5[Au(STm)_2]$ [47]. However, commercial and medicinal preparations of myochrisine are more complex, due to the use of excess thiomalate ligand over the 1 : 1 ratio. In solutions of the drug, tetramers (ESI-MS) and open-chain structures (XAS) have been detected [47]. Solganol and myochrisine are water soluble but are insoluble in hydrophobic environments and thus have to be administered intramuscularly in order to prevent hydrolysis in acidic gastric juice. The lipophilic auranofin (19.14), on the other hand, can be administered orally with about 25% resorption. After long-term therapy with myochrisine and auranofin, constant levels of 30–50 mg/ml blood are attained; in the blood, myochrisine is mainly bound to albumin in the serum, while auranofin is equally distributed between serum and erythrocytes.

The increasingly common therapy with gold compounds to retard active rheumatic processes has some not uncommon side effects resembling allergic reactions on skin and mucous membranes, as well as gastrointestinal and renal problems. These effects restrict the gold therapy to only about two-thirds of patients; it is assumed that some are due to the formation of Au^{III} compounds. Since Au^{I} forms thermodynamically stable complexes with sulfur ligands, the addition of sulfur-containing chelating agents such as penicillamine or dimercaprol (2.1) and of antihistamines or adrenocorticosteroids might reduce the toxicity of the gold drugs.

Arthritis is an inflammation of the tissue which surrounds the joints. It is assumed that the damage is caused by the action of hydrolytic enzymes from lysosomes (rheumatoid arthritis as autoimmune reaction). Examinations of the tissue show that gold is preferentially accumulated in the joints and is stored in the lysosomes of macrophages, forming "aurosomes". Inhibition of the lysosomal enzyme activity can be rationalized by assuming a coordination of gold to the thiolate groups, RS^-, present in the enzymes. *In vitro*, disodiumgold(I) thiomalate (myochrisine) readily reacts with other thiolates, RS^-, under release of thiomalate. This process is analogous to the rapid exchange reaction of the similar, linearly coordinated and thus kinetically labile Hg^{II} thiolate complexes (associative substitution mechanism; see Section 17.5).

According to a different hypothesis, Au^{I} compounds are able to inhibit the formation of undesired antibodies in the collagen region.

Yet another hypothesis suggests that irreversible damage of the joints may result from lipid oxidation with subsequent degradation of proteins by the free radicals formed (compare Section 16.8). Superoxide ions, $O_2{}^{\bullet -}$, which can be produced by activated phagocytes (compare Section 10.5) play an important role in this process. It has been shown that several oxidants can relatively easily convert $O_2{}^{\bullet -}$ to reactive, not spin-inhibited singlet oxygen, $^1O_2$ (5.3). In this context, it is important to note that Au^{I} compounds should be able to deactivate this excited singlet state of dioxygen, due to the particularly high spin–orbit coupling constant of this heavy element (intersystem crossing) [48].

19.4.2 Lithium in Psychopharmacologic Drugs

For several decades, manic–depressive (bipolar) psychoses have been treated with lithium salts, often in the form of exactly measured lithium carbonate. The Li^+ ion is therapeutically valuable because it counteracts both phases in the typically cyclical course of this disorder [49–51].

Difficulties in the therapy with lithium compounds result from the relatively high toxicity of the metal and the resulting very small therapeutic window. While a concentration of about 1 mmol Li^+/l (1 mM) blood is necessary for a successful treatment, a 2 mM concentration can cause toxic side effects, particularly in the renal and nervous systems (tremor). Concentrations of 3 mM and higher may eventually be lethal for the patient. For this reason, lithium carbonate and other salts of Li^+ are administered orally in several carefully controlled doses per day. On acute poisoning, the blood should be purified using Na^+-containing dialysis fluids.

Li^+ is relatively rare in the earth's crust and in sea water. It is also a lighter and, with an ionic radius of 60 pm for the coordination number 4, significantly smaller and thus more polarizing homologue of Na^+, a fact that might directly explain the neurological (side) effects (compare Chapter 13). On the other hand, Li^+ shares a diagonal relationship and the same physiologically important affinity towards phosphate ligands with the slightly larger Mg^{2+}. There are several hypotheses regarding the specific antipsychotic mode of action for lithium in which the effect on the cellular information system (via binding to phosphates) is a focal point; inhibitions of the inositol/phosphate metabolism, of an adenyl cyclase and of a guanine nucleotide binding protein, a "G-protein", are under discussion [49–51].

19.4.3 Bismuth Compounds against Ulcers

Bismuth compounds have been used for more than 200 years for the treatment of gastric ulcers. Administered Bi^{III} compounds include the nitrate, salicylate (both soluble) and basic colloidal bismuth subcitrate (CBS), which forms colloids in water [3,4,52,53]. The structures of the *in vivo*-generated compounds are rather speculative, since the coordination number of Bi^{III} in its compounds is very flexible and ranges from 3 to 10, exhibiting frequently irregular polyhedra. *In vitro* studies reveal that polymeric aggregates such as the polyanionic chain $[Bi_2(cit)_2]^{2-}$ (cit^{4-} = citrate tetraanion) (19.20) are formed, probably covering the surface of the ulcer [52,53].

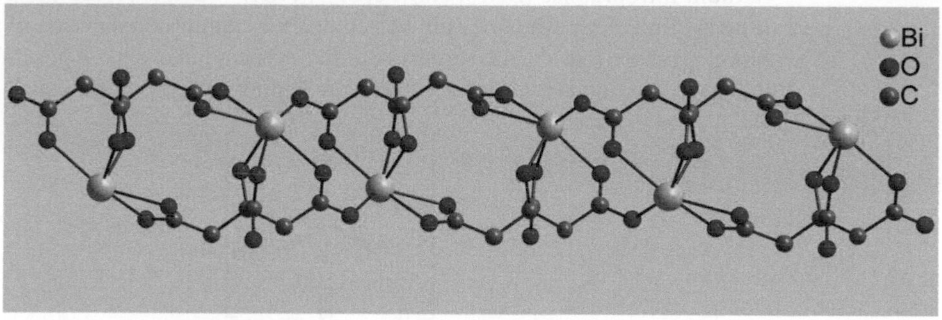

(19.20)

The mode of action of Bi^{III} is assumed to be mainly antimicrobacterial activity against *Helicobacter pylori* (compare Section 9.2). Typical side effects are nausea and vomiting. The chemically very related Sb^{III} tartrate (trade names: Tartar Emetic Stibophen, Astiban) and Sb^{V}-containing compounds such as sodium stibogluconate (Pentostam) and meglumine antimonate (Glucantime) are used as antiparasitic agents (e.g for the treatment of leishmaniasis) [4,52,53]. Sb^{V} compounds are generally regarded as prodrugs; they are reduced to Sb^{III} (considered to be more toxic) *in vivo*. The precise molecular structures of stibogluconate and meglumine antimonate have not been unambiguously determined (compare Salvarsan; Section 19.1).

19.4.4 Vanadium-containing Insulin Mimetics and V-containing Anti-HIV Drugs

Pentavalent vanadium, V^V, as vanadate (VO_4^{3-}), and tetravalent V^{IV}, as vanadyl $[VO]^{2+}$, are in use in insulin-mimetic drugs. They can imitate the function of insulin and stimulate the uptake of glucose. At the same time, many V-containing compounds exhibit cell toxicity. Vanadium complexes with organic ligands are usually less toxic and exhibit higher lipophilicity. The complex bis(maltolato)oxovanadium(IV) (19.21) is more effective than $VOSO_4$ [3–5].

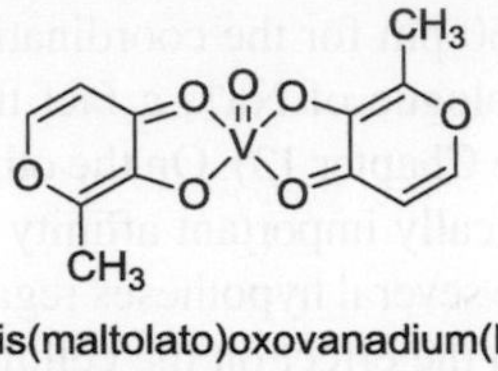

(19.21)

Oxovanadium(IV) complexes with thiourea or porphyrin ligands, vanadium-substituted polyoxotungstates and the vanadocene complex $[V(Cp)_2(acac)]$ (acac = acetylacetonate) (19.22) have been reported to exhibit anti-HIV properties [8,42].

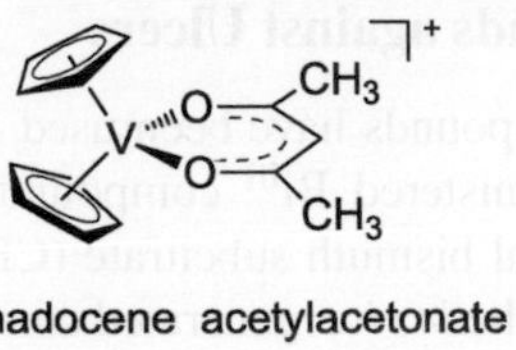

(19.22)

19.4.5 Sodium Nitroprusside

Sodium nitroprusside (SNP), $Na_2[Fe(CN)_5(NO)] \times 2H_2O$ (19.23), is a nitrosyl complex, long in clinical use. The low-spin Fe^{II} complex is administered in case of suddenly elevated blood pressure (e.g. during surgery) or after myocardial events. After infusion of the drug, the desired reduction of the blood pressure begins within 1–2 minutes.

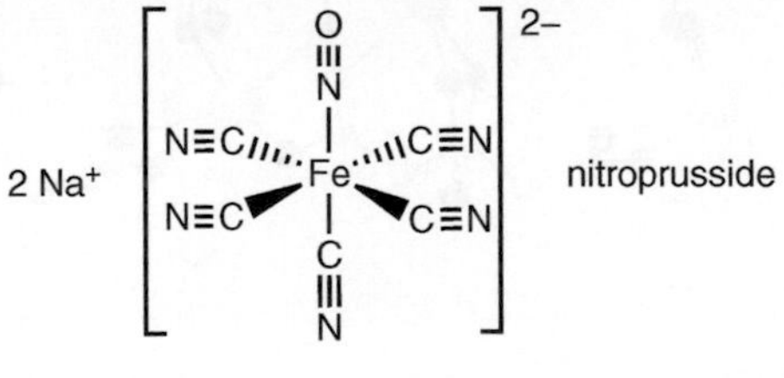

(19.23)

It is assumed that *in vivo* nitroprusside $[Fe(CN)_5(NO)]^{2-}$ is reduced to $[Fe(CN)_5(NO)]^{3-}$, which is labile and loses CN^-. Subsequently, $[Fe(CN)_4(NO)]^{2-}$ releases NO (Section 6.5), which leads to a relaxation of the smooth muscles of the blood vessels. This is the reason for the therapeutic effects [3–5,36,54].

Alternatively, nitroglycerine can be administered in case of angina pectoris. Sildenafil citrate (19.24), sold as Viagra, Revatio and under various other trade names, is used to treat erectile dysfunction and pulmonary arterial hypertension (PAH). It was initially studied for use in hypertension (high blood pressure) and angina pectoris (a symptom of ischemic heart disease). Phase I clinical trials suggested that the drug had little effect on angina, but that it could induce marked penile erections. Sildenafil acts by inhibiting a cGMP-specific phosphodiesterase type 5, an enzyme that regulates blood flow (see also Section 14.2).

The company Pfizer therefore decided to market it for erectile dysfunction, rather than for angina pectoris.

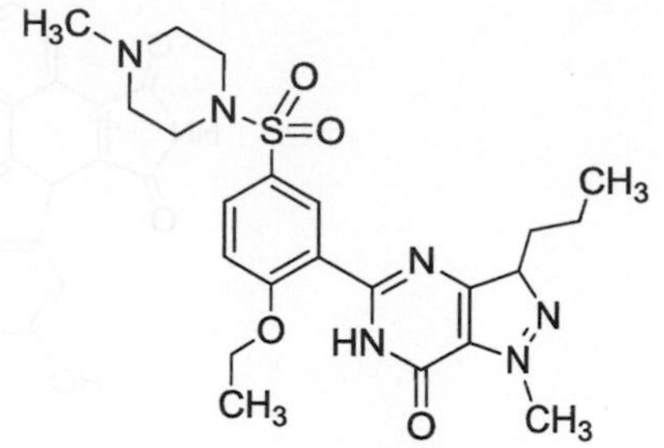

(19.24)

Sildenafil

In conjunction with studies on other NO-releasing complexes, carbon monoxide, CO, has come into the spotlight. Although generally toxic, the controlled *in vivo* release of CO has very similar consequences for blood vessels to that of NO, and corresponding CO-releasing molecules (CORMs) have been the subject of increasing interest in the last 10 years (see Sections 6.5 and 19.5) [55]. Importantly, the CO molecule is far more inert than the radical $NO^\bullet$, and their methods of binding to metalloproteins (preferably heme) are markedly different.

19.5 Bioorganometallic Chemistry of Nonessential Elements

According to the established definition, organometallic compounds contain at least one covalent metal–carbon bond. In many cases, this bond is labile to hydrolysis (compare Chapter 9) or weak; therefore, bioorganometallic chemistry appears to a be a limited scientific subdiscipline. However, in previous chapters we have met a number of organometallic species with importance in biology. Methylcobalamin contains a discrete CH_3–Co bond (Chapter 3) and is able to biomethylate Hg^{II} and Sn^{IV} (Chapters 3 and 17). Metal–carbon-bonded species also occur in a number of nickel-dependent enzymes, such as CODH/ACS, the [NiFe] hydrogenases and the methyl-coenzyme M reductase (MCR) (Chapter 9). While the modeling of these naturally occurring bioorganometallic molecules forms part of the investigation of the corresponding enzymes, the newly evolving field of bioorganometallic chemistry is additionally motivated by medical applications of ferrocene derivatives, group 4 metallocenes, CO-releasing compounds, ruthenium-arene compounds and other organometallic systems [41–43]. Examples of organometallic anticancer drugs were presented in Section 19.3, along with non-organometallic drugs, and a few aspects can be extracted which reveal why organometallic compounds are interesting.

1. Organic ligands can be varied to a large extent. For instance, complexes carrying bulky ligands with hydrophobic properties can be as rationally designed as highly water soluble molecules. Substitution using electron-withdrawing or electron-donating groups can further vary the electronic and structural properties of such ligands and their complexes. Many biologically active molecules have been derivatized in order to modify their physiological performance. Covalent binding to a metal can be regarded as substitution using a more or less electron-withdrawing group. Examples are the ferrocene-conjugates

(19.10), (19.11) and the protein kinase inhibitor (19.25), which show an affinity two orders of magnitude higher than the corresponding organic framework alone [42,43].

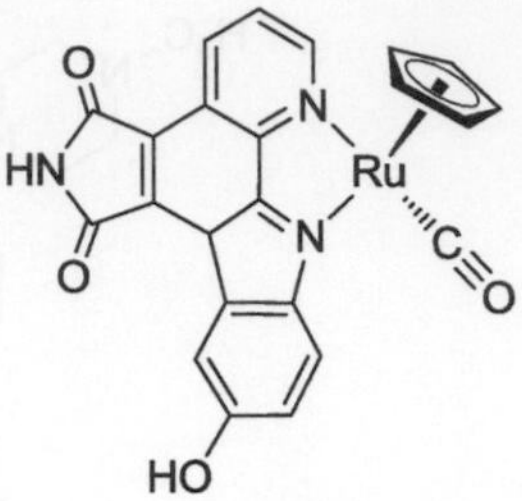

(19.25)

2. Metal–carbon bonds are often thermodynamically and hydrolytically stable for the noble metals, such as Pt, Rh, Ru (cytotoxic RAPTA complexes (19.13)) and Au (cytotoxic carbene complexes (19.15)). Apart from Cu in the 3d series, Co (B_{12}) and to a lesser extent Ni (CODH/ACS) are also suitable. The outstanding stability of ferrocene in biological environments is due to the multidentate (η^5) binding of the two cyclopentadienido ligands (Cp^-) to the configurationally stable low-spin Fe^{II}, which effectively blocks the attack of water.
3. For the heavy metals and metalloids Hg, Sn, As and Se, the metal–carbon bond is less thermodynamically stable and bond homolysis occurs, leading to biologically unwanted radicals ($M{-}C \rightarrow M\cdot + \cdot C$). Hydrolysis is relatively slow. The antibacterial agent mercurochrome (19.26) is hydrolyzed very slowly, resulting in a slow release of the metal ion Hg^{2+} (or $[Hg(OH)]^+$), which inhibits thiol functions in bacterial cell walls (compare Chapter 17) [3].

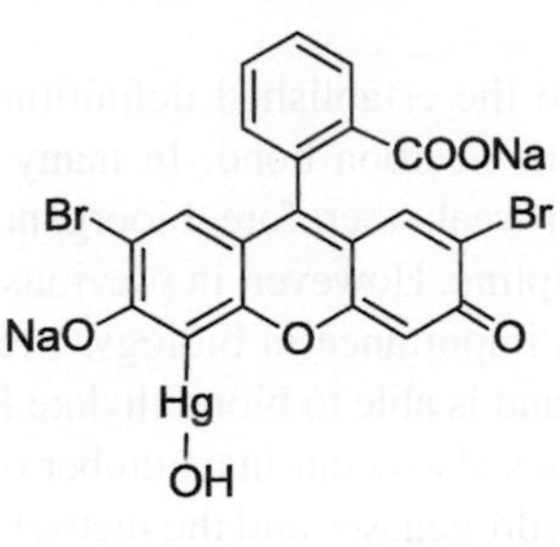

(19.26)

4. For organometallic complexes of the more electropositive metals of groups 4–6, the hydrolysis may be slowed down by employing *ansa* systems (19.27), which involves chelating bis-cyclopentadienide ligands that are less prone to hydrolysis. Some of these compounds exhibit *in vitro* anticancer activity at concentrations one order of magnitude lower than titanocene dichloride [42].

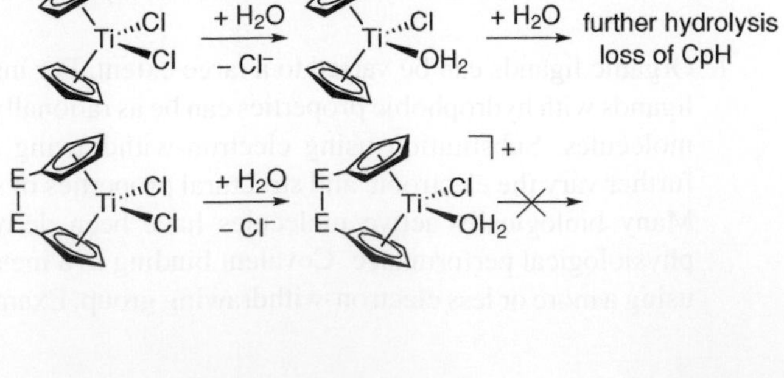

(19.27)

5. Some ligands, such as arenes and CO, are not hydrolyzed in aqueous media, their replacement depending on the presence or absence of potentially stronger ligands. Therefore, very stable complex fragments of the nonessential elements chromium $[Cr(CO)_3]$, rhenium or technetium $[M(CO)_3X]$ (M = Tc (see Chapter 18.1.5) or Re; X: halide) can be used to modify biomolecules without release of the toxic CO molecule [42,55]. Other carbonyl complexes, such as $[Ru(CO)_3Cl(glycinato)]$ (19.28), are able to release CO in a controlled way and can be used for the modulation of coronary vasoconstriction [42,55].

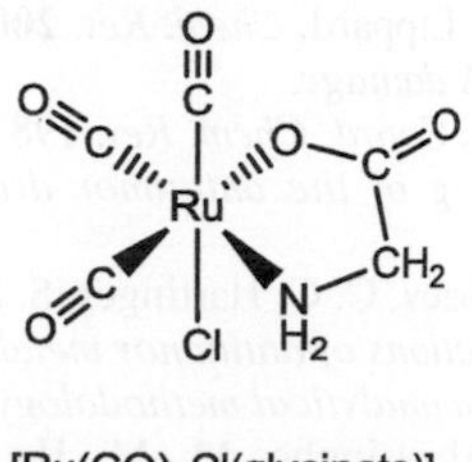

(19.28)

Further Reading

E. Alessio, *Bioinorganic Medical Chemistry*, Wiley-VCH, Weinheim, **2011**.
J. C. Dabrowiak, *Metals in Medicine*, John Wiley & Sons, Chichester, **2009**.
C. Jones, J. Thornback, *Medicinal Applications of Coordination Chemistry*, RSC Publishing, Cambridge, **2007**.

References

1. N. C. Lloyd, H. W. Morgan, B. K. Nicholson, R. S. Ronimus, *Angew. Chem. Int. Ed. Engl.* **2005**, *44*, 941–944: *The composition of Ehrlich's salvarsan: resolution of a century-old debate.*
2. C. Biot, W. Castro, C. Y. Botté, M. Navarro, M., *Dalton Trans.* **2012**, *41*, 6335–6349: *The therapeutic potential of metal-based antimalarial agents: implications for the mechanism of action.*
3. N. Farrell, *Metal complexes as drugs and chemotherapeutic agents*, in J. A. McCleverty, T. J. Meyer (eds.) *Comprehensive Coordination Chemistry II, Vol. 9*, Elsevier, Amsterdam, **2003**.
4. Z. Guo, P. J. Sadler, *Angew. Chem. Int. Ed. Engl.* **1999**, *38*, 1512–1531: *Metals in medicine.*
5. S. P. Fricker, *Dalton Trans.* **2007**, 4903–4917: *Metal based drugs: from serendipity to design.*
6. K. H. Thompson, C. Orvig, *Dalton Trans.* **2006**, *6*, 761–764: *Metal complexes in medicinal chemistry: new vistas and challenges in drug design.*
7. L. Ronconi, P. J. Sadler, *Coord. Chem. Rev.* **2007**, *251*, 1633–1648: *Using coordination chemistry to design new medicines.*
8. R. W.-Y. Sun, D.-L. Ma, E. L.-M. Wong, C.-M. Che, *Dalton Trans.* **2007**, *43*, 4884–4892: *Some uses of transition metal complexes as anti-cancer and anti-HIV agents.*
9. (a) J. D. Hoeschele, *Dalton Trans.* **2009**, *48*, 10 648–10 650: *In remembrance of Barnett Rosenberg*; (b) B. Rosenberg, L. van Camp, T. Krigas, *Nature (London)* **1965**, *205*, 698–699: *Inhibition of cell division in E. coli by electrolysis products from a platinum electrode*; (c) B. Rosenberg, L. van Camp, J. E. Trosko, V. H. Mansour, *Nature (London)* **1969**, *222*, 385–386: *Platinum compounds: a new class of potent antitumor drugs.*
10. J. Suryadi, U. Bierbach, *Chem. Eur. J.* **2012**, *18*, 12 926–12 934: *DNA metalating-intercalating hybrid agents for the treatment of chemoresistant cancers.*
11. B. Lippert, *Cisplatin, Chemistry and Biochemistry of a Leading Anticancer Drug*, Wiley-VCH, Weinheim, **1999**.

12. R. C. Todd, S. J. Lippard, *Metallomics* **2009**, *1*, 280–291: *Inhibition of transcription by platinum antitumor compounds.*
13. L. Kelland, *Nat. Rev. Cancer* **2007**, *7*, 573–584: *The resurgence of platinum-based cancer chemotherapy.*
14. A. M. Montaña, C. Batalla, *Curr. Med. Chem.* **2009**, *16*, 2235–2260: *The rational design of anticancer platinum complexes: the importance of the structure-activity relationship.*
15. J. Reedijk, *Eur. J. Inorg. Chem.* **2009**, 1303–1312: *Platinum anticancer coordination compounds: study of DNA binding inspires new drug design.*
16. T. Gianferrara, I. Bratsos, E. Alessio, *Dalton Trans.* **2009**, *37*, 7588–7598: *A categorization of metal anticancer compounds based on their mode of action.*
17. Y. Jung, S. J. Lippard, *Chem. Rev.* **2007**, *107*, 1387–1407: *Direct cellular responses to platinum-induced DNA damage.*
18. P. Umapathy, *Coord. Chem. Rev.* **1989**, *95*, 129–181: *The chemical and biological consequences of the binding of the antitumor drug cisplatin and other platinum group metal complexes to DNA.*
19. A. R. Timerbaev, C. G. Hartinger, S. S. Aleksenko, B. K. Keppler, *Chem. Rev.* **2006**, *106*, 2224–2248: *Interactions of antitumor metallodrugs with serum proteins: advances in characterization using modern analytical methodology.*
20. (a) P. M. Abeysinghe, M. M. Harding, *Dalton Trans.* **2007**, *32*, 3474–3482: *Antitumour bis(cyclopentadienyl) metal complexes: titanocene and molybdocene dichloride and derivatives*; (b) T. Schilling, K. B. Keppler, M. E. Heim, G. Niebch, H. Dietzfelbinger, J. Rastetter, A.-R. Hanauske, *Invest. New Drugs* **1996**, *13*, 327–332: *Clinical phase I and pharmacokinetic trial of the new titanium complex budotitane.*
21. A. Levina, A. Mitra, P. A. Lay, *Metallomics* **2009**, *1*, 458–470: *Recent developments in ruthenium anticancer drugs.*
22. I. Ott, *Coord. Chem. Rev.* **2009**, *253*, 1670–1681: *On the medicinal chemistry of gold complexes as anticancer drugs.*
23. S. M. Aris, N. P. Farrell, *Eur. J. Inorg. Chem.* **2009**, 1293–1302: *Towards antitumor active trans-platinum compounds.*
24. C. Metcalfe, J. A. Thomas, *Chem. Soc. Rev.* **2003**, *32*, 215–224: *Kinetically inert transition metal complexes that reversibly bind to DNA.*
25. B. M. Zeglis, V. C. Pierre, J. K. Barton, *Chem. Commun.* **2007**, 4565–4579: *Metallo-intercalators and metallo-insertors.*
26. E. Freisinger, R. K. O. Sigel, *Coord. Chem. Rev.* **2007**, *251*, 1834–1851: *From nucleotides to ribozymes – a comparison of their metal ion binding properties.*
27. B. Lippert, D. Gupta, *Dalton Trans.* **2009**, *24*, 4619–4634: *Promotion of rare nucleobase tautomers by metal binding.*
28. M. Ljungman, *Chem. Rev.* **2009**, *109*, 2929–2950: *Targeting the DNA damage response in cancer.*
29. R. R. Iyer, A. Pluciennik, V. Burdett, P. L. Modrich, *Chem. Rev.* **2006**, *106*, 302–323: *DNA mismatch repair: functions and mechanisms.*
30. M. A. Fuertes, C. Alonso, J. M. Pérez, *Chem. Rev.* **2003**, *103*, 645–662: *Biochemical modulation of cisplatin mechanisms of action: enhancement of antitumor activity and circumvention of drug resistance.*
31. S. E. Sherman, D. Gibson, A. H. J. Wang, S. J. Lippard, *J. Am. Chem. Soc.* **1988**, *110*, 7368–7381: *Crystal and molecular structure of cis-$Pt(NH_3)_2$\{d(pGpG)\}], the principal adduct formed by cis-diamminedichloroplatinum(II) with DNA.*
32. P. M. Takahara, C. A. Frederick, S. J. Lippard, *J. Am. Chem. Soc.* **1996**, *118*, 12 309–12 321: *Crystal structure of the anticancer drug cisplatin bound to duplex DNA.*
33. V. Sharma, D. Piwnica-Worms, *Chem. Rev.* **1999**, *99*, 2545–2560: *Metal complexes for therapy and diagnosis of drug resistance.*
34. P. D. W. Eckford, F. J. Sharom, *Chem. Rev.* **2009**, *109*, 2989–3011: *ABC efflux pump-based resistance to chemotherapy drugs.*
35. P. J. Barnard, S. J. Berners-Price, *Coord. Chem. Rev.* **2007**, *251*, 1889–1902: *Targeting the mitochondrial cell death pathway with gold compounds.*
36. T. W. Hambley, *Dalton Trans.* **2007**, *21*, 4929–4937: *Developing new metal-based therapeutics: challenges and opportunities.*

37. A. Vessières, S. Top, W. Beck, E. Hillard, G. Jaouen, *Dalton Trans.* **2006**, *4*, 529–541: *Metal complex SERMs (selective oestrogen receptor modulators). The influence of different metal units on breast cancer cell antiproliferative effects.*
38. N. J. Farrer, L. Salassa, P. J. Sadler, *Dalton Trans.* **2009**, *48*, 10 690–10 701: *Photoactivated chemotherapy (PACT): the potential of excited-state d-block metals in medicine.*
39. K. M. Buettner, A. M. Valentine, *Chem. Rev.* **2012**, *112*, 1863–1881: *Bioinorganic chemistry of titanium.*
40. C. G. Hartinger, S. Zorbas-Seifried, M. A. Jakupec, B. Kynast, H. Zorbas and B. K. Keppler, *J. Inorg. Biochem.* **2006**, *100*, 891–904: *From bench to bedside – preclinical and early clinical development of the anticancer agent indazolium trans-tetrachlorobis(1H-indazole) ruthenate(III)] (KP1019 or FFC14A).*
41. D. R. van Staveren, N. Metzler-Nolte, *Chem. Rev.* **2004**, *104*, 5931–5985: *Bioorganometallic chemistry of ferrocene.*
42. C. G. Hartinger, P. J. Dyson, *Chem. Soc. Rev.* **2009**, *38*, 391–401: *Bioorganometallic chemistry – from teaching paradigms to medicinal applications.*
43. G. Jaouen, (ed.) *Bioorganometallics: Biomolecules, Labeling, Medicine*, John Wiley & Sons, Chichester, **2005**.
44. J. Chen, J. Stubbe, *Nat. Rev. Cancer* **2005**, *5*, 102–112: *Bleomycins: towards better therapeutics.*
45. L. Que Jr., *J. Biol. Inorg. Chem.* **2004**, *9*, 684–690: *The oxo/peroxo debate: a nonheme iron perspective.*
46. C. Glidewell, *J. Chem. Educ.* **1989**, *66*, 631–633: *Ancient and medieval Chinese protochemistry.*
47. C. F. Shaw III, *Chem. Rev.* **1999**, *99*, 2589–2600: *Gold-based therapeutic agents.*
48. E. J. Corey, M. M. Mehrotra, A. U. Khan, *Science* **1987**, *236*, 68–69: *Antiarthritic gold compounds effectively quench electronically excited singlet oxygen.*
49. N. J. Birch, *Chem. Rev.* **1999**, *99*, 2659–2682: *Inorganic pharmacology of lithium.*
50. D. M. De Freitas, M. M. C. A. Castro, C. F. G. C. Geraldes, *Acc. Chem. Res.* **2006**, *39*, 283–291: *Is competition between Li^+ and Mg^{2+} the underlying theme in the proposed mechanisms for the pharmacological action of lithium salts in bipolar disorder?*
51. R. S. B. Williams, A. J. Harwood, *Lithium metallotherapeutics*, in M. Gielen, E. R. T. Tiekink (eds.), *Metallotherapeutic Drugs and Metal-based Diagnostic Agents: The Use of Metals in Medicine*, John Wiley & Sons, Chichester, **2005**.
52. R. Ge, H. Sun, *Acc. Chem. Res.* **2007**, *40*, 267–274: *Bioinorganic chemistry of bismuth and antimony: target sites of metallodrugs.*
53. N. Yang, H. Sun, *Coord. Chem. Rev.* **2007**, *251*, 2354–2366: *Biocoordination chemistry of bismuth: recent advances.*
54. L. E. Goodrich, F. Paulat, V. K. K. Praneeth, N. Lehnert, *Inorg. Chem.* **2010**, *49*, 6293–6316: *Electronic structure of heme-nitrosyls and its significance for nitric oxide reactivity, sensing, transport and toxicity in biological systems.*
55. (a) C. C. Romao, W. A. Blättler, J. D. Seixas, G. J. L. Bernardes, *Chem. Soc. Rev.* **2012**, *41*, 3571–3583: *Developing drug molecules for therapy with carbon monoxide*; (b) R. Alberto, R. Motterlini, *Dalton Trans.* **2007**, 1651–1660: *Chemistry and biological activities of CO-releasing molecules (CORMs) and transition metal complexes.*

Index

Bioinorganic Chemistry: Inorganic Elements in the Chemistry of Life – An Introduction and Guide, Second Edition.
Written and Translated by Wolfgang Kaim, Brigitte Schwederski and Axel Klein.

Printed and bound by CPI Group (UK) Ltd, Croydon, CR0 4YY
10/10/2024
14571756-0001